Communications in Computer and Information Science 2590

Rationale

The CCIS series is devoted to the publication of proceedings of computer science conferences. Its aim is to efficiently disseminate original research results in informatics in printed and electronic form. While the focus is on publication of peer-reviewed full papers presenting mature work, inclusion of reviewed short papers reporting on work in progress is welcome, too. Besides globally relevant meetings with internationally representative program committees guaranteeing a strict peer-reviewing and paper selection process, conferences run by societies or of high regional or national relevance are also considered for publication.

Topics

The topical scope of CCIS spans the entire spectrum of informatics ranging from foundational topics in the theory of computing to information and communications science and technology and a broad variety of interdisciplinary application fields.

Information for Volume Editors and Authors

Publication in CCIS is free of charge. No royalties are paid, however, we offer registered conference participants temporary free access to the online version of the conference proceedings on SpringerLink (http://link.springer.com) by means of an http referrer from the conference website and/or a number of complimentary printed copies, as specified in the official acceptance email of the event.

CCIS proceedings can be published in time for distribution at conferences or as post-proceedings, and delivered in the form of printed books and/or electronically as USBs and/or e-content licenses for accessing proceedings at SpringerLink. Furthermore, CCIS proceedings are included in the CCIS electronic book series hosted in the SpringerLink digital library at http://link.springer.com/bookseries/7899. Conferences publishing in CCIS are allowed to use Online Conference Service (OCS) for managing the whole proceedings lifecycle (from submission and reviewing to preparing for publication) free of charge.

Publication process

The language of publication is exclusively English. Authors publishing in CCIS have to sign the Springer CCIS copyright transfer form, however, they are free to use their material published in CCIS for substantially changed, more elaborate subsequent publications elsewhere. For the preparation of the camera-ready papers/files, authors have to strictly adhere to the Springer CCIS Authors' Instructions and are strongly encouraged to use the CCIS LaTeX style files or templates.

Abstracting/Indexing

CCIS is abstracted/indexed in DBLP, Google Scholar, EI-Compendex, Mathematical Reviews, SCImago, Scopus. CCIS volumes are also submitted for the inclusion in ISI Proceedings.

How to start

To start the evaluation of your proposal for inclusion in the CCIS series, please send an e-mail to ccis@springer.com.

Alexandra I. Cristea · Erin Walker · Yu Lu ·
Olga C. Santos · Seiji Isotani
Editors

Artificial Intelligence in Education

Posters and Late Breaking Results, Workshops and Tutorials, Industry and Innovation Tracks, Practitioners, Doctoral Consortium, Blue Sky, and WideAIED

26th International Conference, AIED 2025
Palermo, Italy, July 22–26, 2025
Proceedings, Part I

Springer

Editors
Alexandra I. Cristea
Durham University
Durham, UK

Erin Walker
University of Pittsburgh
Pittsburgh, PA, USA

Yu Lu
Beijing Normal University
Beijing, China

Olga C. Santos
UNED
Madrid, Spain

Seiji Isotani
University of São Paulo
São Carlos, Brazil

ISSN 1865-0929 ISSN 1865-0937 (electronic)
Communications in Computer and Information Science
ISBN 978-3-031-99260-5 ISBN 978-3-031-99261-2 (eBook)
https://doi.org/10.1007/978-3-031-99261-2

This Springer imprint is published by the registered company Springer Nature Switzerland AG
The registered company address is: Gewerbestrasse 11, 6330 Cham, Switzerland

Preface

Welcome to the 26th International Conference on Artificial Intelligence in Education, taking place from July 22–26, 2025 in the vibrant city of Palermo, Italy. AIED 2025 is ranked A in ICORE[1], in the top 14.92% of all 784 ranked venues, and clearly growing and attracting more submissions, this year with over 800 initial submissions. Thus, as we celebrate over a quarter of a century of advancing the science and engineering of intelligent human-technology ecosystems that support learning, this edition marked a significant evolution in our conference structure, with the introduction of a dedicated track-based system. This new format was designed to capture the growth of the area, as well as the dynamic and multifaceted nature of research in educational AI, reflecting the rapid pace and expanding impact of the field.

The theme for AIED 2025 — "AI as a Catalyst for Inclusive, Personalised, and Ethical Education: Empowering Teachers and Students for an Equitable Future" — embodies our vision for a transformative future in education. Artificial Intelligence is progressing at an unprecedented rate, with generative AI and related technologies spearheading this transformative shift. As these innovations reshape both teaching and learning processes, they unlock novel opportunities for creating personalised and dynamic learning experiences. Even as traditional AI techniques continue to evolve and enrich our understanding, true change lies in leveraging these advancements to provide educators with powerful tools for deeper insights, tailored instruction, and increased student engagement.

At the heart of this transformation is the commitment to equity and ethical practice. By harnessing AI's capacity to personalise learning experiences, we can address the diverse needs of students and reduce educational disparities, especially among underserved and marginalised communities. Equally important is the need to integrate AI responsibly, ensuring that fairness, transparency, and privacy remain paramount in our educational practices. This dual commitment—to innovation and to ethical inclusion—was the cornerstone of our conference theme.

The conference program comprised seven thematic tracks:

- **Track 1 – AIED Architectures and Tools:** This track highlighted innovative engineering advances, from multi-agent systems and tangible interfaces to scalable, domain-specific applications across a variety of educational contexts.
- **Track 2 – Machine Learning and Generative AI:** Emphasising data-driven techniques and generative models, this track explored novel algorithms and computational approaches that enhance predictive and adaptive functionalities in educational environments.
- **Track 3 – Learning, Teaching, and Pedagogy:** Focusing on empirical and theoretical work, this track examined the impact of AI on teaching practices, learner engagement, and the evolution of both formal and informal learning experiences.

[1] https://portal.core.edu.au/conf-ranks.

- **Track 4 – Human-Centred Design and Design-Based Research:** This track addressed the design principles and interactive strategies that place educators and learners at the centre, fostering collaborative human-AI partnerships to promote agency and effective decision-making.
- **Track 5 – Teaching AI:** Dedicated to advancing AI literacy, the research in this track explored methods and technologies that demystify AI, empowering individuals to understand, utilise, and critically evaluate its capabilities.
- **Track 6 – Ethics, Equity, and AIED in Society:** This track delved into the societal dimensions of AI in education, with an emphasis on fairness, transparency, privacy, and strategies to mitigate educational disparities.
- **Track 7 – Theoretical Aspects of AIED and AI-Based Modeling for Education:** This track emphasised conceptual and theoretical contributions providing deeper insights into cognitive, motivational, and socio-cognitive processes, underpinning rigorous models of learning and teaching.

AIED 2025 attracted an exceptionally broad range of submissions. We received over 800 submissions across all tracks, of which 711 met the eligibility criteria and were accepted for review. From the full paper submissions, 130 were accepted as full papers, and an additional 129 were accepted as short papers. The acceptance rates for both full and short papers were 18%, respectively.

Every paper and presentation at AIED 2025 underwent a rigorous, double-blind peer-review process designed to ensure the highest scholarly standards and fair evaluation, with an average of three reviews per submission. This meticulous process—drawing on the expertise of leading researchers, seasoned reviewers, and senior committee members—culminated in an outstanding program that spanned the full spectrum of AIED research and practice.

Beyond the scholarly sessions and paper presentations, which are published in the LNCS volumes of the conference, AIED 2025 offered a rich program of keynote addresses, panel discussions, interactive demonstrations, and specialised sessions which are compiled in these CCIS volumes. The specialised tracks included **Late Breaking Results (LBR)**, **Doctoral Consortium**, **BlueSky**, **WideAIED**, and **Practitioners, Industry, and Policy (PIP)**. Furthermore, 14 **workshops** and 2 **tutorials** were held during AIED 2025.

This year had both a general AIED panel as well as a joint panel with sister conferences, Educational Data Mining (EDM 2025) and ACM Learning @ Scale (L@S 2025), which were co-located in Palermo with AIED 2025. Attendees had the opportunity to participate in the Doctoral Consortium, Workshops and Tutorials, Late-Breaking Results sessions, and dedicated tracks for practitioners, industry, and policy-makers. Notably, the BlueSky track invited visionary reflections on past progress and future trajectories in AIED, while the WideAIED track amplified perspectives from under-represented regions and communities. Additionally, the conference offered presentations on the latest publications in IJAIED2, the journal associated with the conference. As a member of the IAALDE[2] alliance, AIED 2025 hosted, last but not least, presentations of the best papers from other IAALDE conferences (in particular, ICCE 2024 and LAK 2025).

[2] https://alliancelss.com/.

This conference would not have been possible without the unwavering commitment of the International Artificial Intelligence in Education Society, the Organising Committee, the Program and Senior Program Committees, and the sponsors and partners who support our mission. Their passion and dedication to advancing intelligent educational systems is the foundation of AIED's continued success and innovation.

We invite you to engage wholeheartedly in the exchange of ideas, to challenge conventional paradigms, and to forge new collaborations that will shape the future of education. Together, we will explore how artificial intelligence can act as a true catalyst for creating more inclusive, personalised, and ethical learning environments for teachers and students around the world.

Welcome to AIED 2025 — where the future of education is actively being reimagined, one innovative idea at a time.

July 2025

Alexandra I. Cristea
Erin Walker
Yu Lu
Olga C. Santos
Seiji Isotani

Organization

Conference General Co-chairs

Olga C. Santos	UNED, Spain
Seiji Isotani	University of São Paulo, Brazil

Program Co-chairs

Alexandra I. Cristea	Durham University, UK
Erin Walker	Arizona State University, USA
Yu Lu	Beijing Normal University, China

Local Organising Co-chairs

Davide Taibi	Institute for Educational Technology, National Research Council, Italy
Giusuè Lo Bosco	University of Palermo, Italy

Doctoral Consortium Co-chairs

Zitao Liu	Jinan University, China
Carla Limongelli	University of Roma Tre, Italy
Sergey Sosnovsky	Utrecht University, Netherlands

Workshop and Tutorials Co-chairs

Gabriella Casalino	University of Bari, Italy
Tanja Mitrovic	University of Canterbury, New Zealand
Sebastiane Lalle	Sorbonne University, France

Practitioners, Industry, and Policy Co-chairs

Sreecharan Sankaranarayanan Amazon, USA
Penghe Chen Beijing Normal University, China
Emmanuel G. Blanchard Le Mans University, France
Christian M. Stracke University of Bonn, Germany

Posters and Late Breaking Results Co-chairs

Jingyun Wang Durham University, UK
Ilona Buchem Berliner Hochschule für Technik, Germany

AIED Mentoring Fellowship Co-chairs

Vania Dimitrova University of Leeds, UK
Ig Ibert Bittencourt Universidade Federal de Alagoas, Brazil

Panel Co-chairs

Ryan Baker University of Pennsylvania, USA
Marta Cimitile UnitelmaSapienza, Italy

Blue Sky Co-chairs

Stamos Katsigiannis Durham University, UK
Bruce M. McLaren Carnegie Mellon University, USA

WideAIED Co-chairs

Isabel Hilliger Pontificia Universidad Católica, Chile
Giovanni Fulantelli Institute for Educational Technology, National Research Council, Italy
Ilaria Torre University of Genoa, Italy

Diversity and Inclusion Co-chairs

Valery Psyché	TÉLUQ University, Canada
Kaska Porayska-Pomsta	UCL Institute of Education, UK

Interactive Events Co-chairs

Leonardo B. Marques	Federal University of Alagoas, Brazil
Emmanuel G. Blanchard	LIUM, Le Mans Université, France

Proceedings Co-chairs

Paraskevi Topali	Radboud University, Netherlands
Rafael D. Araújo	Universidade Federal de Uberlândia, Brazil
Elaine Harada	Universidade Federal do Amazonas, Brazil

International Journal of AIED Chair

Conrad Borchers	Human-Computer Interaction Institute, USA

Virtual Experiences Co-chairs

Teng Guo	Jinan University, China
Salvatore Calderaro	University of Palermo, Italy

Publicity Co-chairs

Miguel Portaz	UNED, Spain
Xinyuan Hao	University College London, UK
Mariella Farella	Institute for Educational Technology, National Research Council, Italy

Volunteer Co-chairs

Lele Sha	Monash University, Australia
Domenico Amato	University of Palermo, Italy

Sponsorship Chair

Pietro Paolo Corso	University of Palermo, Italy

Scholarship Chair

Michele Tumminello	University of Palermo, Italy

Website Chair

Andrea Nelson Mauro	Institute for Educational Technology, National Research Council, Italy

Research Track Junior Advisory Board

Conrad Borchers	Human-Computer Interaction Institute, USA
Amine Hatun Ataş	Galatasaray University, Turkey
Krishna Chaitanya Rao Kathala	University of Massachusetts, USA
Guher Gorgun	IPN – Leibniz Institute for Science and Mathematics Education, Germany
Isaac Alpizar-Chacon	Utrecht University, Netherlands

BlueSky Program Committee

Shrikant Malviya	IIT, Patna, India
Ana Serrano-Mamolar	Universidad de Burgos, Spain
Seiji Isotani	Harvard University, USA
Peter Hastings	DePaul University, USA
Kalina Yacef	University of Sydney, Australia
Ig Ibert Bittencourt	Federal University of Alagoas, Brazil
Pablo Arnau-González	Universidad de Valencia, Spain
Vincent Aleven	Carnegie Mellon University, USA

Olga Viberg	KTH Royal Institute of Technology, Sweden
Paulo Carvalho	Carnegie Mellon University, USA
Seyedahmad Rahimi	University of Florida, USA
John Stamper	Carnegie Mellon University, USA
Donald Sturgeon	Durham University, UK
Sunčica Hadžidedić	Durham University, UK
Ivon Arroyo	University of Massachusetts Amherst, USA
Rafael Ferreira Mello	CESAR School, Brazil
Jelena Jovanovic	University of Belgrade, Serbia
Mahboobeh Mehrvarz	Carnegie Mellon University, USA
Benedict du Boulay	University of Sussex, UK
Ryan Baker	University of Pennsylvania, USA

WideAIED Program Committee

Ifeoma Adaji	University of British Columbia, Okanagan Campus, Canada
Gabriel Astudillo	Pontificia Universidad Católica de Chile, Chile
Héctor Ceballos	Tecnológico de Monterrey, Mexico
Paolo Fantozzi	LUMSA, Italy
Rafael Ferreira Mello	CESAR School, Brazil
Giovanni Fulantelli	National Research Council of Italy, Italy
Daniela Giordano	University of Catania, Italy
Isabel Hilliger	Pontificia Universidad Católica de Chile, Chile
Ivana Marenzi	L3S Research Center, Germany
Juan Andrés Talamás	Tecnológico de Monterrey, Mexico
Ilaria Torre	University of Genoa, Italy
Maomi Ueno	University of Electro-Communications, Japan
Giacomo Valente	University of L'Aquila, Italy
Mario Vallarino	University of Genoa, Italy

Practitioners, Industry, and Policy (PIP) Track

Senad Becirovic	PH NOE, Austria
Rajesh Kumar Cherukuri	Amazon, USA
Erwan David	Le Mans Université, France
Annamaria Di Grassi	Università degli studi di Bari, Italy
Guozhu Ding	Guangzhou University, China
Khaoula Ferchichi	Rheinische Friedrich-Wilhelms-Universität Bonn, Germany

Yunlong Huang	Jiangxi University of Science and Technology, China
Sébastien Iksal	Le Mans Université, France
Dora Katsamori	NCSR "Demokritos", Greece
Imran A. S. Khan	Le Mans Université, France
Zhi Liu	Central China Normal University, China
Stefania Massaro	University of Bari, Italy
Julia Maria Mönig	Rheinische Friedrich-Wilhelms-Universität Bonn, Germany
Nurun Nahar	University of Bolton, UK
Abdelkader Ouared	Le Mans Université, France
Dimitra Pappa	NCSR Demokritos, Greece
Yang Pian	Beijing Institute of Education, China
Claudine Piau-Toffolon	Le Mans Université, France
Marjana Prifti Skenduli	Università degli Studi di Bari "Aldo Moro", Italy
Veronica Punzo	Università di Pisa, Italy
Valérie Renault	Le Mans Université, France
Elham Tajik	University of Rochester, USA
Judith Uchidiuno	Georgia Institute of Technology, USA
Tim Walther	Rheinische Friedrich-Wilhelms-Universität Bonn, Germany
Haipeng Wan	Capital Normal University, China
Deliang Wang	University of Hong Kong, China
Qi Wang	Beijing Foreign Studies University, China
Zhifeng Wang	Central China Normal University, China
Lan Wu	Nanjing University of Information Science & Technology, China
Sergej Zerr	Rheinische Friedrich-Wilhelms-Universität Bonn, Germany
Jia Zhu	Zhejiang Normal University, China

Late Breaking Results (LBR) Track

Senior Program Committee Members

Adam Wynn	Durham University, UK
Daner Sun	Education University of Hong Kong, China
Emily Evermann	TU Berlin, Germany
Georges Arnaud Kouayim Bonga	Berliner Hochschule für Technik, Germany
Mandana Sutter	TU Berlin, Germany
Yi Shang	Durham University, UK
Zhongtian Sun	University of Kent, UK

Program Committee Members

Aditi Singh	Cleveland State University, USA
Aditya Joshi	UNSW, Australia
Ajanie Karunanayake	University of Technology Sydney, Australia
Allison Poh	University of Massachusetts Amherst, USA
Almazbek Beishenaliev	Regional Institute of Central Asia, Kyrgyzstan
Álvaro Sobrinho	Federal University of the Agreste of Pernambuco, Brazil
Andres Felipe Zambrano	University of Pennsylvania, USA
Andy Smith	North Carolina State University, USA
Ange Adrienne Nyamen Tato	Université Laval, Canada
Angelo Rega	Pegaso Telematic University, Italy
Anita Dąbrowicz-Tlałka	Gdańsk University of Technology, Poland
Anna Korchak	Higher School of Economics, Moscow, Russia
Anna Rafferty	Carleton College, USA
Anuradha Singh	Banaras Hindu University, India
Atsushi Shimada	Kyushu University, Japan
Bailing Lyu	University of Utah, USA
Batuhan Sayis	Universitat Pompeu Fabra/Universidad de Granada, Spain
Beate Grawemeyer	Coventry University, UK
Ben Nye	University of Southern California, USA
Boxuan Ma	Kyushu University, Japan
Bradford Mott	North Carolina State University, USA
Candace Walkington	Southern Methodist University, USA
Chaewon Kim	Florida State University, USA
Changhao Liang	Kyoto University, Japan
Chaohua Ou	Georgia Institute of Technology, USA
Cheng Tang	Kyushu University, Japan
Chenjia Zhu	University of Hong Kong, China
Christian M. Stracke	University of Bonn, Germany
Chu Hui-Chun	東吳大學, Taiwan
Chung Yiu Richard Yeung	Education University of Hong Kong, China
Cunling Bian	Ocean University of China, China
Dan Sun	Hangzhou Normal University, China
Dan R. Kohen-Vacs	Holon Institute of Technology, Israel
Daniel Alberto Perez Perez	Durham University, UK
Deliang Wang	Beijing Normal University, China
Di Zhang	Zhejiang Normal University, China
Diana-Laura Borza	Babeş-Bolyai University, Romania
Duc Nguyen	Ho Chi Minh University of Technology, Viet Nam

Eduardo Oliveira	University of Melbourne, Australia
Ehsan Latif	University of Georgia, USA
Emanuel Queiroga	Federal University of Pelotas, Brazil
Emmanuel G. Blanchard	LIUM, Le Mans Université, France
Fahim Ahmed	University of Dhaka, Bangladesh
Fatiha Sadat	UQAM, Canada
Federica Illuzzi	Università degli Studi di Bari Aldo Moro, Italy
Franziska Bickel	North Carolina State University, USA
Gabriel Astudillo	Pontificia Universidad Católica de Chile, Chile
German Capdehourat	Ceibal, Uruguay
Gila Kurtz	Holon Institute of Technology, Israel
Gongqi Lin	Victoria University, Australia
Hai Li	University of Florida, USA
Hannes Rosenbusch	University of Amsterdam, Netherlands
Hanqi Li	New York University, USA
Hao Tang	City University of New York, USA
Hideaki Yoshida	Morioka University, Japan
Hiu Ching Hung	Friedrich-Alexander University of Erlangen-Nuremberg
Hongming Li	University of Florida, USA
Huiying Cai	Jiangnan University, China
Huiyong Li	Kyushu University, Japan
Ifeoma Adaji	University of British Columbia, Okanagan Campus, Canada
Isabela Drămnesc	West University of Timisoara, Romania
Jacek Stańdo	Technical University of Lodz, Poland
Jarne Thys	Hasselt University, Flanders Make, Belgium
Jaroslaw A. Chudziak	Warsaw University of Technology, Poland
Jasbir Singh	University of Otago, New Zealand
Jennifer Kleiman	University of Georgia, USA
Jiahui Wu	Beijing University of Posts and Telecommunication, China
Jianwei Li	Beijing University of Posts and Telecommunications, China
Jiayu Zheng	Johns Hopkins University, USA
Jie Gao	McGill University, Canada
Jing Zhang	Dalian University of Technology, China
Jingxi Liu	Education University of Hong Kong, China
John Hollander	Arkansas State University, USA
Jonathan Kummerfeld	University of Sydney, Australia
Junyoung Lee	Nanyang Technological University, Singapore
Kevin Huang	WestEd, USA

Koichi Yamashita	Tokoha University, Japan
Krishna Chaitanya Rao Kathala	University of Massachusetts Amherst, USA
Leisi Pei	Education University of Hong Kong, China
Lena Hollenstein	St. Gallen University of Teacher Education, Switzerland
Liang-Yi Li	National Taiwan Normal University, Taiwan
Liangyu Chen	East China Normal University, China
Ling Li	Zhejiang Normal University, China
Lingyun Huang	Education University of Hong Kong, China
Liu Tianrui	University of Sydney, Australia
Luca Marconi	University of Milano-Bicocca, Italy
Luis P. Prieto	Universidad de Valladolid, Spain
Machi Shimmei	Tohoku University, Japan
Manikandan Ravikiran	Hitachi India Pvt. Ltd., India
Manpreet Singh	Education University of Hong Kong, China
Maomi Ueno	University of Electro-Communications, Japan
Margarida Romero	Université Côte d'Azur, France
Maria Marina	AIRI, Skoltech, Russia
Mariana Silva	University of Illinois Urbana-Champaign, USA
Marius Vogt	St. Gallen University of Teacher Education, Switzerland
Martin Ruskov	University of Milan, Italy
Matthieu Branthôme	Université de Bretagne Occidentale, France
Michael Liut	University of Toronto Mississauga, Canada
Michael Neumann	Hochschule Hannover, Germany
Min Lan	Zhejiang Normal University, China
Min Lu	Akita University, Japan
Min-Chi Chiu	嶺東科技大學資訊管理系, Taiwan
Mohammed Seyam	Virginia Tech, USA
Morgan Lee	Worcester Polytechnic Institute, USA
Mourya Teja Kunuku	Kennesaw State University, USA
Naiming Liu	Rice University, USA
Nasrin Dehbozorgi	Kennesaw State University, USA
Nghia Duong-Trung	German Research Centre for Artificial Intelligence, Germany
Nina Zhou	A*STAR, Singapore
Noshin Tahsin	Kishoreganj University, Bangladesh
Nuodi Zhang	Florida State University, USA
Ottavio G. Rizzo	Università degli Studi di Milano, Italy
Pascal Muam Mah	AGH University of Krakow, Poland
Pei-Chen Wu	National Taichung University of Education, Taiwan

R. Yamamoto Ravenor	Ochanomizu University, Japan
Rezaul Tutul	Berliner Hochschule für Technik, Germany
Rohit Saluja	IIT Mandi, India
Rositsa V. Ivanova	University of St. Gallen, Switzerland
Ruiwei Xiao	Carnegie Mellon University, USA
Ryan Zhang	Johns Hopkins University, Canada
Sabine Seufert	University of St.Gallen, Switzerland
Sabrina Eimler	Hochschule Ruhr West, Germany
Shan Zhang	University of Florida, USA
Shen Ba	Education University of Hong Kong, China
Shinobu Hasegawa	Japan Advanced Institute of Science and Technology, Japan
Shotaro Naganuma	Kyushu University, Japan
Siyuan Liu	Nanyang Technological University, Singapore
Songhee Han	Florida State University, USA
Soujatya Sarkar	Indian Institute of Technology, Mandi, India
Susan Beudt	Deutsches Forschungszentrum für Künstliche Intelligenz, Germany
Tadachika Ozono	Nagoya Institute of Technology, Japan
Uwe Maier	University of Education Schwäbisch Gmünd, Germany
Vasile Rus	University of Memphis, USA
Victor Sotelo	UNICAMP, Brazil
Vijay Prakash	Indian Institute of Technology Bombay, India
Wangda Zhu	Cornell University, USA
Xiangyu Tan	Shanghai Open University, China
Xianwen Zheng	Japan Advanced Institute of Science and Technology, China
Xinying Hou	University of Michigan, USA
Xuansheng Wu	University of Georgia, China
Yael Feldman-Maggor	KTH Royal Institute of Technology, Sweden
Yan Zhou	Zhejiang Normal University, China
Yannis Dimitriadis	University of Valladolid, Spain
Yaping Xu	Beijing Normal University, China
Yasuhiro Noguchi	Shizuoka University, Japan
Yiling Dai	Kyoto University, Japan
Yin Nicole Yang	Education University of Hong Kong, China
Yo Ehara	Tokyo Gakugei University, Japan
Yongnam Kim	Seoul National University, South Korea
Yoshimitsu Miyazawa	National Center for University Entrance Examinations, Japan
Yu Xiong	Chongqing University of Posts and Telecommunications, China

Yu-Kai Huang	Carnegie Mellon University, USA
Zechu Feng	University of Hong Kong, China
Zhi Liu	Central China Normal University/National Engineering Research Center for E-Learning, China
Zhizi Zheng	Education University of Hong Kong, China
Zifeng Liu	University of Florida, USA
Ziwei Wang	University of Sydney, Australia
Żywilla Fechner	Lodz University of Technology, Poland
昂轩 陈	北京大学, China

Doctoral Consortium Mentors

Andrew Lan	University of Massachusetts Amherst, USA
Bita Akram	North Carolina State University, USA
Bruce M. McLaren	Carnegie Mellon University, USA
Carla Limongelli	Roma Tre University, Italy
Carrie Demmans Epp	University of Alberta, Canada
Christian M. Stracke	University of Bonn, Germany
Gautam Biswas	Vanderbilt University, USA
Giora Alexandron	Weizmann Institute of Science, Israel
Ig Ibert Bittencourt	Federal University of Alagoas, Brazil
Jionghao Lin	University of Hong Kong, China
John Stamper	Carnegie Mellon University, USA
Judy Kay	University of Sydney, Australia
Kasia Muldner	Carleton University, Canada
Lauri Malmi	Aalto University, Finland
Luc Paquette	University of Illinois Urbana-Champaign, USA
Maria Mercedes Rodrigo	Ateneo de Manila University, Philippines
Michael Yudelson	Carnegie Mellon University, USA
Miguel L. Bote-Lorenzo	Universidad de Valladolid, Spain
Mingyu Feng	WestEd, USA
Neil Heffernan	Worcester Polytechnic Institute, USA
Penghe Chen	Beijing Normal University, China
Peter Brusilovsky	University of Pittsburgh, USA
Rawad Hammad	University of East London, UK
Roger Azevedo	University of Central Florida, USA
Ryan Baker	University of Pennsylvania, USA
Seiji Isotani	University of São Paulo, Brazil
Sergey Sosnovsky	Utrecht University, Netherlands
Sharon Hsiao	Santa Clara University, USA

Teng Guo	Jinan University, China
Danielle McNamara	Arizona State University, USA
Thomas Breideband	University of Colorado Boulder, USA
Tobias Ley	Danube University Krems, Austria
Valery Psyché	Université TÉLUQ, Canada
Zitao Liu	Jinan University, China

International Artificial Intelligence in Education (IAIED) Society

IAIED Management Board

President

Olga C. Santos	UNED, Spain

President-Elect

Seiji Isotani	University of São Paulo, Brazil

Secretary

Vania Dimitrova	University of Leeds, UK

Journal Editors

Vincent Aleven	Carnegie Mellon University, USA
Cristina Conati	University of British Columbia
Marcus Specht	Delft University of Technology

Finance Chair

Benedict du Boulay	University of Sussex, UK

Membership Chair

Benjamin D. Nye	University of Southern California, USA

IAIED Officers

Conrad Borchers	Carnegie Mellon University, USA
Xinyuan Hao	University College London, UK
Jionghao Lin	University of Hong Kong, China
Miguel Portaz	UNED, Spain

IAIED Executive Committee

Akihiro Kashihara	University of Electro-Communications, Japan
Amruth Kumar	Ramapo College of New Jersey, USA
Christothea Herodotou	Open University, UK
Jeanine A. Defalco	CCDC-STTC, USA
Judith Masthoff	Utrecht University, Netherlands
Maria Mercedes T. Rodrigo	Ateneo de Manila University, Philippines
Ning Wang	University of Southern California, USA
Olga C. Santos	UNED, Spain
Rawad Hammad	University of East London, UK
Zitao Liu	Jinan University, China
Bruce M. McLaren	Carnegie Mellon University, USA
Cristina Conati	University of British Columbia, Canada
Diego Zapata-Rivera	Educational Testing Service, Princeton, USA
Erin Walker	University of Pittsburgh, USA
Seiji Isotani	University of São Paulo, Brazil
Tanja Mitrovic	University of Canterbury, New Zealand
Noboru Matsuda	North Carolina State University, USA
Min Chi	North Carolina State University, USA
Alexandra I. Cristea	Durham University, UK
Neil Heffernan	Worcester Polytechnic Institute, USA
Andrew M. Olney	University of Memphis, USA
Irene-Angelica Chounta	University of Duisburg-Essen, Germany
Beverly Park Woolf	University of Massachusetts, USA
Ig Ibert Bittencourt	Universidade Federal de Alagoas, Brazil

International Alliance to Advance Learning in the Digital Era (IAALDE)

The IAIED Society is a member of the International Alliance to Advance Learning in the Digital Era (IAALDE) and belongs to its board. IAIED is currently represented at IAALDE by Olga C. Santos, as president of IAIED.

AIED 2025 Sponsors

Platinum

Google

Gold

Gates Foundation

Silver

Hewlett Packard Enterprise

Bronze

Eedi

VitalSource

duolingo english test

Best Paper Award Sponsorship

Springer

Interactive Events Track at AIED 2025

Emmanuel G. Blanchard[1] and Leonardo B. Marques[2]

[1] Université du Mans, France
emmanuel.blanchard@univ-lemans.fr
[2] Federal University of Alagoas, Brazil
leonardo.marques@cedu.ufal.br

Presentation

The Interactive Events (IE) Track at AIED 2025 was designed to offer a dynamic platform for showcasing innovations at the intersection of education, interactive systems, and artificial intelligence. It welcomed researchers, practitioners, and industry professionals from both within and beyond the AIED community to present interactive systems and demonstrations that address a wide range of educational challenges, with the goal of fostering collaboration and advancing the field. Authors of accepted papers or posters in the main conference or other AIED tracks were particularly encouraged to participate, as the IE Track provided an additional opportunity to showcase their systems.

Contributions

IE Track contributions had to present interactive systems operating across various platforms, such as standalone applications, web-based tools, mobile devices, and more. Below is a non-exhaustive list of possible contributions:

- virtual/augmented reality environments for learning
- adaptive educational systems
- serious/educational games
- intelligent tutoring systems
- agent-based learning environments
- data-driven and/or analytic-based systems for education
- LLM-based educational technologies
- authoring tools for AIED systems
- any other interactive tool relevant to the AIED community

Contributions were required to focus on systems that had already been implemented, and to demonstrate functional features that clearly illustrated the primary educational benefits described in the submission. A selection of them was invited to be featured on the AIED Society's showcase page: https://iaied.org/showcase.

AIED 2025 Keynotes

How Generative AI Has Changed Everything and Nothing in Education

Kristen DiCerbo

Chief Learning Officer at Khan Academy

Abstract. What changes in a world with generative AI in education and what remains the same? This talk will examine this question from multiple perspectives, addressing both how students learn and how we know if they have learned something. There are fundamental, well-replicated findings about how students learn that new technology does not change. However, new technology can change the emphasis on what knowledge, skills, and attributes we deem important for students to learn. Generative AI also changes the capabilities we have at our disposal to support learning. Does it help us address some of the seemingly intractable challenges that education technology has faced in learning and assessment? Or do those challenges remain unsolved? What new challenges does generative AI introduce? I will explore all of this using the lens of lessons learned in the development, implementation, and evaluation of Khanmigo, an AI-powered tutor for students and assistant for teachers.

Bio. Dr. Kristen DiCerbo is the Chief Learning Officer at Khan Academy, where she leads the content, assessment, design, product management, and community support teams. Time magazine named her one of the top 100 people influencing the future of AI in 2024. Dr. DiCerbo's career has focused on embedding insights from education research into digital learning and assessment experiences. Prior to her role at Khan Academy, she was Vice-President of Learning Research and Design at Pearson, served as a research scientist supporting the Cisco Networking Academies, and worked as a school psychologist. Kristen has a Ph.D. in Educational Psychology from Arizona State University.

The Technopolitics of AIED: A Developing Country's Perspective

Maria Mercedes T. Rodrigo

Ateneo de Manila University

Abstract. In the IJAIED opinion piece "Is the AIED Conundrum a First-World Problem?", I argued that educational systems in developing countries such as the Philippines could greatly benefit from AI-based teaching and learning because such technologies can augment these resource-poor environments and provide students with consistent quality of instruction. The rise of generative AI and the increasing availability of extended reality applications and technologies has increased this potential for benefit. However, equity continues to be a problem. In this keynote, I will discuss the GenAI and XR developments in the Asia-Pacific, showing how these boost learning and the learning experience. I will note how innovation and use differs among developed vs. developing nations. I will discuss challenges to sustainability and scaling, e.g. a lack of hardware, software, connectivity, and teacher training as well as an inability to keep pace with changing software authoring tools' pricing models. Finally, I will discuss the implications of LLM diplomacy on developing countries. The launch of DeepSeek and similar open, free applications provides entry points for AI participation to those of us who are resource-constrained. However, we are also aware that these platforms are values-laden, and that using them for education specifically gives their creators influence over students' and teachers' values and world views. Developing nations have to balance economic and political pressures against opportunities for technological progress, deciding what tradeoffs they are willing to make in order to expand their participation in AIED.

Bio. Maria Mercedes (Didith) T. Rodrigo is a professor at the Department of Information Systems and Computer Science. Her research interests include learning analytics, artificial intelligence in education, technology in education, and educational games. She is the head of the Ateneo Laboratory for the Learning Sciences and co-lead of the Ateneo Virtual, Augmented, and Mixed Reality Laboratory. Her research work focuses on artificial intelligence in education, learning analytics, and educational games. In collaboration with colleagues and students, Dr. Rodrigo's publications focus on artificial intelligence in education, learning analytics, and educational games. She serves on the editorial boards of several high-impact journals including Artificial Intelligence in Education, the Internet and Higher Education, and Research and Practice in Technology-Enhanced Learning. Dr. Rodrigo is on the Executive Committee of the Artificial Intelligence Education Society. In 2021, Dr. Rodrigo received the Distinguished Researcher Award from the Asia-Pacific Society for Computers in Education (APSCE). She is President of APSCE. Outside of her academic work, she writes books for children. She won the National Children's Book Award Kids' Choice Award for Made Perfect in Weakness in 2016. In

2018, she won the 12th Cardinal Sin Catholic Book Awards for Cave Dweller. Her book, Blueplate, won the 2024 National Children's Book Award Kids' Choice Award and was a finalist for the Cardinal Sin Catholic Book Awards.

Beyond the Algorithm: Participatory approaches to the use of AI to support inclusive and ethical education

Francesca Pozzi

Director of CNR – Institute for Educational Technology

Abstract. The evolution of technology has always gone hand in hand with the evolution of learning theories. Artificial intelligence is no exception, as it emerges at the intersection of learning theories and – going beyond the algorithm in itself – is neither inherently 'good' nor 'bad', rather, it is the way we use it that can produce either positive or negative effects. The debate on AI highlights the potential of this technology in terms of innovation, while also warning us about the possible risks of a tool that does not necessarily provide an accurate reflection of reality and whose responses tend to reflect the values and norms of some communities, rather than others. The focus of this talk, therefore, is to reflect on how AI should be used in schools and universities to maximize the common good rather than exacerbating inequalities for the exclusive benefit of some minorities. Specifically, I argue that by promoting a participatory use of AI based on social constructivist approaches and leveraging AI interactions to encourage student exchange and discussion, we can foster the construction of new, shared, and equitable meanings. This approach can also help students reflect on and recognize the ethical risks embedded in AI systems. Promoting the use of AI as a sparring partner rather than as an expert that completes the task can lead to the enhancement of cognitive skills rather than their replacement. Discussing AI-generated results with others can encourage reflection on the structure and value of the output and can be used to stimulate students to do better than the algorithm. Additionally, it helps foster critical thinking about how misinformation can be created. However, to achieve this, educational systems need to be radically rethought and transformed in their organizational, structural, curricular, and methodological dimensions to return agency and centrality to learners and teachers need to be trained to shift their role and guide this process with due awareness. The alternative approach, consisting of simply shutting the door to this technology, would lead to schools that are even further than today from the external reality and to students prone to all the risks as soon as they leave the protected and artificial, although not intelligent, school environment.

Bio. Francesca Pozzi is Director of the Institute for Educational Technology of the National Research Council (ITD-CNR). She received her PhD in Languages, Cultures and Information and Communication Technologies from the University of Genoa and – since 1998 – she has been working as a researcher at ITD-CNR. Over the years she has coordinated and participated in numerous national and international projects, in which she has dealt with various aspects of Educational Technology. She is author of 180+ publications and is co-editor since 2016 of the journal 'Italian Journal of Educational

Technology (IJET)' (formerly 'Tecnologie Didattiche'), the first Italian scientific journal dedicated to this topic. She is also a member of Editorial Boards of several international journals and is annually involved in the scientific committees of 20+ international conferences. She is a member of the Teaching Board of the Doctorate in Digital Humanities at the University of Genoa and has been Chair of several scientific events and member of the organising committees of several scientific events.

Contents – Part I

WideAIED

Doctoral Consortium

Contents – Part II

Late Breaking Results

Contents – Part III

Workshops

BlueSky

CLJ4AI: Citizen Learning Journey for AI

Luis Felipe Coimbra Costa[1,2], Carlos Eduardo Barbosa[2,3](✉), Henrique de Andrade[2], Yuri Lima[2], Eduardo Carvalho Nunes[4], Luciana Maria Azevedo Nascimento[2], Ana Moura Santos[5], and Geraldo Xexéo[2]

[1] NUCLIO - Núcleo Interativo de Astronomia e Inovação em Educação, 2785-817 São Domingos de Rana, Portugal
luisfcosta@nuclio.org

[2] Universidade Federal do Rio de Janeiro, 21941-450 Rio de Janeiro, Brazil
{eduardo,handrade,yuriodelima,lmnascimento,xexeo}@cos.ufrj.br

[3] Centro de Análises de Sistemas Navais, Marinha do Brasil, 20091-000 Rio de Janeiro, Brazil

[4] Universidade de Trás-os-Montes e Alto Douro, 5000-801 Vila Real, Portugal
eduardonunes@utad.pt

[5] Instituto Superior Técnico, Universidade de Lisboa, 1649-004 Lisbon, Portugal
ana.moura.santos@tecnico.ulisboa.pt

Abstract. Science, Technology, Engineering, and Mathematics education faces challenges in maintaining student engagement, reducing gender disparities, and fostering inclusive learning experiences. At the same time, the growing enthusiasm for Generative Artificial Intelligence among young learners presents an opportunity to address these issues by integrating AI literacy into K-12 curricula. We propose the Citizen Learning Journey for AI project, aimed at equipping students with essential digital literacy and AI competencies while promoting AI's responsible and ethical use in their future careers and daily lives. The project leverages GenAI using UNESCO's seven roles to motivate students in STEM subjects, integrating the UNESCO Competency Framework for Students and employing Educational Design Research. It adapts the Heroine's Learning Journey (HLJ) model to create a dynamic, narrative-driven approach that enhances student engagement. Athena, an AI Virtual Assistant, built, is at the initiative's core. Athena incorporates key elements—Challenge, Mission, Story, and Aid—that are aligned with HLJ stages. These elements support competency development across four domains: Human-Centered Mindset, Ethics of AI, AI Techniques and Applications, and AI System Design. The project provides educational materials reinforcing learning, including interactive tutorials, stage videos, quizzes, and multimedia resources emphasizing AI's social and ethical dimensions. The project aligns with the values of the European Union and the UN Sustainable Development Goals, advancing sustainable development through an educational approach integrating AI, equity, and inclusion.

Keywords: STEM · Generative AI · Citizen Learning · Competency Framework for Students · Heroine's Learning Journey · Education

A. I. Cristea et al. (Eds.): AIED 2025, CCIS 2590, pp. 3–16, 2025.
https://doi.org/10.1007/978-3-031-99261-2_1

1 Introduction

Science, Technology, Engineering, and Mathematics (STEM) education continues to face persistent challenges, including low student engagement (SDG 4.4), gender disparities (SDG 4.5, 5.B), and a lack of inclusive learning practices for diverse student populations. Conventional pedagogical strategies often fail to connect abstract concepts to students' lived experiences, resulting in diminished interest and participation in STEM learning pathways [14].

Recent studies have demonstrated a significant rise in interest among adolescents and young adults in Generative Artificial Intelligence (GenAI) technologies [20]. This growing enthusiasm offers a valuable opportunity to rethink how STEM education can be made more engaging and relevant. By aligning students' curiosity about GenAI with formal educational objectives, it becomes possible to integrate digital and AI literacy—particularly within the field of Artificial Intelligence in Education (AIED)—in ways that foster both cognitive development and social responsibility.

However, the incorporation of GenAI into education raises significant concerns. Issues related to its accuracy, ethical implications, and cultural biases have been widely discussed. Moreover, the COVID-19 pandemic has accelerated digital transformation in education, further exposing technological inequities and highlighting the need for flexible, student-centered learning environments [35]. In this context, it becomes essential to explore frameworks that address both pedagogical shortcomings and the ethical integration of AI tools.

International policy developments reflect growing attention to these challenges. For instance, the OECD's 2025 report calls for coordinated efforts to build capacity for AI adoption in education, emphasizing evidence-based approaches [31]. Similarly, the AAAI 2025 Presidential Panel Report [32] emphasizes the rapid evolution of GenAI research and the urgency of implementing reliable assessment and governance structures in education. Countries such as the United Kingdom and Spain have already incorporated GenAI into their national educational strategies, focusing on safe and ethical implementation [15,25]. In parallel, China's Ministry of Education promotes intelligent technologies as central to education reform [28].

From a labor market perspective, demand for AI-related competencies continues to grow. Reports show that 28% of European workers currently engage with some form of AI in their jobs [6]. At the same time, projections suggest the EU will need 1.2 million ICT professionals by 2030 [9,16] and The Future of Jobs Report projects that AI will drive demand for digital and STEM skills [39]. In response, the European Union intends to position STEM as a strategic education [8], and the European Union's AI Act has mandated AI literacy initiatives as a public priority [5]. These developments underline the need to equip students with foundational AI competencies in alignment with both national and international strategies.

Despite these encouraging efforts, a key challenge remains: the lack of inclusive, adaptive learning frameworks that effectively engage diverse learners. GenAI technologies such as ChatGPT offer new possibilities for dynamic, interactive learning, yet their deployment requires careful educational design to ensure equitable outcomes and responsible use [3]. This article addresses this challenge by introducing the Citizen Learning Journey for AI (CLJ4AI). This framework integrates UNESCO's Competency Framework for Students (CFS) [36] and the Heroine's Learning Journey (HLJ) model [11] to create a narrative-driven, ethically grounded approach to AI literacy in K-12 education, in alignment with EU 2030 STEM goals [18].

In the following section, we examine related work to contextualize this proposal within ongoing academic and policy discussions on GenAI in education, identifying both opportunities and unresolved gaps.

2 Related Work

The applications of AI in education are diverse [24], including adaptive learning and the use of virtual assistants for personalized learning [34]. These applications have evolved very rapidly, drawing on a large volume of work and diverse perspectives as indicated by Wang et al.'s review [38], allowing research opportunities in this rapidly changing field. One of the opportunities that opens up is the study of GenAI.

The emergence of GenAI technologies has significantly influenced educational methodologies, particularly by enabling personalization, enhancing engagement through narrative techniques, and supporting novel pedagogical practices. Advances in large language models (LLMs) and multimodal systems have broadened the scope of educational applications [29].

One of the most studied benefits of GenAI in education is its potential to support adaptive learning environments. Intelligent tutoring systems powered by GenAI can dynamically adjust content and scaffolding based on learners' needs, thereby offering personalized educational trajectories [29]. Such systems have shown promise in facilitating differentiated instruction across cognitive and developmental levels. Moreover, bibliometric analyses reveal a growing interest in GenAI for automating assessments and generating customized content [26].

In addition to personalization, GenAI enhances engagement through storytelling and multimodal interaction. AI-generated narratives, incorporating text, voice, and visual elements, have been shown to foster creativity and curiosity among students—especially in early education settings [4]. Systematic reviews highlight how these approaches can stimulate interdisciplinary thinking and deepen conceptual understanding. However, the cultural adaptability of such systems remains a subject of active inquiry [40].

Another promising application of GenAI is its role in the learning-by-teaching paradigm. In this approach, students interact with teachable AI agents, which act as novice learners. This configuration allows students to reinforce their own knowledge by instructing the AI, particularly in technical domains such as programming and mathematics [7]. Although promising, this strategy raises questions about academic integrity, especially in the context of AI-assisted writing and assessment generation [26].

Despite the benefits, the limitations of GenAI-based educational tools must be addressed. Accessibility disparities, especially across socio-economic or geographic lines, may exacerbate the digital divide. Concerns about data privacy and bias in model outputs also warrant ongoing scrutiny [40]. To mitigate these issues, researchers advocate for ethical guidelines, transparent algorithm design, and robust oversight mechanisms in AI deployment [22].

Together, these studies underscore the need for an educational framework that leverages GenAI's adaptive and narrative potential while addressing its ethical, cultural, and technological challenges.

3 Theoretical Background

This section presents the conceptual foundations that support the framework of CLJ4AI. It examines two main models: UNESCO's CFS, which guides the integration of AI literacy into school curricula, and the HLJ, which addresses gender disparities in STEM education through narrative-based methodologies.

3.1 UNESCO AI Competency Framework for Students

UNESCO has been carrying out an important leadership mission in promoting educational innovation, which is aligned with the Education Agenda 2030 and Sustainable Development Goal 4. In addition to various documents and guides [37], as part of this mission, UNESCO introduced the CFS [36], providing a structured model for incorporating AI literacy into formal education.

The CFS defines twelve essential competencies, organized across four dimensions: *Human-Centred Mindset*, *Ethics of AI*, *AI Techniques and Applications*, and *AI System Design*. Each dimension addresses specific aspects of AI literacy, ranging from ethical considerations to technical proficiency. The framework's design aims to equip students with both the operational skills to use AI tools and the critical reasoning skills to evaluate the social impacts of AI.

To aid progressive learning, the CFS outlines three levels of mastery: (a) *Understand*, (b) *Apply*, and (c) *Create*. At the fundamental "Understand" level, students develop a basic awareness of AI concepts and their ethical implications. At the "Apply" level, they are expected to engage with AI tools in real-world contexts actively. Finally, at the "Create" level, students innovate by designing AI solutions based on ethical and technical rigor.

Beyond serving as a curricular guide, the CFS functions as a diagnostic and prescriptive tool, offering detailed learning objectives that can be adapted to diverse educational systems. By defining clear outcomes at each progression level, the framework aims to form responsible and critically engaged citizens capable of navigating an increasingly AI-driven society.

3.2 Heroine's Learning Journey

HLJ was developed to address the persistent underrepresentation of women in STEM fields, particularly in computing and engineering, and uses structured narratives to foster motivation and self-confidence among learners [11].

The HLJ is organized into three acts and twelve stages, each representing a critical phase in a student's learning process. Within each stage, four elements are emphasized: Challenge (the task or difficulty faced), Mission (the objective or learning goal), Story (an immersive narrative framing the experience), and Aid (external support or mentoring), as illustrated in Fig. 1. This structure enables a balance between emotional engagement and skill development, encouraging students to see themselves as protagonists in their educational journey.

Technologically, the HLJ model is supported by the Heroine's Learning System (HeLaS) [21], a scalable digital platform that facilitates implementation across varied educational contexts. Empirical applications of the HLJ in MOOCs, particularly at the Instituto Superior Técnico (IST) of the University of Lisbon, have demonstrated notable successes: an increase in female enrollment from 37.3% to 59.2% and heightened self-reported motivation among participating students [11].

The HLJ model has been evaluated by educational experts from EU, earning recognition for enhancing female participation in STEM [12,13] and also received international recognition, including awards from the ACM SIGHPC Education Committee [1] and the Brazilian Computer Society [10]. These acknowledgments highlight its effectiveness in promoting gender equity in STEM education through innovative pedagogical approaches.

While initially focused on addressing gender disparities, the HLJ's adaptable narrative framework offers broader applications for promoting diversity, engagement, and resilience across various student populations. Its emphasis on structured progression through challenges, stories, and support systems makes it particularly well-suited for integration with digital and AI literacy initiatives.

Together, the UNESCO CFS and the HLJ model offer complementary foundations for designing an AI literacy framework that is both technically robust and socially inclusive. In the following section, we detail how CLJ4AI synthesizes these two models, creating a dynamic and ethical learning pathway for K-12 students in the era of Generative AI.

4 Framework Integration: CFS and HLJ for AI Literacy

The CLJ4AI project articulates an integrated framework that brings together CFS and HLJ. In this approach, the HLJ stages, organized into elements such as Challenge, Mission, Story, and Aid, are mapped onto the key competency aspects of the CFS, creating personalized learning journeys that foster technical proficiency and ethical, responsible citizenship.

As illustrated in Fig. 2, this integration is achieved by aligning each HLJ stage with a central competency domain from the CFS. For instance, the stage "Learning About the Heroine's World (1)" paired with the act "The Light of Knowledge (6)" corresponds to the *Human-centred mindset* dimension, with progression levels ranging from fostering human agency to encouraging human accountability and nurturing citizenship in the era of AI. Similarly, other stages—such as "Malala's Call (2)", "External and Internal (3)", and "Facing a Storm (4)" are

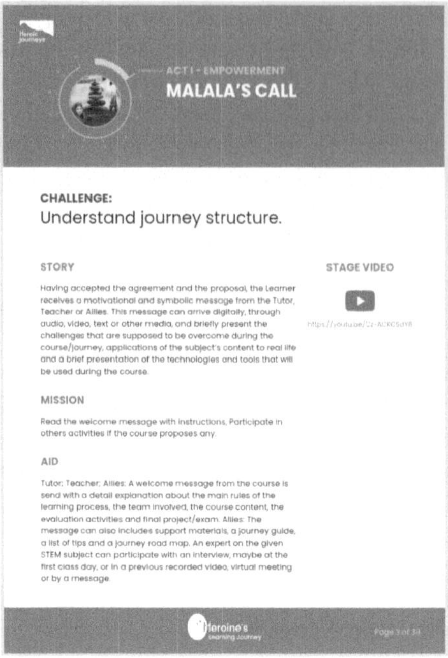

Fig. 1. HLJ Conceptual Representation.

systematically related to the dimensions *Ethics of AI*, *AI techniques and applications*, and *AI system design*, respectively, each unfolding across the progression levels of Understand, Apply, and Create. The stage "Transforming Fear into Strength (5)" is further developed through a sequence of the acts "Self-Regulated Learning (10)", "System Reboot (11)", and "Becoming a STEM Heroine (12)", which extend the narrative and emphasize the continuous evolution of the learning journey, even without a direct mapping to an additional CFS aspect.

4.1 System Architecture: Modular Design of CLJ4AI

In addition to the conceptual mapping, Fig. 3 presents a technical diagram of the system architecture and workflow of the CLJ4AI. This architecture integrates the HeLaS platform with Athena, the AI Virtual Assistant, to deliver the educational journey through interactive tutorials, multimedia content, quizzes, and narrative-based challenges. Based on UNESCO's guidelines [33], Athena is designed to assume multiple roles (e.g., idea generator, Socratic interlocutor, personalized tutor). The implementation leverages the HeLaS platform as its technological backbone. The architecture is organized into three interconnected modules: Knowledge Base (KB), Artificial Intelligence (AI), and User Interface (UI).

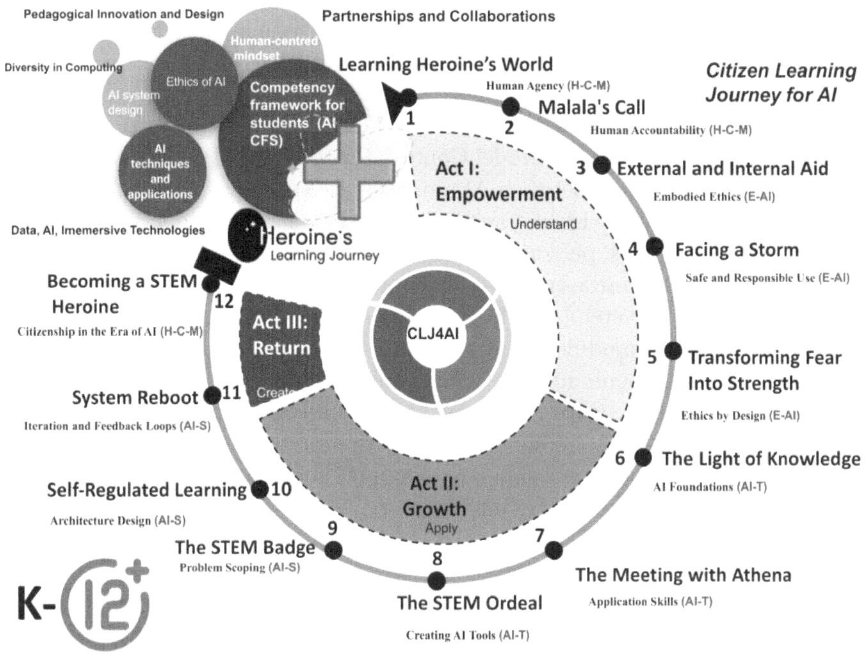

Fig. 2. CLJ4AI framework with 12 HLJ stages and 12 CFS competencies.

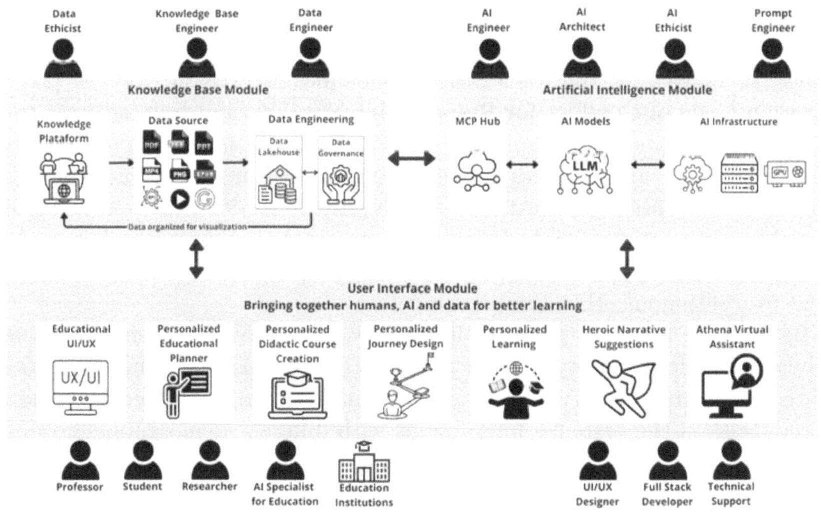

Fig. 3. Technical diagram of the CLJ4AI system architecture, showing the integration of the HeLaS platform with Athena and supporting educational modules.

1. The Knowledge Base aggregates domain-specific resources in AI, STEM education, and digital literacy. Educational content is managed through structured sources such as APIs, documents, SQL databases, and multimedia files. Data engineering techniques support the maintenance of a Data Lakehouse with rigorous data governance protocols to ensure integrity and accessibility.
2. The AI module includes the Model Context Protocol (MCP) Hub, which standardizes the interaction between AI models and external systems [23]. It hosts large language models (LLMs) and specialized generative models tailored for educational narratives, personalization, and adaptive learning strategies. Scalable computational infrastructures, including GPU resources and cloud services, support the system's performance and reliability.
3. The User Interface module focuses on delivering a seamless, learner-centered experience. Components include a Personalized Educational Planner, Didactic Course Creation tools, Journey Design modules based on HLJ, and dynamic Heroic Narrative Suggestions. Athena, embedded within this UI layer, offers continuous personalized support, serving various roles such as tutor, motivator, and Socratic interlocutor.

Stakeholders involved in the development and refinement of the platform include educators, students, researchers, UI/UX designers, developers, and technical support teams, ensuring a participatory, iterative design process. Data flows iteratively between the Knowledge Base, AI models, and User Interface components, consistent with the principles of Educational Design Research (EDR) methodology [27].

4.2 Ethical and Legal Considerations

The CLJ4AI framework is deeply grounded in ethical standards and legal compliance. It aligns with the EU's Ethics Guidelines for Trustworthy AI [22], the European Code of Conduct for Research Integrity [2], and fully adheres to the General Data Protection Regulation (GDPR) [19].

Transparency and fairness are operationalized through explainability-by-design features embedded in Athena, which provide students and educators with accessible justifications for AI recommendations. An independent Ethics Advisory Panel regularly audits Athena's outputs and publishes transparency reports to ensure continuous ethical improvements.

To mitigate bias and promote inclusivity, educational content is systematically audited for stereotypes and representation gaps. Training datasets are curated to ensure diversity and periodic evaluations are conducted to assess any discrepancies in the system's interactions with different demographic groups.

Privacy protection is reinforced through local data processing where feasible, pseudonymization of personal data, and granular consent management mechanisms. Students and guardians maintain control over data usage, with clear options for opting out or revoking permissions, and data access is strictly monitored through automated audits.

It is important to highlight that ethical education is not limited to backend compliance but is embedded within the student learning experience itself. Specific modules will guide students through ethical dilemmas related to AI, fostering reflective and values-based decision-making. Athena also simulates real-world scenarios to encourage responsible and critical thinking about AI use.

Educators are supported through an Ethics Dashboard that enables them to monitor AI outputs, flag concerns, suggest revisions, and access continuous professional development on AI ethics. These measures ensure that CLJ4AI promotes a culture of responsibility, inclusivity, and transparency, consistent with international best practices.

Ethical approval for the first phase of this project was already obtained from the Ethics Committee for the Monitoring of School Surveys under the Directorate- General for Education and Science Statistics within the Ministry of Education, Science, and Innovation of Portugal [17].

Building upon this integrated conceptual and technical foundation, the following section presents the evaluation methodology employed to assess the educational impact and iterative refinement of the CLJ4AI framework through empirical studies.

5 Evaluation Methodology

CLJ4AI project employs EDR methodology [27] to develop, test, and refine its educational framework iteratively. This approach is particularly appropriate given the project's dual aim: to create practical educational solutions and to generate new theoretical insights in the fields of AI literacy and STEM education.

EDR integrates research with real-world practice, emphasizing systematic cycles of design, implementation, evaluation, and redesign. In CLJ4AI, these cycles align the CFS with the narrative-driven structure of the HLJ, ensuring that each educational element supports both technical competency development and ethical awareness.

Each iteration of the EDR cycle begins by identifying specific challenges in STEM and AI education, designing targeted interventions, and implementing them in authentic educational settings, as shown in Fig. 4. Subsequent phases involve collecting stakeholder feedback, analyzing outcomes, and refining both the content and delivery mechanisms based on empirical evidence. This process promotes continuous improvement, ensuring that CLJ4AI remains responsive to learner needs and educational best practices.

The evaluation strategy for CLJ4AI closely follows the methodology previously applied in assessing the HLJ framework [12]. A panel of educators from partner institutions will participate in the evaluation, providing structured assessments of the newly developed narrative stages and competency mappings.

Evaluation instruments include detailed materials outlining each stage's narrative, learning objectives, and targeted competencies. Educators evaluate these materials using a mixed-methods approach, combining qualitative feedback (via open-ended responses) with quantitative evaluations (via Likert-scale ratings).

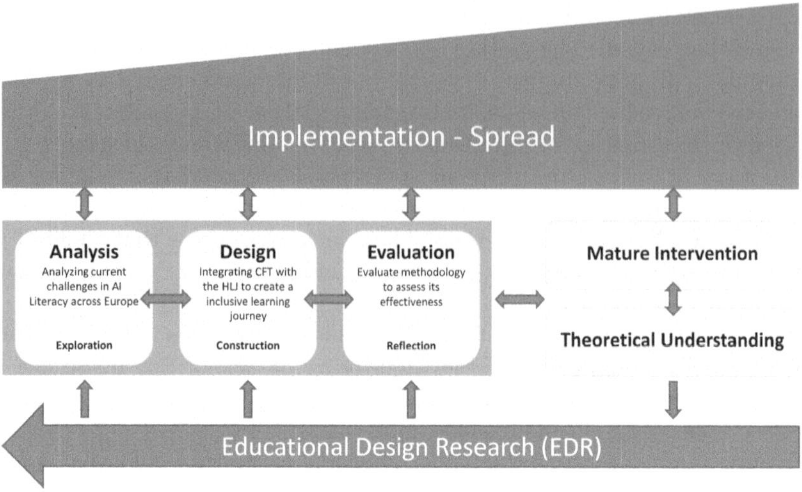

Fig. 4. Educational Design Research framework applied in CLJ4AI.

The evaluation criteria focus on pedagogical coherence, the relevance of AI competencies addressed, and the overall capacity of the learning journey to foster responsible AI usage among students. In addition to educator evaluations, CLJ4AI incorporates participatory design methodologies involving students directly in the development process. Following the principles of user-centered design [30], co-design workshops are conducted with diverse groups of students, utilizing techniques such as user journey mapping and rapid prototyping. These workshops ensure that the system's interface, narratives, and learning activities are intuitive, engaging, and culturally sensitive.

Through iterative usability studies, prototypes of the Athena system will be tested in authentic educational environments. Student feedback will be collected using structured questionnaires, interviews, and focus groups. This feedback will be systematically analyzed and integrated into subsequent EDR cycles, facilitating incremental improvements in both functionality and educational effectiveness. The evaluation methodology combines educator assessments, participatory design with students, and iterative field testing to ensure that CLJ4AI is rigorously grounded in empirical evidence. It also maintains a strong commitment to inclusivity, ethical engagement, and adaptability across diverse educational contexts.

6 Discussion

The CLJ4AI framework emerges from integrating the UNESCO CFS and HLJ model, offering a approach to STEM education focusing on AI. The CFS provides a robust structure for embedding AI literacy into K-12 curricula through four key domains: Human-Centered Mindset, Ethics of AI, AI Techniques and Applications, and AI System Design. In parallel, the HLJ model leverages a narrative-driven methodology to enhance student motivation, particularly among underrepresented groups such as young women in STEM.

By merging these frameworks, CLJ4AI creates a holistic educational paradigm that transcends traditional technical instruction, cultivating AI proficiency, critical ethical reasoning, and civic responsibility. This synergy addresses the pressing need for innovative pedagogies that prepare students for a technological future while aligning with global educational priorities, such as the UN Sustainable Development Goals.

A defining feature of CLJ4AI is its capacity to tailor learning experiences to students' diverse developmental stages and needs. The framework aligns the HLJ's 12 narrative stages—such as "Learning About the Heroine's World" and "Malala's Call"—with the CFS's competency progression levels (Understand, Apply, Create), enabling a scaffolded journey from foundational AI concepts to advanced problem-solving and innovation. For example, early stages introduce students to ethical considerations and human agency in AI, while later stages guide them toward designing AI systems with real-world applicability. This staged approach ensures that learners, regardless of prior knowledge or pace, can engage with content that evolves in complexity and relevance. By personalizing education in this manner, CLJ4AI promotes inclusivity and bridges gaps in STEM participation, particularly for girls and students from diverse backgrounds.

Athena is central to the CLJ4AI framework, to be developed on the HeLaS platform and designed to embody UNESCO's seven roles for AI in education. Athena enhances the learning experience by delivering real-time, adaptive support through interactive tutorials, quizzes, and narrative-driven resources that reinforce technical skills and ethical principles. For instance, it can generate challenges and missions aligned with HLJ stages while addressing CFS competencies, such as evaluating the societal impact of AI or applying machine learning techniques. Athena not only augments student engagement but also alleviates educators' workload, enabling them to focus on mentorship and curriculum design.

Implementing CLJ4AI and Athena poses challenges. Its technological demands may deepen inequities, requiring strong infrastructure. Curriculum adaptation and teacher training need time, funding, and support. Privacy concerns highlight the need for strict data protection that is aligned with ethical and legal standards.

7 Conclusions

STEM education faces persistent challenges in engagement, equity, and inclusivity. Still, the growing interest in GenAI among youth creates an opportunity to leverage AIED to foster cognitive development and social responsibility.

This work presents the CLJ4AI framework, which integrates the UNESCO CFS and the HLJ model to foster AI literacy in K-12 education. Supported by the AI Virtual Assistant Athena, CLJ4AI combines competency development with narrative-driven learning to enhance engagement, ethical reasoning, and technical skills. It offers personalized, inclusive pathways in STEM education while addressing challenges of motivation, equity, and the ethical use of GenAI.

CLJ4AI offers educators innovative tools to teach AI literacy by combining ethical reflection, technical skills, and adaptive support through Athena, thus reducing instructional burdens. For students, the framework delivers dynamic, narrative-based learning pathways that foster critical thinking, creativity, and responsible AI use. Policymakers can leverage CLJ4AI as a strategic model aligned with the UN SDGs and EU digital strategies to promote equity, innovation, and digital empowerment in national education policies. Companies benefit from the preparation of a new generation of professionals equipped with both technical AI competencies and strong ethical foundations. For society at large, CLJ4AI fosters a digitally literate society prepared for an AI-driven future.

However, important challenges remain. CLJ4AI's technological demands raise concerns about digital equity, while the adaptability of narrative frameworks like HLJ across cultures requires further study. Data privacy, ethical compliance, and infrastructure accessibility must remain priorities in future deployments. Future work will validate CLJ4AI's impact on AI literacy, STEM motivation, and ethical awareness while enhancing Athena's natural language, multilingual, and feedback capabilities.

Acknowledgments. This work was funded by *Coordenação de Aperfeiçoamento de Pessoal de Nível Superior* (CAPES) under Grant Finance Code 001. We acknowledge using AI generative tools to support text processing tasks: translation, rewriting to improve clarity, and condensing text to meet page limits. We reviewed all outputs to ensure alignment with our original intent.

Disclosure of Interests. The authors declare that they have no competing interests.

References

1. ACM SIGHPC Education Committee: Acm sighpc education award 2024 announcement (2024). https://sighpceducation.acm.org/events/award24_announce/
2. ALLEA: The European Code of Conduct for Research Integrity (2023). https://allea.org/wp-content/uploads/2023/06/European-Code-of-Conduct-Revised-Edition-2023.pdf
3. Almassaad, A., Alajlan, H., Alebaikan, R.: Student perceptions of generative artificial intelligence: investigating utilization, benefits, and challenges in higher education. Systems **12**, 385 (2024). https://doi.org/10.3390/systems12100385
4. Arif, S., Arif, T., Khan, A.J., Haroon, M.S., Raza, A.A., Athar, A.: The Art of Storytelling: Multi-Agent Generative AI for Dynamic Multimodal Narratives. arXiv preprint (2024). https://arxiv.org/abs/2409.11261
5. Artificial Intelligence Act Initiative: Article 4 - Artificial Intelligence Act (2025). https://artificialintelligenceact.eu/article/4/

6. Cedefop: Skills Empower Workers in the AI Revolution: First findings from Cedefop's AI skills survey. Technical report, European Centre for the Development of Vocational Training (Cedefop), Thessaloniki, Greece (2025). https://doi.org/10.2801/6372704
7. Chen, A., Wei, Y., Le, H., Zhang, Y.: Learning-by-Teaching with ChatGPT: The Effect of Teachable ChatGPT Agent on Programming Education (2024). https://arxiv.org/abs/2412.15226
8. Commission, E.: STEM Education Strategic Plan: Skills for Competitiveness and Innovation (2025). https://ec.europa.eu
9. Commission, E.: Union Skills and EU Competitiveness (2025). https://commission.europa.eu/topics/eu-competitiveness/union-skills_en
10. COPPE: Project aimed at increasing female representation in STEM areas has won an award (2024). https://coppe.ufrj.br/en/planeta-coppe/project-aimed-at-increasing-female-representation
11. Costa, L.F.C., et al.: Heroine's learning journey: motivating women in STEM online courses through the power of a narrative. IEEE Access **12**, 20103–20124 (2024). https://doi.org/10.1109/ACCESS.2024.3360376
12. Costa, L.F., et al.: Stem education for young women via heroic narratives: evaluating the heroine's learning journey. In: Simpósio Brasileiro de Educação em Computação (EDUCOMP), pp. 52–62. SBC (2024)
13. Costa, L., et al.: Societal impact of the heroine's learning journey: inspiring women in STEM through the power of role models. Societal Impacts **5**, 100100 (2025)
14. Council, N.R., et al.: National Science Education Standards. National Academies Press (1995)
15. Department for Education: Generative Artificial Intelligence (AI) in education (2025). https://www.gov.uk/government/publications/generative-artificial-intelligence-in-education/generative-artificial-intelligence-ai-in-education
16. DIGITALEUROPE: Position paper on bridging the digital divide (2022). https://cdn.digitaleurope.org/uploads/2022/09/DIGITALEUROPE-position-paper-on-bridging-the-digital-divide.pdf
17. Direção-Geral de Estatísticas da Educação e Ciência (DGEEC): Inquérito no Sistema de Monitorização de Inquéritos em Meio Escolar (MIME), ID 14933. https://mime.dgeec.mec.pt/InqueritoConsultar.aspx?id=14933
18. European Commission: 2030 digital compass: The european way for the digital decade (2021). https://eufordigital.eu/wp-content/uploads/2021/03/2030-Digital-Compass-the-European-way-for-the-Digital-Decade.pdf
19. European Parliament and Council of the European Union: General data protection regulation (GDPR) (regulation (EU) 2016/679) (2016). https://eur-lex.europa.eu/eli/reg/2016/679/oj
20. Harvard Graduate School of Education: Teen and Young Adult Perspectives on Generative AI (2024). https://digitalthriving.gse.harvard.edu/wp-content/uploads/2024/06/Teen-and-Young-Adult-Perspectives-on-Generative-AI.pdf
21. Heroic Journeys: Heroic Journeys Framework. https://heroicjourneys.life/
22. High-Level Expert Group on Artificial Intelligence: Ethics Guidelines for Trustworthy AI (2019). https://digital-strategy.ec.europa.eu/en/library/ethics-guidelines-trustworthy-ai
23. Hou, X., Zhao, Y., Wang, S., Wang, H.: Model Context Protocol (MCP): Landscape, Security Threats, and Future Research Directions (2025). arXiv preprint arXiv:2503.23278. https://doi.org/10.48550/arXiv.2503.23278
24. Huang, J., Saleh, S., Liu, Y.: A review on artificial intelligence in education. Acad. J. Interdisc. Stud. **10**(3) (2021)

25. INTEF: Inteligencia Artificial en el aula. https://descargas.intef.es/cedec/proyectoedia/guias/contenidos/inteligencia_artificial/index.html
26. Liu, J., et al.: A bibliometric analysis of generative AI in education: current status and development. Asia Pac. J. Educ. **44**(1), 156–175 (2024). https://doi.org/10.1080/02188791.2024.2305170
27. McKenney, S., Reeves, T.: Conducting Educational Design Research. Routledge, Abingdon (2018)
28. Ministry of Education of the People's Republic of China: Press release - ministry of education, prc (2024). http://en.moe.gov.cn/news/press_releases/202404/t20240422_1127028.html
29. Mittal, U., Sai, S., Chamola, V., Sangwan, D.: A comprehensive review on generative AI for education. IEEE Access (2024). https://doi.org/10.1109/ACCESS.2024.3468368
30. Norman, D.A., Draper, S.W.: User centered system design; new perspectives on human-computer interaction. L. Erlbaum Associates Inc. (1986)
31. OECD: Everybody Cares About Using Education Research Sometimes: Perspectives of Knowledge Intermediaries. Educational Research and Innovation, OECD Publishing, Paris (2025). https://doi.org/10.1787/5ef88972-en
32. Rossi, F., Members, A.P.P.: AAAI 2025 Presidential Panel on the Future of AI Research (2025). https://aaai.org/wp-content/uploads/2025/03/AAAI-2025-PresPanel-Report-FINAL.pdf
33. Sabzalieva, E., Valentini, A.: ChatGPT and Artificial Intelligence in Higher Education: Quick Start Guide. Technical report, IESALC, Paris, France and Caracas, Venezuela (2023). https://unesdoc.unesco.org/ark:/48223/pf0000385146
34. Sajja, R., Sermet, Y., Cikmaz, M., Cwiertny, D., Demir, I.: Artificial intelligence-enabled intelligent assistant for personalized and adaptive learning in higher education. Information **15**(10), 596 (2024)
35. Stéphan, V.L.: Educational Research and Innovation Measuring Innovation in Education 2023 Tools and Methods for Data-Driven Action and Improvement. OECD Publishing (2023)
36. UNESCO: AI Competency Framework for Students (2024). https://unesdoc.unesco.org/ark:/48223/pf0000391105
37. UNESCO: Artificial Intelligence in Education (2025). https://www.unesco.org/en/digital-education/artificial-intelligence
38. Wang, S., Wang, F., Zhu, Z., Wang, J., Tran, T., Du, Z.: Artificial intelligence in education: a systematic literature review. Expert Syst. Appl. **252**, 124167 (2024)
39. World Economic Forum: The Future of Jobs Report 2025 (2025). https://www.weforum.org/publications/the-future-of-jobs-report-2025/
40. Yusuf, A., Pervin, N., Román-González, M., Noor, N.M.: Generative AI in education and research: a systematic mapping review. Rev. Educ. **12** (2024). https://doi.org/10.1002/rev3.3489

Why the Future of AIED is Causal: Arguments for Creating a Tradition Based on Causal Thinking

Lea Cohausz(✉)

University of Mannheim, Mannheim, Germany
lea.cohausz@uni-mannheim.de

Abstract. Artificial Intelligence in Education (AIEd) differs from other application areas of AI by having a vulnerable user group supposed to benefit from the research in the area and by frequently asking more of machine learning (ML) models than only to produce predictions. The first difference is apparent in the comparatively strong research tradition of AIEd concerning fairness and explainability. The second difference is pronounced in the motivation for AIEd research and tools; e.g., the aim is not to predict who is likely to drop out but to intervene to avoid dropout. In this paper, we argue that these differences lead to causal thinking being particularly valuable for AIEd compared to other application areas of artificial intelligence and argue in favor of establishing a tradition surrounding causal thinking. We state that a focus on causality can further increase the area's existing strengths while simultaneously leading to new insights. We discuss that by incorporating causal thinking, AIEd becomes more accurate, fairer, more explainable, achieves better interventions, makes better recommendations, and can even make a more convincing case for its importance. Moreover, establishing a tradition around causal thinking naturally fosters interdisciplinary research and the inclusion of stakeholders such as teachers and students.

Keywords: causality · XAI · AI fairness · interventions · recommendation · stakeholder inclusion

1 Introduction

Research on Artificial Intelligence in Education (AIEd) has led to many advances and breakthroughs, especially in recent years, as has the discipline of AI in general [25]. AIEd, more so perhaps than any other application area within AI, is also a driving force in discussions that are recently gaining more and more attention in AI in general: discussions around fairness [4,24], interpretability and explainability [5,33], inclusion of stakeholders and experts [2,26,28], and many other topics that make AI more responsible and less volatile [20].

The reason for AIEd's strong focus on responsibility is, of course, straightforward: those who are supposed to directly or indirectly benefit from it are mostly

A. I. Cristea et al. (Eds.): AIED 2025, CCIS 2590, pp. 17–31, 2025.
https://doi.org/10.1007/978-3-031-99261-2_2

children, teenagers, and young adults – a population especially in need and with a special right of protection as evidenced by the Sustainable Development Goal IV [34]. Moreover, by impacting their education, AIEd has lasting consequences concerning their social, economic, mental, and even physical well-being [31]. As such, focusing on responsible and equitable AI is consequential and absolutely necessary.

Nevertheless, we argue that for AIEd to realize its potential of producing beneficial, fair, and reliable tools, one way of thinking has to be established more. We believe that for AIEd to overcome obstacles in the way of achieving responsible solutions, it is necessary to move away from association-based thinking (i.e., thinking based on the correlations in the data) and instead move towards a causal perspective. Association-based thinking is almost ubiquitous in Artificial Intelligence (AI): most Machine Learning (ML) methods rely on correlation, eXplainable AI (XAI) methods construct their explanations on the basis of the correlation-based models, and AI fairness is mostly assessed with association-based notions. While current ML methods are powerful and employing XAI and fairness notions is certainly better than not doing so, we argue that the goals of AIEd align poorly with association-based decision-making. In AIEd, we are typically concerned not only with predicting dropout but also with understanding why someone is likely to drop out and intervening. We are not concerned with recommending some exercise based on what other students did but with recommending an exercise that will actually foster understanding. These objectives cannot be answered with correlation alone; they require causal thinking. While other application areas of AI also benefit from a causal perspective, many application areas that are heavily researched do not require it as much: for autonomous driving, it does not matter why a camera detects a street; for Netflix recommendations, it does not matter how Alice and Bob liking a movie affects whether Charlie will like a movie. In this paper, we will argue that it matters for AIEd and that taking a causal perspective is important. We will also argue that the focus on causality can make AIEd better in more ways than one may initially think: it may make models fairer, provide better insights, interventions, and recommendations, allow for the natural inclusion of researchers of other disciplines as well as of teachers and students, and even allows us to make better arguments for why AIEd tools are valuable.

Figure 1 shows different steps within an AIEd project; we argue that each of the mentioned steps benefits from causal reasoning directly while seamlessly including knowledge from other disciplines and domain experts. Other steps within ML projects (e.g., model selection) are influenced indirectly by these steps in the figure and their relation to causal thinking. Of course, not all AIEd projects will include all steps. The paper is organized by moving from the inner to the outer squares. For each step, we will highlight why causality will lead to advantages and how to incorporate causality. We will also briefly look at current and future research endeavors and discuss how practitioners in the industry may include causality. Our goal is not to provide deep introductions into each step (indeed, papers exist that do a great job at that) but to showcase the general importance of causal thinking across an AIEd project. We are aware that meaningful causal-based work already exists in the AIEd community [7,9]. We are also not the first to call for more causal thinking in areas of AIEd [38], but we take a more holistic view, highlighting its advantages in more than one step of AIEd tool development and research. Before starting with the different steps, we will provide some general background on causality.

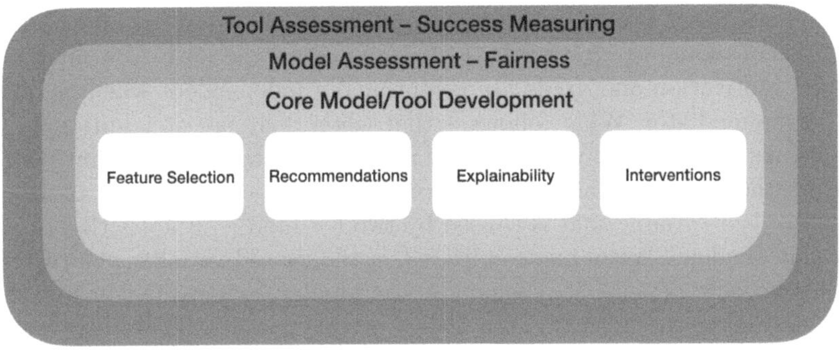

Fig. 1. The different areas within AI tool development and assessment for which causality may immediately be helpful.

2 Causality

When we talk about causality, we mean that the value of one variable influences the value of another variable [27]. For example, if Alice studied more for her math class, then she would get a better math grade on the next test. A causal influence may happen directly (as in the example) or indirectly (if Alice has more motivation to study math, she will study more and pay more attention in class, and therefore, get a better grade). Generally, verifying whether a causal relationship between variables exists in a setting can best be tested using randomized experiments where the variable believed to be a cause of another variable is the treatment variable, i.e., gets intervened on [27]. For example, to test how much a reading tutor improves students' literacy, one can randomly assign tutors to some students and not others. However, for many relationships and settings, randomized experiments are ethically concerning, impossible (e.g., one cannot randomize gender), or challenging to unfeasible for other reasons (e.g., complexity, number of interesting variables involved). In such cases, the explicit modeling of the data-generating mechanism based on observational data and domain knowledge is the best option and allows for the subsequent analysis of causal effects without experiments under certain assumptions[1] [17,27,38].

[1] For an introduction regarding the connection between the assumptions in causal graphs and causal inference consult Pearl [27].

To graphically model causal relationships, one often chooses Directed Acyclic Graphs (DAGs), with vertices (or nodes) representing variables and edges representing causal relationships [27]. Figure 2 shows examples of DAGs. DAGs provide information concerning conditional independencies: every variable is conditionally independent of all other variables given its parents in the graph. This independence information, in principle, allows the computation of causal effects [27]. Some important terms concerning specific (conditional) independence relationships of variables are mediators, colliders, and confounders. As depicted in Fig. 2a), a mediator is a variable on the path from one variable to another variable [27]. For example, study time is a mediator on the path from motivation to grade. Motivation and grade are conditionally independent when one conditions on the mediator. What follows is that when there are no additional direct relationships between two variables, the mediator captures all relevant information from the variable that is a cause of the mediator. A collider, as depicted in Fig. 2b), is a variable that is caused by two (or more) variables that are otherwise independent. When one conditions on the caused variable, the two other variables become dependent, though. A confounder, as depicted in Fig. 2c), is a variable that is the mutual cause of two other variables. These two variables become independent once one conditions on the variable causing them.

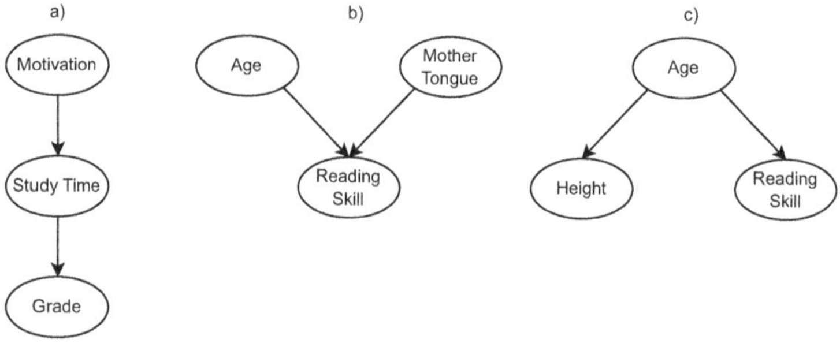

Fig. 2. The figure displays different small DAG structures as examples. The left one shows a chain structure with *Study Time* being a mediator. The middle one shows a collider structure, with *Reading Skill* being the collider. The right one shows a fork structure, with *Age* being a confounder.

To mathematically represent causal relationships, two frameworks, which are generally equivalent, exist: the structural causal model (SCM) and the potential outcome framework [19,27]. Both require background information concerning the causal relationships, i.e., for the SCM, we need the DAG and, additionally, the structural functions that connect the variables. Typically, this causal knowledge can stem from theory, domain experts (e.g., teachers), data (via data-driven structure learning algorithms), or a combination of all of them [9]. More on modeling causal relationships will follow in Sect. 7.

3 Developing an AIEd Tool

Even before we start developing an AIEd tool, causal thinking is very important. Suppose we are aware of a problem that we believe can be mitigated by developing an AIEd tool. In that case, we should first critically consider what the tool will impact and whether it will likely have the intended effect. This assessment may benefit from asking domain experts (students or teachers) about their opinions, problems, and ideas. If we believe a tool will impact whatever we aim to improve, we can continue with the following steps.

3.1 Feature Selection

Causality can already come into play when selecting features for training a Machine Learning (ML) model [15,40]. Feature selection is generally an important step in the pre-processing pipeline, helping the Machine Learning model to focus on relevant features, leading to easier training and more stable models [40]. Causal feature selection is a relatively well-known concept where the idea is only to include variables that are directly impacting the target variable, i.e., that are its parents (direct causes) [30,40]. Typically, causal-based feature selection is performed by learning the whole DAG from observational data or learning the Markov Blanket of the target variable (its parents, children, and children's parents) [40]. Then, only the parents are selected.

The advantage of only including causally relevant features can be divided into three aspects: performance, actionable variables, and fairness. By including causally relevant features, all information that can be extracted from the known variables will be used, whereas all variables that contain superfluous information (for example, those connected to the target through a confounder which would be included using correlation-based feature selection methods) or are completely disconnected are discarded; as such, more accurate models can be learned faster while requiring less data [15,40]. Consider Fig. 3 as a small example. We aim to predict the reading skills of elementary school children. In our example, math skills and motivation will be discarded because they contain no extra information not present in the reading experience. If regular correlation-based approaches were chosen, it is likely that the other variables would also be included – despite them not providing a real benefit. Moreover, we will automatically generate actionable insights by including only causally relevant features. For example, suppose we use an interpretable model or employ XAI on an ML model and observe that reading experience is important for the prediction of a student's reading skills. In that case, we know that increasing reading experience is likely to help the student increase their skills. Suppose we included the variable math skills in the training data, and this was important for the model (again, ML models are association-based, so the model may place a weight on it). In that case, we may be inclined to conclude that increasing the math skills will increase the reading skills. However, as evidenced by the DAG, trying to advance math skills would not be a good tactic to improve reading skills. It is important to highlight that, of course, not every included variable may directly be an actionable one. More on interventions will follow in Sect. 3.3. Finally, as Saarela [30] shows, by additionally excluding sensitive features, i.e., features we deem unfair, such as gender, age, and other socio-demographic features, we achieve fairer models. On the one hand, fairness increases because of the exclusion of those features (although we do not need causal feature selection for that); and on the other hand, fairness increases as features caused by the sensitive features but not causing the target feature will not be included which limits the sensitive information available for the model. An alternative approach would be to select only those features as allowed training data that are not descendants of sensitive features, which would be guaranteed to achieve fair models but would potentially severely limit model accuracy as, practically, most features will be influenced by sensitive features [21] – in particular in education. More on fairness will follow in Sect. 4.

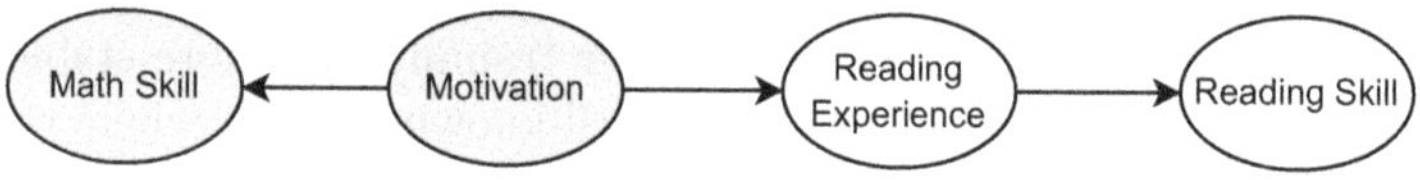

Fig. 3. A small example for feature selection.

Current and Future Research: In the general AI literature, causal feature selection has received steady attention, and recent arguments in its favor include the advantages regarding fairness and actionability [30,40]. Although Saarela [30] even evaluated causal feature selection with educational datasets, the topic has not received much attention in AIEd yet. In particular, it may be interesting to see how well causal feature selection works when comparing the utility and fairness of different bias mitigation approaches, especially for educational datasets. Moreover, it would be interesting to see whether the success regarding the utility of causal feature selection is dependent on data size and, if so, to what extent.

Industry Notes: We advise practitioners to include causal feature selection as one option in their pre-processing pipeline, especially when one aims to generate insights from the models. As Yu et al. [40] developed an open-source package for causal feature selection, the inclusion of causal feature selection should be relatively straightforward.

3.2 Explainability

As mentioned, eXplainable AI (XAI) has received much attention in AIEd and is believed to have distinct requirements when employed for AIEd compared to other AI application areas [20]. The area is challenging in many ways as different XAI techniques can lead to different results for the same model [33] and as it is uncertain which explanations are preferred by users [32]. We argue, however, that it is also an area susceptible to causal conclusions where often only association-based conclusions are justified – and where inappropriate causal thinking may initially even hurt. As XAI insights are typically directly taken from the model, they do ideally explain how the model reaches its decision. However, this conclusion is not transferable to real-life relationships, and it is important to stress the distinction. Considering our example in Fig. 3 again, if we observe using an XAI technique like, e.g., SHAP that more reading experience or a higher math skill would lead to the model predicting better reading skills, a user may intuitively believe that this connection also holds in real life. While this may be true for the reading experience, it is not true for math skills, as these are only related to reading skills through a mutual confounder. The ML model cannot learn real-world causal relationships, but human interpretation may go in this direction [7,33]. What follows is that we need to find ways to highlight that only association-based conclusions are justified or, even better, to make explanations more causal in nature. One way to achieve this may be to additionally create a DAG and use this, in combination with the XAI results, to explain not only how the model reaches the decision but also what the real-world data-generating mechanism looks like. This procedure may also lead to more comprehensible and trustworthy explanations.

Current and Future Research: While XAI in AIEd has been researched quite heavily and in many directions, including (appropriate) causality into the explanations is not typically done [5,20,29]. As the value of explanations directly depends on the users' responses to them [32], user studies could be conducted both evaluating how often users misinterpret explanations causally and how helpful, trustworthy, and comprehensible they find explanations that include real causal knowledge.

Industry Notes: Practitioners including XAI in their tools should be aware of potential misunderstandings and take action to avoid them. An easy remedy may be to explain to users that ML models are association-based and do not reflect causal relationships. As research surrounding XAI and causality evolves, practitioners may want to think about including causal knowledge in explanations.

3.3 Creating Interventions

Tightly connected to the area of explainability is that of interventions. As already mentioned, the ultimate goal of AIEd prediction models is typically not only to achieve accurate predictions but to create interventions out of those predictions. Usually, in case of an unfavorable outcome, the aim is to determine which variable's value would need to change to produce a different result (also known as contrastive samples) [7,8]. However, as discussed in Sects. 3.1 and 3.2, important variables for a model or important variables according to correlation analyses do not reflect variables that truly causally impact the target variable. Again, considering the example, it would be poor advice to state that a student should increase their math skills to improve their reading skills. Yet, pure correlation is often used as the basis for interventions [3].

Achieving causally sound interventions could be approached similarly to achieving causally sound explanations. Generally, in case of an unfavorable outcome for a person according to an ML model, one could choose one of two approaches. Either, one could learn an SCM on the data and infer which intervention may be particularly helpful, keeping everything else equal[2]. Alternatively, one could identify all possible (but maybe minimal) subsets of features for whom a change in value changes the prediction [8] and then identify those subsets only containing ancestors (i.e., nodes on a direct path towards the target variable) of the target according to the DAG. These are then recommended as interventions if they are actionable (i.e., not unchangeable in the context).

Current and Future Research: Some attempts to include causality into interventions exist [7,29,37]. However, these have not been evaluated on a larger scale or in real applications, which may be an interesting future endeavor and necessitate stakeholder inclusion. Moreover, different procedures could be investigated, including the different results for different procedures (such as the ones described above).

Industry Notes: If practitioners aim to construct interventions, they should consider whether causally sound interventions are important. Currently, there is no standard procedure to achieve causally sensible interventions. However, at the very least, it may be worthwhile to consult domain experts such as teachers or students to ask them which changes may causally affect the target variable.

3.4 Creating Recommendations

Exercise, course, and content recommendation is a big topic in AIEd, with the aim to keep students motivated and engaged and to keep them from being frustrated or dropping out [1,35]. However, not considering the causal mechanisms behind the patterns leading to recommendations can lead to poor recommendations. For example, if we use collaborative filtering, i.e., information from successful other students, to recommend a course order to a struggling student, this recommendation may not work as intended. It may be the case that successful students study courses in a specific order because they are successful students, not because that order makes them successful. Therefore, we need to consider how varying variables (e.g., study success, completed courses, or exercises) interact with each other.

As there are many different goals of recommendation (e.g., exercise, content, course, study program) and many different methods for recommendation (e.g., content-based recommendation, collaborative filtering, item response theory, knowledge tracing), there is no single procedure that can account for causal relationships [1,14,35].

[2] Again, whether the inference is possible depends on certain assumptions [27].

Current and Future Research: Luo et al. [22] and Gao et al. [13] surveyed how causal recommendation approaches can generally be classified – though not for the education use case. In education, several approaches implicitly or sometimes explicitly counter problems arising from a pure association-based view. For example, Wagner et al. [35] construct recommendations by using information from students that are also similar to the student to recommend regarding their grades and previous course patterns. Yang et al. [39] explicitly model how courses change skills and how this affects the results of future enrollments via knowledge graphs. Explicit causal recommendation approaches are rare in education and should be investigated more, but some approaches specifically for educational settings exist [14]. In particular, it may be interesting for future research to observe whether and how recommendations change between "regular" and causal recommendation methods.

Industry Notes: Practitioners should be aware that by making recommendations that are not necessarily causal, the recommendations may not work as intended and could even have contrary effects. Several implementations are available as causal recommendation methods are comparatively well-researched (albeit not in the educational context). Gao et al. [13] and Luo et al. [22] gathered a repository of links leading to implemented approaches.

4 Fairness

One of the areas of ML in which the causal perspective has recently gained a lot of attention is that of ML fairness [23]. The key argument is that only by understanding the data-generating mechanism depicted by the DAG can one adequately say whether the data or the ML model resulting from the data is fair and can mitigate potential biases [9,23]. A prominent example showcasing the value of causal thinking in fairness is Simpson's paradox [23]. The example concerns admission to the University of Berkeley, which seems to be unfairly biased against women. Prominent fairness metrics like demographic parity would then suggest that the data or models resulting from it are unfair. However, in reality, women apply for more competitive departments, and department choice influences the probability of admission. If one conditions on department choice, then the model no longer exhibits a bias against women [23]. In general, not considering the underlying causal mechanisms of the data used for ML models can lead to wrong conclusions, unfair models, and an unclear and often unjustified selection of applied fairness notions. Because of these problems, causal fairness notions have been developed that take the underlying data generating mechanism into account [23]. Most of the causal fairness notions require a fully defined structural causal model or at least its graphical part, the DAG [23]. Moreover, the estimation of causal biases typically requires different kinds of independencies or equalities to hold [23]. Based on the causal notions of fairness, different mitigation strategies exist. A major distinction between different causal fairness notions and corresponding mitigation techniques is whether they allow sensitive features to impact the predicted feature through specified other predictive features. In Simpson's paradox, e.g., we may say that department choice is a feature through which information from gender is allowed to pass [23].

Current and Future Research: Much research in the general AI literature has recently described different causal fairness notions and how to mitigate unfairness detected by those notions [6,18,23]. In AIEd, some recent work also highlighted the importance of the causal perspective, showing that different underlying DAGs and different application scenarios lead to very different fairness conclusions [9,10] – we encourage future discussion in this direction. The use of causal fairness notions has hardly been considered in AIEd, however, and may be an interesting endeavor with a special focus on how the results differ from employing other, non-causal fairness notions. Similarly, evaluating how causal bias mitigation changes prediction (compared to the unfair baseline and regular bias mitigation methods) may be interesting.

Industry Notes: Practitioners who aim to use causal fairness notions should be aware of the sometimes very strong causal assumptions necessary for inferring unfairness [23]. This limitation should not discourage them from using these notions but only make them cautious. Moreover, practitioners should discuss with domain experts whether there are features through which sensitive information is allowed to pass and, if so, through which – neglecting to do so may either lead to overly strict or still unfair models. This distinction also determines the notion used.

5 Measuring Success

Regardless of what we do in AIEd, ultimately, we want the tools developed to be helpful. As such, we need to test whether they are helpful in the intended way. Ideally, this involves randomized experiments or quasi-experiments (when true randomization is not feasible) to test the success. For example, we could test whether the personalized recommendation of math exercises improves students' grades. For this, we need a control group whose members studied without such personal recommendations and a treatment group whose members received them [11]. Additionally, we need a clear idea of what "success" would look like and an operationalization for this (e.g., math skills operationalized by grades)—which may vary by country and context [36]. Experiments are already a causal endeavor, but we believe that deeper causal thinking will allow us to improve our analysis and provide more insights. For example, it may be valuable to look at how helpful the scrutinized tool is for different subgroups of students. However, we can only theorize which subgroups may be interesting to look at when thinking causally. For example, we may have the following idea: An automatic feedback system for essay writing may help students whose essays are already on a high level more because they have a better grasp on the topic, allowing them to incorporate the feedback better. We should then look at each subgroup (in this case, the subgroups are determined by the essay writing level) separately, which may provide interesting insights and make newly emerging fairness problems apparent.

Current and Future Research: We believe that AIEd has a relatively strong tradition of measuring success with experimental settings [12,16]. We argue in favor of developing a general framework for experiments that helps practitioners and researchers alike answer relevant questions when conducting experiments (such as how to measure success and which groups to look at). By having convincing experimental set-ups, the area of AIEd can only make itself stronger and convince important stakeholders of the value of such tools. How to communicate the results to political stakeholders should become a much more prominent discussion in the research community, but causal modeling will certainly allow for better understandable arguments.

Industry Notes: We can only emphasize the strength experiments for practitioners and encourage them to focus on particularly sensitive groups. However, experiments should be carried out carefully and with ethical considerations in mind. Part of these considerations should already be whether a tool is even expected to have a positive impact, as mentioned in the beginning.

6 Joining Forces with Other Disciplines and Experts

The need to model causal relationships apparent in several of the discussed steps naturally invites the perspectives of scholars of other disciplines (such as sociologists, psychologists, or education researchers) but also stakeholders (such as teachers or students), and we believe that this is another important strength of including causality in AIEd. It is not just AI researchers who can benefit from such a collaboration. As Weidlich et al. [37,38] and Hicks et al. [17] write, by working together to create DAGs, we force everyone to make causal assumptions explicit and the analysis of the resulting DAG may allow us to test theories with observational data where otherwise experimental data would be necessary—something of interest to other researchers as well. As such, interdisciplinary research and research involving stakeholders may become very natural and even lead to a lasting collaboration: Social scientists may help construct graphs, AI researchers may help with testing theories, and social scientists and domain experts may point to AI applications that they believe are promising, AI researchers may provide new insights (e.g., using XAI) to other scholars and suggestions for interventions to stakeholders, stakeholders may demand explainability and fairness, forcing AI researchers to advance research in this direction. We are certainly not the only ones believing that the inclusion of other disciplines and stakeholders is important for the future of AIEd; however, we are convinced that a focus on causality is a gateway for mutual, lasting exchange on equal footing. Moreover, we want to highlight that causal models that may be accurate for one setting and context may not be for others. Hence, it is always of crucial importance to include stakeholders and discuss how contexts may impact causal relationships, which may lead to more equitable outcomes for people from poorer countries, as well as insights into how contexts shape education. How the collaboration among the different groups should be organized, and especially how to make the mentioned methods applicable to the classroom, should be part of future research and an ongoing discussion.

7 The Elephant in the Room: Creating Causal Models

We firmly believe that the causal perspective can only be advantageous for AIEd and can make many of the strengths that AIEd already has more pronounced. However, one problem with causality has frequently been mentioned throughout the paper: the need to know something about the underlying data-generating mechanisms and, in some cases, even the need for a complete SCM. Assuming that we know the underlying causal model – partially or completely – is a strong assumption that is made by many papers introducing techniques using the causal perspective [6,23]. While it is likely that a combination of data-driven structure learning and expert knowledge will provide us at least with a partially correct causal model, it is unlikely that we will know the whole causal model[3] To what extent incorrect or partial causal models hamper the results of using causality in AIEd may depend on what exactly we aim to do and is another avenue of research we want to highlight. However, we argue that, in general, causal thinking will always improve what we do. By thinking causally, we will have thought about why we include which features, why a model works or is fair, why interventions or recommendations make sense, and why a tool is helpful. Even if our exact causal models are wrong, explicitly stating our assumptions is probably better than not thinking causally.

8 Conclusion

In this paper, we argued in favor of developing a tradition of including causal thinking in AIEd. We argue that the application area of education is special in its needs, which has already led to pronounced strengths (e.g., XAI, fairness) compared to other areas of AI and that AIEd will benefit further from causal thinking. In isolation, the idea of including causality in model development and assessment steps is not novel. However, we aim to provide a complete picture of the advantages, not of the advantages of a concrete step alone. Hence, we showed that causal thinking can help with many things, from finding ideas concerning helpful AIEd tools to evaluating a tool's actual helpfulness and many steps along the pipeline.

[3] Data-driven structure learning methods, e.g., are known to be error-prone in certain settings.

References

1. Abdelrahman, G., Wang, Q., Nunes, B.: Knowledge tracing: a survey. ACM Comput. Surv. **55**(11), 1–37 (2023)
2. Alfredo, R., et al.: Human-centred learning analytics and AI in education: a systematic literature review. Comput. Educ. Artif. Intell. 100215 (2024)
3. Alturki, S., Alturki, N.: Using educational data mining to predict students' academic performance for applying early interventions. J. Inf. Technol. Educ. JITE. Innov. Pract. IIP **20**, 121–137 (2021)
4. Baker, R.S., Esbenshade, L., Vitale, J., Karumbaiah, S.: Using demographic data as predictor variables: a questionable choice. J. Educ. Data Mining **15**(2), 22–52 (2023)
5. Chaushi, B.A., Selimi, B., Chaushi, A., Apostolova, M.: Explainable artificial intelligence in education: a comprehensive review. In: World Conference on Explainable Artificial Intelligence, pp. 48–71. Springer, Heidelberg (2023). https://doi.org/10.1007/978-3-031-44067-0_3
6. Chiappa, S.: Path-specific counterfactual fairness. In: Proceedings of the AAAI Conference on Artificial Intelligence, vol. 33, pp. 7801–7808 (2019)
7. Cohausz, L.: Towards real interpretability of student success prediction combining methods of xai and social science. Int. Educ. Data Mining Soc. (2022)
8. Cohausz, L.: When probabilities are not enough-a framework for causal explanations of student success models. J. Educ. Data Mining **14**(3), 52–75 (2022)
9. Cohausz, L., Kappenberger, J., Stuckenschmidt, H.: What fairness metrics can really tell you: a case study in the educational domain. In: Proceedings of the 14th Learning Analytics and Knowledge Conference, pp. 792–799 (2024)
10. Cohausz, L., Tschalzev, A., Bartelt, C., Stuckenschmidt, H.: Investigating the importance of demographic features for edm-predictions. Int. Educ. Data Mining Soc. (2023)
11. Dennis, A.R., Valacich, J.S.: Conducting experimental research in information systems. Commun. Assoc. Inf. Syst. **7**(1), 5 (2001)
12. Essel, H.B., Vlachopoulos, D., Tachie-Menson, A., Johnson, E.E., Baah, P.K.: The impact of a virtual teaching assistant (chatbot) on students' learning in Ghanaian higher education. Int. J. Educ. Technol. High. Educ. **19**(1), 57 (2022)
13. Gao, C., Zheng, Y., Wang, W., Feng, F., He, X., Li, Y.: Causal inference in recommender systems: a survey and future directions. ACM Trans. Inf. Syst. **42**(4), 1–32 (2024)
14. Gilbert, J.B., Miratrix, L.W., Joshi, M., Domingue, B.W.: Disentangling person-dependent and item-dependent causal effects: applications of item response theory to the estimation of treatment effect heterogeneity. J. Educ. Behav. Stat. **50**(1), 72–101 (2025)
15. Guyon, I., Aliferis, C., et al.: Causal feature selection. In: Computational Methods of Feature Selection, pp. 79–102. Chapman and Hall/CRC (2007)
16. Han, J.W., Park, J., Lee, H.: Analysis of the effect of an artificial intelligence chatbot educational program on non-face-to-face classes: a quasi-experimental study. BMC Med. Educ. **22**(1), 830 (2022)
17. Hicks, B., Kitto, K., Payne, L., Buckingham Shum, S.: Thinking with causal models: a visual formalism for collaboratively crafting assumptions. In: LAK22: 12th International Learning Analytics and Knowledge Conference, pp. 250–259 (2022)
18. Hort, M., Chen, Z., Zhang, J.M., Harman, M., Sarro, F.: Bias mitigation for machine learning classifiers: a comprehensive survey. ACM J. Resp. Comput. **1**(2), 1–52 (2024)

19. Imbens, G.W., Rubin, D.B.: Causal Inference in Statistics, Social, and Biomedical Sciences. Cambridge university press, Cambridge (2015)
20. Khosravi, H., et al.: Explainable artificial intelligence in education. Comput. Educ. Artif. Intell. **3**, 100074 (2022)
21. Kusner, M.J., Loftus, J., Russell, C., Silva, R.: Counterfactual fairness. Adv. Neural Inf. Process. Syst. **30** (2017)
22. Luo, H., et al.: A survey on causal inference for recommendation. Innov. **5**(2) (2024)
23. Makhlouf, K., Zhioua, S., Palamidessi, C.: Survey on causal-based machine learning fairness notions. arXiv preprint arXiv:2010.09553 (2020)
24. Memarian, B., Doleck, T.: Fairness, accountability, transparency, and ethics (fate) in artificial intelligence (ai) and higher education: a systematic review. Comput. Educ. Artif. Intell. **5**, 100152 (2023)
25. Mittal, U., Sai, S., Chamola, V., et al.: A comprehensive review on generative ai for education. IEEE Access (2024)
26. Nguyen, A., Ngo, H.N., Hong, Y., Dang, B., Nguyen, B.: Ethical principles for artificial intelligence in education. Educ. Inf. Technol. **28**(4), 4221–4241 (2023)
27. Pearl, J.: Causality. Cambridge university press, Cambridge (2009)
28. Perrotta, C., Selwyn, N.: Deep learning goes to school: toward a relational understanding of ai in education. Learn. Media Technol. **45**(3), 251–269 (2020)
29. Rai, D., Beck, J.E.: Causal modeling of user data from a math learning environment with game-like elements. In: Biswas, G., Bull, S., Kay, J., Mitrovic, A. (eds.) AIED 2011. LNCS (LNAI), vol. 6738, pp. 528–530. Springer, Heidelberg (2011). https://doi.org/10.1007/978-3-642-21869-9_91
30. Saarela, M.: On the relation of causality-versus correlation-based feature selection on model fairness. In: Proceedings of the 39th ACM/SIGAPP Symposium on Applied Computing, pp. 56–64 (2024)
31. Schuller, T.: The Benefits of Learning: The Impact of Education on Health, Family Life, and Social Capital. Psychology Press, Hove (2004)
32. Swamy, V., Du, S., Marras, M., Kaser, T.: Trusting the explainers: teacher validation of explainable artificial intelligence for course design. In: LAK23: 13th International Learning Analytics and Knowledge Conference, pp. 345–356 (2023)
33. Swamy, V., Radmehr, B., Krco, N., Marras, M., Käser, T.: Evaluating the explainers: black-box explainable machine learning for student success prediction in moocs. arXiv preprint arXiv:2207.00551 (2022)
34. United Nations: Sustainable development goals (2015). https://sdgs.un.org/goals. Accessed 30 Apr 2025
35. Wagner, K., Merceron, A., Sauer, P., Pinkwart, N.: Can the paths of successful students help other students with their course enrollments?. Int. Educ. Data Mining Soc. (2023)
36. Weatherton, M., Schussler, E.E.: Success for all? A call to re-examine how student success is defined in higher education. CBE—Life Sci. Educ. **20**(1), es3 (2021)
37. Weidlich, J., Gašević, D., Drachsler, H.: Causal inference and bias in learning analytics: a primer on pitfalls using directed acyclic graphs. J. Learn. Anal. **9**(3), 183–199 (2022)
38. Weidlich, J., Hicks, B., Drachsler, H.: Causal reasoning with causal graphs in educational technology research. Educ. Tech. Res. Dev. **72**(5), 2499–2517 (2024)

39. Yang, T., Ren, B., Gu, C., Ma, B., He, T., Konomi, S.: Towards better course recommendations: integrating multi-perspective meta-paths and knowledge graphs. In: Proceedings of the 15th International Learning Analytics and Knowledge Conference, pp. 137–147 (2025)
40. Yu, K., et al.: Causality-based feature selection: methods and evaluations. ACM Comput. Surv. (CSUR) **53**(5), 1–36 (2020)

Pedagogy-Driven Evaluation of Generative AI-Powered Intelligent Tutoring Systems

Kaushal Kumar Maurya(✉) and Ekaterina Kochmar

EduNLP Lab, Mohamed bin Zayed University of Artificial Intelligence, Abu Dhabi, UAE
{kaushal.maurya,ekaterina.kochmar}@mbzuai.ac.ae

Abstract. The interdisciplinary research domain of Artificial Intelligence in Education (AIED) has a long history of developing Intelligent Tutoring Systems (ITSs) by integrating insights from technological advancements, educational theories, and cognitive psychology. The remarkable success of generative AI (GenAI) models has accelerated the development of large language model (LLM)-powered ITSs, which have potential to imitate human-like, pedagogically rich, and cognitively demanding tutoring. However, the progress and impact of these systems remain largely untraceable due to the absence of reliable, universally accepted, and pedagogy-driven evaluation frameworks and benchmarks. Most existing educational dialogue-based ITS evaluations rely on subjective protocols and non-standardized benchmarks, leading to inconsistencies and limited generalizability. In this work, we take a step back from mainstream ITS development and provide comprehensive state-of-the-art evaluation practices, highlighting associated challenges through real-world case studies from careful and caring AIED research. Finally, building on insights from previous interdisciplinary AIED research, we propose three practical, feasible, and theoretically grounded research directions, rooted in learning science principles and aimed at establishing fair, unified, and scalable evaluation methodologies for ITSs.

Keywords: Intelligent Tutoring Systems (ITSs) · Generative AI (GenAI) · Large Language Models (LLMs) · Evaluation · Pedagogy

1 Introduction

Over the past decades, Artificial Intelligence (AI) has undergone multiple paradigm shifts, each bringing new opportunities and challenges. However, the emergence of large language models (LLMs) in the current era of generative AI (GenAI) marks the most notable breakthrough [69], with transformative impact across numerous domains. One of the most pressing global challenges is ensuring high-quality and equitable education – a cornerstone of societal progress. The educational domain is now actively exploring ways to harness the power of LLMs to bridge this gap, aiming for universal access to personalized and effective learning experiences. AI methods have been part of AIED research for many

A. I. Cristea et al. (Eds.): AIED 2025, CCIS 2590, pp. 32–46, 2025.
https://doi.org/10.1007/978-3-031-99261-2_3

years [44]. Recent advances in GenAI have pushed the boundaries of AI-driven educational applications [66], including automated student feedback generation, intelligent tutoring systems (ITS), lesson planning, and more. Among these, ITSs, also referred to as AI Tutors, have garnered significant attention. The rise of conversational LLMs closely aligns with the long-standing vision of ITSs which are interactive, adaptive, and scalable learning environments. This has led to the development of *educational dialogue-based ITSs* like LearnLM [56], Khanmigo [2], and similar.

Despite the active development of ITSs, their true progress and impact remain difficult to quantify due to the absence of universally accepted pedagogical principles, robust evaluation frameworks, and comprehensive benchmarks [26]. One of the primary challenges lies in the inherently multidisciplinary nature of educational research, where existing studies in cognitive and learning sciences are often conducted on small, homogeneous populations, primarily from WEIRD (Western, Educated, Industrialized, Rich, and Democratic) countries, with variable implementation parameters [44]. This has led to the development of subjective, personalized, or limited-scale evaluation frameworks and benchmarks that lack generalizability and broad acceptance [54,55]. For instance, most traditional assessments of ITSs rely on *self-reported* outcomes [19], which are inadequate for modern, interactive, and on-the-fly GenAI-based ITSs. Additionally, common natural language generation (NLG) metrics are not designed to capture pedagogical values of interactions, often requiring an optimal single reference and being susceptible to adversarial inputs [54]. Finally, efforts to develop evaluation models that capture pedagogical effectiveness are either incomplete, failing to encompass the full range of pedagogical nuances, limited to subjective and non-scalable human evaluations, or restricted to narrow domains and specific use cases [26]. Addressing these limitations is crucial for advancing the reliability and scalability of ITS evaluation methodologies.

The lack of appropriate evaluation protocols affects both human tutoring and ITSs. In the "No Child Left Behind Act" (NCLB) in the United States, overemphasis on standardized test scores led to shallow learning, reduced engagement, and limited teaching methods [7,22,25]. Similarly, Automated Essay Scoring (AES) systems driven by LLMs focus on surface-level features like grammar and vocabulary, often neglecting critical thinking and conceptual understanding [23,40,41]. This limits teachers' insight into students' reasoning and hinders personalized instruction [64]. Finally, a field experiment in a Turkish high school involving nearly 1,000 students from grades 9 to 11 examined the impact the best state-of-the-art GenAI model (GPT-4) as tutor on math learning [52]. Over four 90-minute sessions covering 15% of the curriculum, students engaged in reviewing lectures, AI-assisted practice, and unassisted exams. While AI-supported students solved 48% more practice problems correctly, they scored 17% lower on unassisted tests, suggesting that reliance on AI during practice may have impeded the development of independent problem-solving skills. These cases highlight the need for evaluation metrics that assess conceptual clarity, engagement, pedagogical effectiveness, and critical thinking alongside human-like responses.

This work makes several key contributions, listed below:

1. We present *the current state of evaluation methodologies* for ITSs by providing a comprehensive review of existing AIED evaluation research and pedagogical principles. This review covers a broad spectrum, ranging from evaluation approaches based on self-reports for traditional, static and rule-based ITSs, to NLG-based and pedagogically oriented evaluation for advanced deep learning and GenAI-based ITSs.
2. We pinpoint *key challenges* in the evaluation of ITSs by highlighting the diverse pragmatics of evaluation and the lack of unification in the existing pedagogical principles, evaluation methodologies, and benchmarks.
3. Finally, we propose *three feasible research directions* grounded in learning science principles and previous research in AIED, as a step towards mitigating the key challenges. These directions aim to build unified, scalable evaluation approaches that help track the progress and quality of GenAI-powered ITSs.

2 Current State of Evaluation for ITS

In this section, we present a comprehensive review of existing evaluation methodologies for ITSs, which broadly fall into three categories: *traditional*, *NLG-based*, and *pedagogy-oriented* evaluation methods.

2.1 Traditional Evaluation Approaches

Traditional educational research on evaluating teacher effectiveness has a well-established and extensive history, as detailed by [19] and [37]. This includes a range of methodological approaches for assessing teachers' practices, such as the analysis of classroom artifacts (e.g., evaluation of teacher assignments and student work), teaching portfolios, teacher self-reports of instructional practices via surveys, teaching logs, and interviews, as well as student evaluation of teacher performance. Pre-GenAI ITSs, including AutoTutor [20], DeepTutor [47], iDRIVE [50], and Writing Pal (W-Pal) [36] utilize rule-based and deep learning models to simulate human tutor behaviors, thereby supporting learning in areas such as STEM subjects, language acquisition, and computer literacy. The pedagogical effectiveness of these systems has been evaluated through role-simulation reports [20] and empirical or extrinsic studies measuring learning gains [38]. For instance, previous research demonstrates that students interacting with AutoTutor show improvements in solving deep reasoning problems compared to those participating in traditional classroom-based learning [38].

The AIED community are active contributors to emerging methodologies for measuring the effectiveness of human tutors and ITSs [4,5,16,49]. For example, [31] developed self-assessment for tutors in extrinsic studies based on self-regulatory learning theories. [12] developed an automated computational tool, Coh-Metrix, to evaluate the Writing Pal (W-Pal) ITS based on text cohesion. While these approaches have been effective in evaluating traditional human instructors and early AI-driven ITSs, their applicability to generative AI-powered tutoring systems raises significant concerns [30]. Unlike their predecessors, generative AI models exhibit complex, context-dependent behaviors that pose unique challenges in evaluation. Existing methodologies, which often rely on static performance metrics or predefined criteria, may fail to capture the dynamic and adaptive nature of generative AI tutors [34]. This gap highlights the need for new, more nuanced approaches that account for the variability and learning capabilities inherent in generative systems. Additionally, the reliance on traditional extrinsic frameworks may overlook critical aspects such as ethical considerations, bias in AI responses, and the alignment of AI-generated content with educational goals [66]. These concerns compel us to critically reassess the evaluation of GenAI-powered ITS effectiveness.

2.2 NLG-Based Evaluation

In recent years, general Natural Language Generation (NLG)-based domain-agnostic automated evaluation metrics have been leveraged to assess the quality of responses from ITSs [26]. These metrics rely on lexical and semantic matching to validate coherence, fluency, and human-likeness of the generated responses. For instance, [32] employed BLEU [39] and BERTScore [68] to measure response quality of a generative AI-based tutor. The BEA 2023 shared task [54] incorporated additional metrics like DialogRPT [17] alongside BLEU and BERTScore to automatically evaluate ITS' responses developed by different teams. However, these metrics are not specifically designed to capture the pedagogical values inherent in effective tutoring [55]. Furthermore, these approaches require a gold reference to compute evaluation scores, which is often unavailable or non-unique [3]. Additionally, prior research has demonstrated the susceptibility of these models to superficial responses, where simple and generic outputs like "Hello" or "teacher" can misleadingly achieve high evaluation scores, thus undermining the robustness of these metrics [57].

2.3 Pedagogically Oriented Evaluation

As discussed in the previous two subsections, traditional evaluation approaches are primarily based on *self-report* assessments, while NLG-based models often fall short in capturing pedagogical nuances. In this section, we review approaches that aim to capture the pedagogical values of GenAI-powered ITSs. The most reliable evaluation method remains human assessment; however, human evaluation is prone to subjectivity and personalized biases [18]. In the context of ITSs, additional challenges arise: the AIED and educational NLP communities accept multiple definitions of learning science principles [26, 63], leading to the absence of a unified and widely accepted protocol for pedagogical strategies and their associated definitions. For instance, [55] introduced an AI Tutor test considering only 3 pedagogical values, while [29] presented 30 pedagogical strategies, and more recently, [26] proposed 28 distinct pedagogical strategies. Efforts have been made to unify these pedagogical frameworks, but they remain limited to specific tasks, such as single-turn mistake remediation [35], although they establish a good footprint for future research. Automated evaluation metrics designed to capture pedagogical values often focus on a narrow subset of strategies, such as coherence in uptake [15], or rely on training discriminator models [33] to score pedagogical values; such approaches are frequently unreliable and incomplete. Specifically, *meta-cognition* [1,44,45] and *active learning* [10,46] have emerged as key dimensions for evaluating the performance of ITSs. However, both automated and human evaluation approaches currently lack reliability, scalability, and adaptability, making it difficult to consistently and efficiently assess the pedagogical effectiveness of these systems across diverse educational contexts. More recently, LLMs have been actively used as proxy judges [21], though their evaluations remain largely restricted to metrics like helpfulness and human-likeness, often failing to capture the full spectrum of rich pedagogical values [35].

3 Key Challenges in Evaluation of ITS

The previous section outlined existing efforts and briefly touched upon the limitations of ITS evaluation. In this section, we will dive deeper and focus on two key challenges that hinder AIED progress in ITS evaluation methodologies.

3.1 Challenge #1: Diverse Pragmatics of Evaluation

One of the major challenges in evaluating GenAI-powered ITS lies in the *diverse pragmatics of evaluation* [26], which arise from the wide spectrum of design choices involved in benchmark creation and assessment methodologies.

A fundamental consideration is the **type of participants** involved in benchmark creation [8]. Depending on whether participants are experts, novices, real learners, role-playing participants, or researchers, the quality and nature of the benchmark data can vary significantly. Real learners bring authenticity to the evaluation process, reflecting genuine learning behaviors and challenges, while role-playing participants or researchers may offer more controlled and diverse scenarios, albeit with potential biases. Another crucial factor is the **structure of the evaluation benchmark** itself [34,35]. Benchmarks can be designed as single-turn or multi-turn interactions, each capturing different aspects of pedagogical effectiveness. Single-turn benchmarks focus on evaluating isolated responses, which can provide clarity but may overlook the dynamics of an evolving tutoring session. In contrast, multi-turn benchmarks capture the continuity and coherence of dialogue, which is essential to assess the system's ability to adapt and respond to ongoing learner needs, but may overlook atomic utterance level responses. Moreover, the benchmarks creation can itself be **guided and unguided**. An unguided benchmark creation allows for open-ended interactions that better reflect natural conversations, while scenario-guided benchmarks offer structured contexts that help standardize evaluation and ensure coverage of specific learning objectives. This distinction affects the system's ability to handle spontaneous and varied learner inputs as opposed to its proficiency in following defined pedagogical strategies.

As discussed in Sect. 2, **evaluation methods** themselves differ significantly [34], ranging from human judgment to automated metrics. Human evaluation provides context-sensitive feedback but can be subjective and inconsistent, while automated evaluation offers scalability but often misses nuanced pedagogical values, requires good references and is not always reliable. Evaluation can be conducted at the single-turn level, focusing on immediate response quality, or at the conversation level, assessing the overall effectiveness and flow of the tutoring interaction. Conversation-level evaluation is particularly valuable for understanding long-term engagement and pedagogical impact. Finally, **comparative evaluation** introduces additional complexity, as systems can be evaluated one at a time, offering a focused assessment but risking inconsistent judgments, or side-by-side, allowing direct comparisons that highlight relative strengths and weaknesses. Each approach offers trade-offs between depth of evaluation and comparative clarity, underscoring the need for careful methodological choices when designing and interpreting ITS evaluation frameworks.

3.2 Challenge #2: Lack of Evaluation Unification

The implications of diverse pragmatics of evaluation lead to the *absence of unification* across several critical dimensions. This lack of standardization impedes the systematic assessment and cross-comparison of different models, ultimately limiting progress in the development of effective and generalizable tutoring systems. We identify four primary categories contributing to this challenge.

Lack of Unified Learning Science and Pedagogical Principles: While different ITSs may have specialized goals and may justifiably focus on distinct learning science principles tailored to their specific use cases, there is a notable lack of uniformity even among systems designed for similar educational objectives [26]. Different systems leverage varying cognitive, constructivist, and behaviorist models [6,42], leading to inconsistencies in pedagogical strategies [59]. As a result, this heterogeneity in both theoretical and pedagogical foundations hampers the reproducibility, scalability, and pedagogy-oriented evaluation of ITSs, ultimately limiting their effectiveness and comparability.

Lack of Unified Evaluation Metrics: As noted in Sect. 2, ITS evaluation uses automated methods, which are scalable but often unreliable, and human evaluation, which is reliable but costly and non-scalable [11]. This dichotomy results in ambiguous interpretations of ITSs performance and undermines the reliability of comparative analyses. Standardized metrics are crucial for establishing clear benchmarks and tracking incremental improvements. Efforts of [35] provide a foundation and promise for unification in this direction.

Lack of Unified Evaluation Benchmarks: Benchmark datasets are essential for the validation and generalization assessment of models, particularly in ITSs. However, there is a notable dearth of comprehensive and diverse datasets specifically designed for ITS applications, which significantly hinders progress in this area [34]. The current benchmarks often fall short in terms of domain coverage and interaction depth, limiting their ability to accurately capture the intricate dynamics of real-world educational dialogues. This gap impedes the development of pedagogically sound, robust, and scalable ITS design. Table 1 provides an overview of some of the popular datasets used in ITS evaluation, highlighting their respective domains, dialogue characteristics, and grounding mechanisms. Despite their utility, these datasets fail to provide the necessary breadth and complexity required for a holistic assessment of ITS.

Table 1. Overview of benchmarks used in ITS evaluation

Dataset	Synthetic	Domain	# Dialogues	# Moves	Grounding	Setting
CIMA [51]	No	Language	391	max 5	Image, answer	1:1
Bridge [60]	No	Math	700	max 10	Image, confusion	1:1
MathDial [32]	Mixed	Math	2861	max 4	Confusion, answer	1:1
TSCC [9]	No	Language	102	max 5	None	1:1
TalkMoves [53]	No	Science	567	max 10	None	Classroom
NCTE [14]	No	Math	1660	–	None	Classroom
MRBench [35]	Mixed	Math	200	max 10	Confusion, answer	1:1

4 Path Forward

In the previous two sections, (1) we reviewed the current state of evaluation from different perspectives and (2) pinpointed the key challenges that hinder the progress of evaluation methodologies for ITSs. These highlight multiple research opportunities that help bridge the ITS evaluation gap. In this section, we take a *step forward* and propose three key research directions that can help mitigate the existing challenges and provide a framework for the holistic evaluation of ITSs. These directions are inspired by the gaps observed in the ITS evaluation literature and are grounded in previous research in AIED and learning science principles. However, we acknowledge that this is just a preliminary effort toward addressing numerous open problems, and we call for the AIED research community to join hands in this endeavor. As discussed in Sect. 2, *meta-cognition* and *active learning* are well-established key principles of ITS design. Building on these foundations, this proposal aims to develop automated, scalable, and reliable evaluation methodologies to assess these two critical dimensions. Although the proposed research directions are targeted for the evaluation of GenAI-powered ITSs, it is also applicable to human tutors.

4.1 Research Line #1: Evaluation Unification

The challenges mentioned in Sect. 3.2 highlight that the lack of a unified evaluation framework is one of the key obstacles to the progress in ITS development. While we acknowledge that a unified framework may introduce social biases related to different learning goals, subjects, and contexts within ITSs, we still believe that establishing a standardized approach led by the AIED community would provide a strong foundation for future research and facilitate easier adaptability. As an initial step, for focused educational use cases such as student mistake remediation and feedback generation, a unified framework should be developed based on existing work. This framework should incorporate key learning science principles, effective pedagogical strategies, standardized evaluation metrics, and well-defined benchmarks.

Research Statement: Current evaluation practices for ITSs often focus on isolated aspects of system performance, resulting in fragmented insights. We hypothesize that a unified evaluation framework, bringing together AIED researchers to develop widely accepted evaluation protocols and benchmarks for similar use cases, will enable a more holistic assessment of ITSs.

Potential Approach: Let's consider a case study from the recent work by [35] to convince readers and assess the feasibility of this research direction. In [35], the authors proposed a unified evaluation taxonomy to systematically assess the ITSs' and human tutor's current response quality given the conversation history for the student mistake remediation (SMR) task. They extensively reviewed the existing literature and concluded that several existing works only partially address the essential pedagogical properties required for effective tutoring. Based on this analysis, they proposed an eight-dimensional evaluation taxonomy grounded in established learning science principles. Furthermore, they annotated and released the `MRBench` benchmark, developed based on this taxonomy, to quantitatively measure the pedagogical effectiveness of ITSs. This effort, though focused on the SMR task in educational settings, presents a strong proof of concept for the unification of evaluation frameworks. Future research can develop automated evaluation metrics for each dimension to enhance scalability and reliability. This effort lays a foundation for future work toward a unified evaluation paradigm, which should include a *comprehensive review of the existing literature*, *high-quality dataset creation*, the *formulation of an evaluation taxonomy* aligned with pedagogical principles, *rigorous training of annotators*, *detailed annotation processes*, and the *development of robust automated evaluation metrics* for large-scale deployment.

4.2 Research Line #2: Measuring Pedagogical Guidance

For any learner, successful learning enables the generalization of skills beyond the given context and fosters deep, reflective thinking. To cultivate these skills, tutors often provide pedagogical guidance in the form of hints, examples, explanations, acknowledgments, and constructive feedback. This approach is rooted in the principle of metacognition [1,13] from learning sciences. By offering timely and context-aware support, tutors help students navigate their learning journey by promoting mistake discovery and providing constructive feedback (both positive and negative), among others.

Research Statement: This research direction aims to develop a robust quantitative framework to assess the appropriateness and richness of the pedagogical guidance provided by GenAI-powered ITSs. Specifically, given a partial conversation between a student and a human tutor or ITS, we propose methods to systematically evaluate the effectiveness of the tutor's next response in supporting the student's metacognitive development.

Potential Approach: In existing AIED research [1,44,45], metacognition is one of the most frequently cited capabilities of ITSs. However, there is a lack of scalable and automated methodologies to assess how well a tutor's guidance aligns pedagogically with metacognitive principles. Building on prior research [26,27], we consider a tutor's next response appropriate if it provides *pedagogical guidance*, is *factually accurate*, and remains *contextually relevant*. See Fig. 1 (left) for a visual illustration of these principles.

A modular approach to modeling each of the three aforementioned properties offers a promising research direction. These modules can collectively help to develop a holistic discriminator reward model to distinguish between *good* and *bad* responses. *Pedagogical guidance* can be approached as identifying the student's current state (e.g., a confusion) and selecting the most appropriate pedagogical strategy (e.g., a hint, examples, etc.) based on that state. This can be achieved by leveraging prior work [60,61], where human tutor models and language model-based classifiers were employed. For *contextual relevance*, sentiment embedding-based approaches can be adapted as proposed by [15,68]. Finally, the most challenging aspect is the identification of *factual correctness*, which can apply at the local level (where the current response contradicts statements in the conversation history) or at the global level (inconsistencies with established factual knowledge). This issue is particularly prevalent in GenAI-powered ITSs, as they are prone to hallucinations [24]. At the local level, this can be framed as a textual entailment problem, while at the global level, prior research on factuality verification [62] offers valuable insights and a path forward.

4.3 Research Line #3: Measuring Active Learning

Active learning is a critical component of effective tutoring, allowing students to engage actively in the learning process through critical thinking and reflection. This approach aligns with well-established learning science principles, such as constructivism and inquiry-based learning [43,59]. Effective tutors encourage active engagement by posing thought-provoking questions, guiding students

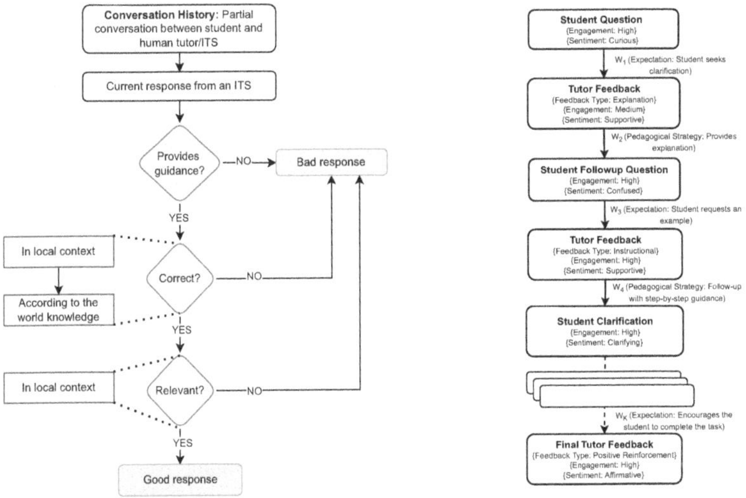

Fig. 1. (Left) Flowchart illustrating good and bad tutor responses to measure pedagogical guidance. (Right) Conversation flow graph to measure the student's active learning, where nodes represent different utterance features and edge weights indicate the intensity of conversational dependency.

without offering immediate answers, and maintaining open-ended conversations to promote reflective thinking.

Research Statement: This research direction should aims to develop a systematic framework for evaluating and measuring the active learning capabilities of ITSs. Specifically, for an ongoing dialogue between a student and an ITS, we propose methods to assess how effectively the tutor stimulates *active engagement*, *adapts to the student's emotional and cognitive states*, and provides *encouraging feedback*.

Potential Approach: Previous research in AIED [10,46] has emphasized the importance of active learning strategies. However, scalable and automated methods to measure these strategies remain underexplored. We believe that a graph-based modeling approach is a promising direction for assessing the active learning capabilities of tutoring systems. Within this approach, the dialogue between the student and tutor is represented as a dynamic, directed *conversation flow graph* [48,65], where each node represents an individual utterance, and edges capture conversational dependencies, such as follow-up questions, clarifications, or feedback. Figure 1 (right) presents a sketch of the flow graph. Node features encode engagement signals, sentiment, and feedback types, while edge weights W_i represent the strength of interactions.

The conversation flow graph can be modeled using a graph neural network (GNN) [28,58], enabling the prediction of engagement, adaptive support, and encouraging feedback from its structure. By capturing both content and conversational dynamics, the model can identify student engagement patterns like sustained reflective dialogue and effective adaptation to student states. This approach allows for a more nuanced analysis of the tutoring process, distinguishing between productive and unproductive interactions. Furthermore, the GNN framework can generalize across diverse ITS datasets, making it a scalable solution for evaluating different tutoring strategies. Additionally, attention mechanisms in GNNs [65] can highlight influential interactions, offering insights into tutor behavior that drive active learning and promote student understanding.

Other modeling approaches beyond GNNs can be used to represent the conversation flow graph. For instance, Graph Attention Networks (GATs) [58] can be employed to dynamically assign importance weights to neighboring nodes, allowing the model to focus on critical conversational turns such as key feedback or pivotal clarifications. Additionally, Graph Transformers [67] offer a promising alternative by leveraging global attention mechanisms to capture long-range dependencies across the dialogue, thus enabling a more holistic understanding of complex tutoring interactions. These advanced architectures could further enhance the model's ability to identify nuanced patterns of student engagement and tutor support strategies.

4.4 Implications of Proposed Research Directions

If developed and deployed, these research directions could significantly enhance ITSs. A *unified evaluation framework* would standardize assessments, promoting consistency and transparency across systems. While it may oversimplify the diverse needs in different educational contexts, careful design can mitigate the risk of bias and ensure that specific learning goals are not overlooked. Furthermore, taxonomy-driven evaluation serves as a foundational step toward systematically assessing ITSs. At the same time, it offers limited reflection of real-world tutoring quality, where student interactions are often complex and non-linear. The next step should involve conducting real-world randomized controlled experiments to rigorously evaluate the effectiveness of any tutoring system. Finally, unification efforts must strike a balance: evaluation should not be overly fine-grained—such as requiring a separate taxonomy for each utterance—nor overly coarse, where a tutor is assumed to be holistically optimal. This level of granularity in evaluation design is critical for the effectiveness of taxonomy- and benchmark-driven research in ITSs. *Measurement of pedagogical guidance* would improve the adaptability of ITS, ensuring that guidance aligns with students' metacognitive needs, fostering deeper engagement. Although assessing the subtleties of student cognition and emotions is challenging, thoughtful integration of adaptive mechanisms could enhance the accuracy and appropriateness of interventions. *Measuring active learning* would allow systems to better stimulate critical thinking, adapt to student states, and provide timely, context-sensitive feedback, enhancing overall learning experiences. While graph-based models may face scalability issues and complexity in capturing dynamic interactions, ongoing advancements in GenAI-based computational models can address these challenges. Together, these advancements could lead to more effective, scalable, and personalized ITSs, transforming educational practices.

5 Conclusion

The remarkable performance of GenAI models has led to active development and new opportunities for dialogue-based ITSs in the AIED field. However, the true progress of ITSs remains unclear due to the lack of reliable and robust evaluation measures. This challenge arises from the diverse pragmatics of evaluation and the non-uniform adoption of subjective and personalized protocols and benchmarks. This work highlights this gap by presenting the current state of ITS evaluation from multiple perspectives, identifying key challenges, and proposing three potential and feasible research directions rooted in pedagogical principles and existing AIED research. These directions pave the way towards the establishment of reliable, scalable, and robust evaluation frameworks for ITSs. We believe that this work provides a comprehensive overview of the existing landscape and offers valuable insights to advance the evaluation of ITSs.

Acknowledgments. We are grateful to Google for supporting this research through the Google Academic Research Award (GARA).

References

1. Abdelshiheed, M., Hostetter, J.W., Barnes, T., Chi, M.: Leveraging deep reinforcement learning for metacognitive interventions across intelligent tutoring systems. In: AIED, pp. 291–303. Springer, Heidelberg (2023). https://doi.org/10.1007/978-3-031-36272-9_24
2. Academy, K.: Khanmigo: AI-powered tutoring (2023). https://www.khanacademy.org
3. Al-Hossami, E., Bunescu, R., Smith, J., Teehan, R.: Is a dialogue based tutoring system that emulates helpful co-constructed relations during human tutoring effective? Experiments with code debugging. In: ACM TSCSE, pp. 53–59 (2024)
4. Albacete, P., Jordan, P., Katz, S.: Is a dialogue-based tutoring system that emulates helpful co-constructed relations during human tutoring effective? In: Conati, C., Heffernan, N., Mitrovic, A., Verdejo, M.F. (eds.) AIED 2015. LNCS (LNAI), vol. 9112, pp. 3–12. Springer, Cham (2015). https://doi.org/10.1007/978-3-319-19773-9_1
5. Allen, L.K., Crossley, S.A., Snow, E.L., Jacovina, M.E., Perret, C., McNamara, D.S.: Am i wrong or am i right? gains in monitoring accuracy in an intelligent tutoring system for writing. In: Conati, C., Heffernan, N., Mitrovic, A., Verdejo, M.F. (eds.) AIED 2015. LNCS (LNAI), vol. 9112, pp. 533–536. Springer, Cham (2015). https://doi.org/10.1007/978-3-319-19773-9_55
6. Anderson, J.R.: Cognitive Psychology and its Implications, 4th edn. W. H. Freeman (1995)
7. Au, W.: High-stakes testing and curricular control: a qualitative metasynthesis. Educ. Res. **36**(5), 258–267 (2007)

8. Brown, P.C., Roediger III, H.L., McDaniel, M.A.: Make it Stick: The Science of Successful Learning. Harvard University Press, Cambridge (2014)
9. Caines, A., et al.: The teacher-student chatroom corpus. arXiv. (2020)
10. Castro-Schez, J.J., Glez-Morcillo, C., Albusac, J., Vallejo, D.: An intelligent tutoring system for supporting active learning: a case study on predictive parsing learning. Inf. Sci. **544**, 446–468 (2021)
11. Chi, M., Wylie, R.: The ICAP framework: linking cognitive engagement to active learning outcomes. Educ. Psychol. **49**(4), 219–243 (2008)
12. Crossley, S.A., Varner, L.K., Roscoe, R.D., McNamara, D.S.: Using automated indices of cohesion to evaluate an intelligent tutoring system and an automated writing evaluation system. In: Lane, H.C., Yacef, K., Mostow, J., Pavlik, P. (eds.) AIED 2013. LNCS (LNAI), vol. 7926, pp. 269–278. Springer, Heidelberg (2013). https://doi.org/10.1007/978-3-642-39112-5_28
13. Dehaene, S.: How we learn: Why brains learn better than any machine... for now. Penguin (2021)
14. Demszky, D., Hill, H.: The NCTE transcripts: a dataset of elementary math classroom transcripts. arXiv preprint arXiv:2211.11772 (2022)
15. Demszky, D., et al.: Measuring conversational uptake: a case study on student-teacher interactions. arXiv preprint arXiv:2106.03873 (2021)
16. Fang, Y., He, B., Liu, Z., Liu, S., Yan, Z., Sun, J.: Evaluating the design features of an intelligent tutoring system for advanced mathematics learning. In: AIED, pp. 338–350. Springer, Heidelberg (2024). https://doi.org/10.1007/978-3-031-64302-6_24
17. Gao, X., Zhang, Y., Galley, M., Brockett, C., Dolan, B.: Dialogue response ranking training with large-scale human feedback data. arXiv:2009.06978 (2020)
18. Gehrmann, S., Clark, E., Sellam, T.: Repairing the cracked foundation: a survey of obstacles in evaluation practices for generated text. arXiv (2020)
19. Goe, L., Bell, C., Little, O.: Approaches to evaluating teacher effectiveness: a research synthesis. National Comprehensive Center for Teacher Quality (2008)
20. Graesser, A.C., Chipman, P., Haynes, B.C., Olney, A.: AutoTutor: an intelligent tutoring system with mixed-initiative dialogue, pp. 612–618. IEEE (2005)
21. Gu, J., et al.: A Survey on LLM-as-a-Judge. arXiv preprint arXiv:2411.15594 (2024)
22. Hargreaves, A.: Teaching in the Knowledge Society: Education in the Age of Insecurity. Teachers College Press (2003)
23. Haswell, R.: Automated essay scoring and the future of writing assessment. Coll. Compos. Commun. **58**(1), 36–63 (2006)
24. Huang, L., et al.: A survey on hallucination in large language models: principles, taxonomy, challenges, and open questions. ACM TIS, 1–55 (2025)
25. Hursh, D.: Assessing no child left behind and the rise of neoliberal education policies. Am. Educ. Res. J. **44**(3), 493–518 (2007)
26. Jurenka, I., et al.: Towards responsible development of generative AI for education: an evaluation-driven approach. arXiv:2407.12687 (2024)
27. Khuwaja, R., Desmarais, M., Cheng, R.: Intelligent guide: combining user knowledge assessment with pedagogical guidance. In: Frasson, C., Gauthier, G., Lesgold, A. (eds.) ITS 1996. LNCS, vol. 1086, pp. 225–233. Springer, Heidelberg (1996). https://doi.org/10.1007/3-540-61327-7_119
28. Kipf, T., Welling, M.: Semi-Supervised Classification with Graph Convolutional Networks. In: ICLR (2017). https://arxiv.org/abs/1609.02907
29. Koedinger, K.R., Booth, J.L., Klahr, D.: Instructional complexity and the science to constrain it. Science **342**(6161), 935–937 (2013)

30. Lin, C.C., Huang, A.Y., Lu, O.H.: Artificial intelligence in intelligent tutoring systems toward sustainable education: a systematic review. Smart Learn. Environ. **10**(1), 41 (2023)
31. Long, Y., Aleven, V.: Skill diaries: improve student learning in an intelligent tutoring system with periodic self-assessment. In: AIED, pp. 249–258 (2013)
32. Macina, J., et al.: Mathdial: a dialogue tutoring dataset with rich pedagogical properties grounded in math reasoning problems (2023). arXiv:2305.14536
33. Macina, J., Daheim, N., Hakimi, I., Kapur, M., Gurevych, I., Sachan, M.: Mathtutorbench: a benchmark for measuring open-ended pedagogical capabilities of LLM tutors. arXiv:2502.18940 (2025)
34. Macina, J., et al.: Opportunities and challenges in neural dialog tutoring. In: Vlachos, A., Augenstein, I. (eds.) EACL, pp. 2357–2372 (2023)
35. Maurya, K.K., Srivatsa, K., Petukhova, K., Kochmar, E.: Unifying AI tutor evaluation: an evaluation taxonomy for pedagogical ability assessment of LLM-powered AI tutors. Proc. NAACL **2025**, 1234–1251 (2024)
36. McNamara, D.S., et al.: The writing-pal: natural language algorithms to support intelligent tutoring on writing strategies. In: K-12 Education: Concepts, Methodologies, Tools, and Applications (2014)
37. Muijs, D.: Measuring teacher effectiveness: some methodological reflections. Educ. Res. Eval. **12**(1), 53–74 (2006)
38. Nye, B.D., Graesser, A.C., Hu, X.: AutoTutor and family: a review of 17 years of natural language tutoring. Int. J. Artif. Intell. Educ. **24**(4), 427–469 (2014). https://doi.org/10.1007/s40593-014-0029-5
39. Papineni, K., Roukos, S., Ward, T., Zhu, W.J.: Bleu: a method for automatic evaluation of machine translation. In: ACL 2002, pp. 311–318 (2002)
40. Perelman, L.: Construct validity, length, and automated essay scoring. Educ. Leadersh. **70**(4), 44–49 (2012)
41. Perelman, L.: Critiquing automated essay scoring systems: a challenge for educational assessment. J. Writ. Assess. **7**(1), 1–15 (2014)
42. Piaget, J.: The Language and Thought of the Child. Routledge and Kegan Paul (1959)
43. Piaget, J.: The Theory of Stages in Cognitive Development. McGraw-Hill (1972)
44. Pinkwart, N.: Another 25 years of AIED? Challenges and opportunities for intelligent educational technologies of the future. IJAIED **26**, 771–783 (2016)
45. Porayska-Pomsta, K.: AI as a methodology for supporting educational praxis and teacher metacognition. IJAIED **26**, 679–700 (2016)
46. Rouzegar, H., Makrehchi, M.: Generative AI for Enhancing Active Learning in Education: A Comparative Study of GPT-3.5 and GPT-4 in Crafting Customized Test Questions. arXiv:2406.13903 (2024)
47. Rus, V., Niraula, N., Banjade, R.: DeepTutor: an effective, online intelligent tutoring system that promotes deep learning. In: AAAI 2015, vol. 29 (2015)
48. Scarselli, F., Gori, M., Tsoi, A.C., Hagenbuchner, M., Monfardini, G.: The graph neural network model. IEEE Trans. Neural Netw. **20**(1), 61–80 (2008)
49. Schmucker, R., Xia, M., Azaria, A., Mitchell, T.: Ruffle&Riley: insights from designing and evaluating a large language model-based conversational tutoring system. In: AIED, pp. 75–90. Springer, Heidelberg (2024)
50. Spreng, M.: iDrive-The New Interaction Concept for Automotive Cockpits. Technical report, SAE Technical Paper (2002)
51. Stasaski, K., Kao, K., Hearst, M.A.: CIMA: a large open access dialogue dataset for tutoring. In: Proceedings of BEA 2020, pp. 52–64 (2020)

52. Süngü, A., et al.: Generative AI Can Harm Learning. SSRN (2024). https://www.ssrn.com/abstract=4895486
53. Suresh, A., Jacobs, J., Harty, C., Perkoff, M., Martin, J.H., Sumner, T.: The TalkMoves dataset: K-12 mathematics lesson transcripts annotated for teacher and student discursive moves. arXiv:2204.09652 (2022)
54. Tack, A., Kochmar, E., Yuan, Z., Bibauw, S., Piech, C.: The BEA 2023 shared task on generating AI teacher responses in educational dialogues. In: Proceedings of BEA 2023, pp. 785–795 (2023)
55. Tack, A., Piech, C.: The AI teacher test: measuring the pedagogical ability of blender and GPT-3 in educational dialogues. arXiv:2205.07540 (2022)
56. Team, L., Modi, A., Veerubhotla, A.S., Rysbek, A., Huber, A., Wiltshire, B., et al.: LearnLM: improving gemini for learning. arXiv:2412.16429 (2024)
57. Vasselli, J., Vasselli, C., Nohejl, A., Watanabe, T.: NAISTeacher: a prompt and rerank approach to generating teacher utterances in educational dialogues. In: Proceedings of BEA 2023, pp. 772–784 (2023)
58. Velickovic, P., et al.: Graph attention networks. In: ICLR (2018)
59. Vygotsky, L.S.: Mind in Society: The Development of Higher Psychological Processes. Harvard University Press, Boston (1978)
60. Wang, R., Zhang, Q., Robinson, C., Loeb, S., Demszky, D.: Bridging the novice-expert gap via models of decision-making: a case study on remediating math mistakes. In: Proceedings of NAACL 2024, pp. 2174–2199 (2024)
61. Wang, R.E., Ribeiro, A.T., Robinson, C.D., Loeb, S., Demszky, D.: Tutor CoPilot: a human-AI approach for scaling real-time expertise. arXiv:2410.03017 (2024)
62. Wang, Y., et al.: Factuality of large language models: a survey. arXiv:2402.02420 (2024)
63. Weinstein, Y., Madan, C.R., Sumeracki, M.A.: Teaching the science of learning. Cogn. Res. Princip. Implicat. **3**, 1–17 (2018)
64. Whitlock, A.: The limitations of automated feedback in educational assessment. Educ. Technol. Rev. **18**(2), 122–137 (2020)
65. Wu, Z., Pan, S., Chen, F., Long, G., Zhang, C., Yu, P.S.: A comprehensive survey on graph neural networks. IEEE TNNLS 4–24 (2020)
66. Yan, L., et al.: Practical and ethical challenges of large language models in education: a systematic scoping review. BJET 90–112 (2024)
67. Yun, S., Jeong, M., Kim, R., Kang, J., Kim, H.J.: Graph transformer networks. In: NeurIPS, vol. 32 (2019)
68. Zhang, T., Kishore, V., Wu, F., Weinberger, K.Q., Artzi, Y.: Bertscore: evaluating text generation with bert. arXiv:1904.09675 (2019)
69. Zhao, W.X., et al.: A Survey of Large Language Models. arXiv (2023)

AI in Educational Digital Escape Rooms: State of the Art and Perspectives

Jakub Swacha[1(✉)], Karolina Muszyńska[1], Jose Maria Font Fernandez[2], Selay Arkün Kocadere[3], Ricardo Queirós[4], Robertas Damaševičius[5], and Rytis Maskeliūnas[5]

[1] Department of IT in Management, University of Szczecin, Szczecin, Poland
{jakub.swacha,karolina.muszynska}@usz.edu.pl

[2] Game Lab, Department of CS and MT, Malmö University, Malmö, Sweden
jose.font@mau.se

[3] Department of CEIT, Hacettepe University, Ankara, Türkiye
selaya@hacettepe.edu.tr

[4] ESMAD, Polytechnic of Porto & CRACS INESC TEC, Porto, Portugal
ricardoqueiros@esmad.ipp.pt

[5] CRTCS, Kaunas University of Technology, Kaunas, Lithuania
{robertas.damasevicius,rytis.maskeliunas}@ktu.lt

Abstract. Artificial Intelligence (AI), in particular Generative Artificial Intelligence (GenAI), is a quickly developing field capable of revolutionizing educational digital escape rooms. Traditionally reliant on static content, these immersive environments have faced limitations in adaptability, replayability, and personalization. However, recent advancements in AI and GenAI enable dynamic puzzle generation, adaptive storytelling, and AI-driven non-player characters (NPCs) with agentic AI, allowing for highly responsive and personalized experiences. This paper reviews the state-of-the-art in integrating AI (with the focus on GenAI) into educational digital escape rooms, integrating interdisciplinary insights from cognitive science, game design, and machine learning, and showing how AI can improve engagement, scalability, and content diversity, but also indicates challenges related to ethical AI use, bias in algorithmic decision-making, and the need for robust evaluation frameworks to assess player satisfaction and learning outcomes.

Keywords: Artificial Intelligence · GenAI · Large Language Models · Virtual Escape Rooms · Educational Escape Rooms

1 Introduction

Educational Escape Rooms (EERs) and Educational Digital Escape Rooms (EDERs) [50] represent a significant shift in pedagogical design, combining immersive storytelling, collaborative problem-solving, and gamified learning to enhance engagement. Rooted in experiential learning theories [39,53], these environments embed educational content into time-bound, narrative-driven challenges where participants solve interconnected puzzles. Studies highlight their

A. I. Cristea et al. (Eds.): AIED 2025, CCIS 2590, pp. 47–60, 2025.
https://doi.org/10.1007/978-3-031-99261-2_4

effectiveness in boosting cognitive engagement and knowledge retention across domains such as STEM, healthcare, and soft skills training [69]. They can also integrate machine learning to analyze participant behavior in real time, enabling adaptive difficulty levels and personalized feedback to enhance learning outcomes [22].

Despite their demonstrated advantages, they pose serious challenges such as the costly physical setups, the static content limitations of traditional designs, and the need for robust evaluation frameworks to measure long-term impact. EERs and EDERs encounter significant limitations that impede their scalability, adaptability, and long-term pedagogical efficacy [20]. Traditional EERs are inherently resource-intensive, often necessitating substantial investments in physical infrastructure, specialized equipment, and manual content design. The static nature of these environments poses a critical barrier to replayability, as participants who have previously engaged with a room may perceive subsequent attempts as predictable or redundant, thereby diminishing engagement and learning outcomes. EDERs, although they address some logistical constraints through digital platforms, are not immune to analogous challenges [48]. Current implementations frequently rely on labor-intensive manual content creation, resulting in finite and pre-scripted experiences that lack the flexibility to adapt dynamically to individual learner profiles or evolving educational objectives [52]. The absence of sophisticated personalization mechanisms often leads to formulaic gameplay structures, where puzzles and narratives fail to account for variations in cognitive abilities, cultural contexts, or learning styles. These limitations are exacerbated by the growing demand for scalable solutions capable of accommodating diverse user populations and contexts, from K-12 education to professional training. Consequently, there is an urgent need for innovative methodologies that can overcome these barriers by introducing dynamism, adaptability, and scalability into escape room design [72].

Artificial Intelligence (AI), and especially Generative Artificial Intelligence (GenAI), provide techniques capable to address static content limitations in EDERs, such as Procedural Content Generation (PCG), Natural Language Processing (NLP), Reinforcement Learning (RL), and Generative Adversarial Networks (GANs) to create adaptive learning experiences [58]. They can also be used to foster personalized education and promote equity, inclusivity, and accessibility for students with various physical, emotional, or cognitive needs with adaptive mechanisms that tailor challenges to each student and by generating culturally sensitive narratives [21,26,77].

This paper aims to explore the intersection of AI and EDERs by performing a critical review of prior work within five technological domains: PCG, NLP, RL, GANs, and Multimodal AI, as well as regarding ethical concerns, including algorithmic bias and data privacy. Its results are reported in Sect. 2, whereas Sect. 3 synthesizes the findings into a conceptual framework of the use of AI in EDERs indicating the perspectives for future research, and Sect. 4 elaborates on the key takeaways.

2 State of the Art Review

2.1 Procedural Content Generation (PCG) in EDERs

Procedural Content Generation (PCG) is revolutionizing EDERs by automating puzzle and environment creation, enhancing adaptability, and improving engagement [75]. By dynamically generating content, PCG overcomes static design limitations like limited replayability and lack of personalization, also enabling adaptive difficulty by ensuring that puzzles match the cognitive abilities of the learners. Several approaches [4,68] use evolutionary algorithms to balance puzzle complexity in real time, optimizing challenge levels and fostering deeper understanding through a fitness function that iteratively modifies puzzle design parameters. Graph-based PCG approaches [54] create branching narrative puzzles using directed acyclic graphs (DAGs), encouraging strategic exploration while ensuring varied yet coherent experiences.

Using constraint-based systems with PCG ensures that the generated content aligns with pedagogical goals and narrative coherence. Naswir et al. [59] propose a hybrid framework combining rule-based constraints and machine learning to create STEM-focused puzzles. Pittman [63] suggests using ontologies to encode learning objectives and reinforcement learning to optimize puzzle variability. Kuka et al. [44] advocate dual-layered approaches to balance educational benchmarks while retaining the unpredictability necessary to sustain engagement across repeated interactions. Constraint-based PCG helps address cultural and contextual factors by integrating sociocultural metadata, such as regional education priorities and linguistic preferences [12].

Design metaphors play a key role [35], while NLP techniques generate inclusive and accessible puzzles and narratives [46]. Hybrid PCG approaches now combine multiple algorithms for greater flexibility [14]. GANs generate photorealistic textures and 3D assets [37], enhancing the aesthetic and immersive quality of EDERs [55]. PCG also fuses with real-time behavioral analytics. Cutting [16] highlights analyzing attention and stress levels to adjust puzzle complexity, ensuring engagement while avoiding cognitive fatigue.

PCG enhances replayability [67] of EDERs, addressing a key challenge in traditional escape room design by generating new content with each iteration. This is crucial in education, where repeated exposure can lead to disengagement. PCG also improves scalability by automating content creation, reducing reliance on manual design, and enabling rapid EDER deployment in educational settings, where educators can customize escape room scenarios to match learning objectives, democratizing content creation to empower educators to use EDERs as versatile tools for both formal and informal learning, making EDERs more accessible as dynamic learning tools [24].

2.2 Natural Language Processing (NLP) in EDERs

Natural Language Processing (NLP) is transforming EDERs by enabling adaptive narratives, responsive dialogues, and immersive storytelling [18]. Advanced transformer-based models, such as GPT-4 and BERT, allow EDERs to generate context-sensitive content that evolves dynamically based on user interactions, enhancing engagement and facilitating user-tailored learning pathways.

Bathnagar [7] used fine-tuned GPT models for culturally sensitive narratives. Trimboli [74] highlights how dynamic adjustments in character backstories and dialogue align with regional educational priorities, improving inclusivity and engagement. Multilingual NLP models support cross-cultural EDERs. Pakray [61] combined mBERT with rule-based translation to generate multi-language narratives while preserving cultural nuances.

NLP-driven dialogue systems have been instrumental in simulating intelligent non-player characters (NPCs) with real-time feedback and hints that enrich the pedagogical value of EDERs, e.g. in conversational agents for medical training [1] and in an NLP-powered mentor NPC for leadership training [41]. NLP can provide customized feedback that reinforces correct decisions and corrects miscalculations in real time [38], increasing authenticity and the development of critical thinking and decision-making skills.

Sentiment analysis and discourse modeling improve emotionally resonant dialogues. Sentiment-aware transformers tailor NPC interactions in soft skills training [64], and Guan [23] found that analyzing emotional cues fosters empathy.

Recent research integrates NLP with multimodal data streams—physiological responses, gaze tracking, and facial expressions —to enhance EDER immersion. Krzeminska's multimodal approach [43] combined textual analysis with biometric data for adaptive narratives, while Iavarone [31] used stress and attention patterns to adjust pacing and complexity, ensuring engagement without overload. NLP also merges with AR/VR for immersive storytelling. Pilai [62] used NLP-driven voice recognition in VR, enabling spoken interactions.

2.3 Reinforcement Learning in EDERs

Reinforcement Learning (RL), lets optimize gameplay mechanics and personalization in EDERs by adapting game dynamics, difficulty, and feedback to individual learning styles and cognitive abilities [50]. Kabanda et al. [33] demonstrated RL-driven difficulty tuning in a cybersecurity-themed EDER. RL modifies puzzle complexity in real-time based on response times and errors [6], easing challenges when needed by adjusting constraints or providing hints [56], thus balancing engagement and mitigating frustration.

RL frameworks help optimize collaborative learning in EDERs. Yang et al. [82] modeled participants as independent agents, analyzing task completion, communication, and interactions to adjust task allocations and resolve collaboration issues. RL uses real-time behavioral analytics to enhance adaptive gameplay in EDERs [18]. Integrated with multimodal data—physiological responses, gaze tracking, and interaction patterns—RL improves personalization and immersion [40]. De Lima [49] used a multimodal RL framework to optimize narrative pacing and puzzle design by analyzing stress, attention, and cognitive load, ensuring engagement without overload. RL also optimizes reward structures in EDERs

[18]. Snyman [70] applied intrinsic motivation models, tailoring rewards to exploration or mastery-driven learners. Educators can deploy personalized EDERs [10], while transfer learning enables pre-trained RL models to generalize across domains, reducing domain-specific training [78].

2.4 Multimodal Data Integration in EDERs

Using wearable devices, biometric sensors, and computer vision techniques, real-time data on learners' emotional states, stress levels, cognitive loads, and behavioral patterns can be captured and then processed by AI models to dynamically adapt content, pacing, and feedback, ensuring highly immersive and personalized learning experiences that cater to individual needs [42]. Multimodal data fusion can create adaptive environments that respond intelligently to learners' physiological and cognitive states [13]. Horves et al. [28] suggested incorporating electrodermal activity (EDA) sensors to measure arousal levels. EDA data can be used to detect moments of increased stress or frustration, triggering interventions such as hints or simplified tasks to restore equilibrium, which could not only enhance the accessibility of EDERs, but also ensure that learners remained motivated and focused throughout the experience.

Computer vision techniques have played a transformative role in enabling emotionally responsive narratives and interactions within EDERs [85]. Facial expression analysis and gesture recognition can be used to infer learners' emotional states, such as frustration, confusion, or satisfaction [32]. Based on that, the NPCs can be developed to provide empathetic responses and adaptive feedback that aligned with the learners' emotional contexts [30]. For example, if education game detects signs of frustration, the NPCs can offer supportive encouragement or simplified explanations, fostering a positive and inclusive learning environment, which can not only enhance the emotional resonance of the experience, but also promote empathy and interpersonal understanding among the participants. Wang [79] suggest, that emotion recognition with natural language processing (NLP) can create emotionally intelligent dialogue systems.

Another significant advancement lies in the use of multimodal data to model learners' behavioral patterns and optimize interaction design [84] and to create multisensory feedback systems that enhance the immersiveness [19]. Haptic feedback systems that synchronized tactile stimuli with visual and auditory cues can simulate realistic environmental interactions [34]. Therefore, learners' physiological responses to adjust the intensity and timing of haptic feedback, ensuring a seamless and engaging sensory experience [81] and such multisensory approaches not only would enrich the realism of EDERs but also facilitate deeper immersion and emotional engagement. Augmented reality (AR) and multimodal AI can be used to create context-sensitive environmental adaptations [29]. Gaze tracking and gesture recognition can be used to modify virtual objects and spatial layouts in real time, allowing learners to interact with dynamic and responsive environments [8].

Despite these advancements, several challenges remain in the integration of multimodal data into GenAI-driven EDERs and educational games [57]. Key issues include ensuring data privacy and ethical use of biometric data, addressing algorithmic biases in emotion recognition and behavioral modeling, and developing robust validation frameworks to assess the pedagogical effects.

2.5 GANs in EDERs

GANs can be used in the development of EDERs to create photorealistic assets, interactive 3D models, and dynamic environments [5]. They redefine immersive learning by improving aesthetics and amplifying educational impact through visually rich, contextually and culturally relevant settings [56].

Villegas et al. [76] used StyleGAN to create dynamic classrooms with emotional cues, while latent space interpolation enabled seamless cultural transitions [51]. High-resolution textures and real-time adjustments in lighting and spatial layouts enhance immersion [3]. Beyond visuals, GANs improve pedagogical impact by aligning generated environments with educational goals and learner profiles, ensuring content remains meaningful and scenario-specific [2,47]. GANs have also been applied to 2D to 3D image conversion [37] , enabling educators to quickly prototype and deploy scenarios tailored to specific learning objectives, including immersive and experiential learning experiences in Virtual and Augmented Reality [17,25].

2.6 Ethical Considerations in EDERs

While the integration of AI into Educational Digital Escape Rooms (EDERs) offers transformative benefits, it also raises critical ethical and pedagogical concerns that warrant careful consideration.

One of the most pressing ethical concerns in the use of AI for EDER development is the potential for algorithmic bias, which is common to machine-learning based approaches such as those in GenAI. Yu studies ethical challenges in AI [83], important for AI-driven EDERs, emphasizing the risk of biased output in content generation processes. For example, if the training datasets used to develop GenAI models are not representative of diverse populations, the resulting narratives, puzzles, or environments may inadvertently favor certain cultural, linguistic, or cognitive profiles while marginalizing others [66].

To mitigate these risks, researchers have advocated the development of fairness-aware AI systems [45]. For example, Chinta proposed a framework for auditing GenAI-generated content to identify and address biases related to gender, race, socioeconomic status, and learning styles [15]. Priyadarshini [65] suggested employing adversarial debiasing techniques to ensure equitable outcomes. Such measures are critical to foster inclusive EDERs that cater to the needs of all learners, regardless of their background or abilities.

The potential for overreliance on AI-generated content represents another critical pedagogical challenge [71]. Although AI can automate content creation and adaptation, excessive dependence on these systems can undermine educators' roles and diminish opportunities for human-centered instructional design [86]. Bozkurt et al. [9] cautioned against viewing GenAI as a substitute for thoughtful pedagogical planning, emphasizing the need for educators to retain oversight and control over the learning experience. Hensai et al. developed a collaborative platform that allowed educators to customize and curate GenAI-generated content, ensuring alignment with curriculum standards and learning goals [27], empowering educators to use AI as a creative partner in designing engaging and meaningful learning experiences.

Ensuring transparency and accountability in AI-driven EDERs is essential to build trust and foster learner agency [11]. For example, learners should be able to understand why certain puzzles were presented or why specific feedback was provided, enabling them to critically evaluate and engage with the material [36].

Ensuring pedagogical effectiveness of GenAI-generated content remains a key priority [73] to assess content quality, adaptability, and alignment with educational objectives, using both quantitative metrics (e.g., knowledge retention rates) and qualitative insights (e.g., learner feedback). For example, Wang et al. conducted a longitudinal study to evaluate the efficacy of a GenAI in STEM education [80]. Their findings revealed improvements in conceptual understanding and problem-solving skills, but also identified areas for improvement, such as the need for more nuanced scaffolding mechanisms. Nyaabba et al. [60] underscored the importance of continuous evaluation and refinement to ensure that GenAI-driven EDERs deliver meaningful and impactful learning experiences.

3 Synthesis of Findings

The results of the review can be summarized as the Conceptual Framework of AI in Educational Digital Escape Rooms (EDERs) (Fig. 1), which provides a structured overview of how AI, and in particular GenAI, enhances adaptive gameplay, personalization, and immersive learning. At its core, user interactions—including player decisions and actions—serve as inputs for AI processing, which dynamically adjusts puzzles, narratives, and feedback to optimize engagement and learning. This process is powered by five key AI techniques: Procedural Content Generation (PCG) for dynamic puzzle design, Natural Language Processing (NLP) for interactive NPCs and story adaptation, Reinforcement Learning (RL) for real-time difficulty adjustments, Generative Adversarial Networks (GANs) for realistic and immersive environments, and Multimodal AI, which integrates behavioral and biometric data for emotion-aware adaptation. The AI-driven outputs—dynamic puzzles, story-driven challenges, and personalized feedback—ensure that the learning experience remains engaging, scalable, and tailored to individual needs.

The core of the framework lies in the AI Processing, responsible for producing dynamic content tailored to individual player (learner) experiences. The Procedural Content Generation (PCG) generates puzzles, riddles, and challenges, ensuring that no two gameplay experiences are identical. Natural Language Processing (NLP) enhances narrative depth by generating adaptive storylines and interactive dialogues with non-player characters (NPCs), making conversations feel natural and responsive to player choices. Reinforcement Learning (RL)

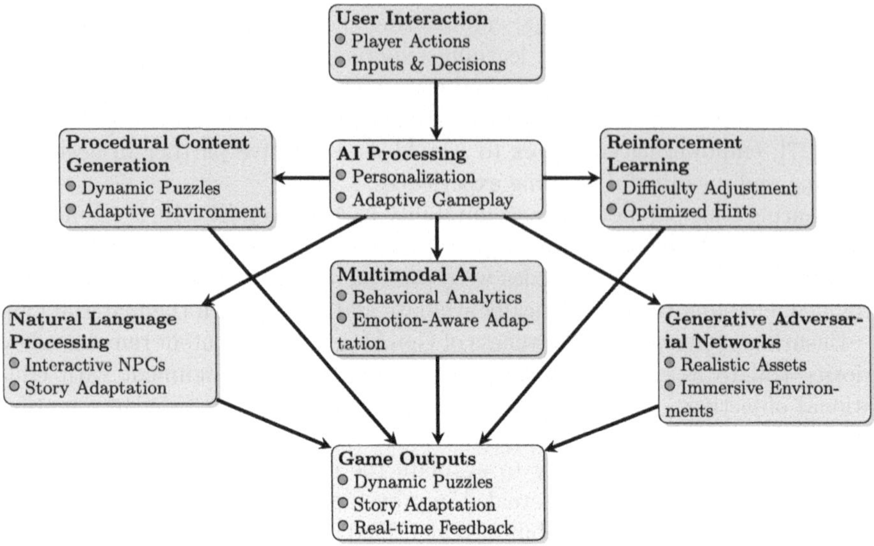

Fig. 1. Conceptual Framework of Generative AI in EDERs

dynamically adjusts the difficulty of the game and provides personalized feedback by analyzing the performance of the player, ensuring an optimal balance between challenge and engagement. GANs create visually rich environments and interactive 3D assets, allowing for realistic settings and immersive worlds. Finally, Multimodal AI processes diverse data inputs such as gaze tracking, speech patterns, and physiological responses, enabling the game to respond intelligently to player emotions and cognitive states.

Using AI in EDERs enhances game-based learning by dynamically adjusting puzzle complexity, storytelling, and progression to support **adaptability**. It **personalizes** content based on user behavior, cognitive ability, and learning goals, ensuring a tailored experience. AI-driven narratives and immersive interactions boost **engagement**, motivation, and retention. By automating content creation, AI improves **scalability** and replayability while also offering **intelligent feedback** through real-time hints and performance analytics.

There are nevertheless relevant challenges inherent to these technologies that can't be overlooked. **Algorithmic bias** can lead to content that reinforces stereotypes or overlooks diverse learning needs. **Data privacy** is a major concern, especially when collecting and processing sensitive biometric and behavioral information. There's also a risk that over-reliance on AI for content generation could **undermine educators' control** over pedagogical strategies. Additionally, the high computational demands of AI-driven personalization may pose **infrastructure barriers**, potentially limiting access and scalability.

Future research in AI-driven EDERs should focus on enhancing transparency, inclusivity, and adaptability to create more effective and ethical learning environments. Explainable AI (XAI) helps make AI-driven adaptations transparent, enabling educators and learners to understand personalized elements. Research into fairness-aware AI can reduce bias and support culturally inclusive content. Multimodal learning analytics, using biometric and behavioral data, allow dynamic personalization based on cognitive and emotional states. Hybrid human-AI approaches ensure co-creative balance, educator involvement, and authorship alongside automation. Finally, AI-powered collaborative learning using multi-agent reinforcement learning (MARL) can enhance teamwork and problem-solving, making digital escape rooms more adaptive and impactful in formal and informal education.

4 Discussion and Conclusion

This paper provided a survey of AI and, in particular, GenAI methods in Educational Digital Escape Rooms (EDERs), synthesizing current research across procedural content generation, natural language processing, reinforcement learning, multimodal data integration, and generative adversarial networks. It underscores how these methods collectively enhance the adaptability, personalization, and overall effectiveness of digital educational experiences. A central contribution of this paper is the proposed conceptual framework, which integrates diverse AI technologies for the use in EDERs, providing a structured roadmap for future research and practical implementations.

Several research questions remain unanswered, highlighting future study directions: balancing narrative coherence with adaptive flexibility, ensuring unbiased and equitable AI-driven content, and developing robust frameworks for evaluating pedagogical effectiveness and long-term learning outcomes. Also several barriers and limitations persist, including computational complexity, data privacy concerns, ethical implications related to sensitive biometric data usage, and algorithmic bias. Practical challenges such as educator acceptance and the integration of advanced AI technologies into existing educational systems also present hurdles.

In conclusion, while AI and GenAI advance educational digital escape rooms, realizing their full potential will require overcoming identified barriers through interdisciplinary collaboration among AI specialists, educators, researchers, and policymakers. The proposed framework may serve as a pivotal reference point for future advancements in educational technology.

Acknowledgments. The presented work has been co-financed by the Minister of Science of Poland under the Regional Excellence Initiative program.

Disclosure of Interests. The authors have no competing interests to declare relevant to this article's content.

References

1. Babu, A., Boddu, S.B.: BERT-based medical chatbot: enhancing healthcare communication through natural language understanding. Exploratory Res. Clin. Soc. Pharmacy **13**, 100419 (2024)

2. Bahroun, Z., Anane, C., Ahmed, V., Zacca, A.: Transforming education: a comprehensive review of generative artificial intelligence in educational settings through bibliometric and content analysis. Sustainability **15**(17), 12983 (2023)
3. Bansal, G., Nawal, A., Chamola, V., Herencsar, N.: Revolutionizing visuals: the role of generative AI in modern image generation. ACM Trans. Multimed. Comput. Commun. Appl. **20**(11), 1–22 (2024)
4. Bari, A.G., Gaspar, A., Wiegand, R.P., Albert, J.L., Bucci, A., Kumar, A.N.: Evoparsons: design, implementation and preliminary evaluation of evolutionary parsons puzzle. Genet. Program Evolvable Mach. **20**, 213–244 (2019)
5. Basyoni, L., et al.: Generative AI-driven metaverse: the promises and challenges of AI-generated content (2025), https://www.techrxiv.org/doi/pdf/10.36227/techrxiv.174062953.37029023
6. Bechtold, O.: AI Cruciverbalist-artificial intelligence (Machine Learning and Constraint Satisfaction Programming) for grid based word puzzle generation. Master's thesis, Graz University of Technology (2020)
7. Bhatnagar, D.: Fine-tuning large language models for domain-specific response generation: a case study on enhancing peer learning in human resource. Master's thesis, Dublin, National College of Ireland (2023)
8. Bhowmik, A.K.: Virtual and augmented reality: human sensory-perceptual requirements and trends for immersive spatial computing experiences. J. Soc. Inform. Display **32**(8), 605–646 (2024)
9. Bozkurt, A., et al.: The manifesto for teaching and learning in a time of generative AI: a critical collective stance to better navigate the future. Open Praxis **16**(4), 487–513 (2024)
10. Carvalho, C., Teran, L., Mota, M., Pereira, R.: A systematic mapping study on digital game adaptation dimensions. In: Proceedings of the 21st Brazilian Symposium on Human Factors in Computing Systems, pp. 1–14 (2022)
11. Celestin, M., Vasuki, M., Kumar, A.D., Sujatha, S.: The future of education. DK Int. Res. Found. (2024)
12. Chan, G., Banire, B., Ataguba, G., Frempong, G., Orji, R.: A panoramic view of socio-cultural sensitivity in digital technologies: a comprehensive review and future directions. Int. J. Hum.-Comput. Interact. **41**(4), 1917–1945 (2025)
13. Chango, W., Lara, J.A., Cerezo, R., Romero, C.: A review on data fusion in multimodal learning analytics and educational data mining. Wiley Interdisciplinary Rev. Data Min. Knowl. Disc. **12**(4), e1458 (2022)
14. Chen, T.: Optimization-based approaches for structured content creation in game design. Ph.D. thesis, University of Minnesota (2021)
15. Chinta, S.V., et al.: FairAIED: navigating fairness, bias, and ethics in educational AI applications (2024), https://arxiv.org/pdf/2407.18745?
16. Cutting, J., Deterding, S.: The task-attention theory of game learning: a theory and research agenda. Hum.-Comput. Interact. **39**(5–6), 257–287 (2024)
17. Damaševičius, R., Sidekerskiene, T.: Designing metaverse escape rooms for microlearning in STEM education. In: Fostering Pedagogy Through Micro and Adaptive Learning in Higher Education: Trends, Tools, and Applications, pp. 192–211. IGI Global (2023). https://doi.org/10.4018/978-1-6684-8656-6.ch009
18. Damaševičius, R., Sidekerskienė, T.: Digital escape room designs in education. IGI Global (2025). https://doi.org/10.4018/979-8-3693-4219-0
19. Dawkins, O., Young, G.W.: Engaging place with mixed realities: Sharing multisensory experiences of place through community-generated digital content and multimodal interaction. In: International Conference on Human-Computer Interaction, pp. 199–218. Springer (2020)

20. Dritsas, E., Trigka, M.: Methodological and technological advancements in e-learning. Information **16**(1), 56 (2025)
21. Fotaris, P., Mastoras, T., Lameras, P.: Designing educational escape rooms with generative AI: a framework and ChatGPT prompt engineering guide. In: Proceedings of the European Conference on Games-Based Learning, pp. 180–189 (2023)
22. Gm, D., Goudar, R., Kulkarni, A.A., Rathod, V.N., Hukkeri, G.S.: A digital recommendation system for personalized learning to enhance online education: a review. IEEE Access **12**, 34019–34041 (2024)
23. Guan, J.Q., Ying, S.F., Zhang, M.L., Hwang, G.J.: From experience to empathy: an empathetic vr-based learning approach to improving EFL learners' empathy and writing performance. Comput. Educ. **220**, 105120 (2024)
24. Guerrero, M.J.S., Calderón-Lesano, D.: Escape room strategy to motivate the learning of healthy habits in high school students. Investigación Y Desarrollo **18**(1) (2023)
25. Gupta, A., Mohummed, F.: Role of generative AI in augmented reality (AR) and virtual reality (VR) application testing. JAIMLD **1**, 426–30 (2023)
26. Hadinezhad, S., Garg, S., Lindgren, R.: Enhancing inclusivity: Exploring AI applications for diverse learners. In: Kourkoulou, D., Tzirides, A.O., Cope, B., Kalantzis, M. (eds.) Trust and Inclusion in AI-Mediated Education: Where Human Learning Meets Learning Machines, pp. 163–182. Springer, Cham (2024). https://doi.org/10.1007/978-3-031-64487-0_8
27. Han, A.: StoryAI: designing, developing, and evaluating generative AI-powered story-authoring platform for young learners. Ph.D. thesis, University of California, Irvine (2024)
28. Horvers, A., Tombeng, N., Bosse, T., Lazonder, A.W., Molenaar, I.: Detecting emotions through electrodermal activity in learning contexts: a systematic review. Sensors **21**(23), 7869 (2021)
29. Hu, Y., et al.: Vision-based multimodal interfaces: a survey and taxonomy for enhanced context-aware system design. In: Proceedings of the 2025 CHI Conference on Human Factors in Computing Systems (CHI 2025), article 758, pp. 1–31. ACM, New York (2025)
30. Huyen, N.T.: Fostering design thinking mindset for university students with npcs in the metaverse. Heliyon **10**(15), e34964 (2024)
31. Iavarone, B.: Understanding emotive response to textual stimuli: a multimodal approach. Ph.D. thesis, Scuola Normale Superiore (2024)
32. Imani, M., Montazer, G.A.: A survey of emotion recognition methods with emphasis on e-learning environments. J. Netw. Comput. Appl. **147**, 102423 (2019)
33. Kabanda, G., Chipfumbu, C.T., Chingoriwo, T.: A reinforcement learning paradigm for cybersecurity education and training. Orient. J. Comput. Sci. Technol. **16**(1), 12–45 (2023)
34. Kang, N., Sah, Y.J., Lee, S.: Effects of visual and auditory cues on haptic illusions for active and passive touches in mixed reality. Int. J. Hum Comput Stud. **150**, 102613 (2021)
35. Khaled, R., Nelson, M.J., Barr, P.: Design metaphors for procedural content generation in games. In: Proceedings of the SIGCHI Conference on Human Factors in Computing Systems, pp. 1509–1518 (2013)
36. Khosravi, H., et al.: Explainable artificial intelligence in education. Comput. Educ. Artif. Intell. **3**, 100074 (2022)
37. Koh, A., Tan, S.Y., Nasrudin, M.F.: A systematic literature review of generative adversarial networks (GANs) in 3d avatar reconstruction from 2d images. Multimedia Tools Appl. **83**(26), 68813–68853 (2024)

38. Kökver, Y., Pektaş, H.M., Çelik, H.: Artificial intelligence applications in education: natural language processing in detecting misconceptions. Educ. Inf. Technol. **30**, 3035–3066 (2024)
39. Kolb, D.A., Boyatzis, R.E., Mainemelis, C.: Experiential learning theory: previous research and new directions. In: Perspectives on Thinking, Learning, and Cognitive Styles, pp. 227–247. Routledge (2014)
40. Koorathota, S.: Multimodal deep learning systems for analysis of human behavior, preference, and state. Ph.D. thesis, Columbia University (2023)
41. Kostilainen, S.: Next generation of NPC dialogue: creating responsive NPCs (Non-Player Characters) with Retrieval-Augmented Generation and real-time player data. Master's thesis, Jyväskylä: Jamk University of Applied Sciences (2024)
42. Kourtesis, P.: A comprehensive review of multimodal xr applications, risks, and ethical challenges in the metaverse. Multimodal Technol. Interact. **8**(11), 98 (2024)
43. Krzemińska, I.: Multimodal recognition of users states at human-AI interaction adaptation. Technium **26**, 102–140 (2025)
44. Kuka, L., Sabitzer, B.: Navigating educational frontiers in the AI era: a teacher's autoethnography on AI-infused education. In: Proceedings of the 16th International Conference on Computer Supported Education (CSEDU 2024), vol. 1, pp. 355–362 (2024)
45. Le Quy, T., Roy, A., Iosifidis, V., Zhang, W., Ntoutsi, E.: A survey on datasets for fairness-aware machine learning. Wiley Interdisciplinary Rev. Data Min. Knowl. Disc. **12**(3), e1452 (2022)
46. Lee, A., Poch, R., Shaw, M., Williams, R.: Engaging diversity in undergraduate classrooms: a pedagogy for developing intercultural competence: ASHE Higher Education Report, vol. 38, no. 2. Wiley (2012)
47. Lester, J.C., Ha, E.Y., Lee, S.Y., Mott, B.W., Rowe, J.P., Sabourin, J.L.: Serious games get smart: intelligent game-based learning environments. AI Mag. **34**(4), 31–45 (2013)
48. Lim, I.: Unlocking the potential of educational escape rooms in higher education: theoretical frameworks and pathways ahead. In: Fostering Pedagogy Through Micro and Adaptive Learning in Higher Education: Trends, Tools, and Applications, pp. 141–167. IGI Global (2023)
49. de Lima, E.S., Silva, B.M., Galam, G.T.: Adaptive virtual reality horror games based on machine learning and player modeling. Entertain. Comput. **43**, 100515 (2022)
50. Makri, A., Vlachopoulos, D., Martina, R.A.: Digital escape rooms as innovative pedagogical tools in education: a systematic literature review. Sustainability **13**(8), 4587 (2021)
51. Manaswi, N.K.: Generative adversarial networks with industrial use cases: learning how to build GAN applications for retail, healthcare, telecom, media, education, and HRTech. BPB Publications (2020)
52. McGeary, B., Guder, C., Ganeshan, A.: Opening up educational practices through faculty, librarian, and student collaboration in oer creation: moving from labor-intensive to supervisory involvement. Partnership **16**(1), 1–27 (2021)
53. Miettinen, R.: The concept of experiential learning and john dewey's theory of reflective thought and action. Int. J. Lifelong Educ. **19**(1), 54–72 (2000)
54. Mikkola, E.J., Perälä, M., Hotulainen, R., Salmenkivi, E., Kallioniemi, A.: Causal variables in the community of inquiry: Creating a directed acyclic graph of the effectiveness of the philosophy for children program. Int. J. Educ. Res. **127**, 102414 (2024)

55. Miroshnichenko, A.: On the intersection of theory and practice: hybrid multi-modal game design as a medium of affective digital communication. Ph.D. thesis, York University Toronto (2020)
56. Mittal, U., Sai, S., Chamola, V., et al.: A comprehensive review on generative AI for education. IEEE Access **12**, 142733–142759 (2024)
57. Moon, J., et al.: Generative artificial intelligence in educational game design: nuanced challenges, design implications, and future research, pp. 1–13. Technology, Knowledge and Learning (2024)
58. Moundridou, M., Matzakos, N., Doukakis, S.: Generative AI tools as educators' assistants: designing and implementing inquiry-based lesson plans. Comput. Educ. Artif. Intell. **7**, 100277 (2024)
59. Naswir, A.F.: The future of artificial intelligence in 2d game development. J. Data Sci. Technol. Comput. Sci. **4**(2), 1–9 (2024)
60. Nyaaba, M., Zhai, X., Faison, M.Z.: Generative AI for culturally responsive science assessment: a conceptual framework. Educ. Sci. **14**(12), 1325 (2024)
61. Pakray, P., Gelbukh, A., Bandyopadhyay, S.: Natural language processing applications for low-resource languages. Nat. Lang. Process. **31**(2), 183–197 (2025)
62. Pillai, A.S.: Advancements in natural language processing for automotive virtual assistants enhancing user experience and safety. J. Comput. Intell. Robot. **3**(1), 27–36 (2023)
63. Pittman, R.E.: Taxonomy of Learning. PediaPress (2020)
64. Priya, P., Firdaus, M., Ekbal, A.: Computational politeness in natural language processing: a survey. ACM Comput. Surv. **56**(9), 1–42 (2024)
65. Priyadarshini, A., Gago-Masague, S.: Fair evaluator: an adversarial debiasing-based deep learning framework in student admissions. In: 2024 IEEE 6th International Conference on Cognitive Machine Intelligence (CogMI), pp. 152–161. IEEE (2024)
66. Register, Y.: The future of AI can be kind: strategies for embedded ethics in AI education. Ph.D. thesis, University of Washington (2024)
67. Saffari, S., Dorrigiv, M., Yaghmaee, F.: Enthusiasms, challenges, and future trends in procedural content generation in games using machine learning. Model. Simul. Electr. Electron. Eng. **3**(4), 7–22 (2024)
68. Scirea, M.: Adaptive puzzle generation for computational thinking. In: International Conference on Human-Computer Interaction, pp. 471–485. Springer (2020)
69. Sidekerskienė, T., Damaševičius, R.: Out-of-the-box learning: digital escape rooms as a metaphor for breaking down barriers in STEM education. Sustainability **15**(9), 7393 (2023). https://doi.org/10.3390/su15097393
70. Snyman, C., et al.: Exploring escape rooms as a game-based learning approach to promote student motivation. Ph.D. thesis, University of Pretoria (2023)
71. Stuchlikova, L., Weis, M.: From information to insight: reimagining critical thinking pedagogy in the age of artificial intelligence. In: 2024 International Conference on Emerging eLearning Technologies and Applications (ICETA), pp. 591–598. IEEE (2024)
72. Tawafak, R.M., Mathew, R., Alfarsi, G., AlSideiri, A., Malik, S.I.: Escape rooms and e-learning unite for smart cities education entertainment. In: 2024 25th International Arab Conference on Information Technology (ACIT), pp. 1–6. IEEE (2024)
73. Tillmanns, T., et al.: Mapping tomorrow's teaching and learning spaces: a systematic review on GenAI in higher education. Trends High. Educ. **4**(1), 2 (2025)
74. Trimboli, D.: Mediating Multiculturalism: Digital Storytelling and the Everyday Ethnic. Anthem Press (2020)

75. Vaghani, V., Oyibo, K.: A scoping review of genetic algorithms in serious games: applications, challenges, and future directions. Priprints.org (2024)
76. Villegas-Ch, W., Navarro, A.M., Mera-Navarrete, A.: Using generative adversarial networks for the synthesis of emotional facial expressions in virtual educational environments. Intell. Syst. Appl. **25**, 200479 (2025)
77. Vučković, A., Sikimić, V.: Global justice and the use of AI in education: ethical and epistemic aspects. AI Soc. **40**, 3087–3104 (2024). https://doi.org/10.1007/s00146-024-02076-x
78. Wang, B.: Domain adaptation in reinforcement learning: Approaches, limitations, and future directions. J. Inst. Eng. (India): Ser. B **105**(5), 1223–1240 (2024)
79. Wang, J., Wang, J., Dai, S., Yu, J., Li, K.: Research on emotionally intelligent dialogue generation based on automatic dialogue system (2024), https://arxiv.org/pdf/2404.11447
80. Wang, K.D., Wu, Z., Tufts II, L., Wieman, C., Salehi, S., Haber, N.: Scaffold or crutch? examining college students' use and views of generative AI tools for STEM education. In: 2025 IEEE Global Engineering Education Conference (EDUCON), pp. 1–10. IEEE (2025)
81. Wee, C., Yap, K.M., Lim, W.N.: Haptic interfaces for virtual reality: challenges and research directions. IEEE Access **9**, 112145–112162 (2021)
82. Yang, J., Li, A., Farajtabar, M., Sunehag, P., Hughes, E., Zha, H.: Learning to incentivize other learning agents. Adv. Neural. Inf. Process. Syst. **33**, 15208–15219 (2020)
83. Yu, P.K.: The algorithmic divide and equality in the age of artificial intelligence. Florida Law Rev. **72**, 331 (2020)
84. Yusuf, A., Noor, N.M., Bello, S.: Using multimodal learning analytics to model students' learning behavior in animated programming classroom. Educ. Inf. Technol. **29**(6), 6947–6990 (2024)
85. Zeng, H., He, X., Pan, H.: Implementation of escape room system based on augmented reality involving deep convolutional neural network. Virtual Reality **25**(3), 585–596 (2021)
86. Zhai, C., Wibowo, S., Li, L.D.: The effects of over-reliance on AI dialogue systems on students' cognitive abilities: a systematic review. Smart Learn. Environ. **11**(1), 28 (2024)

On the Role of Domain Experts in Creating Effective Tutoring Systems

Sarath Sreedharan[1(✉)], Kelsey Sikes[1], Nathaniel Blanchard[1], Lisa Mason[2], Nikhil Krishnaswamy[1], and Jill Zarestky[3]

[1] Department of Computer Science, Colorado State University, Fort Collins, USA
{sarath.sreedharan,Kelsey.Sikes,nathaniel.blanchard, nikhil.krishnaswamy}@colostate.edu

[2] Colorado State University Extension, Fort Collins, USA
lisa.mason@colostate.edu

[3] School of Education, Colorado State University, Fort Collins, USA
jill.zarestky@colostate.edu

Abstract. The role that highly curated knowledge, provided by domain experts, could play in creating effective tutoring systems is often overlooked within the AI for education community. In this paper, we highlight this topic by discussing two ways such highly curated expert knowledge could help in creating novel educational systems. First, we will look at how one could use explainable AI (XAI) techniques to automatically create lessons. Most existing XAI methods are primarily aimed at debugging AI systems. However, we will discuss how one could use expert specified rules about solving specific problems along with novel XAI techniques to automatically generate lessons that could be provided to learners. Secondly, we will see how an expert specified curriculum for learning a target concept can help develop adaptive tutoring systems, that can not only provide a better learning experience, but could also allow us to use more efficient algorithms to create these systems. Finally, we will highlight the importance of such methods using a case study of creating a tutoring system for pollinator identification, where such knowledge could easily be elicited from experts.

Keywords: Explainable AI · Concept-Based Teaching · Concept Highlighting · Feature Localization

1 Introduction

Artificial Intelligence (AI) as a field is currently in the midst of a transformative moment, marked by advances and achievements happening at a pace nearly unthinkable just a few years ago [20]. Rapid advancement has also brought with it some implicit and explicit notions of how to build successful AI systems. Chief among those notions is the push to downplay the need for, and in some cases completely reject, the use of any expert-specified knowledge [25]. While AI for education, as a subfield, has always embraced the use of expert knowledge,

A. I. Cristea et al. (Eds.): AIED 2025, CCIS 2590, pp. 61–68, 2025.
https://doi.org/10.1007/978-3-031-99261-2_5

we feel in this moment of the extreme proliferation of end-to-end trained AI systems; it is worth re-emphasizing the need and advantage of using expert-specified knowledge. Through this position paper, we would like to go one step further and call for additional efforts to more effectively (a) gather, (b) represent, and (c) utilize experts in and for AI systems.

Our primary argument relates to the positioning of education in the larger landscape of possible AI applications and how it differs from the kind of problems on which the broader AI community typically focuses. To begin, many recent successes in AI relate to tacit knowledge tasks [7], like vision [26] and robotics [1], where it is unclear how relevant expert knowledge could even be gathered. The other domain where there is a lot of recent excitement is the development and deployment of large-language models (LLM) [2]. These models operate on such large scales that gathering expert knowledge would never be cost-effective. One could argue that LLM fine-tuning through processes like RLHF [18] acts as a proxy for expert feedback. However, these systems are able to get by with feedback from non-expert users, given that these systems are mostly deployed in low-stakes everyday use cases.

Education, on the other hand, can be a high-stakes application. For example, we do not want our systems to hallucinate and provide the learners with misleading or incorrect lessons; factually correct, appropriately scaffolded lessons that are responsive to individual learner needs are beyond the current capabilities of AI systems. While the state-of-the-art methods may not be able to provide the level of robustness and efficiency we would require for authentic or impactful educational settings [5], there are ways we can accommodate the system's shortcomings by leveraging existing human expertise. We have an extensive knowledge base on creating effective lessons and supporting learners across disciplinary domains (e.g. mathematics or history) and institutional contexts (e.g. elementary school or workplace learning). We also have experts who have the training to help craft approaches. While we want to minimize the demands placed on the experts, the solution should not be to ignore them. We should build systems that leverage expert input to help more learners with a wider variety of needs [3].

To demonstrate our argument, we present two examples where expert-specified knowledge can be combined with state-of-the-art AI methods to generate better learning outcomes. First, we will show how explainable AI methods could be extended and augmented with expert-specified rules to generate lessons automatically. Second, we will examine how one could combine a hierarchically structured curriculum with a POMDP-based adaptive tutoring system to effectively generate lessons. To demonstrate the viability of these approaches, we ground them in an important citizen science learning problem of pollinator identification.

Tracking pollinator health is a common goal shared by many community science (also known as citizen science) groups across the country [13,17,27,28], particularly those groups that are concerned with issues of environmental sustainability or conservation. Such tracking can provide valuable ecological data pertaining to local environmental health. Learning more about insects and pollinators, including identification and behavior patterns, from experts is generally touted as one benefit of joining such groups. It is also in the interest of group organizers to provide effective educational opportunities for these potential community scientists so they can carry out their scientific mission accurately and reliably. Unfortunately, many of these organizations have capacity limitations, particularly around the availability of content domain experts, and do not have the resources to provide the level of individualized training some learners require. The availability of effective automated instruction systems could make a huge impact on the ability to scale their organizational processes, recruit more participants, and guide those participants appropriately, ultimately enabling such organizations to achieve their scientific or environmental health aims [10,21]. As we will see in the following section, while such organizations may lack the resources to provide extensive training, they are fully capable of providing the level of information our proposed methods would require.

2 Using Explainable AI Methods to Generate Lessons

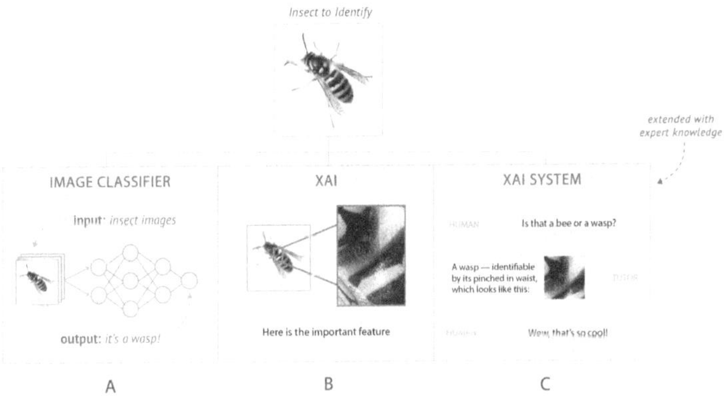

Fig. 1. An example showing the application of our system.

Explainable AI (XAI) as a field emerged out of the realization that as AI systems were becoming more capable, it was not always clear why the system was making the decisions it did [6]. Even in a domain as specific as AI decision-making, the term explanation could include many aspects. However, most works in this space have focused on a few settings that the designers deemed most important. Specifically, one could think of most works in this space as debugging mechanisms that could be used to verify whether the AI system is working the way it was expected. These methods may be meant for the system designer to debug the algorithm/model or for the domain expert to check if it uses the correct problem features to make its decisions [11]. To pick two different XAI techniques, consider feature attribution methods [12] and concept-based explanation methods [9]. In the former, the explanation includes highlighting parts of the input deemed most relevant to the decision. In the case of concept-based explanation, the explanation system can present high-level concepts that might influence the decision.

As a concrete example, consider the task of classifying a wasp presented in Fig. 1. The subfigure (B) presents an example of a feature attribution method showing the image area that determined it was a wasp. A concept-based explanation might explain that the concept "pinched-waist" influenced the decision. An entomologist can examine either explanation and determine that the system is correctly identifying the wasp using a relevant feature.

However, if we wanted to use this tool to help learners understand how to recognize a wasp, we quickly see that the two methods are insufficient. If the learner is not aware of the concept of "pinched-waist" or cannot visually recognize it, then neither of the explanations on its own will teach them about recognizing wasps. On the other hand, these automatically generated pieces of information could be part of a pipeline to generate lessons for the learner. We can create such lessons by using a curated list of rules to recognize wasps, expressed in terms of concepts that the XAI system can identify. Once the AI system has identified an image to contain a wasp, one can iterate over the expert-specified rules provided and see which one can be surfaced to the learner and can be used to demonstrate the fact that the current image contains a wasp. Subfigure (C) in Fig. 1 shows an example of how such a lesson might look for the given image. Note that in this case, we can also account for cases where the AI system may be making incorrect predictions by associating uncertainty with the decision and the presence or absence of concepts. Such approaches (cf. [9,23]) have been previously used to ensure fidelity of generated explanation, in this case, is used to ensure the learner is getting accurate lessons.

3 Expert-Specified Curriculum to Structure Lessons

Partially Observable Markov Decision Processes or POMDPs [22] have been widely discussed as a potential framework to represent adaptive tutoring systems [19]. They are particularly effective as they provide a natural way to encode the system's uncertainty about the learner's current knowledge. POMDPs also provide natural ways to capture actions to query the learner to test for their level of expertise and then provide lessons that might be better suited to it.

However, a naive application of POMDPs to most education settings remains infeasible as generating policies for them is quite computationally expensive [15]. One of the biggest problems is the explosion of belief space that is needed to capture uncertainty related to the potential expertise the current learner might possess. However, the use of a well-structured curriculum, created with the help of domain experts, could be helpful in simplifying the space. In particular, identifying hierarchies between concepts and the order in which they need to be taught could go a long way in simplifying the set of possible expertise levels the system needs to consider at any given point.

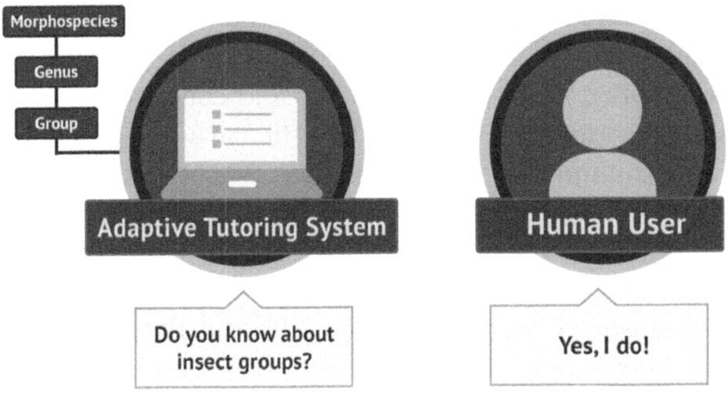

Fig. 2. A hierarchical representation of concepts relevant to pollinator classification.

Going back to pollinator identification tasks, a curriculum could dictate that before they are trained to identify morphospecies, they should be able to identify the genus. Additionally, before they can recognize the genus, they should be able to recognize whether the given insect is a bee, wasp, or a fly [14]. By encoding this order into the POMDP, you can have a policy that focuses on first testing the learner's ability to recognize the three groups before moving on to more granular distinction. Figure 2 shows how a policy produced for such a POMDP might play out based on the feedback given by the learner. This method of structuring belief spaces has shown to provide computational advantages in other settings [24], and we expect it to carry over in this new setting as well.

4 A Potential Tutoring System for Pollinator Identification

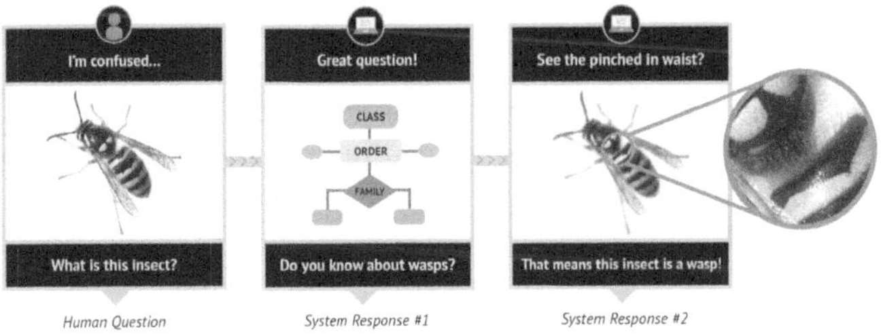

Fig. 3. An overview of our proposed tutoring system.

Putting these two pieces together, one can start seeing the outlines of an effective tutoring system that community scientists can use (visualized in Fig. 3). To build such a system, we would need the domain experts to first build a curriculum that specifies in what order learners should be exposed to or master different concepts. Next, we would also require them to compile a set of rules that the learner can use to recognize different groups of insects. While this may at first glance seem like a lot of information for an expert to provide, this is a set of information that is commonly provided to community scientists as part of their onboarding training.[1] The morphospecies and concept classifiers needed to build the XAI pipeline that was described in the earlier section can be trained using data available from public repositories like iNaturalist [16]. One could also leverage the existing volunteer pool to get additional annotations and examples. We can reuse the organization's existing teaching materials for the lessons and questions needed to build out the POMDP-based tutoring system. The objective function and the transition probabilities used by the POMDP can be initially set to default values and then refined over time based on learner feedback.

Once we have the two pieces in place, we can actually build a tutoring system that will allow the learners to take pictures of insects they see in the wild and ask the system to teach them about them. The POMDP-based system will query them to identify their current level of expertise and then determine the exact level of details that need to be included in the user feedback. This is then used to determine the rules that the XAI component needs to use to identify the feedback to be generated for the current image. The system can also provide them with more resources for them to better learn the target concept (in this case, a group, morphospecies). However, building such a system with the current state-of-the-art AI methods would require the use of expert-specified knowledge. One that we argue is readily available in this case.

5 Recommendations to the Community

To conclude, here are our recommendations to the community.

Building Tools that Empower Experts. Our first recommendation is to build better tooling and interfaces that will empower domain experts, who might not be AI experts, to easily specify their knowledge in forms that AI systems can use. Potential early steps from the AI side in this direction include the use of symbolic interfaces [8] and the use of LLMs as semantic parsers [4]. It is also worth noting that the domain experts themselves may not be education experts, so there is a question about whether we can build tools that can help education experts introduce best practices into the knowledge we collect to support the integration of AI and domain expertise with high-quality education design.

Focusing on interdisciplinary teams. While the AI for education community has always emphasized the importance of building interdisciplinary teams, this trend may erode as more researchers from the broader AI community engage with

[1] Here are slides that were prepared for Native Bee Watch, a citizen science organization, by our co-author Lisa Mason https://rb.gy/r03y0g.

education as their target application. As a community, AI for education needs to emphasize building and supporting interdisciplinary teams.

Broader Public Education. Finally, there is a need to educate the broader AI community, the public, and policymakers about what state-of-the-art AI systems can and cannot do in educational contexts. More work is needed to identify problem spaces where AI might be a valuable tool and where AI technology might still lag behind. We need to perform outreach activities to AI scientists so they better appreciate the challenges related to education and to the policymakers and public to make sure that resources are not being diverted to quixotic projects that are bound to fail. Specifically, in the context of this proposal, we need to do a better job of identifying how expert knowledge can be best utilized and integrated at each level of education, including preK-12, higher ed, workforce development, and informal learning.

Acknowledgments. This study was funded by NSF grant 2303019.

References

1. Ahn, M., et al.: Do as i can, not as i say: grounding language in robotic affordances. arXiv preprint arXiv:2204.01691 (2022)
2. Bubeck, S., et al.: Sparks of artificial general intelligence: early experiments with gpt-4 (2023)
3. D'Mello, S.K., et al.: From learning optimization to learner flourishing: reimagining ai in education at the institute for student-ai teaming (isat). AI Mag. **45**(1), 61–68 (2024)
4. Guan, L., Valmeekam, K., Sreedharan, S., Kambhampati, S.: Leveraging pre-trained large language models to construct and utilize world models for model-based task planning. Adv. Neural. Inf. Process. Syst. **36**, 79081–79094 (2023)
5. Gunasekara, S., Saarela, M.: Explainability in educational data mining and learning analytics: an umbrella review. In: Proceedings of the 17th AIED, pp. 887–892. International Educational Data Mining Society, July 2024
6. Gunning, D.: Explainable artificial intelligence (xai). DARPA, Web **2**(2) (2017)
7. Kambhampati, S.: Polanyi's revenge and ai's new romance with tacit knowledge. Commun. ACM **64**(2), 31–32 (2021)
8. Kambhampati, S., Sreedharan, S., Verma, M., Zha, Y., Guan, L.: Symbols as a lingua franca for bridging human-ai chasm for explainable and advisable ai systems. In: Proceedings of AAAI, vol. 36, pp. 12262–12267 (2022)
9. Kim, B., Wattenberg, M., Gilmer, J., Cai, C., Wexler, J., Viegas, F., et al.: Interpretability beyond feature attribution: quantitative testing with concept activation vectors (tcav). In: International Conference on Machine Learning, pp. 2668–2677. PMLR (2018)
10. Kremen, C., Ullman, K., Thorp, R.W.: Evaluating the quality of citizen-scientist data on pollinator communities. Conserv. Biol. **25**(3), 607–617 (2011)
11. Langley, P.: Varieties of explainable agency. In: ICAPS Workshop on Explainable AI Planning (XAIP), vol. 1365 (2019)
12. Lundberg, S.M., Lee, S.I.: A unified approach to interpreting model predictions. In: Advances in Neural Information Processing Systems, vol. 30 (2017)

13. Mason, L.: Native bee watch community science program (2025), https://arapahoe.extension.colostate.edu/nbw/
14. Mason, L., Arathi, H.: Assessing the efficacy of citizen scientists monitoring native bees in urban areas. Global Ecol. Conserv. **17**, e00561 (2019)
15. Murphy, K.P.: A survey of pomdp solution techniques. Environment **2**(10) (2000)
16. Nugent, J.: Inaturalist. Science. Scope **41**(7), 12–13 (2018)
17. Oregon department of agriculture: oregon bee atlas (2025), https://extension.oregonstate.edu/bee-atlas
18. Ouyang, L., Wu, J., Jiang, X., Almeida, D., Wainwright, C., Mishkin, P., Zhang, C., Agarwal, S., Slama, K., Ray, A., et al.: Training language models to follow instructions with human feedback. Adv. Neural. Inf. Process. Syst. **35**, 27730–27744 (2022)
19. Rafferty, A.N., Brunskill, E., Griffiths, T.L., Shafto, P.: Faster teaching via pomdp planning. Cogn. Sci. **40**(6), 1290–1332 (2016)
20. Rossi, F.: AAAI 2025 presidential panel on the future of AI research (2025), https://aaai.org/wp-content/uploads/2025/03/AAAI-2025-PresPanel-Report-FINAL.pdf
21. Schmeller, D.S., et al.: Advantages of volunteer-based biodiversity monitoring in Europe. Conserv. Biol. **23**(2), 307–316 (2009)
22. Sondik, E.J.: The Optimal Control of Partially Observable Markov Processes. Stanford University (1971)
23. Sreedharan, S., Soni, U., Verma, M., Srivastava, S., Kambhampati, S.: Bridging the gap: Providing post-hoc symbolic explanations for sequential decision-making problems with inscrutable representations. In: International Conference on Learning Representations
24. Sreedharan, S., Srivastava, S., Kambhampati, S.: Using state abstractions to compute personalized contrastive explanations for ai agent behavior. Artif. Intell. **301**, 103570 (2021)
25. Sutton, R.: The bitter lesson. Incomplete Ideas (blog) **13**(1), 38 (2019)
26. Szeliski, R.: Computer vision: algorithms and applications. Springer Nature (2022)
27. Vilen, L., Mason, L., Viders, S., Zarestky, J.: Community science online: building capacity for native bee monitoring. J. Hum. Sci. Extension **11**(2), 12 (2023)
28. Woodard, S.H., et al.: Towards a us national program for monitoring native bees. Biol. Cons. **252**, 108821 (2020)

Practitioners, Industry and Policy

SchrijfBlik: Safeguarding the Validity of Writing Assessment in the Age of AI

Tijmen Poell(✉), Lientje Maas, Marica Balk, and Max van Haastrecht

CitoLab, Stichting Cito, Arnhem, The Netherlands
{tijmen.poell,lientje.maas,marica.balk,max.vanhaastrecht}@cito.nl

Abstract. The rise of generative artificial intelligence (GenAI) in education necessitates new methods for assessing student writing. Teachers struggle to distinguish original work from AI-generated content, which challenges the validity of assessments. This paper introduces `SchrijfBlik`, a prototype designed to help secondary school teachers in the Netherlands gain insight into how students use GenAI for writing assignments. We follow the Double Diamond design process to explore and define our problem setting, and then develop a prototype to address the challenge of safeguarding validity in the age of AI. The results from a questionnaire among roughly 400 teachers and interviews with assessment experts highlighted the importance of visualizing process data on student-AI interactions in an intelligible way. `SchrijfBlik` was iteratively refined to create a user-friendly interface that shows prompts, edits, and reflections on a timeline, thus making the writing process of students more transparent for teachers. Early user tests suggest that `SchrijfBlik` can help to safeguard assessment validity, but further classroom evaluations are necessary to help us gain a better understanding of how we can best help teachers cope with the challenges of today's AI world.

Keywords: generative artificial intelligence · writing assessment · process data · validity

1 Introduction

The proliferation of generative artificial intelligence (GenAI) applications has had a profound impact on education. Writing assignments are a key area where teachers need to rethink their approach. Teachers struggle to distinguish automatically generated texts from their original counterparts [18]. In addition, teachers tend to be overconfident in their ability to detect when students have used AI [7]. This raises questions about the validity of the inferences made by teachers regarding student products.

Several approaches have been suggested to counteract this threat to validity. Teachers can set clear guidelines for students on how GenAI may be used and can require students to submit draft versions to identify inappropriate use of GenAI early on [4]. However, since teachers struggle to distinguish generated

A. I. Cristea et al. (Eds.): AIED 2025, CCIS 2590, pp. 71–80, 2025.
https://doi.org/10.1007/978-3-031-99261-2_6

text from original text, such strategies may create more problems through false accusations than they solve by promoting ethical use [18]. Another approach is to shift to different modes of assessment that limit or exclude the possibility of using GenAI, such as in-class assignments or oral examinations [1]. The downside of such approaches is that they require an extensive time investment by teachers and they ignore the reality of today's world where GenAI can be a valuable addition to students' writing process.

Swiecki et al. (2022) posit that instead "the presence of AI suggests that we should account for the fact that a human-AI team can generate assessment evidence" [16, pp. 8]. It is difficult to discern solely on the basis of the writing product whether the human or AI in the human-AI team produced a written text. The logical place to look for additional evidence is in the writing process. Computer-based assessments offer fertile ground for extracting and analyzing process data [9,10]. In the context of writing assignments, process data can involve how students interact with AI tools, the extent to which they rely on generated content, and how they refine and integrate AI-generated text. By analyzing such data, more meaningful assessments can be developed.

This is not to say that designing meaningful assessment strategies that incorporate process data is a simple task. In fact, it is widely recognized that we are yet to establish appropriate validity theory regarding the use of process data in educational assessments [8,19]. Applications such as automated essay scoring [3,13] and automated feedback generation [15] ingest and create learning process data, but are not designed or theoretically grounded with the purpose of supporting validity arguments. To truly safeguard the validity of writing assessment, we need to conceptualize how process and product data will provide meaningful insight into the achievement of learning goals from the earliest design stages of our assessment [8,11]. Within this work, we describe a prototype designed from the ground up to safeguard validity.

2 SchrijfBlik: Description and Intended Use

To safeguard the validity of writing assessment in the age of AI, we developed the prototype `SchrijfBlik`. `SchrijfBlik`, which can be loosely translated from Dutch as `WriteView`, was designed to cater to the needs of secondary school teachers and students in the Netherlands. Yet, we believe the general principles informing its design can be extended to other contexts and countries.

`SchrijfBlik` allows teachers to administer writing assignments in an online environment where students can use GenAI. With `SchrijfBlik`, data is collected to gain insight into how students interact with GenAI. What prompts do they use? What do they verify? Which parts of the text are edited? Additionally, students are prompted by a chatbot to reflect on their choices. This controlled environment allows us to discern how a student has used GenAI and what was written as original work. The collected process data can help make the writing process more transparent to enable valid inferences regarding students' writing ability.

We used the Double Diamond [5] design process to guide the development of `SchrijfBlik`. The Double Diamond method combines step-by-step and iterative elements to move from a general context description to a contextualized problem definition and finally to a concrete solution. The method can be divided into two focus areas. The first diamond focuses on the exploration of the problem, while the second focuses on the development of the solution. Figure 1 illustrates how the Double Diamond method was used in the development of `SchrijfBlik`. However, this process, like any design process, is iterative and not linear.

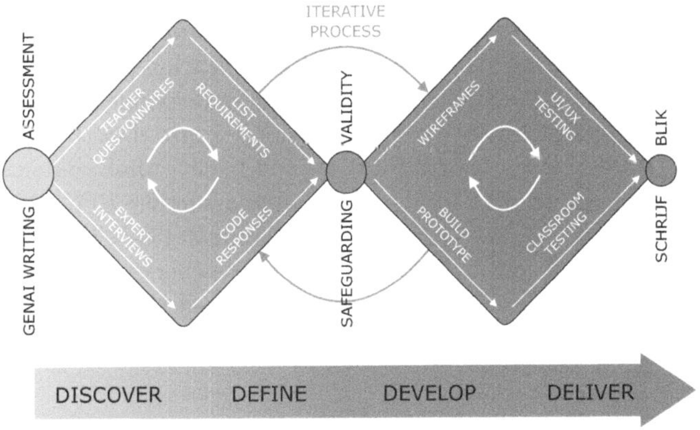

Fig. 1. The Double Diamond [5] process for the SchrijfBlik prototype, inspired by the visualization of Nychka [14].

When further breaking down the double diamond method, four different phases can be identified: discover, define, develop, and deliver. The discover phase involves exploring the boundaries of the problem by collecting data from different perspectives. This phase consisted of expert interviews with assessment specialists and questionnaires distributed to around 400 secondary school teachers in the Netherlands. In the subsequent define phase, the goal is to let the broad understanding converge towards a contextualized problem definition, directly informing the scope for the eventual design. For `SchrijfBlik`, this included creating a list of requirements for the prototype based on the questionnaires and coding the responses from expert interviews to further detail what the ideal solution should look like.

In the develop phase, various options for solutions to the identified problem are explored. The functionality was designed and visualized via wireframes and developers created the alpha version of the prototype. Throughout this phase, we maintained close contact with the assessment specialists to gather feedback. In the final deliver phase, the prototype should be tested in realistic scenarios. We have already conducted the first user experience tests, but broader evaluation in classroom settings is still to be completed, which we planned for 2025.

3 SchrijfBlik: Challenges and Opportunities

This section details the outcomes of the various activities we initiated as part of the Double Diamond design process and describes how these outcomes informed the iterative improvement of the `SchrijfBlik` prototype.

3.1 Discover

To gain a broad understanding of how secondary school teachers in the Netherlands are currently approaching the challenge of safeguarding the validity of writing assessment in the age of GenAI, we distributed three multiple-choice questions to teachers using the Teacher Tapp application [17].

Between 381 and 420 secondary school teachers responded. Figure 2 shows a summary of the results. Only 6% of the teachers indicate that they know exactly who is using GenAI. Nearly half (49%) of them believe that students are using GenAI both as an aid and to evade tasks, with a further 19% indicating that they do not know how students are using GenAI. Just 4% believe their students are not using GenAI. Finally, while 32% of the teachers signaled they had no intention of adapting homework assignments, 46% had already adapted existing exercises (37%) or created completely new assignments (9%).

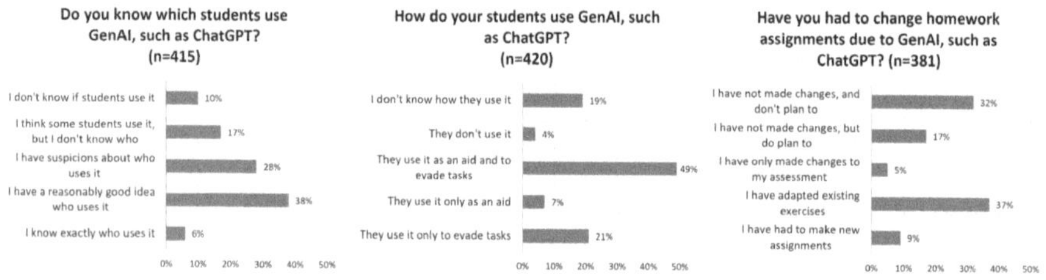

Fig. 2. Answers from secondary school teachers in the Netherlands to three questions, posed via Teacher Tapp, regarding the impact of GenAI on their education [17].

These figures indicate that secondary school teachers in the Netherlands rarely know exactly which students are using GenAI for their assignments, while almost all are convinced that their students are using GenAI. Interestingly, teachers are divided on the need to adapt existing exercises. This can partially be explained by evaluating the results at a more detailed subject level. For example, whereas 83% of the 54 mathematics teachers who responded indicate that they do not plan to adjust exercises, just 11% of the 71 Dutch language teachers and 20% of the 44 English language teachers are of the same opinion. This might be because teachers who rely more heavily on homework assignments involving long written texts are more convinced of the need to adapt to the challenges posed to assessment validity by the introduction of GenAI.

In the next step, expert interviews were arranged with three assessment specialists. The goal of the interviews was to refine our initial conceptualization of objectives and assumptions. For these interviews, we prepared initial sketches that illustrated our vision for the system's design (Fig. 3).

Fig. 3. Initial screen sketches to guide expert interviews with assessment specialists.

The sketches contained three separate tabs. One tab allowed students to work on the assignment, another tab provided access to a GenAI tool, and the last tab was envisioned to be an internet browser. This version already contained the first ideas on tracking the student writing process, for example by registering the copying and pasting of text. Additionally, students would occasionally be prompted to explain why they copied a part of the text. This was intended to encourage them to reflect on proper referencing and on the use of sources.

We interviewed the experts about their general impressions of the sketches. We asked how applicable they thought the system would be in real-world settings and how process data could be used to help with the assessment of writing assignments. Two of the experts saw opportunities and were mostly enthusiastic about the possibility of using process data for assessment. They emphasized that process tracking should be done automatically, meaning it should not be up to students to manually keep track of their steps. This was a crucial and novel system requirement. The experts also indicated that prompted questions could motivate students to reflect on correct source usage or to review their work at the end of the assignment. Finally, they highlighted the importance of clarifying what is being assessed and ensuring that students understand how they are allowed to use GenAI.

3.2 Define

The questionnaire and interview outcomes were used to determine the requirements. We concluded from our initial findings that the goal should be to build a prototype that gives teachers insight into how students use AI as an assistant in the writing process. Such a prototype would give teachers the necessary control mechanisms to know who is using GenAI and how it is being used, where in a traditional assessment setting they are largely kept in the dark. By deliberately avoiding more extreme measures such as banning the use of GenAI, the prototype appeals to more digitally literate teachers who are already adapting their homework assignments and to students who are looking to develop their writing competency as part of a human-AI team.

Feedback from assessment specialists additionally pointed out the need of continuous insight into the student writing process, to avoid having to take cumbersome measures such as forcing students to submit multiple draft versions [4]. Another requirement resulting from our discover phase was that the prototype should expressly not be designed with the idea to 'catch' cheating students. Where comparable tools are built with the intent to detect AI usage, with `SchrijfBlik` we aimed to create an environment where students feel free to use GenAI and discover how to use it responsibly.

3.3 Develop

Figma was used for the first six iterations of development [6]. The wireframes that were designed as part of this process can be found in our Figma file[1]. The first iteration involved the digitization of the sketches constructed during brainstorms and interviews. The second iteration consisted of experimenting with ways of visualizing the different steps of the teacher and student journeys. We started by presenting a list of the steps students took but ended with a continuous and clickable timeline, which allows to present more fine-grained information. This iteration also involved the addition of activating questions for students at several points in the writing process, which are visible in the clickable timeline as well.

The third through fifth iterations involved improvements to the user interface and user experience. Given both technical and ethical limitations, we decided not to actively track internet behavior occurring outside of the `SchrijfBlik` tool. Instead, if students were inactive for a certain period, the chatbot would ask why this occurred. It was up to the students to decide whether they answered this question or not. The other implemented changes of this phase related to personifications applied to the GenAI chatbot incorporated in the prototype.

The final iteration had a shift in focus towards a testable prototype for the initial user tests. We expanded the teacher environment with features to tweak system prompts. This allows teachers to better align the chatbot's communication style with a student's age group. Additionally, this teacher environment facilitates the inclusion of custom chatbot questions to satisfy the requirements for a given assignment.

The `SchrijfBlik` tool, of which Fig. 4 shows an example, was programmed in React [12] and can be viewed via an openly accessible web interface[2].

3.4 Deliver

To evaluate the prototype after the last iteration, we performed user tests with 12 secondary education teachers. Fictional student work was presented to them either with or without the process data from `SchrijfBlik` and they were asked

[1] https://schrijfblik-figma.web.app.
[2] https://schrijfblik-docent.web.app/login.

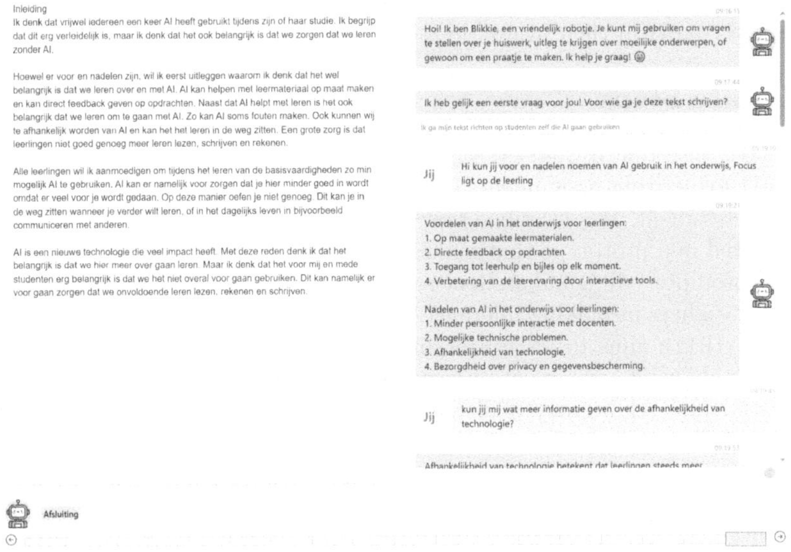

Fig. 4. The teacher environment of the `SchrijfBlik` prototype alpha version.

to score the work on a scale from 1 to 5 for four different categories: argumentation, readability, structure, and word usage. The fictional work consisted of three scenarios, each representing typical student behavior. In scenario 1, we illustrated a student who produced high-quality work but misused GenAI. Scenario 2 illustrated a student who did not deliver a high-quality product but demonstrated a deeper understanding of the subject matter by asking insightful questions to GenAI. Scenario 3 involved a student who followed an appropriate process, resulting in a good product, and utilized GenAI effectively. The results can be found in Fig. 5. Teachers who used `SchrijfBlik` were better at uncovering

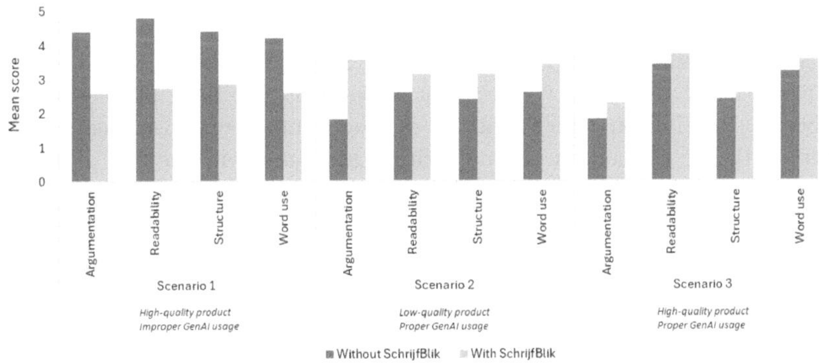

Fig. 5. Average given scores to the fictional student work per scenario.

responsible GenAI usage and, moreover, distinguished between writing ability and use of GenAI in their evaluations.

4 Future Steps

This paper introduced `SchrijfBlik`, a tool to gain insight into students' use of GenAI for writing assignments, with the aim of safeguarding the validity of writing assessment. We described how an iterative design and development process resulted in a prototype that makes the writing process more transparent by showing prompts, edits, and student reflections on a timeline. This transparency supports teachers in making inferences regarding students' writing ability.

`SchrijfBlik` aims to discern original student work from AI-generated content to enable valid assessment of writing skills. However, in addition to writing skills, students' AI literacy skills now play an important role in the writing process. Ideally, we would disentangle the assessment of writing ability and AI literacy within the `SchrijfBlik` environment. This requires adapted assessment frameworks with clear definitions of relevant writing skills and AI literacy skills, along with desired learning outcomes. Such frameworks for human-AI collaborative writing should account for interactions between students and AI, enabling grounded arguments about learning claims based on verifiable evidence [2]. This can aid further development of `SchrijfBlik` by highlighting what process data is relevant to present to teachers and students, and how this information can be effectively presented to support grading and learning.

While offering great opportunities, AI systems also generate risks that must be managed appropriately. In 2024, the AI Act[3] entered into force, introducing a uniform legal framework for any system that includes AI in the EU. The AI Act is structured around a risk-based approach. It categorizes AI systems into four distinct levels based on their potential risks: unacceptable, high, limited, and minimal risk. AI systems intended to be used to evaluate learning outcomes in education are classified as high-risk, also when the outcomes are used solely to direct student learning. Therefore, it is likely that systems like `SchrijfBlik` will be heavily regulated in the future. This entails challenges for the implementation in educational practice, since the system has to be compliant with high-risk AI regulations, for example regarding privacy, explainability, and accountability.

In sum, our first results demonstrate that `SchrijfBlik` is a promising tool to help safeguard the validity of writing assessment. We have planned further classroom evaluations in 2025 to gain more insight into how we can best support teachers with the challenges around writing assessment in the age of AI. We aim to explore what process data teachers find relevant, to what extent they value the process data `SchrijfBlik` provides, and how it impacts their grading decisions.

Acknowledgments. We would like to express our gratitude to the developers of the CitoLab prototyping team for their flexibility, dedication, and insight. Without their efforts and ideas, it would not have been possible to create SchrijfBlik.

[3] https://digital-strategy.ec.europa.eu/en/policies/regulatory-framework-ai.

Disclosure of Interests. The authors have no competing interests to declare that are relevant to the content of this article.

References

1. Bower, M., Torrington, J., Lai, J., Petocz, P., Alfano, M.: How should we change teaching and assessment in response to increasingly powerful generative artificial intelligence? Outcomes of the ChatGPT teacher survey. Educ. Inf. Technol. **29**(12), 15403–15439 (2024). https://doi.org/10.1007/s10639-023-12405-0
2. Cheng, Y., Lyons, K., Chen, G., Gašević, D., Swiecki, Z.: Evidence-centered assessment for writing with generative AI. In: Proceedings of the 14th Learning Analytics and Knowledge Conference, pp. 178–188 (2024). https://doi.org/10.1145/3636555.3636866
3. Conijn, R., Kahr, P., Snijders, C.: The effects of explanations in automated essay scoring systems on student trust and motivation. J. Learn. Anal. **10**(1), 37–53 (2023). https://doi.org/10.18608/jla.2023.7801
4. Cotton, D., Cotton, P.A., Shipway, J.R.: Chatting and cheating: ensuring academic integrity in the era of ChatGPT. Innov. Educ. Teach. Int. **61**(2), 228–239 (2024). https://doi.org/10.1080/14703297.2023.2190148
5. Design council: the double diamond (2005), https://www.designcouncil.org.uk/our-resources/the-double-diamond/
6. Figma, Inc.: Figma: the collaborative interface design tool (2024), https://www.figma.com/
7. Fleckenstein, J., Meyer, J., Jansen, T., Keller, S.D., Köller, O., Möller, J.: Do teachers spot AI? Evaluating the detectability of AI-generated texts among student essays. Comput. Educ. Artif. Intell. **6**, 100209 (2024). https://doi.org/10.1016/j.caeai.2024.100209
8. Goldhammer, F., Hahnel, C., Kroehne, U., Zehner, F.: From byproduct to design factor: on validating the interpretation of process indicators based on log data. Large-scale Assess. Educ. **9**(1), 1–25 (2021). https://doi.org/10.1186/s40536-021-00113-5
9. Goldhammer, F., Naumann, J., Rölke, H., Stelter, A., Tóth, K.: Relating product data to process data from computer-based competency assessment. In: Leutner, D., Fleischer, J., Grünkorn, J., Klieme, E. (eds.) Competence Assessment in Education. MEMA, pp. 407–425. Springer, Cham (2017). https://doi.org/10.1007/978-3-319-50030-0_24
10. Goldhammer, F., Zehner, F.: What to make of and how to interpret process data. Measure. Interdisciplinary Res. Perspect. **15**(3-4), 128–132 (2017). https://doi.org/10.1080/15366367.2017.1411651
11. Lindner, M.A., Greiff, S.: Process data in computer-based assessment. Eur. J. Psychol. Assess. (2023). https://doi.org/10.1027/1015-5759/a000790
12. Meta: react: a javascript library for building user interfaces (2013), https://react.dev/
13. Mizumoto, A., Eguchi, M.: Exploring the potential of using an AI language model for automated essay scoring. Res. Methods Appl. Linguist. **2**(2), 100050 (2023). https://doi.org/10.1016/j.rmal.2023.100050
14. Nychka, J.A.: The materials paradigm and the double diamond design model (2019), https://era.library.ualberta.ca/items/1bbca792-e31a-4e01-b46e-57449266afb8

15. Scarlatos, A., Smith, D., Woodhead, S., Lan, A.: Improving the validity of automatically generated feedback via reinforcement learning. In: Olney, A.M., Chounta, I.A., Liu, Z., Santos, O.C., Bittencourt, I.I. (eds.) Artificial Intelligence in Education, pp. 280–294. Springer, Cham (2024). https://doi.org/10.1007/978-3-031-64302-6_20
16. Swiecki, Z., et al.: Assessment in the age of artificial intelligence. Comput. Educ. Artif. Intell. **3**, 100075 (2022). https://doi.org/10.1016/j.caeai.2022.100075
17. Teacher Tapp Nederland: Monthly overview of results - June 2024 (2024), https://nl.teachertapp.com/juni-2024/
18. Waltzer, T., Pilegard, C., Heyman, G.D.: Can you spot the bot? Identifying AI-generated writing in college essays. Int. J. Educ. Integr. **20**(1), 1–18 (2024). https://doi.org/10.1007/s40979-024-00158-3
19. Zumbo, B.D., Maddox, B., Care, N.M.: Process and product in computer-based assessments. Eur. J. Psychol. Assess. **39**(4), 252–262 (2023). https://doi.org/10.1027/1015-5759/a000748

AMELIA: An AI Teaching Assistant for Equitable STEM Education in Rural Peru

Carla Gamberini[1(✉)], Jorge Paredes[2], and Claudia Roque[3]

[1] MásEducaciónPe, Lima, Peru
carla@mas-educacion.pe
[2] Universidad de Lima, Lima, Peru
[3] MásEducaciónPe, Lima, Peru

Abstract. AMELIA is an AI-powered chatbot designed to improve STEM education for teachers and students in rural Peru, where systemic barriers, gender inequality, and limited digital access restrict learning opportunities. By using AI through WhatsApp, AMELIA offers a low-cost, accessible tool built for low-connectivity environments. It provides learning activities, classroom guidelines, material specifications, and stories of women in STEM, all curated by experts and adapted to the Peruvian context. The program tackles persistent inequalities in STEM education, particularly for girls, who face sociocultural biases, resource scarcity, and underprepared teachers. AMELIA supports teachers with training and implementation guidance, boosts student engagement in STEM, and promotes gender inclusion by offering female role models. Currently active in Cajamarca with 120 teachers reaching 3,000 students, AMELIA has demonstrated strong engagement and received positive feedback for improving teaching practices. It was recently selected by the Ruralia 2024 Accelerator, signaling its scalability and innovation. Continuous improvement is driven by AI refinement, community input, and a monitoring system. Future goals include expanding to other rural regions, enhancing pedagogical resources, and involving families to reduce gender-based barriers. This paper presents AMELIA's design, implementation, outcomes, and scaling strategies, offering a replicable model for leveraging AI to improve teaching practices and reduce gender and geographic disparities in STEM education.

Keywords: AI in Education · Low-Cost EdTech · Women in STEM · Rural Schools · AI for Equity

1 Description of the AIED Implementation, Practice or Policy

1.1 How AMELIA is Addressing Digital, Learning, and Gender Gaps in Rural Areas in Peru

AMELIA is a low-cost AI teaching assistant designed to address the low achievements and gender gap in STEM learning in Peru's rural areas. This tool works as

A. I. Cristea et al. (Eds.): AIED 2025, CCIS 2590, pp. 81–89, 2025.
https://doi.org/10.1007/978-3-031-99261-2_7

a chatbot through WhatsApp, offering teachers a comprehensive suite of learning activities such as games, problems, projects, prototypes with technology simulations, and assessments. Each learning activity includes detailed descriptions, classroom implementation instructions, contextualized material specifications, and time requirements per the teachers' request.

What makes AMELIA unique is that the AI tool integrates into each learning activity an inspiring story of a woman in STEM relevant to the region where the teacher is located and the learning unit that the teacher wishes to teach. Each story highlights the female role models' journey to becoming a successful woman in STEM in order inspire female students to learn that they can pursue a STEM career if they wish to. The learning activities methodology, the STEM content, and women's role model stories have been curated by pedagogical experts, tested with schoolteachers and preservice teachers from the Cayetano University of Peru, as well as aligned with the national curriculum approach to STEM learning.

Education in rural Peru faces substantial challenges. First, rural students consistently underperform in Mathematics compared to their urban counterparts due to limited access to quality resources, inadequate teacher training, and socioeconomic inequalities (Castro & Rolleston, 2018; Castro, 2020). According to the PISA 2022 results, only 11% of students in rural areas reach the minimum satisfactory level in mathematics, compared to 38.7% of students in urban areas. Additionally, Peru ranks 125th out of 137 countries in the World Economic Forum's ranking for mathematics and science.

Second, female students nationwide are particularly underperforming in Mathematics. Only 8% of girls in primary school achieve satisfactory learning outcomes according to the Census Evaluation of Students (2018). In the case of secondary education, PISA 2022 data show that boys outperformed girls in mathematics by nine points. Hence, there is an alarming gender gap in STEM education in Peru that is related to limited access to quality learning resources and teaching, as well as social stereotypes that lead girls to believe they are not capable enough to succeed in STEM subjects (Gamberini, 2023).

Third, there is a massive digital gap in schools in rural areas. According to educational statistics (UE, 2023), educational services in rural regions cater to a total of 1,226,539 students, with 99% of schooling provided by public institutions (Uccelli, Montero; 2023). It should be added that, according to numbers from the 2024 Education Census, there are more than 170 thousand teachers in rural areas, where 50% of rural secondary schools do not have internet access. Hence, teachers and students do not have significant access to integrating digital resources in the teaching and learning process due to the connectivity gap in rural areas.

According to Friedel et al. (2013), digital resources provide educators with access to a wide range of multimedia content, enabling them to create dynamic and interactive lessons that cater to diverse learning styles. Additionally, digital tools increase efficiency by simplifying lesson planning, assessment, and administrative tasks, allowing teachers to dedicate more time to direct teaching. By leveraging these tools, teachers are better equipped to deliver effective instruction and foster meaningful learning experiences in the classroom.

AMELIA aims to bridge these substantial educational challenges by focusing on three key aspects: 1) strengthening teachers' capacities with an AI training program and an AI-based teaching assistant; 2) enhancing STEM learning for students through innovative activities for the classroom; and 3) fighting stereotypes that discourage girls from pursuing STEM fields by integrating women in STEM role models stories to their classrooms. Consequently, AMELIA contributes to addressing the learning crisis in education, closing the digital gap in rural teaching, and the gender gap in STEM education. Currently, AMELIA impacts 120 teachers and 3,000 students from rural schools in Cajamarca, Peru. In this first implementation stage, AMELIA has prioritized targeting secondary school students and teachers. Secondary school, which in Peru goes from 12 to 17 years old, is a critical stage where exposure to STEM subjects can significantly influence future career choices, and it is the stage where the Math learning achievements of female students decrease compared to elementary school achievement (Gamberini, 2023).

Finally, its innovative approach and satisfactory results in Cajamarca have won AMELIA nationwide recognition and an investment fund given by Ruralia, the first rural education accelerator in the country. This prestigious prize validates AMELIA's commitment to bridging educational disparities in rural areas and its potential to sustainably scale to serve more teachers in rural areas. The Ruralia fund will allow the project's expansion to more schools in rural areas between 2025 and 2026.

1.2 The Technology Behind AMELIA

AMELIA's development leverages AI generative technology and WhatsApp to support teachers in low-connectivity environments, making it ideal for rural schools. Its design ensures functionality even with limited digital infrastructure. AMELIA uses Meta's official API to integrate with WhatsApp; Botpress to manage the conversational flow with teachers; and ChatGPT to generate pedagogical activities aligned with national curriculum standards. The chatbot functionality also relies on the management and utilization of diverse key datasets—schools, teachers, and women in STEM stories—stored in a structured SQL database. The school and teacher datasets include comprehensive information on schools and teachers collected during training sessions and user registration. These datasets are periodically updated as usage progresses.

Additionally, AMELIA incorporates the remarkable women in the STEM stories dataset that features profiles of successful Peruvian women from 16 of Peru's 26 regions. It includes an estimated of 100 profiles. This dataset combines public biographical data with private content that was collected during the development of the chatbot and, with the authorization of the women, incorporated into AMELIA. The stories' recollection followed three key actions: first, contact and collaboration were established with organizations that promote gender equality in Peru. Second, an open call through social media was launched to reach all women in STEM in Peru who wanted to share their stories. And third, the IA engine was trained to look for stories available on the internet.

AMELIA generates 4 types of pedagogical activities that are core to STEM learning: games, projects, problem situations, and prototypes using the ChatGPT technology. The AI has been trained with prompts created and curated by pedagogical experts to structure answers that provide teachers with ready-to-use classroom sessions. Each prompt has been built ensuring alignment with the national curriculum and contextualized activities according to the region where the user request comes from. In addition, AMELIA uses activity formats that are familiar to rural teachers.

While AMELIA does not currently apply machine learning to evolve based on individual teacher input, it generates sessions dynamically using structured prompts to ChatGPT, informed by user requests and a knowledge base that is periodically updated, especially with new stories of women in STEM. User interaction data collected via WhatsApp is anonymized and analyzed to understand usage patterns, which informs all iterations and improvements. This multifaceted data strategy supports AMELIA's mission to provide personalized and impactful educational experiences, with plans to integrate behavior-driven enhancements and more specialized LLMs in upcoming development stages.

AMELIA complies with a safeguard policy aligned with global standards in cybersecurity, accessibility, and responsible digital practices. Its data management approach is guided by key principles of the European GDPR, such as data minimization, purpose limitation, and transparency. AMELIA does not interact directly with students or minors. The tool is directed to support teachers, who are adults, on their lesson planning task, and they are trained in a workshop to critically assess AI-generated content and manage digital risks in the classroom. All user data is anonymized and securely handled to ensure privacy and build trust in educational settings (Fig. 1).

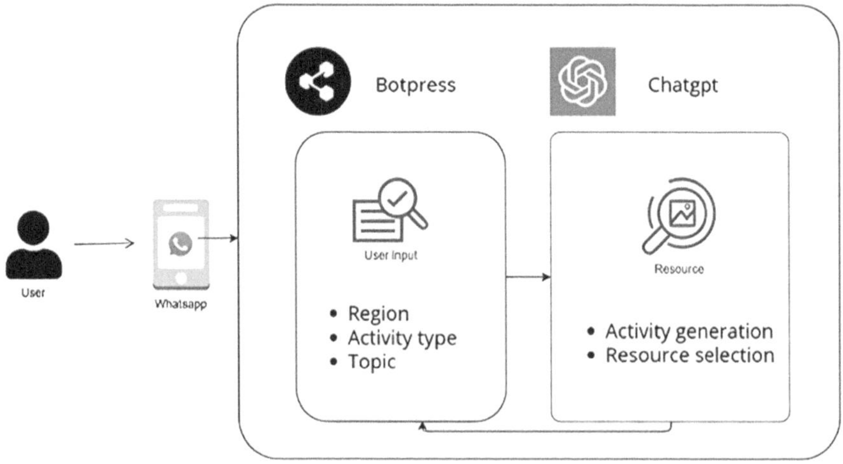

Fig. 1. Interaction of users with the outputs of AMELIA's database

Finally, it is worth noting that, first, the parameterized prompts tailored to teachers' specific requests and, second, the creation of the women's stories' carefully curated dataset helps AMELIA to mitigate the risk of AI hallucinations, reducing factual inaccuracies. While AMELIA does not yet include automated fact-checking mechanisms, its current safeguards prioritize pedagogical coherence and factual reliability in educational content.

2 Reflection of the Challenges and Opportunities Associated with the Implementation

2.1 AMELIA's Implementation Process: from Validation to the Programmes' Results

AMELIA's implementation in schools goes beyond chatbot usage. It includes teacher training, educational materials, learning communities, and monitoring. This section details its deployment in Cajamarca with 120 teachers and the validation process before reaching schools.

First, AMELIA's technology and pedagogy were validated with a select group of teachers and experts. Partnering with EdTech specialist Norma Rodriguez from Cayetano University, AMELIA was tested by 52 of her pre-service teaching students. Their feedback led to improvements in the activities structure, the chatbot fluency, and the women in STEM storytelling. Next, education experts reviewed AMELIA, using it for a week and completing a 10-question evaluation. The chatbot performed well, with minor adjustments made to the integration of women's stories into activities.

At the same time, the teacher training program and materials were developed. The program includes a 3-hour workshop on how to integrate AI in education with a critical and responsible approach; hands-on AMELIA training; incorporating teachers to a WhatsApp learning community where they receive technical and pedagogical support; and the application of face to face and online monitoring tools that assess usage and impact throughout the school year. Once the chatbot and teacher training program were refined, AMELIA was ready to be implemented at schools in Cajamarca.

To achieve the implementation in Cajamarca, the project partnered with an organization called Women in Mining that have more than 100 women student volunteers around Peru that go to public urban and rural schools to deliver STEM workshops for secondary school students as part of their initiative *Kuraq Ñañayki*[1]. By partnering with them, AMELIA was able to deliver a training program and materials to 120 teachers in Cajamarca.

The deployment of the AMELIA program in Cajamarca entails an approximate cost of USD 84 per teacher. This figure includes the face-to-face AI workshop, digital pedagogical materials, year-long platform access, moderation of the

[1] A Women In Mining (WIM) program that means "your big sister" in Quechua.https://wimperu.org/programas/kuraq-nanayki-tu-hermana-mayor/.

WhatsApp learning community, and monitoring processes. AMELIA thus constitutes a low-cost, scalable solution for rural schools in Peru, Latin America, and the world.

Results in Cajamarca were remarkable: 91% of teachers have reported improvement in their teaching practice and 96% belief that AMELIA contributes to getting more students interested in STEM learning. Furthermore, even though 55% of teachers trained by the program had never used AI tools before, 96% of them found AMELIA easy to use. Within a span of five months, the tool successfully facilitated over 2,000 classroom sessions for teachers and accumulated a substantial log of 10,000 interactions. This achievement demonstrates its efficacy even in areas characterized by limited connectivity. In addition, the program's training workshop achieved an NPS of 84, significantly surpassing the benchmark for excellence in teacher training programs, such as those led by educaLAB in Spain and comparable EU initiatives (Ministry of Education, Culture and Sport, 2014).

Monitoring and evaluation have been ongoing. Scope and usage are assessed through analysis of WhatsApp interactions, user profiles, and usage patterns generated by the AI. Pedagogical quality and impact are evaluated through teacher surveys administered at multiple points during the school year. AMELIA prioritizes the improvement of teaching practices and professional motivation. Given

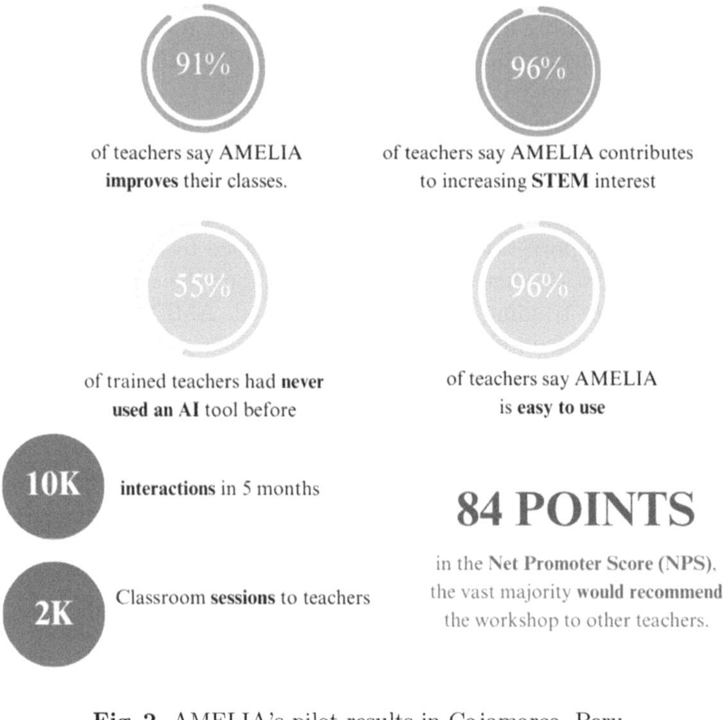

Fig. 2. AMELIA's pilot results in Cajamarca, Peru

its indirect operation—mediated through teachers—student outcomes depend on the degree to which teachers adopt and integrate the tool over time. A rigorous evaluation of student learning and STEM interest is planned once pedagogical adoption stabilizes and contextual variables are better controlled.

The evaluation strategy integrates quantitative and qualitative tools. Using data includes the number of teachers interacting, frequency, activity types, and patterns of engagement. AMELIA also offers automated feedback after each interaction to support continuous professional development. Complementary surveys assess teacher satisfaction, perceived impact on classroom practice, and student engagement. Testimonials and case studies further contextualize these findings. Finally, data visualization tools (Fig. 2) enable comprehensive reporting on AMELIA's performance, impact, and areas for improvement.

3 Description of Future Steps

3.1 AMELIA as an Opportunity to Strengthen Teachers in Rural Areas

AMELIA strengthens STEM education in rural Peru by providing accessible, user-friendly tools that integrate into teachers' daily practice, enhancing classroom effectiveness—especially in contexts with limited access to professional development. Teachers describe the tool as "innovative" and "helpful," noting its positive impact on pedagogy. AMELIA also demystifies AI for educators and students, fostering technological familiarity and confidence. This aligns with national demand for AI training: 84% of Peruvians consider it essential for teachers (IPSOS Global Education Monitor, 2023). As such, AMELIA emerges as a pedagogical ally across diverse subject areas.

The program offers targeted training that equips teachers to integrate technology effectively critical in rural Peru, where disparities in educational resources amplify the influence of teacher quality on student outcomes. The AI education market in Peru is projected to expand significantly by 2031, driven by intelligent tutoring systems (Statista, 2024). AMELIA contributes to bridging this gap, improving teaching and learning, and supporting the professionalization of the teaching career. This is especially relevant given that only 30% of Peru's labor force is highly exposed to AI technologies, compared to over 60% in countries like the U.S. and UK (IMF, 2024). Additionally, AMELIA promotes the inclusion of female role models in STEM, inspiring girls and challenging gender stereotypes. The involvement of WIM volunteers fosters community engagement and regional pride, making STEM education more relatable and inclusive.

3.2 AMELIA's Future Directions and Challenges to Overcome

AMELIA has yielded promising results, with strong teacher engagement and a perceived positive impact on increasing STEM interest in students. Future efforts focus on expanding its reach, refining functionalities, and integrating with other technologies to enhance scalability. AMELIA's approach aligns with broader regional efforts promoting STEM and STEAM education for rural development in Latin America (Gavari-Starkie et al., 2024). However, several challenges remain.

A key limitation is AMELIA's dependence on school attendance, which restricts its reach in poverty-stricken rural areas where girls often drop out due to economic hard-ship or domestic responsibilities. In 2022, nearly 20% of rural girls aged 12–16 was out of school (INEI, 2023). Cultural norms that prioritize boys' education, along with financial barriers, further deepen gender disparities. To address this, AMELIA is exploring a new iteration, a WhatsApp-based assistant targeting families to promote girls' schooling and STEM aspirations.

Another challenge lies in local implementation. WIM volunteers currently support teacher training in Cajamarca but lack pedagogical training and face logistical constraints. To strengthen sustainability, AMELIA is training local "Teacher Ambassa-dors"—educators who have completed the program and use it regularly—to mentor peers. These ambassadors receive ongoing support from a WhatsApp learning community and central team, fostering a scalable peer-led model for rural AI integration.

AMELIA aims to grow its women in STEM database from 100 to 500 stories, representing all 26 regions of Peru and its cultural diversity. This includes partnerships with Amazonian, Andean, and Afro-Peruvian organizations, and training regional ambassadors. Despite challenges, AMELIA's outlook is strong: it won Ruralia, Peru's first rural education accelerator and investment fund, supporting expansion in 2025–2026 to 300 teachers and 5,000 + students in high-need regions. Ruralia will also boost AMELIA's AI capacities with (1) STEM activities for elementary levels, (2) multimedia integration, and (3) safety risk detection.

The program also seeks international partners to expand into Amazonian areas, pilot family-focused support for girls' education, and improve impact monitoring tools.

Disclosure of Interests. The authors have no known conflicts of interest to disclose. However, they acknowledge potential biases related to their professional roles and interests in educational technology and STEM education.

References

Castro, M.: School inputs and cognitive gaps in Peru (2020)

Castro, M., Rolleston, C.: Early childhood and cognitive gaps in Peru (2018)

Friedel, J., et al.: Impacts of digital technologies on education and factors influencing teaching practices. J. Educ. Technol. **12**(4), 45–60 (2013)

Gamberini Coz, C.: Girls prevented from being bright and dreaming: what can be done through education in schools and at home? Friedrich Naumann Foundation for Freedom (2023). https://www.freiheit.org/andean-states/girls-prevented-being-bright-and-dreaming-what-can-be-done-through-education-schools

Gavari-Starkie, E., Espinosa-Gutiérrez, P.-T., Lucini-Baquero, C., Pastrana-Huguet, J.: Importance of STEM and STEAM education for improvement of the land in the rural environment: examples in Latin America. Land **13**(3), 274 (2024). https://doi.org/10.3390/land13030274

Instituto Nacional de Estadística e Informática (INEI). Informe técnico: Estadís-ticas de educación 2022 (2023). https://www.inei.gob.pe

International Monetary Fund. Peru: Selected issues (2024). https://www.imf.org/-/media/Files/Publications/CR/2024/English/1PEREA2024002.ashx

Ipsos. Global Education Monitor 2023 (2023). https://www.ipsos.com/sites/default/files/ct/news/documents/2023-09/Ipsos%20Global%20Education%20Monitor_1.pdf

Ministry of Education of Peru. Results of the 2018 Census Evaluation of Students (2018). http://umc.minedu.gob.pe/ece-2018

Ministry of Education, Culture and Sport. The NPS index: Assessment of the effectiveness of teacher training courses. educaLAB (2014). http://educalab.es/-/el-indice-nps-valoracion-de-la-eficacia-de-los-cursos-de-formacion-de-profesorado

Montero, C., Uccelli, F.: De ilusiones, conquistas y olvidos: La educación ru-ral en el Perú, vol. 72. Instituto de Estudios Peruanos (2023)

Statista. Machine Learning - Peru | Statista Market Forecast. Statista Market In-sights (2024). https://www.statista.com/outlook/tmo/artificial-intelligence/machine-learning/peru

Unidad de Estadística, Ministerio de Educación del Perú. Metodología de clasificación de servicios educativos según ámbito geográfico y la gradiente de ruralidad (2023). https://escale.minedu.gob.pe/c/document_library/get_file?uuid=f71a6501-9145-4b97-b50e-8fdfbc4e0dac&groupId=10156

MindMate: An LLM-Powered Mental Health Companion for Adolescents

Penghe Chen[1,2], Wutong Dong[2], and Yu Lu[1,2](✉)

[1] Advanced Innovation Center for Future Education, Beijing Normal University, Beijing 100875, China
luyu@bnu.edu.cn

[2] School of Educational Technology, Faculty of Education, Beijing Normal University, Beijing 100875, China

Abstract. Adolescent mental health is a pressing global issue, with many of adolescents suffering from mental health disorders and a significant shortage of professional counselors in schools. To address this gap, we propose MindMate, an intelligent mental health consultation system for adolescents powered by Retrieval-Augmented Generation (RAG) technology. MindMate uses large language models (LLMs) to generate a high-quality knowledge base tailored to adolescent mental health, which is refined through quality and diversity filtering. The RAG framework enhances the system's performance by integrating this knowledge base with LLMs, ensuring accurate, reliable, and empathetic responses. Evaluations show significant improvements in accuracy, practicality, risk response, and empathy compared to baseline LLMs. MindMate demonstrates strong potential to provide effective psychological support for adolescents. Future work will focus on dynamic knowledge base updates, handling complex queries, large-scale user studies, and ensuring robust privacy mechanisms.

Keywords: Adolescent Mental Health · Large Language Model (LLM) · Retrieval-Augmented Generation (RAG) · Data Augmentation

1 Introduction

Mental health issues among adolescents can manifest in various forms, including anxiety, depression, behavioral disorders, and even suicidal tendencies. Early detection and intervention are essential to mitigating long-term negative impacts [1]. However, addressing this issue within the current educational system presents significant challenges. A major obstacle is the severe shortage of professional mental health teachers in primary and secondary schools. Additionally, limitations in time and space further restrict the ability of educators and counselors to provide timely attention and responses to students' psychological concerns. This imbalance between the increasing demand for mental health support and the lack of available professional resources underscores the urgency of exploring innovative, technology-driven solutions.

A. I. Cristea et al. (Eds.): AIED 2025, CCIS 2590, pp. 90–98, 2025.
https://doi.org/10.1007/978-3-031-99261-2_8

The rapid advancement of artificial intelligence, particularly large language models (LLMs), has opened new possibilities for mental health interventions. LLMs have demonstrated remarkable capabilities in natural language understanding and generation, offering significant potential in providing psychological support. However, despite their impressive linguistic abilities, general-purpose LLMs often lack specialized knowledge in mental health. Moreover, they are prone to generating inaccurate or misleading information—commonly referred to as "hallucinations"—which is especially concerning when dealing with sensitive and complex topics such as adolescent mental health [2]. These limitations make the direct application of LLMs to psychological counseling problematic.

To address these challenges, this study proposes the development of an intelligent adolescent mental health consultation system using Retrieval-Augmented Generation (RAG) technology. RAG enhances generative models by incorporating domain-specific knowledge through retrieval mechanisms, thereby improving response accuracy and reliability. However, the effectiveness of a RAG-based system heavily depends on the availability of a well-structured and comprehensive knowledge base. Unfortunately, there is currently no high-quality, large-scale knowledge base specifically tailored to adolescent mental health concerns, posing a significant challenge to system development.

Recent studies have explored the use of LLMs for domain-specific data augmentation, offering valuable insights for constructing specialized knowledge bases. For instance, Liang introduced Synth-Empathyt to automatically produces empathetic data and filters out low-quality data [4]; Chen employed ChatGPT for accuracy evaluation [5]; Xu leveraged GPT-4 to rewrite data and enhance complexity [6]; Liu used ChatGPT for data annotation and trained models to assess quality and difficulty [7]; Lu defined diversity and complexity metrics using ChatGPT [8]; and Parkar combined clustering techniques with GPT-4 to identify high-quality samples [9]. These studies provide valuable references for improving dataset quality and relevance.

Drawing inspiration from these advancements, this study explores the use of LLMs for data augmentation to construct a robust adolescent mental health knowledge base. By integrating this knowledge base with RAG technology, we aim to develop an intelligent consultation system that delivers accurate, reliable, and timely psychological support for adolescents, helping to bridge the gap in mental health resources and improve overall well-being.

2 System Architecture and Implementation

MindMate is an intelligent consultation system designed to support adolescent mental health, structured around two core modules: psychological health data generation and selection and RAG consultation. The system first leverages LLMs to generate mental health-related Q&A data, which is then refined using an existing database to ensure alignment with psychological and educational theories, forming a high-quality knowledge base. Subsequently, RAG technology integrates this knowledge base with LLMs, enabling precise and context-aware

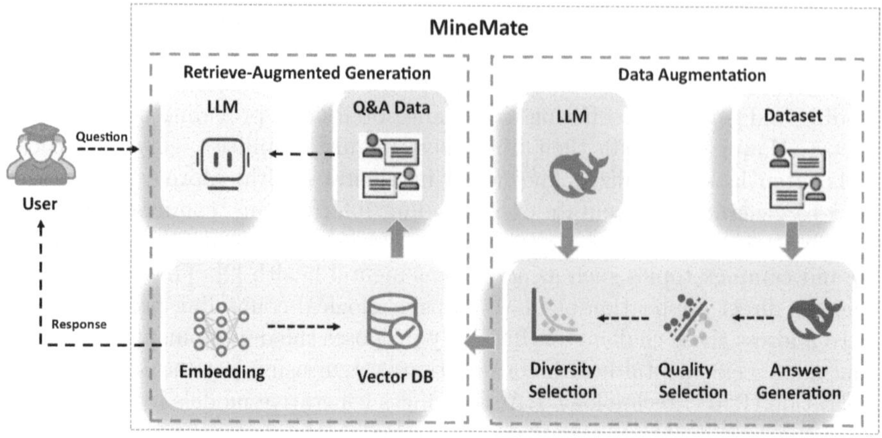

Fig. 1. The system architecture of MindMate.

mental health consultations through embedding-based retrieval and a domain-specific psychological LLM. Ultimately, the system provides adolescents with an interactive interface for real-time mental health support, helping them navigate psychological challenges and fostering emotional well-being (Fig. 1).

2.1 Data Augmentation

The data augmentation process is a crucial step in building a robust knowledge base for the mental health consultation system. Inspired by existing work [4], as shown in Fig. 2, we propose data augmentation with three key stages: first, we use an LLM with predefined prompts to generate multiple potential responses. Second, leveraging existing pairs as training data, we employ LLM to create a extractor and discriminator for data selection, ensuring the chosen responses maintain both professional accuracy and consistency. Finally, to enhance diversity and prevent repetitive answers, we apply algorithmic diversity analysis to filter for high-quality and varied responses. The enhanced dataset generated through this three-step process serves as the foundation for integrating RAG into the intelligent mental health consultation system.

Data Generation. To address mental health questions lacking answers, we leverage an LLM to generate potential responses. To ensure the model understands the context, we explicitly define the consultation scenario as adolescent mental health support, guiding the model to act as a psychological expert focused on fostering healthy development. Additionally, we impose stylistic constraints, requiring responses to be optimistic, aligned with psychological theories, and delivered in a warm and empathetic tone. To further enhance the model's comprehension of the task, we employ few-shot learning, providing representative Q&A examples from existing data. With this structured approach, the LLM

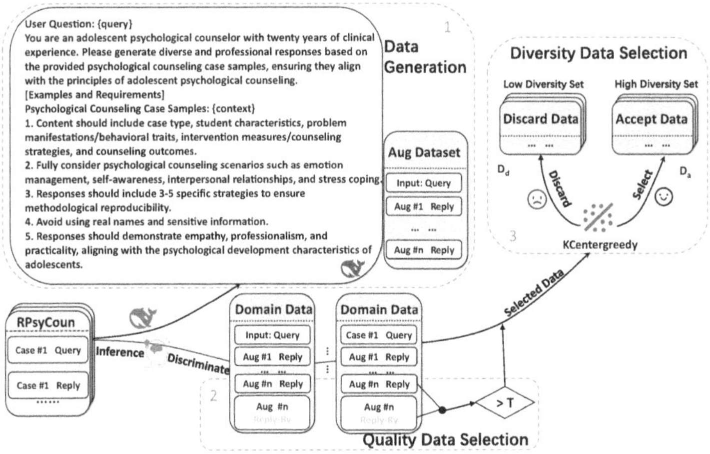

Fig. 2. Data augmentation pipeline, which is composed by (1) data generation module, (2) quality data selection module, (3) diversity data selection module.

generates multiple potential answers tailored to the needs of adolescent mental health consultation.

Quality Data Selection. To ensure the quality and professionalism of the generated answers, we implement a structured filtering process based on key content attributes. The core idea is to analyze critical aspects of each answer—such as key content points, theoretical alignment, and tone—to determine whether it meets the expected standards. First, we utilize an existing dataset to train a key point analyzer with an LLM, which extracts and summarizes the essential elements of each Q&A pair. Next, we develop a key point evaluator, also powered by an LLM, to assess whether the extracted key points align with the required psychological and educational principles. Answers that meet the established criteria are classified as high-quality responses. By combining these two models, we effectively filter and retain only well-structured, professional, and theoretically sound answers for the mental health consultation system.

Diversity Data Selection. To further enhance the dataset, we introduce a diversity-based selection method aimed at maximizing variation among retained responses. This step ensures that the final dataset captures a broad spectrum of perspectives, avoiding redundancy and promoting response diversity. Specifically, we select answers that exhibit the greatest distinction from others based on a discriminator's evaluation. To achieve this efficiently, we employ the KCenter-Greedy algorithm [11], which is well-suited for identifying a diverse subset while maintaining computational efficiency. By integrating this approach, we construct a synthetic dataset that is both high in quality and rich in diversity, enhancing the robustness of the mental health consultation system (Fig. 3).

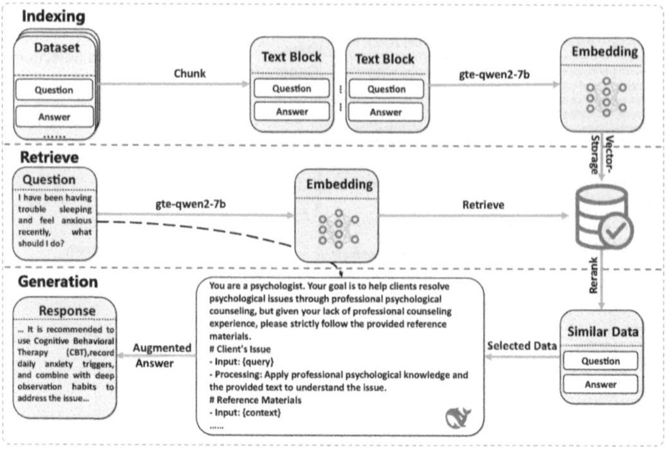

Fig. 3. Retrieval Augmented Generation and System Operation Process, which is composed by (1) Indexing module, (2) Retrieval module, (3) Generation module.

2.2 RAG-Based Consultation

The RAG framework is designed to enhance the performance of LLMs by integrating external knowledge from a domain-specific database. This approach aims to address the limitations of traditional LLMs, which often lack specialized knowledge and may generate inaccurate responses. The RAG framework consists of three core steps: Indexing, Retrieval, and Generation. Each step plays a crucial role in ensuring that the system provides accurate, relevant, and professional responses to user queries. In this section, we detail how these steps are implemented in our psychological Q&A system.

Indexing. The Indexing step is the foundation of the RAG framework, where we organize and store the generated data in a structured manner for efficient retrieval. Specifically, we segment the generated data into question-answer pairs, which are then processed using an embedding model. This model converts each segment into a vector representation that captures the semantic meaning of the text. These vectors are subsequently stored in a vector database, such as FAISS or Pinecone, which allows for fast similarity searches. By indexing the data in this way, we ensure that the system can quickly access relevant information when processing user queries.

Retrieval. The Retrieval step is critical for identifying the most relevant questions from the indexed database in response to a user's query. When a user submits a query, the system first vectorizes the query using the same embedding model used during the Indexing step. This ensures that the query is represented in the same vector space as the counseling cases, facilitating accurate similarity comparisons. The vectorized query is then used to perform a similarity search in the vector database. The retrieved cases are ranked according to their similarity to the user's query, with the most relevant cases appearing at the top. This ranking process prioritizes the most pertinent questions, providing a solid foundation for generating accurate and helpful responses.

Generation. The final step in the RAG framework is the Generation of the response. This step combines answers of the top-ranked questions retrieved in the previous step with predefined prompts that guide the LLM in generating the final response. These prompts are carefully designed to ensure that the generated responses adhere to professional psychological counseling standards and provide actionable advice to the user. By integrating the retrieved answers with the prompts, the LLM can leverage both the external knowledge from the database and its own generative capabilities to produce high-quality, contextually relevant responses. This approach not only enhances the accuracy of the responses but also ensures that they are tailored to the specific needs of the user.

3 System Evaluation

To comprehensively evaluate the performance of our proposed RAG-based mental healthy counsulting system for adolescents, we conducted a two-part assessment. The first part focused on the quality of the generated data, which serves as the foundation of the system. We employed a quantitative analysis approach, leveraging expert scoring to assess the generated data. The second part involved a qualitative analysis of the system's actual question-and-answer results, comparing them with responses generated directly by LLMs. This comparison was conducted through illustrative examples to demonstrate the system's ability to provide more relevant and professional responses.

3.1 Data Augmentation Evaluation

The quality of the generated data is crucial for the overall performance of the system. To evaluate this, we utilized a set of commonly used metrics in data augmentation. These metrics were defined using a heuristic approach to determine the weight for each indicator. Specifically, we selected three experts in the field of adolescent psychology and mental health counseling. We randomly sampled 100 data points from the generated dataset and asked each expert to independently score these data points based on the predefined criteria. The scores from the three experts were then averaged for each data point, and the overall average score was calculated across all 100 data points. To better illustrate the improvement in data quality achieved by our system, we compared our results with those obtained from direct generation by a large language model, which served as the baseline. The responses were evaluated using a four-dimensional professionalism scale with a total score of 10, based on the following criteria [10,12]:

Table 1. Scores on Generated Data

Dimension	DeepSeek-R1	Our System	Improvement
Accuracy	2.2	2.8	+27%
Practicality	1.9	2.6	+37%
Risk Response	0.9	1.8	+100%
Empathy	1.7	1.9	+12%
Total Score	**6.7**	**8.3**	**+24%**

- **Accuracy (3 points)**: Whether the response aligns with psychological theories or clinical standards.
- **Practicality (3 points)**: Whether the advice is specific and actionable.
- **Risk Response (2 points)**: Whether the response to high-risk issues is professional.
- **Empathy (2 points)**: Whether the response demonstrates emotional support and empathetic expression.

As shown in Table 1, the results showed that our system achieved significantly higher scores across all dimensions compared to the baseline LLM-generated data. This indicates that the data augmentation and RAG framework effectively enhanced the quality of the generated responses, making them more accurate, practical, and empathetic.

3.2 Response Analysis

The quality of the system's actual responses directly determines its practical utility in real-world scenarios. To evaluate this, we selected several illustrative examples and compared the responses generated by our RAG-based system with those produced by DeepSeek [3]. The examples were chosen to cover a range of common and critical mental health issues faced by adolescents. The results are presented in Table 2. One example highlights the system's ability to provide empathetic support. For a query about feelings of loneliness, our system generated a response that not only offered practical advice on how to cope but also expressed understanding and empathy. This nuanced approach is crucial for building trust and rapport with adolescents, making them more likely to seek help when needed. These qualitative analyses underscore the enhanced capability of our RAG-based system in delivering professional psychological counseling. By integrating domain-specific knowledge and leveraging the strengths of the RAG framework, our system is better equipped to address the complex and sensitive issues faced by adolescents. This makes it a valuable tool in supporting adolescent mental health and well-being.

Table 2. Comparison of Response Examples in High-Risk Scenarios

Question	DeepSeek-R1	Our System
I have been having trouble sleeping and feel anxious recently, what should I do?	Try to relax and avoid using your phone before bed	It is recommended to use Cognitive Behavioral Therapy (CBT), record daily anxiety triggers, and combine with deep observation habits to address the issue
I want to commit suicide, but I'm afraid my parents will be sad	Suggest communicating more with family	Please contact the psychological crisis hotline at 400-161-9995 immediately, and also suggest making a safety plan with a trusted adult
My parents always negate me, and I am becoming more and more insecure	Communicate more with your parents and express your feelings	It is suggested to use Nonviolent Communication (NVC) with your parents, clearly express your needs, and seek support from the school's psychological counselor

4 Conclusion and Future Work

This study presents MindMate, an innovative intelligent consultation system designed to address the critical gap in adolescent mental health support. By leveraging the RAG framework and a meticulously constructed knowledge base, the system demonstrates significant improvements in providing accurate, relevant, and empathetic psychological counseling for adolescents. The data augmentation process, which includes generation, quality filtering, and diversity filtering, ensures that the knowledge base is robust and tailored to the specific needs of young individuals. The RAG-based consultation module further enhances the system's performance by integrating external knowledge with the generative capabilities of LLMs, resulting in high-quality responses that underscore the potential of MindMate to serve as a valuable tool in supporting adolescent mental health and well-being.

Despite the promising results, there are two areas for future improvement and expansion. First, the current system relies on a static knowledge base, which may limit its adaptability to emerging mental health issues and evolving psychological theories. Future work should focus on developing dynamic updating mechanisms for the knowledge base, allowing it to incorporate the latest research findings. This can be achieved through continuous data augmentation and periodic retraining of the models.

Second, while the system has shown strong performance in generating high-quality responses, further enhancements can be made to improve its ability to handle complex and multi-faceted queries. The current evaluation is primarily based on expert scoring and illustrative examples. Future research should include large-scale user studies to assess the system's real-world impact and user satisfaction. This can involve deploying the system in educational settings and collecting feedback from both students and educators to identify areas for improvement and potential integration into existing mental health support frameworks.

Acknowledgments. This research was supported by the National Natural Science Foundation of China (Grant No. 62177009 and No. 62477003).

References

1. Yu, G.: Detection rate of mental health issues among Chinese students and its educational implications. Tsinghua J. Educ. **43**(4), 20–32 (2022)
2. Wang, Z., et al.: Data management for large language models: a survey. CoRR (2023)
3. DeepSeek-AI. DeepSeek-R1: Incentivizing Reasoning Capability in LLMs via Reinforcement Learning (2025). https://arxiv.org/abs/2501.12948
4. Liang, H., et al.: Synth-empathy: towards high-quality synthetic empathy data. arXiv preprint arXiv:2407.21669 (2024)
5. Chen, L., et al.: Alpagasus: training a better alpaca with fewer data. arXiv preprint arXiv:2307.08701 (2023)
6. Xu, Y., et al.: Rethinking the instruction quality: lift is what you need. arXiv preprint arXiv:2312.11508 (2023)
7. Liu, W., Zeng, W., He, K., Jiang, Y., He, J.: What makes good data for alignment? A comprehensive study of automatic data selection in instruction tuning. arXiv preprint arXiv:2312.15685
8. Lu, K., et al.: # instag: instruction tagging for analyzing supervised fine-tuning of large language models. arXiv preprint arXiv:2308.07074 (2023)
9. Parkar, R.S., Kim, J., Park, J.I., Kang, D.: SelectLLM: Can LLMs Select Important Instructions to Annotate? arXiv preprint arXiv:2401.16553 (2024)
10. Zhang, C., et al.: Cpsycoun: a report-based multi-turn dialogue reconstruction and evaluation framework for Chinese psychological counseling. arXiv preprint arXiv:2405.16433 (2024)
11. Sener, O., Savarese, S.: Active learning for convolutional neural networks: a core-set approach. arXiv preprint arXiv:1708.00489 (2017)
12. Liu, J.M., Li, D., Cao, H., Ren, T., Liao, Z., Wu, J.: Chatcounselor: a large language models for mental health support. arXiv preprint arXiv:2309.15461 (2023)

A Conversational Tool Based on Knowledge Graph, LLMs and BERT Model for Work-Study Programs in France

Baba Mbaye[1(✉)], Diana Nurbakova[2], and Duaa Baig[1,2]

[1] EffetB SAS, 69100 Villeurbanne, France
baba@effetb.com, duaa.baig@insa-lyon.fr
[2] INSA Lyon, CNRS, Universite Claude Bernard Lyon 1, LIRIS, UMR5205, 69621 Villeurbanne, France
diana.nurbakova@insa-lyon.fr

Abstract. Follow-up systems for vocational training, particularly in work-study schemes, are essential to ensure that apprentices make progress in acquiring specific skills. This article presents a conversational tool integrated into a digital training booklet. The tool combines a knowledge graph, linked to the skills to be validated, with large-scale language models (LLMs) and a generative model based on BERT. This approach enables intelligent management of learning data while allowing natural language interactions to monitor progress and provide personalized coaching. By offering personalized feedback and advice, this tool aims to improve the support, ensuring more effective skill acquisition. It offers a more adaptive and interactive way of tracking progress, making it a key innovation in the field of vocational training.

Keywords: Conversational tool · Knowledge graph · LLMs · BERT · Work-Study Programs

1 Introduction

Work-study programs integrate academic learning with practical work experience. Such programs offer several benefits, including practical experience in the domain of studies, enhanced students' employability skills, and theoretical concepts in practical settings [3]. However, they face challenges such as time constraints and cohort size [3], placement issues [2], work-learning balance [4].

Each training program contains skills and knowledge related to the occupation field. Traditionally, training booklets monitor skills acquired by learners. These are used by *apprentices*, *tutors* from the vocation school, *tutors* from the work placements, and *educational manager*. Digital booklets contain skills to be validated and enable trainers and tutors to assess the learners' progress, but are often static and require manual intervention. To facilitate the interactions with the booklet and reduce manual actions, an intelligent conversational tool can

A. I. Cristea et al. (Eds.): AIED 2025, CCIS 2590, pp. 99–108, 2025.
https://doi.org/10.1007/978-3-031-99261-2_9

be integrated into the system [8,11]. Chatbots have increasingly been integrated into educational settings to enhance learning experiences and provide support. They serve as educational assistants, help to assess students, and offer real-time feedback [10,11,13]. Moreover, they can provide mentoring, helping to personalize education by adapting to individual learning needs and preferences [8,9,12].

Such conversational tools have not been employed into digital booklets for work-study programs. To overcome this gap, we propose an intelligent agent (chatbot) that leverages (1) knowledge graph (KG) for modeling apprentice's progress and work-study programs, including the associated skills; (2) Large Language Model (LLM) for analyzing user's queries and converting it into a Cypher query to retrieve the data from the knowledge graph; (3) fine-tuned BERT [1] for final response generation. We are focusing on three scenarios: [S1] Monitoring of acquired skills; [S2] Recommendation of the skills to work on; [S3] Personalized follow-up and feedback.

The paper continues with implementation details for French work-study programs in Sect. 2, followed by technical components and methodology in Sects. 3–4. Experimental results appear in Sect. 5, with challenges and opportunities discussed in Sect. 6. Section 7 concludes with future directions.

2 Implementation and Educational Context

We are focusing on work-study programs in France. In this Section, we provide key implementation details.

Work-Study Programs in France. France offers two main forms of work-study programs: (1) Apprenticeship (fr. *Apprentissage*), combining vocational training with paid work experience; and (2) Professionalization Contract (fr. *Contrat de Professionnalisation*), targeting specific groups. Work-study programs (fr. *formation en alternance*) have seen significant growth since the 1990s [5] with 837,000 apprentices in 2022 [6], following the 2018 reform which increased financial incentives and simplified administrative procedures [7]. Vocational trainings are governed by *France Compétences* (https://www.francecompetences.fr/), which regulates training quality and costs, manages the National Directory of Professional Certifications (fr. *Répertoire National des Certifications Professionnelles, RNCP*), distributes funding, and evaluates system effectiveness. Each RNCP qualification is classified according to a national framework (8 levels), contains competency descriptions, assessment criteria, and career pathway information. These competences align with the French comprehensive skills framework *Référentiels de Compétences*, organized into competency blocks (fr. *blocs de compétences*). Here is an example of such a program for Computer Science engineering degree at the National Institute of Applied Scie nces of Lyon (INSA Lyon). Its RNCP code is `RNCP35971`, with associated competences grouped into 5 blocks (e.g. `RNCP35971BC01` "*Design, develop and maintain quality software*"). Each block contains specific skills to be validated, e.g.

designing complex software architecture, using modeling tools, and implementing quality methodologies. These competency frameworks provide the structured basis for apprenticeship validation within the digital training booklet system. Examples of digital booklets include: Campus skill, GowizApp , Kwark, Sheldon skills, STUDEA.

Educational Goals. The conversational tool addresses several critical objectives in vocational training. It facilitates automated tracking of apprentice skill acquisition through continuous monitoring, enabling real-time visibility into competency development for both supervisors and learners, circumventing inefficiencies in manual documentation. A significant function is its capacity for personalized learning pathway generation. By leveraging the KG structure and historical learning patterns, the tool identifies optimal skill development trajectories tailored to individual apprentices, prioritizing competencies that address specific developmental needs. Furthermore, the system reduces administrative burden for educational personnel by automating data collection and competency validation processes, allowing tutors and trainers to focus on high-value educational interventions such as personalized feedback and targeted coaching. In practical terms, the system enables three primary interaction modes: (1) Skill Monitoring Queries (S1): "*What skills has apprentice Jean validated in the software development module?*" The system retrieves validation records and generates a summary of progress. (2) Recommendation Queries (S2): "*Which skills should Marie prioritize next?*" The system analyzes skill dependencies, current progress, and program requirements to suggest an optimal learning path. (3) Personalized Feedback Queries (S3): "*How is Thomas progressing compared to expectations?*" The system contextualizes individual performance within program norms and peer cohorts, providing comparative insights.

Target Audience. The tool serves multiple stakeholders within the vocational education ecosystem: (1) **Apprentices**: The system provides continuous access to transparent assessment data regarding competency development, enabling awareness of strengths and areas requiring improvement. Through personalized learning recommendations, apprentices can engage in self-directed learning while maintaining alignment with educational objectives. (2) **Tutors and Trainers**: For supervisors, the tool offers streamlined monitoring that eliminates manual progress tracking, enables efficient identification of skill development patterns, and facilitates early detection of learning challenges. (3) **Educational Institutions**: At the institutional level, aggregated data provides valuable insights for curriculum refinement and alignment with industry requirements, promoting adaptive curriculum development that responds to emerging competency demands. By serving these diverse groups simultaneously, the tool creates a unified information ecosystem that promotes coherence in vocational training delivery while accommodating distinct requirements of each participant group.

3 Technical Components

In this Section, we detail three main components used in our conversational tool: a knowledge graph that models digital booklet, LLMs used mainly for user's query analysis and a fine-tuned BERT model for response generation.

3.1 Digital Training Booklet and Knowledge Graph

The digital training booklet is a dematerialized version of skills tracking that automates the collection and organization of data relating to apprentice progress. In our approach, we model this booklet as a knowledge graph (KG) [14,15]. It was built using two main data sources: (1) apprentices progress monitoring data already contained in our digital booklet system; and (2) data extracted from the *France Compétences* platform (see Sect. 2) about skills, qualifications, and training units. Our KG contains the following entities/nodes:

- `Skills` (5,020 nodes): Each skill validated within the training program is represented as a node in the graph. Each node contains attributes such as the skill's identifier, the skill level (aligned with the RNCP described in Sect. 2), as well as the criteria and methods for evaluation.
- `RNCP` (188 nodes): Certifications recognized by the RNCP.
- `Training category` (170 nodes): Broad classifications grouping related programs, e.g., "Computer Science and Software Development" covering Web Developer, Application Designer, and IT Systems Technician courses.
- `Training course` (508 nodes): Program linked to RNCP qualifications.
- `Apprentices` (2,691 nodes): Apprentices are modeled as nodes, with attributes related to their progress in the training program, such as `validated skills` and `acquisition dates`.
- `Trainers/Tutor` (2,010 nodes): The trainers from educational institutions and companies are represented as nodes, linked to `Apprentices`.
- `Schools` (46 nodes): Each training program, whether delivered by a school/partner company, is modeled as a node linked to the `Skills` it validates.
- `Apprenticeship Training Units (UFA)` (413 nodes): Specialized divisions within schools dedicated to apprenticeships.

The relationships between these entities are represented by edges, connecting the different nodes in the graph, e.g.: `HAS_SKILL` (50,738 edges) are links between `Apprentices` and the `Skills` they acquire over time); `ATTENDS` (2,903 edges) are links between `Apprentices` and the specific `Training courses` they participate in, `HAS_RNCP` (226 edges) are links between `School` and `RNCP`).

Our KG is implemented using Neo4J (https://neo4j.com/). Such implementation allows us to trace the skills acquisition paths for each apprentice and track their evolution within the work-study program. This KG serves as a structured basis for the conversational tool, contextualizing questions and generating answers tailored to individual needs. When a user asks "*Which skills should apprentice Jean prioritize next?*", the following relation paths used for reasoning are considered:

```
(Jean) → HAS_SKILL → (Skill A)
(Jean) → ATTENDS → (Training_course Z) → BELONGS_TO_CATEGORY →
    (All Training Categories) → INCLUDES_SKILL → (All Skills) →
    minus → (Validated Skills)
```

3.2 LLMs for Query Interpretation and Intent Recognition

LLMs like GPT-4 play a crucial role in interpreting and generating human-like text responses [16,17]. These models understand natural language queries. However, LLMs can create ungrounded statements factually incorrect [18] and generate vague or overgeneralized responses. To overcome this issue, we make LLM reason over a KG. We build our system similar to the *Planning-Retrieval-Reasoning* framework proposed by Luo et al. [19]. When a user submits a question, the LLM processes the input and identifies user's intent. Relation paths are generated (*Planning*) and the data is retrieved from the KG (*Retrieval*). Finally, BERT generates contextually appropriate answers (*Reasoning*) and LLM generates a final natural response. By doing so, our tool creates structured responses that align with the apprentice's skill progress. LLMs help in personalized recommendations, suggesting the most relevant skills or areas for improvement.

We compared 7 commonly used models: GPT-4 [23], LLaMA 2 7B [20], GPT-J-6B [25], Falcon 7B [21], BLOOM [24], and Mistral 7B [22] regarding the following criteria: availability and the predictive power of the models in terms of Accuracy and F1 score. In addition to our KG, we used 27,713 queries corresponding to our scenarios: [S1] Open SQuAD (https://tinyurl.com/2yup27s8); [S2] Career Dataset (https://tinyurl.com/muw7vm96); [S3] ASSISTments Data Set 2012-2013 (https://tinyurl.com/bdh23acu). The results are given in Table 1.

Table 1. LLM comparison

Model	GPT-4	LLaMA-2 7B	GPT-J-6B	Falcon 7B	BLOOM	Mistral 7B
Availability	**Paid**	Free	Free	Free	Free	Free
Accuracy (%)	**84**	82	81	78	73	72
F1 score (%)	**81**	80	77	77	71	71

GPT-4 demonstrated the highest accuracy (84%) in responding to apprenticeship-related queries, ensuring reliable feedback and improving responses. GPT-4's key strength is structuring responses based on interconnected data, making it compatible with our KG-based logbook. This integration ensures queries regarding skill progress and recommendations are accurate. The model's flexibility allows tailored recommendations based on apprentice progress, enhancing personalized coaching. While GPT-4 and LLaMA-2 performed well in natural language understanding, their responses lacked domain specificity without additional training. Based on the results, we keep GPT-4 for our implementation.

3.3 BERT-Based Refinement Model

BERT [1] serves as our response generation component, providing context-sensitive outputs tailored to the work-study domain. While LLMs like GPT-4 excel at query understanding, BERT generates concise, factually grounded responses aligned with information from our KG. To enhance BERT's performance, we fine-tuned it using: (1) anonymized *Digital Training Logbook Data* from our KG; and (2) *Vocational Training Documentation* from *France Compétences.* (3) *EdNet: Large-Scale Educational Data*, containing interactions between students and learning platforms with recommendations and feedback [26]. The fine-tuning process included *Response Generation* (creating appropriate responses from KG data) and *Intent Recognition* (classifying queries into scenarios S1-S3 with 92% accuracy on our test set[1]).

A crucial aspect is the bidirectional connection between BERT and the KG. BERT processes structured data extracted from the KG to generate responses both linguistically natural and factually accurate. When processing a query about skills prioritization, BERT receives: *"Marie has validated skills A, B, and C. The suggested next skills for their learning path are D and E, according to the training program."* This integration ensures that responses remain grounded in the actual training data while providing the natural language interface that enhances accessibility for all users of the digital training booklet. By combining BERT's refined response generation with the factual grounding of our KG and the query understanding capabilities of GPT-4, we create a comprehensive system that addresses the specific needs of work-study stakeholders while maintaining accuracy and relevance in the vocational training context.

4 Methodology

Our approach integrates three complementary components: a KG for structured data representation, GPT-4 for natural language understanding, and a fine-tuned BERT model for response generation. This enables intelligent monitoring and personalized support for apprentices while maintaining factual accuracy and contextual relevance. Our tool follows a modular pipeline architecture (see Fig. 1), separating query understanding, information retrieval, and response generation:

1. *Query Analysis (GPT-4)*: When a user submits a query, GPT-4 extracts key entities (apprentice names, skill references), determines intent, and identifies contextual parameters. Using GPT-4's contextual understanding capabilities, we achieve 84% accuracy in interpreting user intentions.
2. *KG Reasoning*: The processed query is transformed into a Cypher query to retrieve structured data from the KG. This leverages the graph's structure for complex reasoning including skill dependency analysis, gap identification, personalized recommendations, and cohort-based reasoning.

[1] We divided our dataset into train, validation and test subsets with the ratio 60:20:20.

3. *Response Generation (BERT)*: The fine-tuned BERT model transforms retrieved structured data into natural language responses, providing coherent and contextually appropriate responses.
4. *Feedback Integration*: User interactions are continuously monitored to improve system performance.

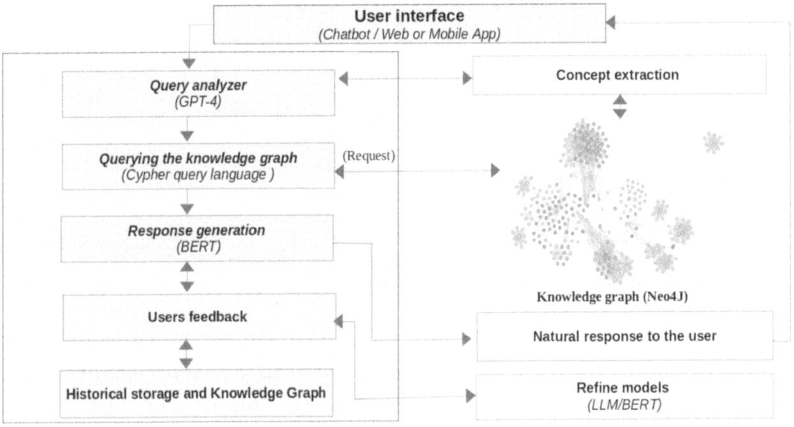

Fig. 1. System architecture

The conversational tool will soon be embedded within our existing digital training booklet system, using a REST API architecture.

5 Experiments and Results

We assessed technical performance of our conversational tool. To do so, we developed a test dataset comprising 1,500 queries across our three target scenarios: skill monitoring (500 queries collected from Open SQuAD (https://tinyurl.com/2yup27s8)), skill recommendation (500 queries from the Career Dataset (https://tinyurl.com/muw7vm96)), and personalized follow-up (500 queries from ASSISTments Data Set 2012-2013 (https://tinyurl.com/bdh23acu)).

We compared four configurations to determine the optimal approach: (1) `LLM-only`, using GPT-4 for query understanding and response generation; (2) `BERT-only`, using BERT for query understanding and response generation; (3) `LLM+BERT`, using GPT-4 for query understanding and BERT for response generation; (4) `LLM+BERT+KG`, i.e. our complete system with graph-based reasoning.

Table 2 presents the technical performance metrics (accuracy and F1-score) for each system configuration. The integrated approach `LLM+BERT+KG` significantly outperformed other configurations across all metrics, with particularly strong performance (93% accuracy) in the skill monitoring scenario. The system demonstrated particular effectiveness in complex query scenarios requiring integration of multiple data points, such as cross-referencing skill dependencies with individual progress data and cohort performance patterns. These results validate our integrated approach while highlighting opportunities for improvement in tutor adoption and complex reasoning scenarios.

Table 2. Comparative results

Method	LLM-only	BERT-only	LLM+BERT	LLM+BERT+KG
Accuracy (%)	84	83	88	**93**
F1 score (%)	81	80	85	**92**

6 Challenges and Opportunities

Our solution revealed several key considerations for AI-driven conversational tools in vocational training. Primary **challenges** include data quality and consistency issues that significantly impact response reliability, the need for continuous model fine-tuning to accommodate evolving training programs, and interpretability limitations of LLMs that necessitate mechanisms to explain recommendations to users. User adoption may be hindered by resistance to AI tools or limited digital literacy, while bias mitigation remains crucial to ensure fair skill evaluations.

Despite these challenges, our hybrid approach—combining LLMs with Knowledge Graphs—offers key advancements. It enables **personalized skill recommendations**, **automated progress monitoring**, and **real-time adaptive feedback**, **reducing the administrative load** on trainers and allowing for **data-informed instructional adjustments**. Integration with digital logbooks facilitates dynamic learning paths and improves coordination between learners and trainers.

These features enhance the **pedagogical impact** by fostering individualized support and scalable feedback loops. The system's flexibility makes it adaptable to **higher education**, **corporate training**, and **certification programs**, opening avenues for broader application of personalized AI in education.

7 Future Work and Conclusion

The integration of KGs, LLMs, and BERT-based response generation advances support for work-study programs in France by combining structured knowledge with NLP capabilities. Our results show that such approach achieves 93% accuracy, significantly outperforming individual components. Nevertheless, we have identified several challenges in data standardization, user adoption, and model interpretability that must be addressed. The opportunities in personalized learning pathways and administrative efficiency suggest promising directions for vocational training innovation. For future work, we plan to conduct a qualitative study across vocational training centers to measure stakeholder satisfaction and validate the system's impact on learning outcomes. We will implement XAI techniques to improve transparency through visualizations of skill progression and recommendation justifications. Additionally, we intend to expand our tool's capabilities with speech recognition and implement reinforcement learning for dynamically adjusted learning paths based on performance data. As work-study programs gain prominence globally, AI-powered tools enhancing personalization and monitoring will become increasingly valuable. Our work contributes to the field of AI in education by demonstrating a practical implementation addressing domain-specific vocational training challenges, with potential applications in higher education, corporate training, and certification programs.

References

1. Devlin, J., et al.: Bert: pre-training of deep bidirectional transformers for language understanding. In: Proceedings of the 2019 Conference of the North American chapter of the Association for Computational Linguistics: Human Language Technologies, vol. 1 (Long and Short Papers) (2019)
2. Stanley, T., Jia, X.: Work-integrated learning in accountancy at Australian universities-forms, future role and challenges. Acc. Educ. **28**(1), 1–24 (2019)
3. Doolan, M., et al.: The benefits and challenges of embedding work integrated learning: a case study in a university education degree program. Aust. J. Teach. Educ. (Online) **44**(6), 91–108 (2019)
4. Uclaray, A., et al.: Social work students' challenges in flexible learning and implications for social work education: a study in Bicol, Philippines. Asean Soc. Work J. 13–27 (2023)
5. Mathieu Le Thuaut. Le boom des formations en alternance. leparisien.fr (2005). https://www.leparisien.fr/seine-et-marne-77/le-boom-des-formations-en-alternance-09-07-2005-2006110150.php. Accessed 13 Mar 2025
6. Ministère du Travail, de la Santé, des Solidarités et des Familles. Les chiffres de l'apprentissage en 2022. Portail de l'Alternance (2024). https://www.alternance.emploi.gouv.fr/actualites/les-chiffres-de-lapprentissage-en-2022. Accessed 13 Mar 2025
7. Assemblée Nationale. Etude d'impact sur le texte, n° 904. Assemblée nationale (2018). https://www.assemblee-nationale.fr/dyn/15/textes/l15b0904_etude-impact. Accessed 13 Mar 2025
8. Wollny, S., et al.: Are we there yet?-a systematic literature review on chatbots in education. Front. Artif. Intell. **4**, 654924 (2021)
9. Alfehaid, A., Hammami, M.A.: Artificial intelligence in education: literature review on the role of conversational agents in improving learning experience. Int. J. **10**(3), 3121–3129 (2023)
10. Georgescu, A.-A.: Chatbots for education–trends, benefits and challenges. In: Conference proceedings of» eLearning and Software for Education «(eLSE), vol. 14, no. 02. Carol I National Defence University Publishing House (2018)

11. Pérez, J.Q., Daradoumis, T., Puig, J.: Rediscovering the use of chatbots in education: a systematic literature review. Comput. Appl. Eng. Educ. **28**(6), 1549–1565 (2020)
12. Ramandanis, D., Xinogalos, S.: Investigating the support provided by chatbots to educational institutions and their students: a systematic literature review. Multimodal Technol. Interact. **7**(11), 103 (2023)
13. Chamorro-Atalaya, O., et al.: Application of the chatbot in university education: a systematic review on the acceptance and impact on learning. Int. J. Learn. Teach. Educ. Res. **22**(9), 156–178 (2023)
14. Hogan, A., et al.: Knowledge graphs. ACM Comput. Surv. (CSUR) **54**(4), 1–37 (2021)
15. Qu, K., et al.: A survey of knowledge graph approaches and applications in education. Electronics **13**(13), 2537 (2024)
16. Wan, Z., et al.: Efficient large language models: a survey. arXiv preprint arXiv:2312.03863 (2023)
17. Kumar, P.: Large language models (LLMs): survey, technical frameworks, and future challenges. Artif. Intell. Rev. **57**(10), 260 (2024)
18. Wang, Y., et al.: Factuality of large language models: a survey. arXiv preprint arXiv:2402.02420 (2024)
19. Luo, L., et al.: Reasoning on graphs: faithful and interpretable large language model reasoning. arXiv preprint arXiv:2310.01061 (2023)
20. Touvron, H., et al.: Llama 2: open foundation and fine-tuned chat models. arXiv preprint arXiv:2307.09288 (2023)
21. Almazrouei, E., et al.: The falcon series of open language models. arXiv preprint arXiv:2311.16867 (2023)
22. Jiang, F.: Identifying and mitigating vulnerabilities in LLM-integrated applications. MS thesis. University of Washington (2024)
23. Achiam, J., et al.: GPT-4 technical report. arXiv preprint arXiv:2303.08774 (2023)
24. BigScience Workshop, et al.: Bloom: a 176b-parameter open-access multilingual language model. arXiv preprint arXiv:2211.05100 (2022)
25. Wang, B.: Mesh-Transformer-JAX: Model-Parallel Implementation of Transformer Language Model with JAX (2021). https://github.com/kingoflolz/mesh-transformer-jax
26. Choi, Y., et al.: EdNet: a large-scale hierarchical dataset in education. In: Bittencourt, I.I., Cukurova, M., Muldner, K., Luckin, R., Millán, E. (eds.) AIED 2020. LNCS (LNAI), vol. 12164, pp. 69–73. Springer, Cham (2020). https://doi.org/10.1007/978-3-030-52240-7_13

From Insights to Impact: An Agentic Learning Intelligence Framework for Student-Centred Success

Joel Weijia Lai(✉), Wei Qiu, and Fun Siong Lim

Institute for Pedagogical Innovation, Research, and Excellence,
Nanyang Technological University, 50 Nanyang Avenue, Singapore 639798, Singapore
{joellai,qiuwei,lim_fun_siong}@ntu.edu.sg

Abstract. Agentic artificial intelligence (AI) in education can reshape student success by fostering adaptive and personalized learning experiences. AI-powered systems are individually built to analyze academic records, career aspirations, and engagement with curricular and co-curricular activities to generate tailored recommendations that dynamically adjust learning pathways. While existing research has explored AI's role in education, its full potential in holistic student development remains under-examined. This practitioner report presents an agentic framework that leverages advances in learning analytics to propose a practical application of agentic AI, emphasizing its future role in integrating academic support, skill-building opportunities, and career preparation. By continuously refining recommendations based on real-time progress, AI-powered systems will enhance student engagement, optimize resource accessibility, and ensure education remains flexible and responsive to evolving career pathways. This ongoing work aims to provide perspective and advance discussions on the future of AI in education, positioning agentic AI as a transformative proponent for student-centered success.

Keywords: Agentic Artificial Intelligence · Learning Analytics · Personalized Pathways · Goal-Oriented Recommendations

1 Introduction

Artificial intelligence (AI) is transforming education by reshaping how students engage with learning and how institutions provide academic support. Learning analytics is one of its most impactful applications, which leverages data to enhance student success [4]. Traditionally, learning analytics has focused on tracking academic performance through structured data, such as grades, attendance, and participation metrics [12,20]. Early warning systems [14], predictive analytics [18], skills identification [7], and adaptive learning platforms [10] have enabled educators to make data-informed decisions to support students at risk

A. I. Cristea et al. (Eds.): AIED 2025, CCIS 2590, pp. 109–118, 2025.
https://doi.org/10.1007/978-3-031-99261-2_10

of failing or dropping out. However, these systems primarily function retrospectively, predicting risks based on past performance, and often do not consider students' personal learning goals. Since education is a dynamic and evolving process, AI-powered learning analytics must move beyond prediction-based intervention models toward a proactive, student-centered approach that continuously adapts to learners' evolving goals and needs.

Generative AI introduces a significant shift in learning analytics by enabling real-time analysis, guidance, and personalization of student experiences. Context-aware large language models and chain-of-thought reasoning play a crucial role in this transformation by synthesizing learning content, adapting instructional materials, and generating tailored goal-oriented recommendations [2]. Despite growing interest in generative AI for education, much existing research has focused on content generation, automated assessment, and personalized tutoring. Learning analytics remains primarily confined to academic monitoring and predictive modeling for student retention [2]. Recent studies have begun to explore generative AI's potential in understanding learning motivation and engagement [3,5,6], but a more comprehensive integration of AI in student development is still lacking.

Agentic AI is an emerging intelligence system demonstrating autonomy, goal-directed behavior, and the ability to take independent actions to achieve objectives. These systems can plan, make decisions, and execute tasks without constant human intervention [1]. Agentic AI expands the possibilities of AI in education by shifting from passive analytics to active guidance, empowering students with personalized learning trajectories that evolve in response to their real-time progress, skills development, and career aspirations. Unlike traditional systems that primarily identify struggling students for remedial interventions, agentic AI dynamically adjusts learning pathways for all students based on academic history, engagement with curricular and co-curricular activities, and emerging career trends. However, existing frameworks often fail to integrate essential non-academic factors such as internships, mentorship programs, and experiential learning, all critical for workforce readiness [13,17,19]. By advancing discussions on agentic AI, this paper highlights its potential to support student success through personalized, adaptive, and holistic learning experiences, positioning AI as a key enabler of lifelong learning and professional growth.

2 Agentic Artificial Intelligence Implementation in Higher Education Practice

Learning analytics at our institution, Nanyang Technological University, Singapore, is performed as a data service through cleaning, engineering, modeling, analysis, visualization, and deployment. This is done by integrating and leveraging leading data science and machine learning technologies, including Denodo, Dataiku, Qliksense, Snowflake, and Sreamlit. These data services cater for the learning needs of 25,000 undergraduate students, pulling information from five key data hubs:

1. The Curriculum Hub contains curricular and co-curricular activity data, offering structured academic content and additional learning experiences beyond the classroom. This information might be used to recommend academic courses, workshops, or extracurricular engagements that align with the user's educational goals.
2. The Education Hub includes diverse continuous learning activities such as student engagement with online interactive exercises, assignments, and announcements, as well as course-specific resources such as videos, textbooks, and PowerPoint slides that would enhance students' understanding of the course.
3. The Career Hub aggregates career preparation opportunities, including internships, mentorship programs, and industry-specific training. This information might help students align their academic experiences with career aspirations and gain exposure to professional environments.
4. The Library/Resource Hub provides access to digital libraries and online learning resources, including books, research articles, and multimedia materials. These resources might support self-directed learning and provide additional context to academic coursework.
5. The Student Academic Hub serves as a personalized knowledge repository that is the foundation of a student profile, which includes information such as their academic background, performance, course registration history, program, and academic requirements. These historical data are essential for refining future recommendations and adapting learning pathways to ensure continuous student growth.

2.1 Existing Work 1: Early Alert for Learning Intervention

This project aims to proactively identify students who may require academic support before the commencement of a new semester. By leveraging machine and deep learning models—specifically SVM, logistic regression, LSTM, and Short-term gated LSTM—our predictive learning analytics model analyzes historical performance data from the aforementioned student academic hub to predict students at risk of failing and pinpoint specific courses where intervention might be necessary [14,15] (See data pipeline Fig. 1). The implementation involves generating a list of students at risk of academic warning. This list is provided to student care managers through a dedicated dashboard. These managers may contact the identified students to offer academic, financial, psychological, or social support. Its primary aim is to analyze student engagement and learning behaviors using data-driven insights, ultimately improving teaching effectiveness and student outcomes by:

- Understanding Student Learning Patterns: By analyzing data such as time spent on course materials, participation in online discussions, and frequency of resource access, the system identifies trends, such as both long-term academic trajectories and short-term performance fluctuations, in how students engage with their studies.

- Providing Timely Interventions: Leveraging predictive data analytics, the system identifies students who may struggle academically before their performance declines significantly. The model detects patterns associated with academic risk by analyzing historical academic records and engagement metrics.

Trialed across six cohorts in five schools (nearly 14,000 students), it achieved over 70% true-positive rates. Reports from student care managers indicate that early intervention has improved academic outcomes, with some students showing significant GPA improvements following personalized support.

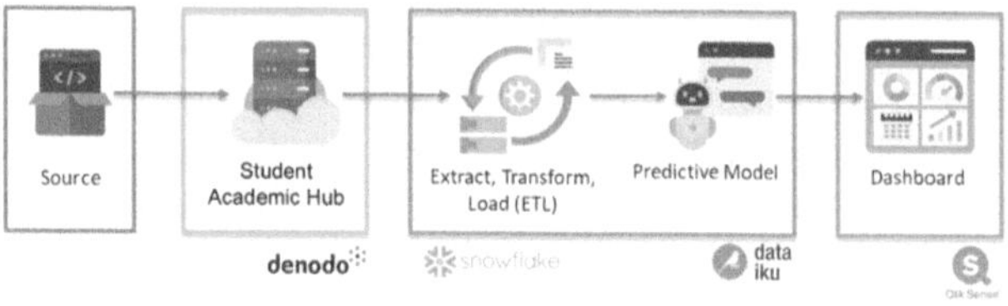

Fig. 1. The automated model pipeline for grade prediction to highlight potential at-risk students. Can be augmented with, and replicated for other forms of academic prediction.

2.2 Existing Work 2: Career Preparation with Skills Profile

Additionally, we have implemented an AI-enabled career preparation platform to enhance university students' awareness of their career readiness and skill development (Fig. 2) based on a needs analysis from a previous study [8]. It acts as an intervention that allows students to self-assess their current competencies and identify gaps in relation to industry requirements. The platform aims to provide students with tailored insights into their suitability for specific roles and to identify skill gaps based on their skill profile and career aspirations. Its key goal is to empower students to map their career goals, develop essential skills, and discover suitable internships, bridging the gap between their current skill set and future aspirations. The platform is a prototype of the career hub and assists students in identifying their skills and skills gaps through:

- AI-powered Skill Profiling: The platform automatically generates a skills profile by analyzing uploaded resumes, showcasing the student's relevant competencies based on AI methods and tagging.
- Comparison with Role Requirements: When students explore internship opportunities, the platform displays the skills required for each role and highlights the skills they already possess. This visual comparison helps students understand which skills they have and which are needed for their desired roles.

- Identification of Skills for Development: The platform automatically shows skills for development. This feature directly addresses the students' skill gaps in relation to their career aspirations.

Fig. 2. Skills in demand for a specific internship, skills possessed by the user are identified from their resume by AI and highlighted in green. This allows the user to identify skill gaps and work on them. (Color figure online)

From our existing works, we know that learning analytics can help inform through predictive means in a student's academic and career-preparation journey. However, they currently operate in silos, answering to specific needs.

We now outline our proposed transformation into an agentic system. By embedding AI-powered learning analytics within student-facing platforms, learners can receive personalized recommendations on course selection, skill development, and career planning based on their learning goals. This shift from an institutional-centric approach to a student-centered model aligns with the vision of agentic AI, empowering students to take ownership of their learning pathways. This expansion will require robust AI-powered interfaces that translate complex learning analytics insights into actionable recommendations tailored to each student. The AI Orchestrator performs this task while handing the agency of goal-setting and learning pathways to the student.

As illustrated in Fig. 3, the agentic AI framework operates as an intelligent intermediary between the user and multiple academic and career-related information hubs. This AI-powered system synthesizes data from various sources to provide highly personalized learning recommendations based on a user's needs, preferences, and academic or career aspirations. At the core of this framework is the AI Orchestrator, which functions as the central intelligence responsible for dynamically processing and generating personalized learning pathways. Students interact with this system via a front-end interface. They are encouraged to select areas they would like to learn and the cadence (such as weekly, fortnightly, or monthly) they want AI recommendations to be delivered to them. In response, the AI Orchestrator will communicate with specialized AI agents

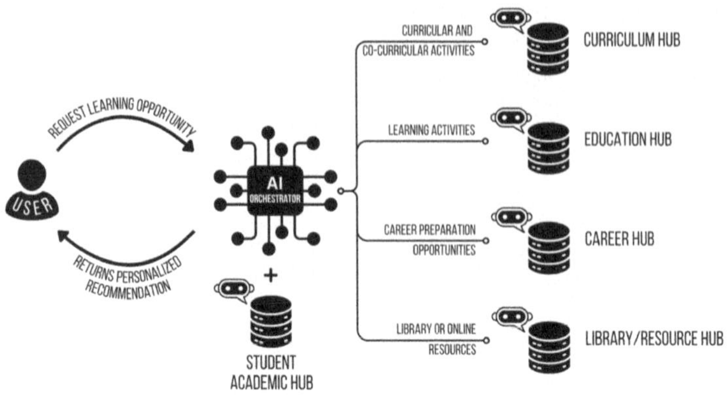

Fig. 3. Agentic AI Framework for Personalized Learning – The AI Orchestrator integrates data from multiple hubs to provide tailored learning recommendations based on user requests.

attached to each hub to fulfill the request. In our proposed architecture, generative AI models embedded within individual hubs enable localized knowledge generation, while the agentic orchestrator synthesizes their outputs to achieve a higher-order, adaptive decision-making capability.

For example, students set learning goals ("I want to be in data science") and when they want recommendations (such as "I want recommendations every fortnight") to be pushed to them. Based on this information and their academic history from the student academic hub, the AI Orchestrator would be activated and seek inputs from the specialized AI agents responsible for each hub. If the request is near the course registration period, the AI agent for the curriculum hub might return information about elective courses that the student can choose to support their learning goals. If the request is during the term, the AI agent curriculum hub could return the available co-curricular activities related to the learning goals instead. Similarly, the AI agent for the education hub may return some practice questions for the courses if the request is near the examination period. If there are some groundbreaking papers on data science, the library hub agent may also return a summary of the latest study in this area. Before the term break, the AI agent of the career hub may alert students of internship opportunities. Taking in all the responses from each hub, the AI Orchestrator collates the responses from each AI agent and then packages the recommendations to the students.

3 Reflection on Challenges and Opportunities Associated with Implementation

Implementing agentic AI in education has several challenges that must be addressed to realize its full potential. One of the most significant hurdles is the labor-intensive process of gathering and integrating data from various sources. Educational institutions collect vast amounts of data from academic records, student engagement metrics, career aspirations, and extracurricular activities. However, this data is often scattered across different systems, managed by various departments, and stored in inconsistent formats. Bringing these disparate data sources into a centralized and structured data warehouse requires extensive collaboration with multiple data owners, which can be time-consuming and complex. Once the data is collected, the next challenge is building systems that can effectively integrate and communicate with an AI Orchestrator. Many institutions rely on legacy systems not designed for interoperability with AI-powered platforms. Developing a seamless integration between learning management systems, student information databases, and career guidance tools demands significant technical investment and careful planning. Ensuring that these systems work together efficiently is critical for AI-powered recommendations to be accurate and relevant.

Beyond technical challenges, ethical considerations and data privacy concerns must be carefully managed. AI-powered education systems depend on vast amounts of student data, raising critical questions about security, confidentiality, and responsible AI use. Institutions must implement strict data governance and use of AI policies to protect student information while ensuring compliance with privacy regulations. Furthermore, trust remains a significant barrier to AI adoption. Some students may be hesitant to rely on AI-generated recommendations due to concerns about bias or a lack of understanding of how the system operates. Henceforth, we need to educate students that they have control over their learning decisions and can give feedback to the system should they find the AI recommendations not helpful.

Despite these challenges, the successful implementation of agentic AI presents significant opportunities for transforming education. Once the data integration and system interoperability issues are resolved, AI can provide a truly personalized learning experience. By analyzing a student's academic history, engagement patterns, and career aspirations, AI can tailor educational content to fit their unique needs. Unlike traditional learning analytics that primarily focus on identifying struggling students for intervention, agentic AI actively supports all learners by dynamically adapting content and learning pathways. Another significant opportunity is the scalability of AI-powered education [16]. Due to resource constraints, personalization has traditionally been challenging to implement at large institutions. However, AI enables scalable, tailored learning experiences, ensuring every student receives support that aligns with their goals [11]. Ultimately, harnessing AI-powered insights allows institutions to create a more equitable and effective learning environment, ensuring that students from all backgrounds have access to high-quality education and support.

4 Future Steps

The next phase of this project will focus on structuring data to enhance the intelligence, adaptability, and privacy of the agentic AI system. Instead of simply integrating and storing data in a centralized warehouse, the emphasis will shift toward organizing information using knowledge graphs within distributed hubs. Each hub will be accessed only by a specialized AI agent, ensuring that data owners maintain control over their respective datasets. This separation of responsibilities enhances data protection and aligns with governance best practices by preventing any single entity from holding comprehensive user profiles. The AI system can better understand relationships between different learning components by structuring academic records, student engagement metrics, career aspirations, and extracurricular activities into localized semantic networks. The orchestrator model will not store or directly manage this data but synthesize insights from multiple agents, each responsible for its domain-specific hub. This approach ensures that data privacy is maintained while allowing for highly contextualized and dynamic recommendations. The system can provide personalized learning journeys without compromising data sovereignty by mapping connections between coursework, skills, career pathways, and real-world job requirements.

We will refine AI models that leverage graph neural networks (GNNs), natural language to SQL generation, and semantic reasoning to make these dynamic recommendations more adaptive and explainable. GNNs are a class of neural models designed to operate on graph-structured data, making them suitable for reasoning over interconnected concepts or relationships [9]. Instead of treating student progress as a linear sequence of courses and grades, the AI will interpret learning as a network of interrelated concepts, competencies, and experiences about their learning goals. This will improve the granularity of goal-oriented recommendations and the system's ability to explain why a particular suggestion is relevant. Enhancing explainability will be a key priority, helping students understand how and why AI-generated guidance is tailored to them while maintaining trust in the system's governance framework. We intend to pilot this with at least one of the hubs and one of the university's colleges in the coming year.

5 Conclusion

Implementing agentic AI represents a transformative shift in higher education, moving beyond traditional learning analytics to offer proactive and personalized student support. By integrating AI-powered learning analytics with career and skill-building recommendations, this system optimizes student engagement and ensures that education remains adaptive to evolving career pathways. While challenges exist, including ethical considerations and the need for institutional integration, ongoing research, and strategic partnerships will be essential to refining the AI framework and expanding its implementation.

Agentic AI has the potential to revolutionize education by empowering students with personalized learning trajectories that evolve in real time. By shifting from reactive intervention models to proactive guidance, AI-powered systems foster a more inclusive, adaptable, relevant, and student-centered approach to learning. As educational institutions embrace this technology, continued research and collaboration will ensure its long-term success and sustainability. By advancing the integration of agentic AI in education, which leverages and agglomerates data and resources uniquely available within the universities, we prepare students for lifelong success and ensure the relevance of university education in an ever-changing world.

Disclosure of Interests. The authors declare that they have no conflict of interest.

References

1. Acharya, D.B., Kuppan, K., Divya, B.: Agentic AI: autonomous intelligence for complex goals–a comprehensive survey. IEEE Access **13**, 18912–18936 (2025). https://doi.org/10.1109/access.2025.3532853
2. Bobula, M.: Generative artificial intelligence (AI) in higher education: a comprehensive review of challenges, opportunities, and implications. J. Learn. Dev. High. Educ. **30** (2024). https://doi.org/10.47408/jldhe.vi30.1137
3. Collie, R.J., Martin, A.J.: Teachers' motivation and engagement to harness generative AI for teaching and learning: the role of contextual, occupational, and background factors. Comput. Educ. Artif. Intell. **6**, 100224 (2024). https://doi.org/10.1016/j.caeai.2024.100224
4. Giannakos, M., et al.: The promise and challenges of generative AI in education. Behav. Inf. Technol. 1–27 (2024). https://doi.org/10.1080/0144929x.2024.2394886
5. Guo, J., Ma, Y., Li, T., Noetel, M., Liao, K., Greiff, S.: Harnessing artificial intelligence in generative content for enhancing motivation in learning. Learn. Individ. Differ. **116**, 102547 (2024). https://doi.org/10.1016/j.lindif.2024.102547
6. Hmoud, M., Swaity, H., Hamad, N., Karram, O., Daher, W.: Higher education students' task motivation in the generative artificial intelligence context: the case of chatgpt. Information **15**(1), 33 (2024). https://doi.org/10.3390/info15010033
7. Lai, J.W., Zhang, L., Chan, Y.S., Sze, C.C., Lim, F.S.: Compelling educational offerings: a study on the efficacy of skills identification platforms with course descriptions. In: INTED2024 Proceedings. INTED2024, vol. 1, pp. 2553–2561. IATED (2024). https://doi.org/10.21125/inted.2024.0709
8. Lai, J.W., Zhang, L., Sze, C.C., Lim, F.S.: Learning analytics for bridging the skills gap: a data-driven study of undergraduate aspirations and skills awareness for career preparedness. Educ. Sci. **15**(1), 40 (2025). https://doi.org/10.3390/educsci15010040
9. Li, X., Tian, Y., Ji, S.: Semantic- and relation-based graph neural network for knowledge graph completion. Appl. Intell. **54**(8), 6085–6107 (2024). https://doi.org/10.1007/s10489-024-05482-2
10. Morze, N., Varchenko-Trotsenko, L., Terletska, T., Smyrnova-Trybulska, E.: Implementation of adaptive learning at higher education institutions by means of Moodle LMS. J. Phys. Conf. Ser. **1840**(1), 012062 (2021). https://doi.org/10.1088/1742-6596/1840/1/012062
11. Ng, S.H.S., Lai, J.W.: AI-augmented heutagogy for higher education (2025). https://doi.org/10.31219/osf.io/nxgau_v1
12. Pei, B., Xing, W., Wang, M.: Academic development of multimodal learning analytics: a bibliometric analysis. Interact. Learn. Environ. **31**(6), 3543–3561 (2021). https://doi.org/10.1080/10494820.2021.1936075

13. Popli, N.K., Singh, R.P.: Enhancing academic outcomes through industry collaboration: our experience with integrating real-world projects into engineering courses. Discov. Educ. **3**(1) (2024). https://doi.org/10.1007/s44217-024-00300-w
14. Qiu, W., Khong, A., Supraja, S., Tang, W.: A dual-mode grade prediction architecture for identifying at-risk students. IEEE Trans. Learn. Technol. **17**, 803–814 (2024). https://doi.org/10.1109/tlt.2023.3333029
15. Qiu, W., Supraja, S., Khong, A.W.H.: Toward better grade prediction via A2GP - an academic achievement inspired predictive model. In: Mitrovic, A., Bosch, N. (eds.) Proceedings of the 15th International Conference on Educational Data Mining, pp. 195–205. International Educational Data Mining Society, Durham, United Kingdom (2022). https://doi.org/10.5281/zenodo.6852984
16. Saragih, R.I.E.: AI-powered education: Transforming learning through personalized and scalable solutions. Int. J. Inf. Syst. Innov. Technol. **3**(2), 1–9 (2024). https://doi.org/10.63322/qm9dk118
17. Sehgal, G.: AI & agentic AI in education: shaping the future of learning (2024). https://www.meratutor.ai/blog/agentic-ai-in-education/. Accessed 12 Feb 2025
18. Sghir, N., Adadi, A., Lahmer, M.: Recent advances in predictive learning analytics: a decade systematic review (2012–2022). Educ. Inf. Technol. **28**(7), 8299–8333 (2022). https://doi.org/10.1007/s10639-022-11536-0
19. Spanjaard, D., Hall, T., Stegemann, N.: Experiential learning: helping students to become 'career-ready'. Australas. Mark. J. **26**(2), 163–171 (2018). https://doi.org/10.1016/j.ausmj.2018.04.003
20. Ye, D.: The history and development of learning analytics in learning, design, & technology field. TechTrends **66**(4), 607–615 (2022). https://doi.org/10.1007/s11528-022-00720-1

Pilot Study of Self-explanation Based Automate Stuck Point Detection and Personalized Feed-Back Recommendation by Educational eXplainable AI Tool in Middle School Math Classes

Kyosuke Takami[1](✉), Ryosuke Nakamoto[2], Yiling Dai[3], Brendan Flanagan[2], and Hiroaki Ogata[2]

[1] Osaka Kyoiku University, Osaka, Japan
takami-k75@cc.osaka-kyoiku.ac.jp
[2] Kyoto University, Kyoto, Japan
[3] Hiroshima University, Hiroshima, Japan

Abstract. In the age of AI in education, the utilization of educational data by AI to provide individually optimized feedback to learners is crucial. Furthermore, understanding how this can be implemented in actual educational settings is essential for envisioning a future where humans co-evolve with AI. In this study, we implemented our recently developed self-explanation based eXplainable AI learning system in real-world settings with first-year middle school students to investigate how AI learning systems can be put into practice. Our system automatically detects stuck points and recommends quizzes based on the learner's self-explanation text data and pen stroke data. Over a period of approximately one month, we gradually introduced this system to 80 students in a mathematics class. The results showed that the system was capable of automatically detecting stuck points from self-explanations and recommending related problems as personalized feed-back.

Keywords: Self-explanation · Explainable AI · K12 · mathematic · Recommendation

1 Introduction

In the age of AI in education, the utilization of educational data by AI to provide individually optimized feedback to learners is crucial. This paper introduces a symbiotic learning system that not only adapts to the learner's performance but also promotes mutual understanding through the explanations by both the learner and the system. We call this system Educational eXplainable AI Tools (EXAIT) [1], and it aims to combine the benefits of both types of explanations in education into a single learning tool that can co-evolve symbiotically through the learner's self-explanation and AI-generated explanation. The student completes a mathematics question, using a stylus pen to input their working out and final answer into the system for evaluation. The system then prompts

A. I. Cristea et al. (Eds.): AIED 2025, CCIS 2590, pp. 119–128, 2025.
https://doi.org/10.1007/978-3-031-99261-2_11

the student to self-explain their answer process by replaying it interactively and annotating points in time to indicate the knowledge applied to overcome sub-problems. Time series analysis is then applied to the self-explanation and answer process data to detect stuck points, that could indicate problems with dependent or related knowledge. This AI system was implemented in a mathematics class for first-year middle school students to assess the types of self-explanatory texts used and the feasibility of providing individualized, optimal feedback.

2 Related Work

2.1 Explainable AI in Education

In recent years, eXplainable AI (XAI) has begun to attract attention in the education field due to emerging concerns about fairness, accountability, transparency, and ethics [2]. XAI is one of the emerging methods for increasing trust in AI systems, which facilitates the use of methods that "enable a human user to understand, appropriately trust, and effectively manage the emerging generation of artificially intelligent partners" [3]. Interpreting the decision process of the model and thereby providing explanations can be expected to have a positive impact on student achievement by increasing student buy-in and trust in the AI. For teachers as well, by presenting an explanation of why the question was recommended to improve academic achievement, they can show what explanatory methods are effective in improving academic achievement.

2.2 Self-explanation

Many studies suggest that self-explanation has learning benefits [4] and is often combined with or an integral part of other instructional designs, such as working examples and control case instruction [5]. However, it has been noted that it is still unclear whether the use of self-explanation is the most effective intervention and use of learners' time [6, 7] in mathematics [8, 9]. To deal with these challenges, educational technology has been applied to self-explanatory learning. For example, the iSTART Tutoring System was developed to support paraphrasing during reading by analyzing students' self-explanations within the context of the target text. iSTART combines NLP methods for self-explanation analysis, scaffolding instructions, and self-explanation practice to develop the necessary skills to it promotes [10]. An automated evaluation algorithm was used to score student self-explanations based on the relevance of the topic to the target text, giving higher scores to those that demonstrated strong consistency and lower scores to those that were off-topic or short in length. This system was found to be effective in supporting self-explanation in texts from a variety of disciplines [11]. In contrast, our Educational eXplainable AI tool (EXAIT) system aims to use self-explanation analysis to inform learner models about their strengths and weaknesses, rather than relying solely on automated evaluation of the content. The analysis of self-explanation is intended to influence the recommendation of learning paths for the student.

2.3 Explainable Recommendation

AI recommender systems are everywhere in our lives. For example, Amazon uses Collaborative Filtering to recommend products [12] and Netflix uses deep learning to recommend movies [13]. In this research field of recommendation mainly in e-commerce, explainable recommendations, which provide explanations about why an item is recommended, have received much attention for improving users' transparency, persuasiveness and trustworthiness [14]. Based on these studies, it seems that in education, explanations from the learning system may provide additional benefits to students who learn from the explanations given by the system. Prior research on intelligent tutoring systems has shown that student motivation in system-based self-regulated learning is improved by prompting and feedback mechanisms, leading to higher achievement [15].

In this research, we proposed a cycle that detects stuck points from self-explanatory text data and pen-stroke data, recommends problems related to those stuck points, and generates explanations of the learner's comprehension of those problems. We conducted evaluations by having first-year middle school students use this cycle in the context of mathematics education. The research questions are as follows:

RQ: *Is it possible to implement an AI system in a middle school educational setting, detect stuck points from accumulated self-explanatory data, recommend related problems with explanations, and provide individualized, optimal feedback?*

3 Methods

3.1 Overview of AI Supported Learning Cycle.

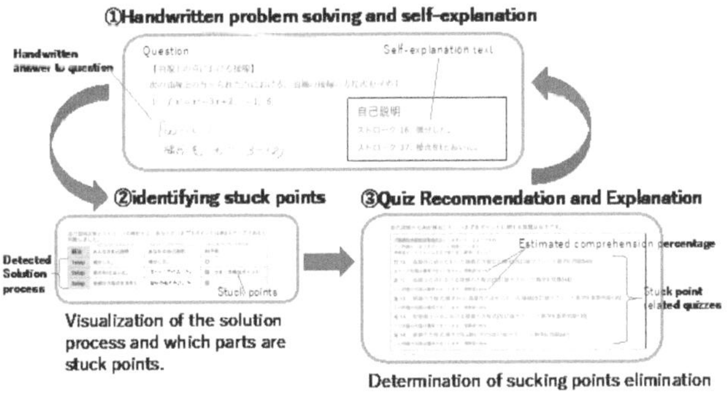

Fig. 1. An overview of the learning cycle proposed in this study.

① Learners use a tablet stylus pen to solve problems on material viewer. Pen strokes are recorded, and the thought process during problem-solving is accumulated as a learning log in the form of videos. Saved videos are played back, and learners provide explanations of their thought processes using a keyboard. This helps improve logical thinking, expression skills, and metacognitive abilities.

② AI analyzes handwritten data and self-explanatory text data to automatically identify stumbling points.
③ Quizzes tailored to the stuck points are recommended. Explanations for why these problems are recommended are also provided, allowing learners to engage with the problems with a sense of understanding.

By repeating the cycle from ① to ③, learners can continuously engage with the material in a way that makes sense to them, leading to an improvement in their understanding. Furthermore, AI learns and evolves from the data generated by learning activities, including text, handwritten data, correct/incorrect answers, thereby enhancing recommendation accuracy and providing more satisfactory explanations. In this way, both AI and humans evolve together (Fig. 1).

3.2 Data Collecting System

The Educational eXplainable AI tools (EXAIT) in this paper was built on the learning system framework, which was developed to support the distribution of learning materials, collection and automated analysis of learning behavior logs in an open and standards-based approach. The main components of the framework are: Moodle LMS which acts as a hub for accessing various courses; the reading system for learning material and quiz exercise distribution; an LRS for collecting learning behavior logs from all of the components; and the learning analytics dashboard to provide feedback to students, teachers and school administrators. This framework enables us to collect and analyze learning behaviors in real time and provide feedback to stakeholders. Quiz books used in the mathematics classes were uploaded to the reading system and multiple-choice quiz questions were created to enable the collection of answers in learning log data.

3.3 Participants

Eighty students from two math classes at a middle school participated for this study. These classes received the same coursework from the same teacher. The teacher assigns a slightly challenging applied problem related to the week's learning content as a self-explanatory task every weekend. Each week, one question is assigned, and students are required to solve it using a stylus pen on the learning system. After solving the problem, they must provide self-explanation by text input, explaining how they arrived at their solution while watching a playback of their solution process video on the system. The first week was dedicated to practice assignments, where students used a stylus pen to solve tasks provided by the system. They practiced inputting self-explanations in text format while watching the solution process video. Starting from the second to the fourth week, regular assignments were given, and the AI's stumbling point detection feature was utilized in the fourth week. Note: Students were not obligated by the teacher to solve recommended quizzes related to detected stuck points.

3.4 Collected Data

We collected the log data from 4-week weekend assignments in middle school mathematics classes during winter semester from January 20th to February 17th, 2023. This log data included students' pen stroke data, self-explanation text data, quiz clicked data and answered data right or wrong to quizzes.

3.5 Data Processing for Stuck Point Related Recommendations

Figure 2 shows the overview of the data processing for stuck points related quiz recommendation. Stuck points are automatically detected through self-explanatory text data and handwritten stroke data by using BERT clustering algorithm [16]. The text of the detected stumbling points is vectorized through morphological analysis. Vectors, weighted by TF-IDF, are computed to represent the relationships between quizzes and the mathematical concepts contained within them [17]. Cosine similarity is calculated from these two vectors, and the top 5 questions with high similarity are recommended. The comprehension level of the recommended questions is estimated based on the student's past quiz performance data (correct/incorrect), and the understanding level of essential knowledge that may be lacking is displayed as recommended questions. Students are encouraged to continue solving these recommended questions until their understanding reaches 100%. For estimation of their comprehension level, we used Python Library of Bayesian Knowledge Tracing Models (BKT) [18]. BKT model has been used to model student knowledge in various educational systems, including tutors for reading skill [19], computer programming [20] and explainable recommendation in mathematics [21, 22].

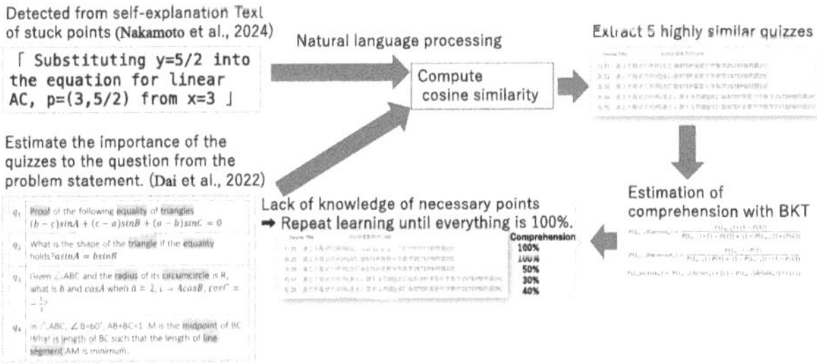

Fig. 2. Data processing for Stuck points related quiz recommendation.

4 Result

4.1 Self-explanation Text Analysis

Table 1 illustrates the total number of characters in the self-explanations for each week. In the week 1, students were trained in using the system during class, guided by the teacher, to solve problems with a stylus pen and enter self-explanations using the keyboard. The reason for the higher word count in self-explanations during the second week is likely due to the complexity of the problem-solving process, which required more words for self-explanation. Alternatively, it's possible that students took more time and care

Table 1. The total number of characters in self-explanations for each week.

	Total self explanation	number of sentences	Mean	Std
Week 1 (Training)	1921	127	15.13	9.394
Week 2 (Assignment 1)	5636	340	16.58	13.62
Week 3 (Assignment 2)	1788	121	14.77	11.29
Week 4 (Assignment 3)	2326	142	16.38	12.05
Whole Weeks	11671	730	15.99	

(AI stuck point detection and recommendation were used in Week 4)

when working on their assignments at home, leading to the inclusion of more extensive self-explanations.

Next, we will examine whether AI was able to detect stuck points in individual learning processes from such diverse self-explanatory texts and whether it could recommend problems related to them.

4.2 Detected Stuck Points

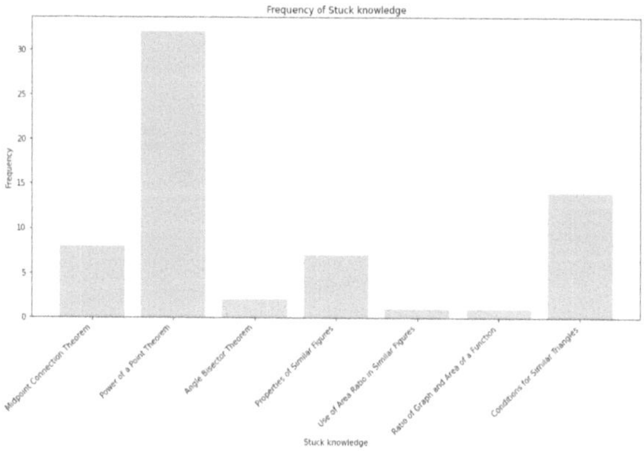

Fig. 3. The diversity of stuck points among individual learners.

During the final week, which was the fourth week, stuck points for each student were detected using self-explanation text and pen stroke data from the self-explanation tasks, and relevant problems were recommended. Thirty students used the AI-based stuck point detection function from self-explanation, and a total of 56 stumbling points were detected (Table 2). Figure 3 shows a histogram of which knowledge points were stuck knowledge blocks. As you can see, the power of a point theorem was the most frequently detected, and the second most frequently detected was conditions for similar triangles, but it is clear that a wide variety of stuck points were detected from the students' various self-explanation data.

Table 2. The statics of detected stuck points

Total detection	Number of user	Mean of detection per one user	Min	Max
56	30	2.13	1	13

In this AI system, it is possible to detect at which step of the answer process the student was struggling. Table 3 provides examples of self-explanation steps where stumbling points were detected. Self-explanations were input at each step, and it was determined at which step the stumbling occurred. Figure 4, which shows a histogram of the steps in the solving process, shows that most students did not detect any stumbling points, but step 3 was the most frequently detected stumbling point in any of the steps 1–5. These results suggest that from the self-explanations, each learner was stumbling at different positions.

Table 3. Examples of self-explanation steps

	Self-explanation
Step 1	Let X-π be A and y-x be B. Find the sum and difference
Step 2	Let the parentheses be A and B, respectively
Step 3	First, solve the equation in (), then move on to the equation for x + y. I'll make it
Step 4	You can use the method of addition and subtraction to explain how to make a combination
Step 5	Align to 7A

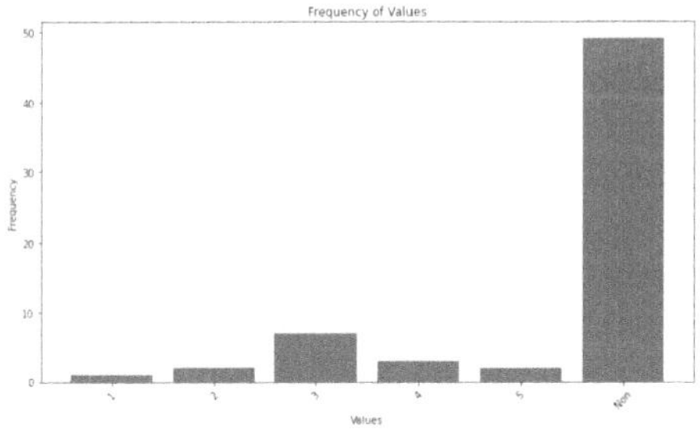

Fig. 4. The diversity of stuck points among individual learners.

4.3 Stuck Point Related Quiz Recommendation

We also explored stuck point related quiz recommendation. In this study, while teachers did not force students to solve the recommended problems, 12 students viewed the recommendation page where quizzes related to their stuck points were displayed, resulting in a total of 16 recommendation page views. Moreover, as illustrated in the example (Table 4), the students' comprehension was estimated using the Bayesian Knowledge Tracing (BKT) based on the data of correct and incorrect answers they had given to the problems up to that point, and this estimation was displayed.

Table 4. Example of stuck point related quizzes with comprehension percentage by BKT

Recommended quiz	Comprehension by BKT
17. Multiplication of monomials [Exercise STEP A Question 17]	1
26. Multiplication of monomials [Exercise STEP B Question 26]	0.770664497
18. Multiplication of monomials [Exercise STEP A Question 18]	1
19. Multiplication of monomials [Exercise STEP A Question 19]	0.765529376
20. Division of monomials [Exercise STEP A Question 20]	0.899723079

5 Discussion and Future Work

This study shows the potential for an Educational eXplainable AI tool (EXAIT) utilizing students self-explanation text and pen stroke data. Results show that based on individual learning logs related to varying self-explanations for each student, it is not only possible to detect specific stumbling points where each learner is facing challenges but also to recommend problems related to those points while providing explanations of how well the learner understands them. However, several challenges remain. One is the difficulty in detecting many stumbling points (Fig. 5, Non-labeled), likely due to the "cold start problem," where little self-explanatory data exists early on. Generating large amounts of such data with LLMs [23]. Second, this study found that students rarely engaged with the recommended problems, as the teacher indicated they were not required. Enhancing teachers' trust in AI and fostering more seamless integration into classroom practices could improve learning outcomes. Notably, as this study was limited to a one-month period, and future research should conduct longer-term evaluations to assess the sustained effectiveness and adaptability of the system. By addressing these challenges, we can create a foundation for effective human-centered AI [24] adoption and envision a future where humans co-evolve with AI.

Acknowledgments. This work was partly supported by JSPS Grant-in-Aid for Scientific Research (B) JP20H01722 and JP23H01001, (Exploratory) JP21K19824, (Early Career) JP23K17012, (A) JP23H00505, and NEDO JPNP20006.

Disclosure of Interests. The authors have no competing interests to declare that are relevant to the content of this article.

References

1. Ogata, H., Flanagan, B., Takami, K., Dai, Y., Nakamoto, R., Takii, K.: EXAIT: educational eXplainable artificial intelligent tools for personalized learning. Res. Pract. Technol. Enhanc. Learn. **19**, 019 (2023)
2. Khosravi, H., et al.: Explainable artificial intelligence in education. Comput. Educ. Artif. Intell. **3**, 100074 (2022)
3. Gunning, D.: Explainable artificial intelligence (XAI). Defense advanced research projects agency (DARPA), nd Web. 2, 1 (2017)
4. Bisra, K., Liu, Q., Nesbit, J.C., Salimi, F., Winne, P.H.: Inducing self-explanation: a meta-analysis. Educ. Psychol. Rev. **30**, 703–725 (2018)
5. Sidney, P.G., Hattikudur, S., Alibali, M.W.: How do contrasting cases and self-explanation promote learning? Evidence from fraction division. Learn. Instr. **40**, 29–38 (2015)
6. McEldoon, K.L., Durkin, K.L., Rittle-Johnson, B.: Is self-explanation worth the time? A comparison to additional practice. Br. J. Educ. Psychol. **83**, 615–632 (2013)
7. Rittle-Johnson, B., Loehr, A.M., Durkin, K.: Promoting self-explanation to improve mathematics learning: a meta-analysis and instructional design principles. ZDM Math. Educ. **49**, 599–611 (2017)
8. Hänze, M., Leiss, D.: Using heuristic worked examples to promote solving of reality-based tasks in mathematics in lower secondary school. Instr. Sci. **50**, 529–549 (2022)
9. Renkl, A.: Learning from worked-examples in mathematics: students relate procedures to principles. ZDM Math. Educ. **49**, 571–584 (2017)
10. Landauer, T.K., McNamara, D.S., Dennis, S., Kintsch, W.: Evaluating self-explanations in iSTART: comparing word-based and LSA algorithms. In: Handbook of Latent Semantic Analysis, pp. 239–254. Psychology Press (2007)
11. Jackson, G.T., Guess, R.H., McNamara, D.S.: Assessing cognitively complex strategy use in an untrained domain. Top. Cogn. Sci. **2**, 127–137 (2010)
12. Smith, B., Linden, G.: Two decades of recommender systems at Amazon.com. IEEE Internet Comput. (2017)
13. Amatriain, X., Basilico, J.: Recommender systems in industry: a netflix case study. In: Ricci, F., Rokach, L., Shapira, B. (eds.) Recommender Systems Handbook, pp. 385–419. Springer, Boston (2015)
14. Zhang, Y., Chen, X.: Explainable recommendation: a survey and new perspectives. Found. Trends® Inf. Retrieval **14**, 1–101 (2020)
15. Duffy, M.C., Azevedo, R.: Motivation matters: interactions between achievement goals and agent scaffolding for self-regulated learning within an intelligent tutoring system. Comput. Hum. Behav. **52**, 338–348 (2015)
16. Nakamoto, R., Flanagan, B., Dai, Y., Yamauchi, T., Takami, K., Ogata, H.: Integrating self-explanation and operational data for impasse detection in mathematical learning. Res. Pract. Technol. Enhanc. Learn. **20**, 19 (2024)
17. Dai, Y., Flanagan, B., Takami, K., Ogata, H.: Design of a user-interpretable math quiz recommender system for Japanese high school students. http://ceur-ws.org/Vol-3120/paper4.pdf. Accessed 08 May 2023
18. Badrinath, A., Wang, F., Pardos, Z.: pyBKT: Python implementation of Bayesian Knowledge Tracing and extensions. Github
19. Beck, J.E., Chang, K.-M.: Identifiability: a fundamental problem of student modeling. Plan. Perspect. **137**, 146 (2007)

20. Corbett, A.T., Anderson, J.R.: Knowledge tracing: Modeling the acquisition of procedural knowledge. User Model. User-Adapt Interact. **4**, 253–278 (1994)
21. Takami, K., Dai, Y., Flanagan, B., Ogata, H.: Educational explainable recommender usage and its effectiveness in high school summer vacation assignment. In: LAK22: 12th International Learning Analytics and Knowledge Conference, pp. 458–464. Association for Computing Machinery, New York (2022)
22. Takami, K., Flanagan, B., Dai, Y., Ogata, H.: Evaluating the effectiveness of Bayesian knowledge tracing model-based explainable recommender. Int. J. Distance Educ. Technol. (2024). https://doi.org/10.4018/ijdet.337600
23. Nakamoto, R., Flanagan, B., Yamauchi, T., Yilling, D., Takami, K., Ogata, H.: Enhancing automated scoring of math self-explanation quality using LLM-generated datasets: a semi-supervised approach (2023)
24. Amershi, S., et al.: Guidelines for human-AI interaction. In: Proceedings of the 2019 CHI Conference on Human Factors in Computing Systems, pp. 1–13. Association for Computing Machinery, New York (2019)

Building Effective Safety Guardrails in AI Education Tools

Hannah-Beth Clark, Laura Benton(✉), Emma Searle, Margaux Dowland, Matthew Gregory, Will Gayne, and John Roberts

Oak National Academy, Manchester, UK
{hannah-beth.clark,laura.benton,emma.searle,eng.mdowland, matthew.gregory,will.gayne,john.roberts}@thenational.academy

Abstract. There has been rapid development in generative AI tools across the education sector, which in turn is leading to increased adoption by teachers. However, this raises concerns regarding the safety and age-appropriateness of the AI-generated content that is being created for use in classrooms. This paper explores Oak National Academy's approach to addressing these concerns within the development of the UK Government's first publicly available generative AI tool - an AI-powered lesson planning assistant (Aila). Aila is intended to support teachers in planning national curriculum-aligned lessons that are appropriate for pupils aged 5–16 years. To mitigate safety risks associated with AI-generated content we have implemented four key safety guardrails: (1) prompt engineering to ensure AI outputs are generated within pedagogically sound and curriculum-aligned parameters; (2) input threat detection to mitigate attacks; (3) an Independent Asynchronous Content Moderation Agent (IACMA) to assess outputs against predefined safety categories; and (4) taking a human-in-the-loop approach, to encourage teachers to review generated content before it is used in the classroom. Through our ongoing evaluation of these safety guardrails, we have identified several challenges and opportunities to take into account when implementing and testing safety guardrails. This paper highlights ways to build more effective safety guardrails in generative AI education tools, including the ongoing iteration and refinement of guardrails, as well as enabling cross-sector collaboration through sharing both open-source code/datasets and learnings.

Keywords: AI · education · safety guardrails · moderation

1 Description of the AIED Implementation

1.1 Background

Over the last five years, the emergence of accessible generative AI has led to an explosion of tools with integrated AI within the education sector. Decisions on which content is appropriate and should or should not be taught in schools are currently set out by the UK Government through the national curriculum objectives and guidance. School leaders interpret and implement this within their schools by creating their own school-based curricula and policies, informing teachers' decisions and lesson planning.

A. I. Cristea et al. (Eds.): AIED 2025, CCIS 2590, pp. 129–136, 2025.
https://doi.org/10.1007/978-3-031-99261-2_12

There has been a massive increase in AI usage amongst teachers, since the launch of accessible generative AI tools like ChatGPT in November 2022, with almost 70% of teachers now using it in their work and one in ten secondary teachers using it during lessons [19]. For AI-generated content to be useful to teachers, it needs to align with the Government and schools' expectations of appropriate content. This is challenging due to the susceptibility of AI tools to bias, hallucinations, and inaccuracies and their ability to produce potentially harmful content [9, 12]. Generative AI will produce different responses each time it is asked a question, which makes testing the outputs of generative AI tools more challenging.

This area of technology is developing rapidly, and both research and legislation must keep up due to the associated risks of using the content produced in schools with children. The UK Government recently published guidance on the safety expectations for using Generative AI [7] in products used in education settings to support this work. However, this is guidance rather than legislation. It is up to individual organisations to interpret and implement effective guardrails within their product based on this guidance. They may also choose to evaluate the effectiveness of their guardrails, but this is not currently compulsory.

1.2 Oak National Academy and Aila

Oak National Academy[1] was established as a response to the COVID-19 pandemic, and became an Arm's Length Body (ALB) of the Department for Education in England in 2022, to improve pupil outcomes and close the disadvantage gap by providing teachers with high-quality curricula and resources. Oak have been working with a range of curriculum partners and their subject experts to create over 10,000 lesson resources for UK teachers of pupils aged 5–16 years, aligned to the national curriculum for England and exam boards. To ensure consistency across multiple lesson creators, we have distilled leading pedagogy and cognitive science research into a detailed curriculum and lesson design rubric. Specific principles, such as '*the prior knowledge starter quiz activates the necessary knowledge required to access the lesson*' or '*explanations are framed around small steps*' from this rubric guide lesson creators to ensure lessons are consistent and high-quality.

Using guidance from a number of sources, including the UK Government, the UK Council for Child Internet Safety, the British Board of Film Classification (BBFC), the BBC and the Pan European Game Information (PEGI) [2, 3, 6, 8, 16, 17], Oak has also created a content guidance rubric that lesson creators can use to highlight to teachers where lessons may require sensitivity (such as lessons on drugs, cyberbullying, domestic abuse, sex education and the Holocaust). This includes four overarching categories ('language and discrimination', 'upsetting content', 'nudity and sex' and 'physical activity or activities requiring equipment'), with a number of clearly defined subcategories in each.

This high-quality corpus of content, along with the codification of our definition of pedagogical quality, puts us in a unique position to innovate with AI, and we launched an AI lesson planning assistant (Aila) in September 2024. We have designed Aila to

[1] https://www.thenational.academy/

co-plan lessons with teachers, supporting them in creating high-quality lesson plans and resources for their classrooms. These are aligned with the national curriculum and can be personalised to their pupils based on locality, needs or interests. More details of our work to ensure the quality of our AI-generated content can be found in our recently published paper [4].

Aila was the first government-funded, publicly available generative AI tool to be launched in the UK, and as it has been designed to produce content for classrooms with pupils aged 5–16 years, safety is imperative. Whilst GPT-4o (the underlying large language model Aila uses at the time of writing) has in-built moderation and filtering mechanisms [15], given the education context of use, we did not feel this provided a sufficient safety net. Therefore, in addition to our existing content guidance rubric, we identified several other key AI safety guidelines which were available at the time of Aila's design [10, 14, 18]. As UNESCO outlines, "unwanted harms (safety risks), as well as vulnerabilities to attack (security risks) should be avoided and should be addressed, prevented and eliminated throughout the life cycle of AI systems to ensure human, environmental and ecosystem safety and security" [18].

Based on these guidelines, we have incorporated four additional safety guardrails. These are designed to block malicious usage and filter any inappropriate content based on a number of categories. These additional guardrails include explicit instructions within Aila's prompt, input threat detection, an independent asynchronous content moderation agent, and the teacher always being the human-in-the-loop. We discuss these in further detail below.

2 Reflection on the Challenges and Opportunities Associated with the Implementation

2.1 Our Four Stages of Safety

Prompt Engineering. We have written an extensive prompt (which is part of our open-source codebase), codifying Oak's definition of best practices in pedagogy and cognitive science according to the latest research in these fields [13]. Within this, we have also outlined what is and is not appropriate for Aila to create. This includes considering the pupil's age, subject, and topic being planned and working to ensure that content is aligned with the national curriculum and appropriate for use in schools. For example, when asked to plan a KS4 lesson on cannabis, Aila draws on the Relationships, Sex and Health Education (RSHE) curriculum [5] to suggest age-appropriate content covering the effects of cannabis and its legal and societal impacts.

Input Threat Detection. We have implemented an input threat detection layer, which monitors user inputs to check for inappropriate, manipulative, or malicious inputs, predominantly through prompt attacks, in order to reduce the risk of jailbreaking. Any input flagged as a threat will not be sent to the large language model (LLM) but will be flagged to our internal moderation systems, and any users who are repeatedly detected will be blocked.

Independent Asynchronous Content Moderation Agent (IACMA). We have built an independent, context-unaware AI agent to assess Aila's outputs. We used Oak's established content guidance rubric to design Aila's 'content guidance' categories and the HarmBench standardised evaluation framework [10] to define Aila's 'toxic' categories.

Our IACMA has no context—either of what Aila has been instructed to do or what the user has inputted—so neither can influence the moderation agent's judgements. The agent categorises content from Aila into three categories, each composed of further sub-categories.

- **Safe**: content is appropriate for use in classrooms with the specified age group;
- **Content guidance**: content is appropriate but may require additional considerations or sensitivity from the teacher when delivering. The topics include 'language and discrimination', 'upsetting content', 'nudity and sex' and 'physical activity or activities requiring equipment'.
- **Toxic**: inappropriate for use in classrooms (highly sensitive or harmful). This includes content on encouraging harmful behaviour (including self-harm and violence), illegal activity, or the creation of weapons or harmful substances.

Our IACMA scores each section of content produced by Aila with a justification against the predefined subcategory descriptions. We use a 5-point Likert scale to evaluate the severity of content within each category.

A score of 5 indicates 'no concern'. Within the 'content guidance' category, a score lower than this triggers the appropriate guidance being shown to the teacher. For any score lower than 5 in a 'toxic' category, the lesson will be blocked, and the user session will be ended. In this instance, an alert is sent to our AI team to notify them of toxic use, it is recorded for reporting purposes, and the lesson is no longer accessible to the user. Repeated attempts to plan toxic lessons will result in a user being blocked from using Aila to plan further lessons.

As we use AI to evaluate and make decisions on AI-generated content, the outcome is not always consistent. Whilst we always have a teacher-in-the-loop, Aila is being used to produce content taught to pupils as young as five, so the appropriateness of the content is imperative. We have designed our IACMA to be oversensitive. This may result in the 'content guidance' warning being shown and lessons being blocked as 'toxic' when not necessary, but we feel that this conservative approach is appropriate. As we continue to evaluate and iterate, we will improve the specificity of this agent to ensure that we minimise false positives that may affect user experience while maintaining high sensitivity, and therefore safety.

Human-in-the-Loop. Our tools and resources are designed for teachers who know their curriculum and their pupils best; therefore, our final safety guardrail is the human teachers themselves. We designed Aila so that the teacher leads the lesson planning process, with the AI suggesting content that they may want to include in their lesson. It is intentionally designed to always have a teacher between the AI-produced content and the pupils. Therefore, they are expected to review the content during the lesson creation and after download to ensure it is aligned with their school's policies and appropriate for their pupils.

2.2 Evaluating the Effectiveness of Our Safety Guardrails

Pre-launch Evaluation. Before launching Aila, we undertook a rigorous testing process to evaluate the initial effectiveness of our safety guardrails. To do this, we used the corpus of Oak lessons created by our expert lesson creators, which had already been assessed for sensitive content against these categories. We put these through our IACMA to ensure that the judgments it made aligned with those made by the humans who created and quality-assured the lesson. We also participated in red-teaming exercises [11] based on the HarmBench standardised evaluation framework [10], working with external experts who tried to use Aila to produce inappropriate content, ensuring our safety guardrails detected and blocked malicious prompts and inappropriate content.

Post-launch Evaluation. We have created an illustrative lesson dataset using Aila to plan lessons in bulk on a number of representative themes that came out of our initial evaluations. We have been able to use these, in addition to the 45,000 lessons that users have created post-launch, to further test our safety guardrails.

We have systematically reviewed the safe, content guidance and toxic lessons from both the user-created and our illustrative lessons with technical and subject experts. We sampled 10,000 lessons across a range of subjects and key stages, reviewing the moderation category and justification, in addition to searching the dataset for known sensitive lesson themes to review their moderation category. This process has highlighted a number of edge cases, such as lessons based on current affairs that occurred after the latest data that the underlying LLM has been trained on (e.g. the 2024 American election and recent conflicts) and lessons on topics outside of the national curriculum but still taught in schools (e.g. lessons on specific educational needs or outdoor skills).

2.3 Challenges and Opportunities

Through our testing, we have identified several challenges and considerations to take into account when testing safety guardrails.

Use of an Illustrative Data Set. We created an illustrative dataset using Aila to create complete lessons without user input. This has allowed us to test our safety guardrails at scale and across different contexts, for example, testing the behaviour of Aila and our moderation agent for a topic when planned for different aged pupils. When using a large data set created in this way, the impact of teacher input is not included, which meant we could not solely rely on this data for testing our safety guardrails.

Due to Aila's design, even lessons created without user input will be the most school and age-appropriate version of a lesson. In the example given earlier, when asked to plan a lesson on 'cannabis for KS4', Aila will suggest a school and age-appropriate lesson to teach on this topic. This means that some user-created lessons may have received a different moderation category than those in our illustrative dataset, even when planned on the same topic, as input from the user could steer the content produced away from being national curriculum aligned and appropriate for the classroom. This highlights the importance of ensuring that real usage data is (where possible) included in the testing process.

Evaluation Process. Aila's moderation agent makes multiple checks during the lesson, checking each lesson section as it is produced and modified. When running the IACMA on the illustrative and user-created lesson sets (where the whole lesson was already created), the moderation category was often misaligned with the original moderation. This was because, in a real usage scenario, the initial sections of a lesson may have received a 'content guidance' or 'toxic' flag before the rest of the content was produced. As more content was created, and if this content was appropriate for use in a classroom, the classification given to the lesson would change. For example, a lesson titled 'weapons of mass destruction' may initially be flagged as being 'toxic' (and be blocked) at the point of initiation. When Aila is given a completed lesson to moderate, the content within the lesson may be for a Year 11 religious studies lesson that focuses on the impact, ethical arguments and religious views on weapons of mass destruction - appropriate for the GCSE religious studies specification [1]. The moderation given to the completed lesson may be downgraded to 'content guidance'. In real usage, this lesson may have received a 'toxic' moderation due to the title and never reached a point where the content was produced, meaning the category given to the real-life lesson and the lesson in the testing dataset would have received different moderation categories.

To overcome the first challenge - the inclusion of teacher input, we tested our safety guardrails on both the illustrative and user-created lesson sets and to address the second challenge of replicating the process used by the IACMA, we applied the moderation tool to sections of the lesson instead of completed lessons. This has provided a more accurate data set that demonstrates the moderation categories that these lessons would actually have produced if they were being created by users in Aila.

3 Description of Future Steps

3.1 Future Steps

This paper has focused on the initial steps we have taken to ensure that Aila produces safe and appropriate content for use in schools. Further work will focus on our steps to address the edge cases (examples mentioned above) and additional tests on these changes.

Generative AI is evolving rapidly, so our approach to safety must continue to be responsive and evolve based on emerging needs. At the time of writing, Aila is unimodal (with only text as the input and output). As it develops to incorporate other mediums (including, for example, imagery and audio), our safety processes will need to evolve to include these.

3.2 Implications for Other Organisations

The use of generative AI in primary and secondary education (aged 5–16 years) is still in its infancy. As such, limited models of effective safety guardrails and datasets exist to test these. To date, this work has resulted in three key contributions to the wider community:

1. Our code, including our prompt, is open source [13], and our content is available under an open licence (OGL). Other organisations can freely access the work we have done to build Aila, and specifically our moderation agent. We do this with a view to enabling other organisations in the sector to have a headstart with building effective safety guardrails for their products.
2. We have created an illustrative dataset of over 1000 lessons (used as part of our guardrail evaluation) representative of the types of lessons created by users. We think that by making this available to other organisations in the sector, we will enable them to test their AI guardrails' effectiveness using a context-specific dataset.
3. Our work has focused specifically on the generation of AI content for use in UK primary and secondary schools. By sharing this example of how a multi-stage, safety-first approach can be applied, we hope to provide a useful starting point for organisations in other sectors or regions aiming to build safe and responsible tools. As part of developing our system, we invested significant time in defining expected behaviour for sensitive themes and topics. This facilitated the creation of context-specific safety categories for our IACMA and enabled us to rigorously test their effectiveness. We believe any organisation adopting this methodology should take a similar approach to defining safety expectations in their own regional and contextual setting.

AI has enormous potential within the education sector to support students and teachers. The emergence of new, innovative projects and tools within this space presents a huge opportunity, but safety must be considered at the forefront of any development. It is imperative that organisations developing generative AI tools think carefully about the audience, use, and risks of their tool in educational contexts, embedding and evaluating appropriate safety guardrails from the start. Governments have a critical role to play in supporting and guiding the safe and equitable implementation of these tools.

This paper presents our initial work in this area, but this is just the beginning. We will continue to reflect and iterate on these processes as we develop Aila further.

Acknowledgements. We are grateful to all of the Oak National Academy curriculum partners and lesson creators and the wider team who have enabled the creation of our corpus of lessons, enabling our work in AI. We would also like to thank Oak's AI squad for contributing to our work on quality and safety within Aila and to teams from the Department for Education and the wider UK Government for their support of the project.

Disclosure of Interests. The authors have no competing interests to declare relevant to this article's content.

References

1. AQA. GCSE Religious Studies A: Specification. https://cdn.sanity.io/files/p28bar15/green/e743211e3d3cc945057c101e7f967c99cefef95f.pdf. Accessed 17 Mar 2025
2. BBC. Section 5: Harm and Offence - Guidelines. BBC Editorial Guidelines. https://www.bbc.com/editorialguidelines/guidelines/harm-and-offence/guidelines. Accessed 11 Mar 2025
3. BBFC. BBFC Classification Guidelines (2024). https://darkroom.bbfc.co.uk/original/ce0ab668a4e5b93e7700a03579b138c6:bebc6c5a654b0e98882bca7014157f55/bbfc-classification-guidelines-new-symbols-2024-aw-v4.pdf. Accessed 11 Mar 2025

4. Clark, H.-B., et al.: Auto-Evaluation: A Critical Measure in Driving Improvements in Quality and Safety of AI-Generated Lesson Resources. The AI + Open Education Initiative (2025)
5. Department for Education. Relationships and sex education (RSE) and health education. https://www.gov.uk/government/publications/relationships-education-relationships-and-sex-education-rse-and-health-education. Accessed 17 Mar 2025
6. Department for Education Keeping children safe in education 2024. https://assets.publishing.service.gov.uk/media/66d7301b9084b18b95709f75/Keeping_children_safe_in_education_2024.pdf. Accessed 11 Mar 2025
7. Department for Education. Generative artificial intelligence (AI) in education. https://www.gov.uk/government/publications/generative-artificial-intelligence-in-education/generative-artificial-intelligence-ai-in-education#using-ai-safely-and-effectively. Accessed 11 Mar 2025
8. Department for Science, Innovation and Technology. Online Safety Act: explainer. https://www.gov.uk/government/publications/online-safety-act-explainer/online-safety-act-explainer. Accessed 12 Mar 2025
9. Farquhar, S., Kossen, J., Kuhn, L., Gal, Y.: Detecting hallucinations in large language models using semantic entropy. Nature **630**(8017), 625–630 (2024)
10. Mazeika, M., et al.: HarmBench: A Standardized Evaluation Framework for Automated Red Teaming and Robust Refusal. https://arxiv.org/abs/2402.04249. Accessed 11 Mar 2025
11. Ministry of Defence. Red Teaming Handbook. https://www.gov.uk/government/publications/a-guide-to-red-teaming. Accessed 06 May 2025
12. MIT Management. When AI Gets It Wrong: Addressing AI Hallucinations and Bias. https://mitsloanedtech.mit.edu/ai/basics/addressing-ai-hallucinations-and-bias/. Accessed 11 Mar 2025
13. Oak National Academy. Aila prompt. https://github.com/oaknational/oak-ai-lesson-assistant. Accessed 06 May 2025
14. Ofsted. Ofsted's approach to artificial intelligence (AI). https://www.gov.uk/government/publications/ofsteds-approach-to-ai/ofsteds-approach-to-artificial-intelligence-ai. Accessed 11 Mar 2025
15. OpenAI. Safety & responsibility. https://openai.com/safety/. Accessed 11 Mar 2025
16. PEGI. PEGI age ratings. https://pegi.info/page/pegi-age-ratings. Accessed 11 Mar 2025
17. UKCCIS. Child Safety Online: A Practical Guide for Providers of Social Media and Interactive Services. https://assets.publishing.service.gov.uk/government/uploads/system/uploads/attachment_data/file/487973/ukccis_guide-final__3_.pdf. Accessed 11 Mar 2025
18. UNESCO. Recommendation on the Ethics of Artificial Intelligence (2022). https://unesdoc.unesco.org/ark:/48223/pf0000381137. Accessed 11 Mar 2025
19. Wespieser, K.: AI teachers, school exclusions and cutting workload. https://teachertapp.com/uk/articles/ai-teachers-school-exclusions-and-cutting-workload/. Accessed 11 Mar 2025

A Gateway for Egalitarian Access to LLM Based Resources

Paul Mithun(✉), Enrique Noriega-Atala, Nirav Merchant, and Edwin Skidmore

Data Science Institute, University of Arizona, Tucson, AZ 85719, USA
{mithunpaul,enoriega,nirav,edwins}@arizona.edu

Abstract. We present AI-VERDE , a unified LLM-as-a-platform service designed to facilitate seamless integration of commercial, cloud-hosted, and on-premise open LLMs in academic settings. AI-VERDE streamlines access management for instructional and research groups by providing features such as robust access control, privacy-preserving mechanisms, native Retrieval-Augmented Generation (RAG) support, budget management for third-party LLM services, and both a conversational web interface and API access. In a pilot deployment at a large public university, AI-VERDE demonstrated significant engagement across diverse educational and research groups, enabling activities that would typically require substantial budgets for commercial LLM services with limited user and team management capabilities. To the best of our knowledge, AI-Verde is the first platform to address both academic and research needs for LLMs within an higher education institutional framework.

Keywords: LLM · VERDE · GenerativeAI

1 Introduction

Large language models (LLMs) such as ChatGPT [17], GPT-4 [15], Mistral [3], Llama [7], etc. have rapidly emerged as transformative tools that demonstrate significant capabilities across a broad spectrum of applications, including natural language processing, content generation, and educational support. Their powerful capabilities have captivated the attention of users and has led to a huge creative exploration and novel applications that span most disciplines.

Due to this accelerated advance and acceptance of the LLM technology by the hoi polloi, Universities and colleges are also currently under pressure to integrate this into academic settings. However, integrating LLM technology into academic settings, especially higher education institutions, faces several unique challenges, such as, the required technological know-how, privacy concerns, limited access to specialized knowledge, intellectual property rights etc. For example, a Professor aiming to incorporate commercial LLMs like ChatGPT into their course work often encounters high costs, as well as difficulties in managing usage across students and generating access tokens. Even for using free open-source alternatives like Llama, technical barriers such as programming knowledge and hosting

A. I. Cristea et al. (Eds.): AIED 2025, CCIS 2590, pp. 137–146, 2025.
https://doi.org/10.1007/978-3-031-99261-2_13

tools are significant, especially for faculty coming from a non-STEM background. Furthermore, even with basic programming knowledge, using an LLM effectively for advanced courses requires fine-tuning or setting up a Retrieval Augmented Generation (RAG) [12] pipeline, both of which involve complex software and hardware requirements. Moreover, issues related to intellectual property, such as the potential public dissemination of copyrighted textbooks, add further challenges to seamless integration.

In this context these are the unique contributions of our work:

1) We present AI-VERDE [1], a first of its kind LLM-platform-as-a-service (LLM-PaaS) solution which is already being incorporated by several professors and researchers at the University of Arizona.
2) We present a detailed survey of several problems that prevent smooth adoption of LLM based technology at Universities and other higher education institutions and show how AI-VERDE addresses these concerns.
3) We do a comparative study with other commercial options and show that AI Verde provides a much lower cost egalitarian gateway to AI tools, and can thus democratize the access to LLMs in colleges and University campuses.

To the best of our knowledge this is the first time, that a survey of such limitations is being done, and a product addressing these limitations, and catering to higher education institutions to holistically address these limitations, as an integrated platform, is being done.

2 AI-VERDE Architecture Components

The primary goal in designing of AI-VERDE was to create egalitarian gateway for members of academia, to smoothly access all facets of AI technology. Hence, while at the core AI-VERDE comprises of a standard RAG pipeline to LLMs, each individual sub-modules and components were made available to be accessed independently as micro-services, in a plug-and-play format.

Specifically, AI-VERDE [2] is a platform designed with the goal of providing seamless access to LLMs (especially RAG if need be) to the academic community through various means like chat interfaces, API etc. To achieve this functional requirement, we tapped into open source technologies to serve LLMs as well as developed custom software to allow the provisioning of a multi-tenant environment to accommodate diverse educational and research groups. In this section we detail several of the major architecture components of AI-VERDE . The individual pieces are deployed and orchestrated with Kubernetes [6].

2.1 Backend: LLM Serving

At the heart of AI-VERDE lies the LLMs to be exposed to our community. We use vLLM [11] to persistently load LLM in a GPU cluster. vLLM allows us to

[1] Video demo: https://youtu.be/hPTiZlcUiZo.

[2] https://datascience.arizona.edu/research/tools/ai-verde.

serve open source models, such as Llama 3.2 [4], Mistral [3] and Phi-3 [1] and provide an API interface. Additionally, we leverage the advanced capabilities offered such as support for numerous model weight formats, integration with the HuggingFace Hub[3], support for LoRA adapters [9], and customized paged attention for increased throughput. Each LLM served through AI-VERDE is associated with a running instance of vLLM.

We use LiteLLM[4] as an *LLM proxy* to provided a unified, managed API access to all the individual vLLM instances. Intuitively, LiteLLM behaves as a *reverse proxy*: It exposes an OpenAI-compliant API access point that routes the requests to the corresponding LLM by its name. LiteLLM enables to implement user access control through and usage metering through the use of API Keys. Additionally, we can also use the LLM proxy functionality to seamlessly meter access commercial LLM API providers such as OpenAI and Anthropic, or other research LLM services such as AnvilGPT[5]. We issue surrogate API keys to allow us to provide a fine-grained managed access and control budgets. To support Retrieval Augmented generation configurations, we host a Weaviate[6] vector database environment. Access to Weaviate is controlled too through the use of API keys.

2.2 Front-End: User Facing Web Interface

We introduce a web-based software interface designed to complement our model serving and storage features. At this interface, the federated login and authentication is achieved through CILogon[7]. For access control, each individual instructional and research groups typically have a different set of requirements. For example one group might want access to different models while some others might want access only to the vector database or only to the RAG pipeline or various combinations thereof. AI VERDE solves this by customizing a pipeline based on the specific requirements of each group. We implement the abstraction of groups as *courses*. Therefore each course is comprised of a list of students and instructors. All users have access to their individual API key. Instructors also have access to the list of students and to the budget information of the course.

We also provide a conversational user interface to enable quick access to an LLM, if that is deemed to be the requirement. The LLM can be configured either in *pass through* or *RAG* mode. The former passes the conversation directly to the LLM configured for the course's interface whereas the later executes a RAG workflow using the vector database assigned to the course. The conversational interface manages the conversation's history and persists and retrieves past conversations in a similar fashion to comparable products such as ChatGPT. Further, As an alternative to the conversational user interface, users are able to

[3] https://huggingface.co/models.
[4] https://github.com/BerriAI/litellm.
[5] https://www.rcac.purdue.edu/news/6826.
[6] https://github.com/weaviate/weaviate.
[7] https://www.cilogon.org/.

programmatically access the LLMs using the API keys provided via the web interface. Programmatic access is enabled thanks to the LiteLLM component and exposes an OpenAI-compliant API interface, which allows the use of popular third-party integration libraries to build AI-powered applications, such as Langchain[8] and Llamaindex[9].

2.3 Document Intake

To support RAG workflows, we developed a document intake service that reads various file formats like MSWord, MSPowerPoint slides, PDF files etc., which are then used to generate a corresponding vector database persisted into AI-VERDE's Weaviate service. The document intake service runs independently of the backend and frontend components. Once a vector database is provisioned, it can be configured as part of a course to enable RAG in the conversational UI.

3 Issues with Commercial LLM Platforms in an Academic Setting

To understand the real-life limitations and challenges of using available LLM-based frameworks (commercial or open source) in academic settings, we first conducted a survey amongst the faculty, staff and students at the University of Arizona. This section outlines and details several of the most prominent concerns identified through this survey, emphasizing the practical and ethical issues that arise when deploying such AI platforms in University campuses. Further, we will discuss how AI-VERDE tackles each of these problems that impede the easy installation and adaptation of LLMs /AI infrastructure on University campuses. Table 1 provides a comparative visual overview of the features mentioned below, to some of the popular commercial and open source alternates.

3.1 Intellectual Property, Privacy and Content Ownership

The primary concern that was raised in the survey, agreed unanimously by all faculty, staff, and students, was the lack of privacy and control over their data. This is because, these platforms often store user data, such as copyrighted course materials or sensitive queries, on corporate servers, where it may be used for downstream model training. This raises issues of data security, intellectual property rights, and confidentiality, particularly in educational settings.

AI-VERDE addresses these concerns by ensuring all data, including user queries and uploaded content, is processed entirely on-premises within a secure infrastructure. Unlike commercial platforms, AI-VERDE does not store or reuse queries for model training and disables personalization features by default to prioritize privacy. These features are only activated with explicit user consent, maintaining strict control over sensitive academic and personal information.

[8] https://www.langchain.com.

[9] https://www.llamaindex.ai.

Table 1. Feature comparison between AI-VERDE and other LLM comparable platforms. ChatGPT and Gemini are representative of commercial offerings, while Anvil is a representative of an academic offering.

Feature	*ChatGPT*	*Gemini*	*Anvil*	*AI-VERDE*
Privacy-Preservation	No	No	No	Yes
Built-in guardrails	Yes	Yes	Unknown	Yes
On-Premises Deployment	No	No	No	Yes
Native RAG support	Limited	Limited	No	Yes
Instructional groups management	No	No	No	Yes
Hosted models customization	No	No	No	Yes

3.2 Limited Access to Specialized Knowledge

General-purpose LLMs platforms, while effective for general information retrieval, often fail to meet domain-specific needs [13]. Specialized chatbots for fields like medicine or law are typically limited to answering frequently asked questions and lack the conversational depth required for complex academic queries [23]. These limitations reduce accuracy and trust, diminishing their utility in research and education.

As mentioned before, AI-VERDE overcomes these challenges by leveraging RAG, an AI methodology that retrieves contextually relevant documents based on user queries and primes the LLM with this information. By indexing specialized knowledge datasets, such as niche research papers or course materials, AI-VERDE enables accurate, context-aware responses without storing or training on user-provided data. This ensures data security while empowering researchers and educators to access precise, domain-specific insights.

3.3 Authorization and Authentication

Another major challenge with commercial chatbots in academic settings is their inability to integrate smoothly with existing university platforms, such as authentication and authorization protocols of the respective learning management systems (LMS) [14]. In the Universities in the United States, for instance, typically proprietary content like grades or course materials in an LMS often requires authentication via dedicated platforms ensuring access is restricted to individuals with university-affiliated email addresses. Furthermore, post-authentication, there are additional authorization requirements, such as ensuring only students registered for a course can access its related chatbot.

Publicly available commercial and open-source platforms typically fail to meet these needs, operating in isolation and lacking flexibility to align with specific university protocols. AI-VERDE addresses these concerns by supporting seamless integration with university authentication platforms through CILogon, allowing registered users of University LMS to access services without added complexity. Additionally, AI-VERDE simplifies inter-university collaboration by offering a plug-and-play authentication interface, enabling smooth integration and secure cross-institutional collaboration.

3.4 Equity and Resource Constraints

Freemium AI platforms often disadvantage financially constrained students and institutions by restricting advanced features to paid subscriptions. Token-based pricing and manual on-boarding processes, even in NSF-funded AnvilGPT[10], hinder scalability, making it difficult for educators to manage large cohorts. These challenges underscore the need for automated enrollment and management solutions in educational LLM applications while addressing financial compliance concerns in cloud-based resources. Additionally, the increasing reliance on cloud-based resources has introduced significant concerns over budget management and financial compliance for academic departments.

AI-VERDE addresses these challenges through open-source components, RAG technology, and partnerships with cost-effective hardware providers like CyVerse [19] and NSF's Jetstream2 [8], ensuring minimal operational costs. It also automates budget management by enabling faculty to allocate class-specific funds and distribute API keys efficiently. Moreover, AI-VERDE reduces the administrative burden by handling routine queries with chatbots, freeing faculty to focus on research, teaching, and impactful academic work, thus providing a literal egalitarian gateway[11].

3.5 Anonymity and Confidentiality of Sensitive Data

Compliance with regulatory standards such as Health Insurance Portability and Accountability Act [2] (HIPAA), Family Educational Rights and Privacy Act [18] (FERPA), and Data Use Agreements (DUAs) in research and education contexts often prohibits the use of external services for handling sensitive data. In several cases respective institutions have to ensure that external service providers do not inadvertently compromise data privacy or violate agreements of the end-users they cater to. For example, many academic research projects involve human subjects, requiring compliance with strict ethical and legal standards like IRB approval to ensure data privacy and confidentiality. However, the usage of commercial large language model (LLM) platforms raises significant concerns due to ambiguous data handling practices [10,21,22], risking the exposure of sensitive information, including, possibly, Personal Health Information (PHI). Additionally, limiting the tracking of individual user activities is also a priority in certain settings and projects.

[10] https://anvilgpt.rcac.purdue.edu/.

[11] It's egalitarian since any person on a University campus has equal access rights to AIVERDE. It is a gateway, since users can access any of their favorite state-of-the-art LLM, picked from the current extremely fast moving LLM production pipeline, while using the same access methodology.

AI-VERDE addresses privacy concerns by operating entirely within a secure, on-premises infrastructure, eliminating risks of external data transfer. It integrates with Soteria[12], a secure data analysis enclave for HIPAA compliance, ensuring secure data processing. Additionally, its gateway model abstracts user interactions, preventing commercial providers from tracking or collecting individual user data, strengthening privacy and supporting ethical data management practices.

3.6 Access to Dedicated Hardware

Effectively harnessing the power of LLMs requires fine-tuning on specialized datasets or integrating them with RAG systems. However, hosting LLMs for a large user base demands dedicated and expensive hardware, with budgets often exceeding $50,000—a prohibitive expense for most campus researchers.

AI-VERDE alleviates these challenges by providing a fully equipped hardware infrastructure with pre-loaded LLMs. Further it features elastic hardware allocation tailored to user needs and integrates seamlessly with NSF Jetstream, CyVerse, and cloud services like AWS[13] and Azure[14], enabling efficient scaling for diverse research and teaching workloads.

While these are the most prominent problems and their solutions through AI-VERDE , there are several more unique contributions of AI-VERDE like reduced hallucination, addressing the steep learning curve of AI by providing a dedicated team of consultants and AI educators etc.

4 Pilot Deployment

Since May 2024, AI-VERDE is deployed as part of a pilot project at The University of Arizona. The purpose of this pilot deployment is to stress test the system as well as understand the feasibility of its maintenance while understanding the needs of the academic community, especially in a live implementation context.

The current users of AI-VERDE at the University of Arizona can be divided into three major user categories: Faculty who use it as a Teaching Assistant chatbot, Researchers who use it for analyzing huge amounts of research papers, and the third are support departments like high performance computing who is using AI-VERDE presented chatbot for letting their users ask query about them on their websites.

4.1 A Quantitative Analysis

During the course of the afore-mentioned pilot deployment of AI-VERDE, 78 different users utilized it for 5 different courses and 10 research projects. The

[12] https://soteria.arizona.edu/.
[13] https://aws.amazon.com/.
[14] https://azure.microsoft.com/.

users represent a population of students, instructors, researchers and staff across multiple academic units of the university, Table 2 shows the token and API consumption statistics of AI-VERDE during a period of six months. In brief, 97 thousand API calls representing more than 110 tokens were passed through the variety of models, both self-hosted and relayed to third-party services. While these metrics are relatively low given the potential user base in an institution of this scale, especially since we are still in the pilot mode, the amount of tokens generate represents significant use and its adoption by an increasing user base and hence its huge potential for adoption in academic settings.

Table 2. AI-VERDE usage metrics during the pilot period of 05/30/2024–11/26/2024. In the top part of the table, the number of tokens transmitted are shown in millions. Specifically, in rows, prompt tokens represent the input data sent to various models and completion tokens represent the response data generated by the models based on the prompts. In columns, the self-hosted and proxy columns represent tokens fed and generated to models hosted in-premises, and relayed to third-party services such as OpenAI and Anvil, respectively. In the bottom part of the table, API calls, represent the total number of requests made to the models either via the conversational interface and through API access.

Token Consumption			
	Self-hosted	*Proxy*	*Total*
Prompt	74.96M	.689M	75.65M
Completion	34.80M	.230M	35.03M
Total	109.76M	.919M	110.68M
API Calls	96,403	1,255	97,658

5 Discussion: Legal Compliance

Note that AI-VERDE was designed to meet legislative requirements like FERPA [20]. For example any users (e.g. a student enrolled in a course) can only see the courses they belong to including their respective instructors, co-instructors, and teaching assistants(TAs) and cannot browse other courses for which they are not registered to. Also the instructor, co-instructors, and TAs are the only users of a course who can view and manage the enrollment for a course. Note that only an email and an institutional log-in are all that is required for an instructor to add students to a course preserving their privacy. Further each individual conversations are only accessible by the user who created the chat, and students have full access to their entire chat history for the duration of the course. To conclude, while FERPA includes a much broader set of privacy, notification, and access controls, AI-VERDE helps address some of the fundamental privacy and access controls, creating a pathway for a more detailed future implementation.

6 Limitations and Future Work

In its current form AI-VERDE has the following limitations. Currently, it has only a limited evaluation of educational impact: that is, there is no evaluation of how AI-VERDE actually impacts teaching effectiveness, student learning outcomes, research productivity, or user satisfaction compared to alternatives or no access. For this purpose, we soon plan on incorporating pre- and post-assesses. This will be in the form of either surveys for users who are actively using the system to compare against commercial offerings like ChatGPTEdu [16], Claude for Education [5] etc. Along with that, we also plan to conduct additional empirical validation and a qualitative assessment of how different educational or research groups use the system, or how the platform supports various pedagogical scenarios.

Furthermore, while AI-VERDE is designed for an institutional setting, the long-term maintenance overhead, cost implications of scaling the hardware on-premises, and support requirements for a growing user base are not fully explored yet. We will include this in our future work, especially as and when our client base increases.

Also in its current form, conversations on AI VERDE chat interface are only accessible by the user who created the chat, and students have full access to their entire chat history for the duration of the course. In the future, we plan to incorporate a feature through which the instructor may be able to see anonymized conversations made by students, along with respective statistics of conversations.

In the future, we plan to integrate AI-VERDE into learning management systems like D2L[15], enabling course-specific instances where faculty can upload materials and students receive tailored interactions throughout the semester. We also aim to develop AI applications for researchers, such as a user interface for proofreading grant proposals against funding agency requirements and providing feedback to improve submissions.

7 Conclusion

This work introduces AI-VERDE, a platform providing LLM services tailored for academic environments, prioritizing privacy, accessibility, and adaptability. Developed with open-source components, it offers capabilities similar to commercial LLMs, with features designed for students, faculty, and researchers. By processing all data within institutional infrastructure, AI-VERDE ensures privacy and mitigates external data exposure risks. Its integration with university systems supports research and teaching, fostering an AI-driven ecosystem that promotes innovation, collaboration, and critical thinking in higher education.

[15] https://www.d2l.com.

References

1. Abdin, M., et al.: Phi-3. arXiv preprint arXiv:2404.14219 (2024)
2. Act, A.: Health insurance portability and accountability act of 1996. Public Law **104**, 191 (1996)
3. Jiang, A.Q., et al.: Mistral 7B. arXiv preprint arXiv:2310.06825 (2023)
4. Touvron, H., et al.: LLaMA 2: open foundation and fine-tuned chat models (2023). https://arxiv.org/abs/2307.09288
5. Anthropic: Claude for education: Partnering with universities on responsible AI (2024). https://www.anthropic.com/education. Accessed 06 May 2025
6. Beda, J., Burns, B., McLuckie, C., et al.: Kubernetes: an open-source container orchestration platform (2014). https://kubernetes.io/. Developed by Google and maintained by the Cloud Native Computing Foundation (CNCF)
7. Dubey, A., et al.: The LLaMA 3 herd of models. arXiv preprint arXiv:2407.21783 (2024)
8. Hancock, D.Y., et al.: Jetstream2: accelerating cloud computing via jetstream. In: Practice and Experience in Advanced Research Computing, pp. 1–8. ACM (2021)
9. Hu, E.J., et al.: LoRA: low-rank adaptation of large language models. In: International Conference on Learning Representations (2022). https://openreview.net/forum?id=nZeVKeeFYf9
10. Jaff, E., Wu, Y., Zhang, N., Iqbal, U.: Data exposure from LLM apps: an in-depth investigation of OpenAI's GPTs. arXiv preprint arXiv:2408.13247 (2024)
11. Kwon, W., et al.: Efficient memory management for large language model serving with pagedattention. In: Proceedings of the ACM SIGOPS 29th Symposium on Operating Systems Principles (2023)
12. Lewis, P., et al.: Retrieval-augmented generation for knowledge-intensive NLP tasks. In: Larochelle, H., Ranzato, M., Hadsell, R., Balcan, M., Lin, H. (eds.) Advances in Neural Information Processing Systems, vol. 33, pp. 9459–9474. Curran Associates, Inc. (2020). https://proceedings.neurips.cc/paper_files/paper/2020/file/6b493230205f780e1bc26945df7481e5-Paper.pdf
13. Minaee, S., et al.: Large language models: a survey. arXiv preprint arXiv:2402.06196 (2024)
14. Oliveira, P.C.D., Cunha, C.J.C.D.A., Nakayama, M.K.: Learning management systems (LMS) and e-learning management: an integrative review and research agenda. JISTEM-J. Inf. Syst. Technol. Manag. **13**(2), 157–180 (2016)
15. OpenAI: GPT-4 technical report. arXiv preprint arXiv:2303.08774 (2023)
16. OpenAI: Introducing chatGPT edu (2024). https://openai.com/index/introducing-chatgpt-edu/. Accessed 06 May 2025
17. OpenAI, T.: ChatGPT: optimizing language models for dialogue. OpenAI (2022)
18. Rights, F.E., Act, P.: Family educational rights and privacy act (FERPA) (2014)
19. Swetnam, T.L., et al.: CyVerse: cyberinfrastructure for open science. PLoS Comput. Biol. **20**(2), e1011270 (2024)
20. U.S. Department of Education: Family educational rights and privacy act (FERPA) (2025). https://studentprivacy.ed.gov/topic/family-educational-rights-privacy-act-ferpa. Accessed 06 May 2025
21. Wang, Z., et al.: Data management for large language models: a survey. CoRR (2023)
22. Yao, Y., Duan, J., Xu, K., Cai, Y., Sun, Z., Zhang, Y.: A survey on large language model (LLM) security and privacy: the good, the bad, and the ugly. High-Confidence Comput. 100211 (2024)
23. Yigci, D., Eryilmaz, M., Yetisen, A.K., Tasoglu, S., Ozcan, A.: Large language model-based chatbots in higher education. Adv. Intell. Syst. 2400429 (2024)

Systematic Control of Multiple-Choice Item Difficulty Through LLM-Based Distractor Generation

Haruki Oka[1,2](✉), Yoshito Tan[1], Tsunenori Ishioka[2], and Kensuke Okada[1]

[1] The University of Tokyo, Tokyo, Japan
oka-haruki715@g.ecc.u-tokyo.ac.jp, {ytan4410,ken}@p.u-tokyo.ac.jp
[2] The National Center for University Entrance Examinations, Tokyo, Japan
tunenori@rd.dnc.ac.jp

Abstract. Advancements in large language models (LLMs) have expanded their use in automatic item generation. In particular, distractor generation (DG), the process of creating distractors for multiple-choice items, has benefited from the flexibility of LLMs to generate test items across various domains. However, despite the growing demand for such applications, systematic methods for controlling difficulty in DG remain underdeveloped. This study explores whether difficulty levels can be systematically adjusted by modifying specific linguistic elements, such as word count, morphosyntactic complexity, and vocabulary complexity, in LLM-generated distractors while keeping the correct answer options unchanged. Furthermore, from the perspective of educational assessment, we evaluated whether the generated items are accurate tools for measuring human achievement by computing item parameters based on item response theory using response data. Our findings confirm that difficulty adjustment in DG can be achieved by systematically modifying distractor elements and that such control is essential for ensuring that items function as intended in real educational contexts. Moreover, our results demonstrate that the items generated using our approach can be validated within the framework of test theory, reinforcing their applicability in educational assessments.

Keywords: Distractor generation · Difficulty control · Automatic item generation · Deep learning · Natural language processing

1 Description of the AIED Implementation, Practice or Policy

1.1 Introduction

Multiple-choice items (MCIs) rely on high-quality distractors to effectively assess examinees' knowledge and understanding [10]. Recent advancements in distractor generation (DG) using computational techniques and AI have enabled the

A. I. Cristea et al. (Eds.): AIED 2025, CCIS 2590, pp. 147–157, 2025.
https://doi.org/10.1007/978-3-031-99261-2_14

automatic creation of natural-sounding distractors [15,21,23]. The advent of large language models (LLMs) has further expanded DG capabilities, allowing test items to meet specific format and domain constraints [2,7,8,20]. Despite these advancements, two critical challenges remain: fine-grained control over difficulty and reliable discrimination of examinees' achievement. In the context of DG, current methods leverage LLMs for natural language output. However, unlike automatic item generation (AIG) approaches that calibrate difficulty through overall item construction, existing DG methods lack explicit mechanisms for controlling item difficulty [3,12,25,26]. Specifically, the direct impact of modifying distractor elements such as word count, morphosyntactic complexity, and vocabulary complexity on item difficulty remains poorly understood [1,11,14]. Moreover, conventional evaluation methods based on linguistic quality or expert judgment do not measure the impact of systematically modifying distractors on the difficulty levels of generated items [2,4,5,7,15,18,20,23,24,29], although item difficulty is one of the most critical factors in assessing the effectiveness of generated distractors [1]. To address these gaps, we systematically manipulated distractor elements and evaluated item difficulty using item response theory (IRT).

Our first contribution is demonstrating systematic difficulty control in LLM-based DG by varying multiple aspects of distractor complexity. Our second contribution is the development of an IRT-based evaluation framework using real response data to support the creation of reliable AI-driven testing tools and further integrating psychometric principles into AIG.

1.2 Method

To systematically examine the impact of the DG on the overall difficulty of MCIs, this study explored different DG methods. Specifically, we evaluated whether modifications to the distractor elements resulted in measurable changes in item difficulty. We focused on English grammar gap-fill items, a common format in MCIs. By generating MCIs with varying distractor elements and analyzing examinee responses, we assessed how these modifications influenced difficulty levels.

Item Generation. The item generation process illustrated in Table 1 was designed to systematically adjust difficulty in DG using GPT-4o in ChatGPT [17]. In Step 1, 30 correct sentences appropriate for a third-year junior high school student were generated, serving as the foundation of the process. In Step 2, these sentences were used to create short-answer items along with their corresponding correct answers using a predefined prompt. In Step 3, the short-answer items from Step 2 were transformed into MCIs through a defined prompt and the DG technique. This transformation involved generating distractors that were semantically similar to the correct answer (referred to as *Baseline*) as well as more complex distractors with increased linguistic difficulty. This study places particular emphasis on Step 3, which is especially suitable for objective testing and offers a solution to the long-standing challenge of difficulty adjustment. Additionally, in alignment with curricular guidelines, the items were designed to reflect topics typically taught to third-year junior high school students, while maintaining a nontrivial difficulty level, even for adults. This ensures the items are appropriate for examinees of all age groups. Furthermore, in Step 3, to examine how variations in distractor complexity influence overall item difficulty, the item stem and correct answer remained constant across all MCIs. Only the baseline distractors were systematically modified according to the following elements:

Table 1. Step-by-step DG for MCIs, illustrating increasing complexity levels

Step 1
We need to finish the project by Friday, so let's start working on it right away
Step 2
We need to finish the project by Friday, so let's ( ) working on it right away
Correct answer: start
Step 3
We need to finish the project by Friday, so let's ( ) working on it right away
Correct answer: start
Distractors 1 (Baseline): begins, commence, initiate
Distractors 2 (Morphosyntactic complexity): starting, starts, started
Distractors 3 (Vocabulary complexity): originate, commence, inaugurate

- Word count: How does increasing the number of words in the distractors affect item difficulty? (Levels: two words and three words)
- Morphosyntactic complexity: How do changes in tense and word forms affect item difficulty? (Levels: moderate and highly complex)
- Vocabulary complexity: How does increasing the complexity of vocabulary in distractors affect item difficulty? (Levels: moderate and highly complex)

For each of these three elements, two difficulty levels were introduced, resulting in six distractor types. Combined with the original, this led to the creation of seven variations for each MCI. After generating 30 MCIs per variation, a manual review was conducted to ensure grammatical correctness and appropriateness. The final dataset was refined to 24 items per variation, from which 10 common items and 14 original items were selected. The seven resulting item sets were constructed so that the common items remained identical across all sets, while the original items varied based on the type of distractors. Short-answer items were systematically distributed across the item sets according to the type of distractor modifications applied. An example MCI created is shown in Table 2.

Item Response Theory (IRT). We employed IRT [22] to analyze the psychometric properties of the generated test items. IRT models the relationship

Table 2. Example of a Generated MCI

Item	Content
Item Stem	The teacher asked us to read the entire chapter before the ____ week
Options	– end of the – reading all books – finish was at – be done on
Element	Three words

between an individual's achievement and item characteristics based on binary responses, allowing for the separate estimation of item parameters and latent achievement. For this study, we used the 2-parameter logistic IRT (2PL-IRT) model, which represents the probability of a correct response as:

$$P(u_{ij} = 1 \mid \theta_i, a_j, b_j) = \frac{\exp(a_j(\theta_i - b_j))}{1 + \exp(a_j(\theta_i - b_j))},$$

where θ_i is the latent achievement parameter for respondent i, a_j is the discrimination parameter for item j and b_j is the difficulty parameter for item j. A higher a_j value suggests that an item j is more effective at distinguishing between different achievement levels, whereas a higher b_j value corresponds to greater item difficulty. We estimated these parameters to examine whether and how distractor manipulation resulted in measurable shifts in item difficulty.

1.3 Experiment

We evaluated whether the generated items matched the expected difficulty levels by conducting an online experiment. Participants completed the test, and the dataset was analyzed using the 2PL-IRT model.

Date. The experiment was conducted in November 2024.

Participants. Data collection targeted Japanese native speakers aged between 18 and 65 years through CrowdWorks, an online crowdsourcing platform. To ensure data quality and exclude non-compliant respondents, we incorporated directed questions scale items [16], which required specific responses as instructed. Respondents who failed to follow the instructions or complete the test in less than five minutes were excluded. A total of 792 participants took part in this study. After applying the exclusion criteria, the final dataset comprised valid responses from 693 participants.

Test Content. Each respondent answered 24 items generated using the GPT-4o. The test employed seven sets of items, with participants randomly assigned to one set. The distribution of respondents across the sets was as follows: Set 1: 98, Set 2: 95, Set 3: 92, Set 4: 101, Set 5: 103, Set 6: 106, and Set 7: 98.

Ethical Considerations. This study was approved by the Ethics Review Committee of The University of Tokyo (Approval Number: 24-408).

Parameter Linking and Analysis of Variance. After estimating the item parameters for each set using the 2PL-IRT model, we performed parameter linking using ten common items. For linking, we applied the Stocking and Lord (SL) method [22], which estimates the linking coefficients by minimizing the difference between the test characteristic curves. Let the test characteristic curve on scale T be $\sum_{j=1}^{J} P_j(\theta_{iT})$, and let the test characteristic curve for scale F, linked to scale T, be $\sum_{j=1}^{J} P_j^*(\theta_{iT})$. Further, let the test characteristic curve on scale T, linked to scale F, be $\sum_{j=1}^{J} P_j^\sharp(\theta_{iF})$. Then, for scales T and F, we define the following divergence measures: $SLC1\left(\theta_{iT}\right) = \left[\sum_{j=1}^{J} P_j\left(\theta_{iT}\right) - \sum_{j=1}^{J} P_j^*\left(\theta_{iT}\right)\right]^2$, $SLC2\left(\theta_{iF}\right) = \left[\sum_{j=1}^{J} P_j\left(\theta_{iF}\right) - \sum_{j=1}^{J} P_j^\sharp\left(\theta_{iF}\right)\right]^2$. Based on these measures, the objective function Q is defined as:

$$Q = \sum_{m=1}^{M} \left(SLC1\left(\theta_{mT}\right) h\left(\theta_{mT}\right) + SLC2\left(\theta_{mF}\right) h\left(\theta_{mF}\right)\right). \quad (1)$$

Here, $h\left(\theta_{mT}\right)$ and $h\left(\theta_{mF}\right)$ denote the probability density functions of the achievement parameter θ_i on scales T and F, respectively, with θ discretized into M quadrature points. In this study, M was set to 30, following the plink package [27] defaults. The linking function that minimized the objective function Q was then determined. After calculating the linking coefficients, the parameters for each unique item were adjusted. The average difficulty parameters were subsequently compared based on the type of distractors using the posterior distributions of the average difficulty parameters and an analysis of variance (ANOVA) [13].

For parameter estimation, we assigned a Normal(0, 100) prior to the mean difficulty parameter of each distractor type, whereas the standard deviations were given non-informative Uniform(0, 10) priors. The average values of the parameters were estimated using the Markov Chain Monte Carlo (MCMC) method. Computations were performed using R version 4.2.0 [19] and cmdstanr package. Four chains, each with a length of 3000 iterations (including a warm-up period of 1000), were generated. The convergence was confirmed when the Gelman-Rubin statistic [9] satisfied $\hat{R} \leq 1.1$ for all parameters. In addition, conventional p-values were calculated to examine the significance of the hypotheses. In our analysis, we conducted ANOVA on three elements—Word count, Morphosyntactic complexity, and Vocabulary complexity. Since we were not concerned with interaction effects among these elements, we examined only the main effects.

1.4 Results

The posterior distributions for each type of distractor, shown in Fig. 1, indicate that increasing the complexity of distractor transformations leads to decreased item difficulty. This trend was pronounced under the vocabulary transformation conditions. To examine statistically significant differences, we conducted a repeated-measures one-way ANOVA for each element: word count transformation, morphosyntactic transformation, and vocabulary transformation.

For the word count transformation condition, the ANOVA revealed a non-significant main effect, $F(2,\ 36) = 0.95$, $p = .396$. Nevertheless, as observed in Fig. 1a, the difficulty levels of the items followed the order: 2-word < 3-word < baseline. This pattern suggests that while the three-word condition might impose an additional cognitive load compared to the two-word condition, the increased word count could also make the distractors appear more unnatural. As a result, the correct answer may become more discernible than in the baseline condition.

For morphosyntactic transformation, the ANOVA yielded $F(2,\ 37) = 0.47$, $p = .631$, indicating no significant effect. The manipulation of verb tense or plural forms in a moderately complex manner did not affect item difficulty, as shown in Fig. 1b. However, if morphosyntactic complexity increases excessively, distractors may appear unnatural or unrealistic, leading test-takers to dismiss them. This could increase the likelihood of selecting the correct answer.

For vocabulary transformation, the ANOVA demonstrated a significant main effect, $F(2,\ 38) = 8.59$, $p < .001$. Post hoc Tukey comparisons revealed that the mean difference between the baseline and moderate complexity conditions was 0.89 (95% CI [0.07, 1.72], $p = .031$) while the difference between the baseline and high complexity conditions was 1.38 (95% CI [0.56, 2.21], $p = .001$). No significant difference was observed between the moderate complexity and high complexity conditions (mean difference ≈ -0.49 (95% CI [−1.30, 0.32], $p = .311$). Figure 1c illustrates that increasing the vocabulary difficulty of the distractors tends to simplify the items. This effect may be due to the increased semantic disparity between the correct answer and the distractors, making the correct response easier to identify. Additionally, test-takers may rely on an exclusion strategy when encountering unfamiliar distractor vocabulary or experience a familiarity effect that makes the correct answer's vocabulary more accessible.

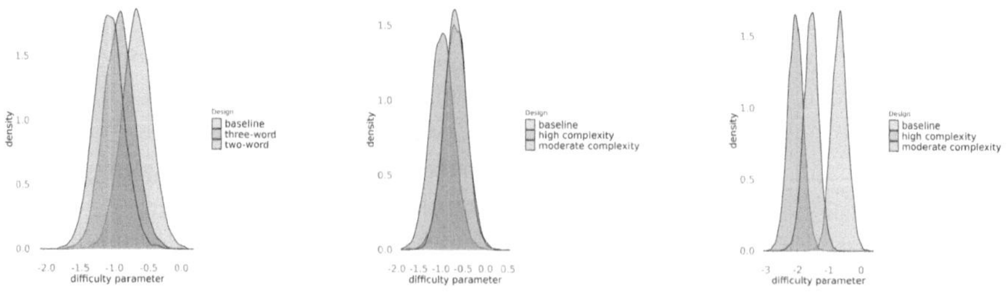

(a) Differences in word count

(b) Differences in morphosyntactic complexity

(c) Differences in vocabulary complexity

Fig. 1. Posterior distributions of the mean difficulty parameters for each type of distractors

In summary, these results indicate that when the correct answer remains unchanged, increasing the complexity of distractors in terms of word count, morphosyntax, or vocabulary, generally reduces item difficulty. This effect likely stems from the heightened contrast between the correct answer and the distractors, making the correct option more conspicuous. These findings highlight the importance of carefully manipulating distractor complexity in practical applications of DG. In particular, overly complex distractors may unintentionally lower item difficulty. Therefore, future DG approaches should incorporate manipulation mechanisms to ensure an appropriate balance of item difficulty.

2 Reflection of the Challenges and Opportunities Associated with the Implementation

2.1 Challenges in Distractor Generation and Difficulty Calibration

In MCIs, the quality of distractors impacts the effectiveness of assessing examinees' knowledge and understanding [10]. Recent advancement have propelled DG into the spotlight as a method for automatically generating these distractors using computational techniques and AI [15,21,23]. The emergence of LLMs has further enhanced DG's flexibility, allowing for the creation of linguistically natural test items while adhering to specific format and domain constraints [2,7,8,20]. However, despite these advancements, two major challenges persist.

First, current DG approaches often lack a precise mechanism for controlling item difficulty. While AIG methods that generate entire test items have made progress in difficulty calibration through overall test construction, controlling difficulty, specifically in DG, remains a challenge [3,12,25,26]. From an educational perspective, it is crucial to precisely control item difficulty [1]. Although some studies have attempted to moderate the discriminative power of DG in relation to examinees' achievement, the intrinsic difficulty calibration of LLM-based DG remains unconsidered in detail [11,14]. Specifically, the direct impact of manipulating distractor elements such as word count, morphosyntactic complexity, or vocabulary complexity on item difficulty has yet to be fully understood.

A second challenge is the absence of a systematic, test-theory-based evaluation of DG-generated items. Traditional evaluation methods have primarily relied on linguistic criteria such as grammatical correctness and fluency, expert judgment from test designers, or student feedback [2,4,5,7,15,18,20,23,24,29]. While these approaches help assess distractor quality, they do not determine whether items effectively differentiate examinees by achievement levels. This limitation raises concerns about their authenticity in real-world educational assessments.

2.2 Opportunities and Proposed Solutions

To address these challenges, our study proposes a twofold approach: systematically manipulating distractor elements and developing an IRT-based evaluation framework. Our method involves altering the linguistic complexity of distractors by increasing the word count, selecting more advanced vocabulary, or introducing additional morphosyntactic complexity while keeping the correct option unchanged. This approach enables us to isolate the impact of distractor manipulation on item difficulty, as demonstrated through English grammar items.

In parallel, we developed an evaluation framework based on IRT. By applying IRT to data from the generated items, we can separate the examinee achievement parameters from the difficulty and discrimination parameters of individual items. This separation facilitates score comparisons that remain independent of specific examinee samples or testing occasions and lays the groundwork for implementing adaptive testing via information functions [28]. Our detailed analysis of examinees' response data provides empirical evidence supporting the effectiveness of our DG approach within a rigorous test-theory framework.

2.3 Related Work and Contextual Comparison

Our study builds on a substantial body of research on AIG and distractor evaluation. Previous studies have explored the application of DG in various domains, including English grammar and vocabulary [2,18,23,29], mathematics [6–8], physics and science [15], and reading comprehension [24]. While these studies provide valuable domain-specific insights, our research focuses on English grammar items, a category with many validated examples, to derive broadly applicable findings.

Additionally, recent AIG studies have explored methods for controlling overall test item difficulty. For example, Cheng et al. [3] and Jiao et al. [12] demonstrated difficulty calibration by controlling various aspects of item construction. Furthermore, Uto et al. [26] and Tomikawa et al. [25] integrated IRT to generate items with predetermined difficulty levels. However, these approaches primarily focus on generating entire items rather than manipulating distractors.

In terms of evaluation, previous research has employed quantitative measures based on linguistic elements and similarity scores [4,5,15,20,23,24,29] and has also relied on manual assessments by experts and students [2,7,18,20,29]. Although these studies have thoroughly examined the relationship between correct answers and distractors, they have not systematically evaluated whether DG-generated items effectively measure examinee achievement, a gap our study aims to fill.

2.4 Summary of Reflections

Overall, our reflections underscore two key contributions of this implementation. First, we demonstrated how systematic manipulation of distractor complexity in LLM-based DG can be employed to control item difficulty. Second, we introduced a practical IRT-based evaluation framework that leverages real examinee responses to verify the effectiveness of the manipulations. These contributions have the potential to assist researchers and practitioners in developing more reliable AI-driven testing tools and integrating psychometric principles into AIG.

3 Description of Future Steps

This study examined whether manipulating distractors could control item difficulty. We increased distractor complexity across three elements—word count, morphosyntactic complexity (e.g., tense, singular/plural), and vocabulary complexity—and found that more complex distractors reduced overall item difficulty by making the correct answer more conspicuous. Future research should refine these modifications to preserve subtle differences while maintaining the intended difficulty levels.

Additionally, we assessed the psychometric properties of the generated items using IRT analysis of human responses, confirming that changes in distractor complexity aligned with test theory. This test-theory-based framework provides a quantitative evaluation method for AI-generated items that is independent of linguistic heuristics or expert judgment and should be further integrated into AIG. Overall, our findings suggest that strategic DG can systematically regulate the difficulty of MCIs. Although this study focused on English grammar, future research should explore other subjects and investigate the combined effects of modifying both distractors and correct answers. Moreover, AIG must tailor items to learner achievement and manage difficulty within specific topic ranges. Future studies should also examine how different LLM architectures influence these processes.

References

1. Alhazmi, E., Sheng, Q.Z., Zhang, W.E., Zaib, M., Alhazmi, A.: Distractor generation in multiple-choice tasks: a survey of methods, datasets, and evaluation. In: Al-Onaizan, Y., Bansal, M., Chen, Y.N. (eds.) Proceedings of the 2024 Conference on Empirical Methods in Natural Language Processing, pp. 14437–14458. Association for Computational Linguistics, Miami (2024). https://doi.org/10.18653/v1/2024.emnlp-main.799
2. Bitew, S.K., Deleu, J., Develder, C., Demeester, T.: Distractor generation for multiple-choice questions with predictive prompting and large language models (2023). https://doi.org/10.48550/arXiv.2307.16338
3. Cheng, Y., et al.: Guiding the growth: difficulty-controllable question generation through step-by-step rewriting. In: Proceedings of the 59th Annual Meeting of the Association for Computational Linguistics and the 11th International Joint Conference on Natural Language Processing (Volume 1: Long Papers), pp. 5968–5978. ACL, Stroudsburg (2021)
4. Chiang, S.H., Wang, S.C., Fan, Y.C.: CDGP: automatic cloze distractor generation based on pre-trained language model. In: Findings of the Association for Computational Linguistics: EMNLP 2022, pp. 5835–5840. ACL, Abu Dhabi (2022). https://doi.org/10.18653/v1/2022.findings-emnlp.429
5. Das, B., Majumder, M., Phadikar, S., Sekh, A.A.: Multiple-choice question generation with auto-generated distractors for computer-assisted educational assessment. Multimed. Tools Appl. **80**(21–23), 31907–31925 (2021)
6. Dave, N., Bakes, R., Pursel, B., Giles, C.L.: Math multiple choice question solving and distractor generation with attentional GRU networks. Int. Educ. Data Min. Soc. (2021)

7. Feng, W., et al.: Exploring automated distractor generation for math multiple-choice questions via large language models. In: Duh, K., Gomez, H., Bethard, S. (eds.) Findings of the Association for Computational Linguistics: NAACL 2024, pp. 3067–3082. ACL, Mexico City (2024)
8. Fernandez, N., Scarlatos, A., Feng, W., Woodhead, S., Lan, A.: DiVERT: distractor generation with variational errors represented as text for math multiple-choice questions. In: Proceedings of the 2024 Conference on Empirical Methods in Natural Language Processing, pp. 9063–9081. ACL, Miami (2024). https://doi.org/10.18653/v1/2024.emnlp-main.512
9. Gelman, A., Carlin, J.B., Stern, H.S., Dunson, D.B., Vehtari, A., Rubin, D.B.: Bayesian Data Analysis, 3rd edn. Chapman & Hall/CRC Texts in Statistical Science. Chapman & Hall/CRC, Philadelphia (2013)
10. Haladyna, T.M.: Developing and Validating Multiple-Choice Test Items, 3rd edn. Routledge (2004). https://doi.org/10.4324/9780203825945
11. Hwang, K., Wang, K., Alomair, M., Choa, F.S., Chen, L.K.: Towards automated multiple choice question generation and evaluation: aligning with bloom's taxonomy. In: Artificial Intelligence in Education, pp. 389–396. Springer (2024)
12. Jiao, Y., Shridhar, K., Cui, P., Zhou, W., Sachan, M.: Automatic educational question generation with difficulty level controls. In: Artificial Intelligence in Education: 24th International Conference, AIED 2023, Tokyo, Japan, 3–7 July 2023, Proceedings, pp. 476–488. Springer, Heidelberg (2023)
13. Kutner, M.H., Nachtsheim, C.J., Neter, J., Li, W.: Applied Linear Statistical Models, 5 edn. McGraw–Hill, Boston (2005)
14. Lee, Y., Kim, S., Jo, Y.: Generating plausible distractors for multiple-choice questions via student choice prediction (2025). https://doi.org/10.48550/arXiv.2501.13125
15. Liang, C., Yang, X., Dave, N., Wham, D., Pursel, B., Giles, C.L.: Distractor generation for multiple choice questions using learning to rank. In: Proceedings of the Thirteenth Workshop on Innovative Use of NLP for Building Educational Applications, pp. 284–290. Association for Computational Linguistics, Stroudsburg (2018)
16. Maniaci, M.R., Rogge, R.D.: Caring about carelessness: participant inattention and its effects on research. J. Res. Pers. **48**, 61–83 (2014)
17. OpenAI: GPT-4o. ChatGPT (May 13 Version) (2024). https://chat.openai.com/. Accessed 02 Nov 2024
18. Palma Gomez, F., Panda, S., Flor, M., Rozovskaya, A.: Using neural machine translation for generating diverse challenging exercises for language learner. In: Proceedings of the 61st Annual Meeting of the Association for Computational Linguistics (Volume 1: Long Papers), pp. 6115–6129. Association for Computational Linguistics, Stroudsburg (2023)
19. R Core Team: R: A language and environment for statistical computing. R Foundation for Statistical Computing, Vienna, Austria (2022). https://www.R-project.org/
20. Scarlatos, A., Feng, W., Smith, D., Woodhead, S., Lan, A.: Improving automated distractor generation for math multiple-choice questions with overgenerate-and-rank. In: Proceedings of the 19th Workshop on Innovative Use of NLP for Building Educational Applications (BEA 2024), pp. 222–231. Association for Computational Linguistics, Mexico City (2024)
21. Shin, J., Guo, Q., Gierl, M.J.: Multiple-choice item distractor development using topic modeling approaches. Front. Psychol. **10**, 825 (2019)
22. Stocking, M.L., Lord, F.M.: Developing a common metric in item response theory. Appl. Psychol. Meas. **7**(2), 201–210 (1983)

23. Susanti, Y., Tokunaga, T., Nishikawa, H., Obari, H.: Automatic distractor generation for multiple-choice English vocabulary questions. Res. Pract. Technol. Enhanc. Learn. **13**(1), 15 (2018)
24. Taslimipoor, S., Benedetto, L., Felice, M., Buttery, P.: Distractor generation using generative and discriminative capabilities of transformer-based models. In: Calzolari, N., Kan, M.Y., Hoste, V., Lenci, A., Sakti, S., Xue, N. (eds.) Proceedings of the 2024 Joint International Conference on Computational Linguistics, Language Resources and Evaluation (LREC-COLING 2024), pp. 5052–5063. ELRA and ICCL, Torino (2024)
25. Tomikawa, Y., Suzuki, A., Uto, M.: Adaptive question-answer generation with difficulty control using item response theory and pretrained transformer models. IEEE Trans. Learn. Technol. **17**, 2186–2198 (2024)
26. Uto, M., Tomikawa, Y., Suzuki, A.: Difficulty-controllable neural question generation for reading comprehension using item response theory. In: Proceedings of the 18th Workshop on Innovative Use of NLP for Building Educational Applications (BEA 2023), pp. 119–129. Association for Computational Linguistics, Toronto (2023)
27. Weeks, J.P.: Plink: an R Package for linking mixed-format tests using IRT-based methods. J. Stat. Softw. **35**(12), 1–33 (2010)
28. Weiss, D.J.: Improving measurement quality and efficiency with adaptive testing. Appl. Psychol. Meas. **6**(4), 473–492 (1982)
29. Yoshimi, N., Kajiwara, T., Uchida, S., Arase, Y., Ninomiya, T.: Distractor generation for fill-in-the-blank exercises by question type. In: Proceedings of the 61st Annual Meeting of the Association for Computational Linguistics (Volume 4: Student Research Workshop), pp. 276–281. Association for Computational Linguistics, Stroudsburg (2023)

Towards Actionable GenAI-Classroom Integration: A Social Role Framework for Teacher's Practice

Xiaoya Yu and Yang Pian(✉)

Faculty of Information Science and Technology Education, Beijing Institute of Education, Beijing, China
bianyang@mail.bnu.edu.cn

Abstract. The integration of Generative Artificial Intelligence (GenAI) in education has reshaped perspectives on teaching, highlighting the need for actionable strategies to support teachers in leveraging AI for instructional design. While existing research underscores the importance of AI literacy, limited guidance exists on how teachers can effectively incorporate GenAI into classroom practice. To address this gap, we propose a GenAI Social Role Framework, which defines three key roles for GenAI in the classroom: Connection Creator (Learn by Connecting), Simulated Learner (Learn by Teaching), and Evaluative Collaborator (Learn by Evaluating). This framework provides structured guidance on how teachers can integrate GenAI into classrooms to facilitate student engagement and deep learning. To illustrate its practical application, we present three real-world classroom scenarios in Beijing, China, demonstrating how GenAI can be embedded into different instructional contexts and classroom activities. These cases highlight the diverse ways in which teachers can design GenAI-supported learning activities, making GenAI a meaningful component of modern pedagogy. Building on educational practice, we further explore the challenges and opportunities in advancing GenAI applications in education and provide insights into future directions for research and implementation. The proposed framework offers a systematic approach to GenAI-enhanced teaching, equipping educators with practical strategies to harness AI for deeper learning experiences.

Keywords: Generative AI · Teacher's Practice · Deep Learning · Social Role Framework · Classroom Integration

1 Introduction

The emergence of Generative Artificial Intelligence (GenAI) has once again shifted the landscape of education, prompting a re-evaluation of teaching practices. Consequently, all stakeholders in education, such as students, teachers, and administrators, must proactively adapt to ensure readiness for the future education [6]. Teachers, as one of the core practitioners in this transformation, must

A. I. Cristea et al. (Eds.): AIED 2025, CCIS 2590, pp. 158–167, 2025.
https://doi.org/10.1007/978-3-031-99261-2_15

not only understand AI but also be able to leverage it to foster deeper learning among students. Up to 2025, China has over 18.8 million teachers serving in its education system. As the front-line implementers of AI-driven educational practices, the AI literacy of this vast professional group holds critical importance for the nationwide advancement of educational modernization. To clarify how teachers should collaborate with AI in this new era, many countries and international organizations - including China's Ministry of Education - have issued guidelines and frameworks on Teacher's AI Literacy [7,14], which outline the key dimensions of AI competencies for educators.

However, these static definitions AI literacy, while valuable, are insufficient to guide teachers in the efficient application of AI tools within their daily teaching practices. For teachers, a theoretical understanding of *"what constitutes an AI-literate individual"* is important but by no means enough. What they need more is the practical guidance on *"how to apply these advanced AI techniques in my teaching"*. In other words, a significant gap persists between AI literacy goals and actionable strategies for classroom implementation, hindering teachers in bridging the "last mile" of real-world AI integration.

To address this gap, this study introduces a **Social Role Framework**, designed as a practitioner-oriented model to assist primary and secondary educators ineffectively integrating Generative Artificial Intelligence (GenAI) into their teaching practices. Grounded in established educational theories and aligned with the specific needs of teachers, the proposed framework positions GenAI as a pedagogical partner across three distinct roles: **Connection Creator**, **Simulated Learner**, and **Evaluative Collaborator**. For each role, the framework provides the corresponding teaching strategies and instructional scaffold design guidance. To illustrate the practical impact and effectiveness of the framework, we present three real-world cases in which the framework was applied by three primary and secondary school teachers in Beijing, China. Guided by the framework, educators designed and implemented GenAI-enhanced activities within their classrooms. Through this approach, our goal is to (1) empower teachers to translate AI literacy from an abstract competency into actionable practice, and (2) foster student deep learning through meaningful engagement with GenAI.

This work contributes to AIED's practitioner discourse by:

- Operationalizing policy mandates into classroom-ready strategies for GenAI integration.
- Democratizing GenAI integration across diverse disciplinary contexts and classroom activities.
- Balancing technological affordances with teacher agency in GenAI-enhanced learning environments.

2 Solution

2.1 Teacher's Needs

To clarify the current state of GenAI-classroom integration, we surveyed 22,964 K-12 teachers in Beijing, China. While 87.53% reported using GenAI tools in teaching, only 48.81% felt confident in their pedagogical use. Additionally, 76.29% were concerned about students becoming overly dependent on AI, potentially undermining the development of higher-order thinking. To further investigate teachers' needs, we conducted in-depth interviews with 26 educators (5 primary, 7 secondaries, and 14 high schools). Thematic coding revealed two key tensions: (1) **AI literacy-actionability paradox**, where teachers understood basic GenAI knowledge and functions (avg = 3.35 out of 5) but had problems in educational practical integration (avg = 2.38 out of 5), and (2) **cognitive depth dilemma**, with 77% interviewees (N = 20) mentioning a decline in students' problem-solving behaviors, and 69% (N = 18) as surface-level output iteration. A teacher's reflection en- capsulated this concern: *"I, myself, use LLMs weekly, but worry if it replaces student critical thinking."*

Based on the needs identified from teachers, we propose two essential conditions for GenAI-classroom integration:

(1) Actionability for Teachers: This necessitates a solution that provides clear, structured steps and appropriate support that enable teachers to seamlessly incorporate GenAI into their classroom practices.
(2) Deep Learning for Students: The integration of GenAI must foster deep learning. It should encourage critical thinking, helping students to develop higher-order cognitive skills such as problem-solving, synthesis, evaluation, and applying knowledge in diverse contexts.

2.2 Theoretical Basis

Sociocultural Theory. Sociocultural theory, proposed by Vygotsky, posits that learning is inherently a social process, achieved through collaboration, discussion, and feedback with others such as teachers and peers [11]. In technology-enhanced learning environments, tools like GenAI can also function as the "others", that is, the interactive partner facilitating deeper interactions between students and the learning contents. This theoretical perspective inspires that educators should shift their focus from merely applying GenAI into classroom activities to **strategically designing GenAI's roles within the activities**. By tailoring the role of GenAI to suit the specific learning content, stages, and student needs, teachers can promote more profound, interactive learning experiences among students.

Deep Learning Theory. Deep learning goes beyond simply understanding the surface content of knowledge [5]. It requires active participation, critical thinking, and ongoing self-reflection throughout the learning process. Based on previous studies, deep learning can be facilitated through three core mechanisms: **(1) Learn by Connecting**, which enables students to integrate new knowledge with prior understanding, fostering deeper comprehension and real-world applicability [1,10]. **(2) Learn by Teaching**, which reinforces knowledge acquisition by engaging students in the process of explaining concepts to others [15]. **(3) Learn by Evaluating**, which cultivates critical thinking by requiring students to assess information for accuracy and relevance, refining their analytical skills [9].

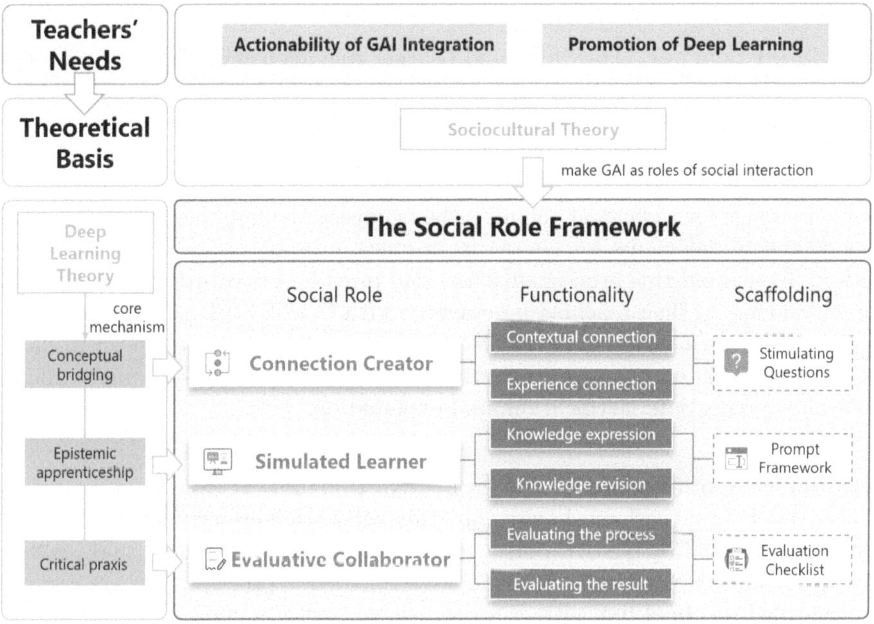

Fig. 1. The social role framework (SRF) for GenAI-classroom integration.

2.3 The Proposed Social Role Framework

Based on the identified teachers' needs and theoretical foundations, we propose the **Social Role Framework (SRF)**, which defines three key roles for GenAI-classroom integration: *Connection Creator*, *Simulated Learner*, and *Evaluative Collaborator*. These roles are designed to foster deep learning by encouraging students to actively engage with GenAI across different learning contexts. To provide clear, actionable guidance, each role is accompanied by functionality descriptions and suggested scaffoldings, which support both teachers and students ineffectively design their own learning activities. Figure 1 presents the detailed information of the SRF.

Connection Creator. The Connection Creator role is grounded in the "Learn by Connecting" mechanism of deep learning. In this role, GenAI helps students relate new content to their prior knowledge or real-world experiences by generating contextualized examples, analogies, or interactive visualizations, making abstract concepts more accessible. In classroom practice, teachers can scaffold this process by using **stimulating questions**, such as *"How does the AI-generated examples connect to your own experiences?"* or *"Can you think of a real-world situation where these examples apply?"* to encourage students to reflect and deepen their understanding.

Simulated Learner. The Simulated Learner role aligns with "Learn by Teaching" mechanism of deep learning. In this role, GenAI functions as a "learner" that requires guidance and instruction to successfully generate specific content. Students take on the role of a "teacher", helping GenAI understand the task by providing clear explanations and refining their input. For example, when generating an image for a classical poem, students must articulate the poem's imagery, mood, and key elements for GenAI to produce an appropriate visual representation. To scaffold this process, teachers can provide a **prompt framework** to guide students in their teaching interactions with GenAI, such as *"What key concepts do you need to explain?"* or *"How can you refine the explanation to make it clearer?"* This encourages students to improve their understanding through continuous reflection, clarification, and explanation.

Evaluative Collaborator. The Evaluative Collaborator role is grounded in "Learn by Evaluating" mechanism. In this role, students critically assess the content generated by GenAI, reflect on its accuracy and relevance, and identify strengths and areas for improvement. To scaffold this process, teachers can use an **evaluation checklist** with criteria such as clarity, accuracy, and real-world applicability. In addition, students can also engage in peer reviews to provide feedback and enhance their evaluative skills.

3 Practical Cases

We have applied our proposed framework in teacher training programs across several primary and secondary schools in Beijing, China. The aim is to help teachers integrate GenAI effectively into their classrooms. After the training, some teachers have already adapted the SRF solutions to designed classroom activities integrating GenAI. In this section, we present three real-world practical cases, one for each role corresponding to SRF. We mainly compare the design of classroom activities before and after applying the SRF to show its effects.

3.1 Case 1: Learn by Connecting in a Science Classroom

Figure 2 illustrates three pedagogical approaches on the same science learning topic *"Natural Phenomena in Our Daily Life"*. **In the traditional task**, without technological aids, students were required to list 20 natural phenomena with scientific explanations within 3 min. This may or may not have elicited cognitive engagement, as such internal activation is implicit and difficult to observe directly; however, the behavioral outcome—classroom silence—suggests a possible disconnect between the knowledge presented and students' prior experience. **In the shallow GenAI-integration task**, students actively repeated AI-generated answers, but lack cognitive engagement due to passive content consumption. **In the SRF-guided task**, teacher employed a two-phase design: GenAI firstly worked as a connection creator, generating 20 natural phenomena. Then students were asked to identify *"Which phenomena align with your experiences?"* (**stimulating question**). This dual structure achieved synchronized cognitive-behavioral engagement through meaningful knowledge bridging.

	Traditional Task	**Shallow GenAI-Integration Task**	**SRF-guided Task**
Technical Tool	None	GenAI	GenAI
Task Design	Please **list 20 natural phenomena** and **explain** them with scientific knowledge **within 3 minutes**.	Please **list 20 natural phenomena** and **explain** them with scientific knowledge **using GenAI**.	Please **list 20 natural phenomena** and **explain** them with scientific knowledge **using GenAI**. **Identify one** natural phenomenon from the list you have **encountered in real life**.
Student Participation	• **cognitive**: engaged • **behavioral**: silence	• **cognitive**: not engaged • **behavioral**: active	• **cognitive**: engaged • **behavioral**: active

Fig. 2. Student cognitive and behavioral engagement in science classroom tasks: comparison of traditional, shallow Gen-AI integration, and SRF-guided pedagogy.

3.2 Case 2: Learn by Teaching in a Geography Classroom

As shown in Fig. 3, in the traditional geography tasks, students identify landforms (*e.g., karst*) from given images. While they can label them correctly, it's unclear if they truly understand the concepts or are just memorizing what the textbook writes. With our SRF-guided solution, students describe karst features in details to a GenAI system (as **simulated learner**), which then creates images based on their descriptions. Even if the images aren't perfectly accurate, the process of explaining helps students clarify their understanding and make knowledge visible.

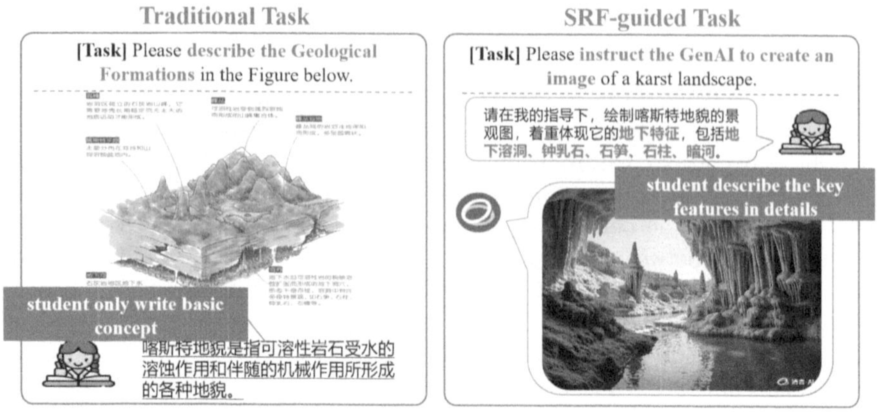

Fig. 3. SRF-guided task (the right panel) to facilitate the externalization of knowledge.

3.3 Case 3: Learn by Evaluating in an English Writing Classroom

Many teachers and parents worry that students might misuse GenAI in writing classes, relying on it to generate essays and leading to shallow learning. In the SRF-guided writing classroom (as shown in Fig. 4), however, GenAI is repositioned as an **Evaluative Collaborator** rather than a replacement. Students are not only required to interact with GenAI to generate content but must also document key interactions, pose meaningful questions, and use an evaluation checklist to justify why their essays meet the acceptable criteria. Through reflective scaffoldings such as *"Did the AI revisions truly improve your essay?"*, students are guided to think deeply about the writing process. As one student reflected, *"This made me take writing more seriously. I had to defend my choices and truly understand the feedback"*.

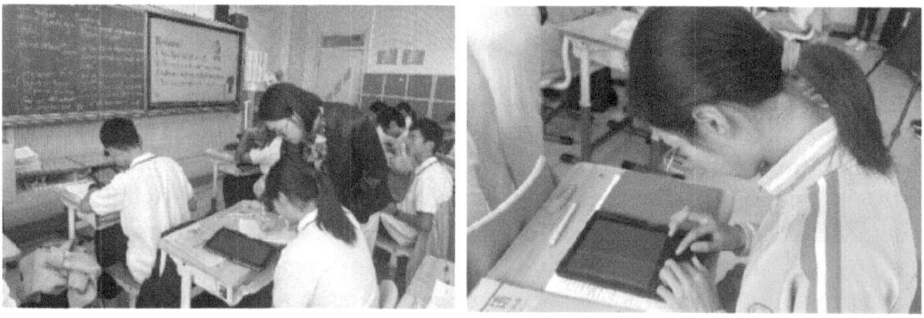

Fig. 4. SRF-guided classroom practice of student writing and evaluating with GenAI.

4 Challenges and Opportunities

While promoting this framework in primary and secondary school teacher training in China, we have observed several challenges and emerging opportunities.

4.1 Challenges

(1) Device Accessibility. A key barrier is the limited access to devices such as tablets or computers. Even in Beijing, not all students have personal equipment, constraining their engagement with GenAI tools. Ensuring hardware accessibility remains fundamental to equitable AI adoption. **(2) Digital Identity Barriers.** Many GenAI platforms require mobile phone verification for registration, posing difficulties for students without smartphones due to financial constraints, parental restrictions, or school policies. While potential workarounds include unified accounts or email-based access, these demand systemic coordination. Localized, offline AI solutions may offer promise but are not yet widely available. **(3) Teacher Resistance and Role Shifts.** Based on our observation during the long-term teacher training practices, teachers often face psychological barriers when adapting to AI-assisted classrooms at the first stage, fearing a loss of authority or relevance as their role shifts from being the primary source of knowledge to a facilitator of student-centered learning. This transition can cause anxiety and resistance [8]. Furthermore, they also worry about there is a risk that students may become overly reliant on AI, out- sourcing critical thinking and decision-making to technology [2]. Balancing the use of GenAI to enhance learning while promoting active student engagement is crucial. Teachers need adequate training and support to navigate this shift and ensure that GenAI complements, rather than replaces, traditional educational practices.

4.2 Opportunities

(1) National Curriculum Reforms. China's recent introduction of a dedicated information technology curriculum, which includes AI as a core component [12], provides a strong foundation for integrating GenAI into education. This reform not only emphasizes the importance of AI literacy but also creates opportunities for schools to align their teaching practices with national educational goals, fostering a generation of students equipped with future-ready skills. **(2) Policy Support and Local Initiatives.** In addition to national policies, cities like Beijing [3] and Shanghai [13] have launched AIED action plans to further drive innovation. These initiatives provide funding, resources, and structured frameworks, enabling schools to experiment with and implement AI-driven teaching tools. Such support fosters a dynamic environment for innovation while ensuring AI integration aligns with broader educational objectives. **(3) Teacher Enthusiasm and Early Successes.** With the emergence of domestic AI models, like DeepSeek [4], ChatGLM [16], and others, teachers are increasingly enthusiastic about incorporating GenAI into their classrooms. Many educators have already begun experimenting with these tools, and early adopters, as demonstrated in this study, have achieved notable successes. These pioneering efforts not only showcase the potential of GenAI but also inspire confidence among other teachers, paving the way for broader adoption and innovation in teaching practices.

5 Future Work

The rapid advancement of artificial intelligence in education presents both opportunities and challenges. With the recent release of Beijing's AI Education Implementation Plan [13], there is a timely opportunity to leverage our role of teacher trainers in promoting AI-driven teaching innovations. The proposed social role framework offers a structured approach to integrating GenAI into classroom practice, equipping educators with practical strategies to enhance deep learning. By embedding this framework in teacher training initiatives, we can foster more effective and scalable AI adoption in educational settings.

Future research will proceed along two key directions. First, empirical studies will be conducted to evaluate the effectiveness of the proposed framework in real classroom environments, focusing on both student learning outcomes and the nature of AI-supported teacher–student interactions. Special attention will be given to how the framework supports teacher autonomy in designing and implementing GenAI-enhanced instruction. Second, we will pursue iterative refinement and large-scale deployment, using teacher feedback to guide continuous improvement and ensure the framework remains adaptable, practical, and grounded in everyday teaching realities.

Acknowledgement. This paper is a staged research outcome of the Educational and Teaching Reform Project of Beijing Institute of Education (Project No. JGYB2024-06).

References

1. Arthur, Y.D., Owusu, E.K., Asiedu-Addo, S., Arhin, A.K.: Connecting mathematics to real life problems: a teaching quality that improves students' mathematics interest. IOSR J. Res. Method Educ. **8**(4), 65–71 (2018)
2. Bastani, H., Bastani, O., Sungu, A., et al.: Generative AI can harm learning. SSRN (2024)
3. Beijing Municipal Education Commission and Others: Notice on issuing the "work plan for artificial intelligence applications in the education sector of Beijing" (2024). https://jw.beijing.gov.cn/xxgk/2024zcwj/2024qtwj/202503/t20250307_4028227.html. Accessed 12 Mar 2025
4. Bi, X., Chen, D., Chen, G., et al.: DeepSeek LLM: scaling open-source language models with longtermism. arXiv preprint arXiv:2401.02954 (2024)
5. Bransford, J.D., Brown, A.L., Cocking, R.R.: How People Learn, vol. 11. National Academy Press, Washington, DC (2018)

6. Chiu, T.K.: The impact of generative AI (GenAI) on practices, policies and research direction in education: a case of ChatGPT and midjourney. Interact. Learn. Environ. **32**(10), 6187–6203 (2024)
7. Cukurova, M., Miao, F.: AI Competency Framework for Teachers. UNESCO Publishing (2024)
8. Ertmer, P.A., Ottenbreit-Leftwich, A.T.: Teacher technology change: how knowledge, confidence, beliefs, and culture intersect. J. Res. Technol. Educ. **42**(3), 255–284 (2010)
9. Forss, K., Cracknell, B., Samset, K.: Can evaluation help an organization to learn? Eval. Rev. **18**(5), 574–591 (1994)
10. King, D., Bellocchi, A., Ritchie, S.M.: Making connections: learning and teaching chemistry in context. Res. Sci. Educ. **38**, 365–384 (2008)
11. Newman, S.: Vygotsky, Wittgenstein, and sociocultural theory. J. Theory Soc. Behav. **48**(3), 350–368 (2018)
12. Ren, Y., Huang, R., Xiong, Z.: From information technology to information science: a dialogue on the compulsory education information science curriculum standards (2022 edition). Curricul. Teach. Mater. Methods **42**(12), 21–31 (2022)
13. Shanghai Municipal Education Commission: Notice on issuing the "action plan for promoting the high-quality development of basic education empowered by artificial intelligence in shanghai (2024–2026)" (2024). https://edu.sh.gov.cn/xxgk2_zdgz_jyxxh_03/20241217/d41f0919ad4446f28076eb15c5669070.html. Accessed 12 Mar 2025
14. Sperling, K., Stenberg, C.J., McGrath, C., Åkerfeldt, A., Heintz, F., Stenliden, L.: In search of artificial intelligence (AI) literacy in teacher education: a scoping review. Comput. Educ. Open 100169 (2024)
15. Wagner, M., Gansemer-Topf, A.: Learning by teaching others: a qualitative study exploring the benefits of peer teaching. Landsc. J. **24**(2), 198–208 (2005)
16. Zeng, A., Xu, B., Wang, B., et al.: ChatGLM: a family of large language models from GLM-130B to GLM-4 all tools. arXiv preprint arXiv:2406.12793 (2024)

SCRIPT: Implementing an Intelligent Tutoring System for Programming in a German University Context

Alina Deriyeva, Jesper Dannath, and Benjamin Paassen(✉)

Bielefeld University, Bielefeld, Germany
bpaassen@techfak.uni-bielefeld.de

Abstract. Practice and extensive exercises are essential in programming education. Intelligent tutoring systems (ITSs) are a viable option to provide individualized hints and advice to programming students even when human tutors are not available. However, prior ITS for programming rarely support the Python programming language, mostly focus on introductory programming, and rarely take recent developments in generative models into account. We aim to establish a novel ITS for Python programming that is highly adaptable, serves both as a teaching and research platform, provides interfaces to plug in hint mechanisms (e.g. via large language models), and works inside the particularly challenging regulatory environment of Germany, that is, conforming to the European data protection regulation, the European AI act, and ethical framework of the German Research Foundation. In this paper, we present the description of the current state of the ITS along with future development directions, as well as discuss the challenges and opportunities for improving the system.

Keywords: Intelligent Tutoring System · Programming Education · Python · Privacy · Regulation

1 Description of the AIED Implementation, Practice or Policy

Intelligent Tutoring Systems (ITSs) have shown their effectiveness in educational scenarios [14,21]. They can help to bridge the gap between demand and supply in education, particularly for skills which require extensive exercise, as well as hints and feedback for these exercises, such as programming [5,15]. We consider the educational context of advanced undergraduate courses, where we introduce Python as a new programming language to experienced computer science students. While we focus on data science education, the system should be sufficiently flexible to support a wide variety of Python programming courses; and teachers should be able to simply upload new course material. In this educational context,

J. Dannath and A. Deriyeva—Both contributed equally.

A. I. Cristea et al. (Eds.): AIED 2025, CCIS 2590, pp. 168–177, 2025.
https://doi.org/10.1007/978-3-031-99261-2_16

we aim to set up an ITS that serves not only as an educational tool, but also as a research platform to develop and investigate different components for ITSs, such as knowledge tracing models or hint and feedback policies based on large language models (LLMs) [18]. Finally, we aim for a system that is compatible with the current, complex regulatory environment of a German university, including the European data protection regulation (GDPR) [2,17], the European AI act [16], and the ethical framework of the German Research Foundation [11]. Starting from these requirements, we found that no pre-existing ITS fits our needs sufficiently.

Accordingly, we are developing SCRIPT (Step-based Coding for Research and Intelligent Programming Tutoring), a novel ITS to support advanced undergrad computer science students in their acquisition of data science and machine learning skills[1]. The ITS functions both as a teaching and as a research tool by enabling to record fine-grained, keystroke-level data and is easily adaptable to novel learner models and pedagogical models as well as randomized experiments. Finally, the ITS is designed to respect the regulatory requirements of Bielefeld University (Germany). The system was already offered as voluntary exam preparation for an "Introduction to Data Mining" course. As a research tool, the ITS has already supported data collection for learner model and feedback research [6–8].

1.1 System Architecture

We adopt a 4-models approach for this ITS [19] including a Pedagogical Model, a Learner Model, a Domain Model, and a User Interface. Figure 1 shows how these

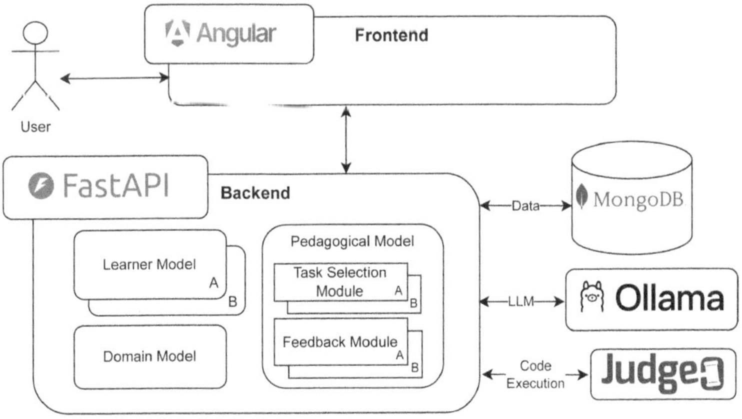

Fig. 1. Technical architecture of the ITS.

[1] The source code is available at https://gitlab.ub.uni-bielefeld.de/publications-ag-kml/script/.

Models are integrated in the technical architecture of the system. The system is provided as a web-application which is hosted in Docker containers on the university's own server infrastructure. The Angular-based[2] frontend (UI) handles all interactions with the user, while all processing is being done server-side. All other system models are encapsulated in the backend. We choose FastAPI[3] for the implementation in order to be able to utilize Python's data science and machine learning stack and use MongoDB[4] as a database in order to store Python objects directly as JSON-documents. The system models are designed to be easily replaceable, the learner model and pedagogical model can be selected individually within each course or can be randomly assigned to users for A/B-testing. All user code is executed with Judge0 [10] to ensure security and allow for fixed compute-resource allocation as well as timeouts. Finally, we use Ollama[5] as a hosting service for open-weight language models, which we use for generating programming hints.

User Interface: The User Interface (UI) connects the other modules to the learner by enabling users to interact with the learning contact and displaying tasks as well as hints to the learner. Figure 2 presents the current interface of the system. The left upper part contains task descriptions, the left bottom part is reserved for displaying different types of feedback (either upon a submission, or whenever a hint is requested).

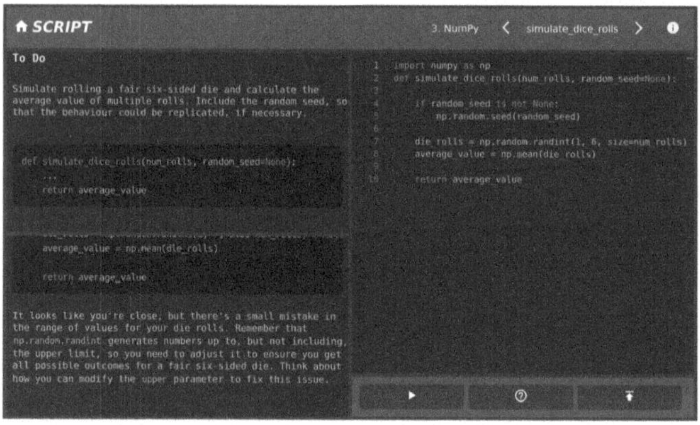

Fig. 2. Screenshot of the current state of the system's UI.

Domain Model: The domain model contains all tasks as well as information that connects them, like a competency model or a list of knowledge components

[2] https://angular.dev/.
[3] https://fastapi.tiangolo.com/.
[4] https://www.mongodb.com/.
[5] https://ollama.com/.

per task (Q-matrix) [3]. Also, it manages and distributes all information related to the difficulty of tasks or knowledge components.

Learner Model: The Learner Model includes all information on the individual learner's competency and mastery of knowledge components. This information is derived via the learner's UI actions and a knowledge tracing algorithm [1].

Pedagogical Model: Finally, we adopted the differentiation of inner and outer loop [20] for the core ITS features. The outer loop is responsible for task selection and the inner loop is responsible for generating feedback to steps inside programming tasks.

More specifically, the outer loop receives information provided by the domain model (esp. task difficulty and task-to-skill relations), and the learner model (esp. ability in each skill). Figure 3 shows the respective flow (currently implemented parts and potential directions for future work are indicated in the figure) of information involved in the outer loop. The baseline version of the outer loop is a pre-defined curriculum.

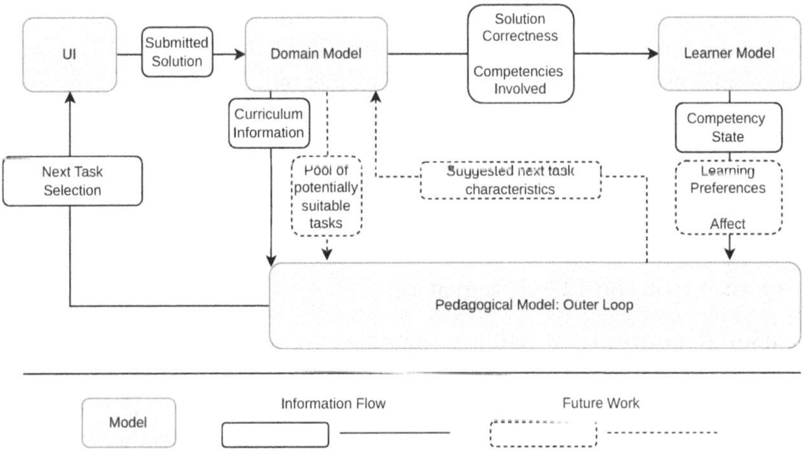

Fig. 3. Schema of the outer loop of the ITS.

For the inner loop (Fig. 4), we use next-step hints as a framework for step-based feedback [15]. This allows to focus on different aspects of feedback, like feedback on mistakes, potential optimizations of user code, or missing pieces of knowledge which can all be the basis for a suggested next step. We decompose the process of hint generation into two modular sub-components, the "Step Generator" and the "Hint Generator", similar to prior work in the ITS literature [4,20]. The "Step Generator" predicts the next program state of the student in the direction of the correct solution. The "Hint Generator" has the goal to translate the predicted step into a textual hint that closes the gap between the learner's current knowledge and the required knowledge to continue with

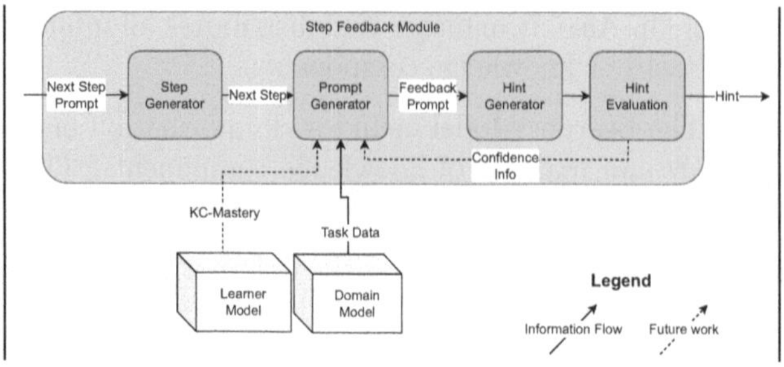

Fig. 4. Schema of the inner loop of the ITS.

the task, without disclosing the complete step. Therefore, the feedback generator should utilize additional information about learner competency as well as task requirements. In a first iteration, we implemented the Step Generator and Hint Generator using different prompt templates for the in-context prompting of LLMs (Fig. 4; all prompts are available in the source code repository), which already incorporate contextual information on tasks from the domain model.

As the system is still in its alpha stage, only contains part of the features described in Figs. 3 and 4 are already implemented. In the future, contextual information from the learner model will be incorporated, as well. Additionally, we plan to include revision steps based on an evaluation of the generated feedback via a certainty metric. In further iterations, the certainty information of the generated LLM output will be used to further improve the feedback generation.

1.2 Context of the ITS Integration

Educational context: SCRIPT is intended to support learning Python by computer science students at the Faculty of Technology at Bielefeld University, Germany. In particular, it was used in context of introductory Python sessions for the courses "Introduction to Machine Learning" and "Introduction to Data Mining", both advanced undergraduate courses for computer science students. Additionally, it was offered as a supplementary exam preparation tool for the "Introduction to Data Mining" course, where the topics were more advanced (refer to Table 1 for a full list of tasks). The current target group predominantly consists of advanced undergraduate students (third year) and early postgraduate students (first year).

Regulatory Context: Within the context of a German university, heightened regulation has to be taken into account. First, because the system processes data of students that could be tied back to individuals (such as IP addresses), it falls within the scope of the EU general data protection regulation (GDPR) [2,17]. Therefore, we strictly separated IP address logs from the actual ITS and instructed students to use pseudonymous usernames in order to avoid any coupling between their data stored in the ITS and their identity. Further, we

Table 1. Tasks with the exam preparation curriculum for Introduction to Data Mining course.

Subdomain	Task Names
Probability Theory	[bayes, dice, independence, mean_and_std]
Statistical Testing	[pearson, pooled_variance, wilcoxon]
PCA	[principal_components, factor_sampling, variance_coverage]
Clustering	[kmeans, ward_linkage, optimal_gmm_clusters]
Logistic Models	[icc, pfa_inference, IRT_logreg]
Markov Models	[most_likely_to_follow, hmm_sampling, bkt]
Deep Models	[dkt_input, training_loop, vae_sampling]
Recommenders	[ndcg_at_k, user_similarity_explicit, user_similarity_cosine]

ensured that no user e-mail addresses are stored beyond the registration process. Our strict adherence to data privacy also meant that we were not yet able to implement the system for mandatory task submissions because that would have required to couple submissions to individual student identities. While this is not strictly forbidden under GDPR, it would mean heightened privacy risks which we wanted to avoid. As such, using the tool had to remain entirely voluntary.

Second, because we want to use the ITS not only as a teaching but also as a research tool, we set the system up to record fine-grained (keystroke-level) data within the ITS, to allow detailed evaluation of the submitted code [12]. However, ethical reasons as well as GDPR forbid us from recording data without explicit consent [11]. As such, we enabled users to use the ITS as a pure teaching tool but also to provide explicit, additional consent to have their data recorded and stored for research purposes. We highlight that fine-grained consent acquisition is a research topic in itself that is far from resolved [13]. Further, because one of the courses is taught by the researchers themselves, it had to be ensured that the researchers had no chance to identify which students used the system – otherwise, students may feel pressured to participate in the research purely to ingratiate themselves with the teachers. Accordingly, in addition to all aforementioned privacy measures, we ensured that the teachers had no access to the servers during the semester but only after another manual anonymization step of all data had been completed and all grades had already been finalized.

Finally, some applications of AI in education are regarded as high risk in the upcoming European AI act [16]. While the precise implications of the EU AI Act for AI in education are not yet fully clear, we aimed to be prepared for the novel regulatory environment and avoided invoking any commercial LLM interface. Instead, we opted for a LLama-70b model, hosted by us or a direct research partner. Regulatory constraints aside, we argue that such a setup is more suited to reduce the environmental impact of ITS operation (due to a smaller model size compared to very large LLMs), enhances scientific reproducibility (because we can exactly specify the model version we are using), and decreases dependency upon commercial LLM providers. Most importantly, self-hosted LLMs can reduce risks in terms of privacy and data security.

2 Reflection of the Challenges and Opportunities Associated with the Implementation

Opportunities: The system is in an ongoing improvement cycle, where we gradually add features to improve the quality of the system. That is, each cycle would consist of integration of new features (in the system or curriculum-wise) and a following semester-long testing in a real world setting. At present, the ITS is only used in a narrow, short-term setting (Python introduction and exam preparation). In the future, our intention is to roll the ITS out as a course-long support tool for studying and solving homework assignments. Further, we also regard the ITS as a training and development tool in study projects, where students can develop novel features and thereby improve the system, themselves.

The ITS also serves as a research platform, in the sense that it allows testing different educational strategies and tools, and allows collecting detailed data about learning interactions of the users (currently at the key-stroke level). This includes, but is not limited to: learner models (what models of knowledge tracing [1] represent the actual performance of users the best), outer loop (what task/intervention suggestion models best support skills acquisition and promote the efficiency of learning), feedback generation (What hints to give in which situation/to different users and how hints should be presented to users in order to achieve ideal learning gains).

Challenges: Developing an ITS from scratch proves to be challenging. As development is mainly guided by two PhD students, the sheer workload of development tasks provides a strain on the time that can be devoted to research. Similarly, keeping up with constantly progressing technologies requires a considerable amount of time. Moreover, as the system is work-in-progress, IT security vulnerabilities are likely: for example, there was a significant vulnerability in the system before the incorporation of judge0 for code execution, which was pointed out by a couple of enthusiastic students.

Complying with the research ethics norms and installing the procedures mentioned in the previous section proved to be additional, substantial challenges. In fact, the roll-out of a semester-long teaching plan via the ITS had to be postponed for a year due to a need to find suitable procedures that comply with all regulations and ensure their approval by the respective committees. While we had anticipated some of these challenges, the sheer complexity of the compliance efforts surprised us. We are convinced that, in the end, these efforts changed our system for the better. However, anyone administering AI systems for education in an EU (in particular, German) context should anticipate this additional effort.

3 Description of Future Steps

The main next milestone is related to the incorporation of the ITS in a semester-long curriculum to support the "Introduction to Data Mining" course. This milestone aims to accomplish both educational (longitudinal support of knowledge acquisition) and research (investigating the systems effectiveness in pre- and post-test setups) goals. However, there are several additional features that we plan to incorporate into the ITS.

Authoring Interface: The current version of the ITS requires manual task file uploads to change the teaching content. Instead, we plan to provide a new component which enables teachers to directly edit tasks, similar to the idea of authoring tools in tutoring systems [9].

Pedagogical Model: We aim to test different recommendation methods for next tasks to investigate their impact onto learning efficiency. Another goal is to enrich the ITS with intervention techniques when a learner struggles to progress within their curriculum; for example, recommending additional materials/hints for a specific skill/topic. This recommendations would follow a pathway similar to the one shown in the Fig. 3. All these variations of the outer loop will also have to be validated in empirical studies. Within the inner loop, the ITS currently allows to request an LLM-generated feedback to the current state of the code of the learner. As displayed in Fig. 4, this feature is planned to be further refined in order to increase feedback reliability. Furthermore, we are investigating the ideal step granularity for next step hints and will incorporate our research results into the ITS in the future [7].

Learner Model: To further support the users of the ITS, we intend to implements a dashboard that would present the findings of the learner model to the users. Additionally, we want to improve the explainability of the learner model's results and employ methods to facilitate interpretability of the presented information.

Evaluation: Once the ITS is used as a semester-long support tool, we plan to use A/B testing with pre- and post-tests to compare skill gains of different pedagogical models.

Open-Source: While we released the current source code of the ITS at https://gitlab.ub.uni-bielefeld.de/publications-ag-kml/script/, the system is still in the alpha stage. As such, we can not yet recommend the system for deployment but only for research use.

Acknowledgments. We gratefully acknowledge the support of the project "SAIL: SustAInable Lifecycle of Intelligent Socio-Technical Systems" project. SAIL is funded by the Ministry of Culture and Science of the State of North Rhine-Westphalia (Germany) (grant no. NW21-059A). We would also like to express our gratitude to Arno Gaußelmann for his contribution to the development of SCRIPT.

Disclosure of Interests. The authors have no competing interests to declare that are relevant to the content of this article.

References

1. Abdelrahman, G.M., Wang, Q., Nunes, B.P.: Knowledge tracing: a survey. ACM Comput. Surv. **55**, 1–37 (2022). https://doi.org/10.1145/3569576
2. Bai, J.Y., Zawacki-Richter, O., Muskens, W.: Re-examining the future prospects of artificial intelligence in education in light of the GDPR and ChatGPT. Turk. Online J. Distance Educ. **25**(1), 20–32 (2024). https://eric.ed.gov/?id=EJ1414336
3. Barnes, T.: The Q-matrix method: mining student response data for knowledge. In: Proceedings of the AAAI 2005 Educational Data Mining Workshop, pp. 1–8 (2005). https://cdn.aaai.org/Workshops/2005/WS-05-02/WS05-02-006.pdf
4. Birillo, A., et al.: One step at a time: combining LLMs and static analysis to generate next-step hints for programming tasks. In: Proceedings of the 24th Koli Calling International Conference on Computing Education Research (Koli Calling) (2024). https://doi.org/10.1145/3699538.3699556
5. Crow, T., Luxton-Reilly, A., Wuensche, B.: Intelligent tutoring systems for programming education: a systematic review. In: Proceedings of the 20th Australasian Computing Education Conference (ACE), p. 53–62 (2018). https://doi.org/10.1145/3160489.3160492
6. Dannath, J., Deriyeva, A., Paaßen, B.: Evaluating task-level struggle detection methods in intelligent tutoring systems for programming. In: Schulz, S., Kiesler, N. (eds.) Proceedings of the German Conference on Educational Technologies (DELFI), Fulda, Germany (2024). https://doi.org/10.18420/delfi2024_07
7. Dannath, J., Deriyeva, A., Paaßen, B.: What is a step? A user study on how to subdivide the solution process of introductory python tasks. In: Mills, C., Alexandron, G., Marras, M., Portaz, M. (eds.) Proceedings of the 18th International Conference on Educational Data Mining (EDM 2025), Palermo, Italy (2025). Accepted
8. Deriyeva, A., Dannath, J., Paassen, B.: Incorporating struggle in performance factors analysis for programming education. In: Bellas, F., Dörpinghaus, J., Fontenla-Romero, O., Tiemann, M. (eds.) Proceedings of the International Workshop on AI in Education and Educational Research (AIEER) at the 27th European Conference on Artificial Intelligence (ECAI) (2024). In press
9. Dermeval, D., Paiva, R., Bittencourt, I.I., Vassileva, J., Borges, D.: Authoring tools for designing intelligent tutoring systems: a systematic review of the literature. Int. J. Artif. Intell. Educ. **28**, 336–384 (2018). https://doi.org/10.1007/s40593-017-0157-9
10. Došilović, H.Z., Mekterović, I.: Robust and scalable online code execution system. In: 2020 43rd International Convention on Information, Communication and Electronic Technology (MIPRO), pp. 1627–1632 (2020). https://doi.org/10.23919/MIPRO48935.2020.9245310
11. German Research Foundation: Guidelines for safeguarding good research practice. Code of conduct (2022). https://doi.org/10.5281/zenodo.6472827
12. Jeuring, J., et al: Towards giving timely formative feedback and hints to novice programmers. In: Proceedings of the 2022 Working Group Reports on Innovation and Technology in Computer Science Education (ITiCSE), pp. 95–115 (2022). https://doi.org/10.1145/3571785.3574124
13. Judel, S., Zeise, C., Schroeder, U.: Informiertes zustimmungsmanagement für learning analytics in lernmanagementsystemen. In: Proceedings of the German Conference on Educational Technologies (DELFI) (2024). https://doi.org/10.18420/delfi2024_06

14. Kulik, J., Fletcher, J.D.: Effectiveness of intelligent tutoring systems: a meta-analytic review. Rev. Educ. Res. **86** (2015). https://doi.org/10.3102/0034654315581420
15. McBroom, J., Koprinska, I., Yacef, K.: A survey of automated programming hint generation: the hints framework. ACM Comput. Surv. **54**(8) (2021). https://doi.org/10.1145/3469885
16. Mueck, M., Gaie, C., Gkikas, D.C.: Introduction to the European artificial intelligence act. In: Mueck, M., Gaie, C. (eds.) European Digital Regulations, pp. 53–90. Springer, Cham (2025). https://doi.org/10.1007/978-3-031-80809-8_3
17. Portela, C., et al.: Bringing AIED, public policy and GDPR to promote educational opportunities in Brazil. In: Olney, A.M., Chounta, I.A., Liu, Z., Santos, O.C., Bittencourt, I.I. (eds.) Proceedings of the 2024 International Conference on Artificial Intelligence in Education (AIED), pp. 74–86. Cham (2024). https://doi.org/10.1007/978-3-031-64315-6_6
18. Roest, L., Keuning, H., Jeuring, J.: Next-step hint generation for introductory programming using large language models. In: Proceedings of the 26th Australasian Computing Education Conference (ACE), Sydney, Australia (2024). https://doi.org/10.1145/3636243.3636259
19. Sottilare, R.A., Graesser, A., Hu, X., Holden, H.: Design Recommendations for Intelligent Tutoring Systems: Volume 1-learner Modeling, vol. 1. US Army Research Laboratory, Orlando (2013)
20. VanLehn, K.: The behavior of tutoring systems. Int. J. Artif. Intell. Educ. **16**, 227–265 (2006)
21. VanLehn, K.: The relative effectiveness of human tutoring, intelligent tutoring systems, and other tutoring systems. Educ. Psychol. **46**(4), 197–221 (2011). https://doi.org/10.1080/00461520.2011.611369

WideAIED

AI Coding Assistants in Competitive Programming: Empirical Studies from India on Human-AI Interactions in Learning, Problem Solving, and Curiosity Development Perspectives

Omkar Joshi(✉), Kushal Mundra, Deepika Yadav, and Syaamantak Das

Centre for Educational Technology, Indian Institute of Technology Bombay, Mumbai, India
omkar.joshi21@gmail.com

Abstract. AI-based coding assistants (AICAs) are increasingly utilized in software development, yet their role in programming education, particularly in competitive programming, remains unclear. This study examines the effectiveness of AICAs in learning environments, focusing on their impact on problem-solving skills, student perceptions, and epistemic curiosity. We conducted two empirical studies involving a total of 10 competitive programming contests and 18 student participants from a prestigious Indian technical institution. In the first study, six AICAs were evaluated in a controlled Kotlin-based competitive programming setting. Based on the performance in the first study, two AICAs were extensively assessed across nine contests on LeetCode, CodeChef, and Codeforces over two months. Data was collected through rubric-based performance evaluations, participant surveys (Likert scale responses), screen recordings, and coding performance rankings. AICAs demonstrated effectiveness in assisting with basic programming tasks, with 59% of students agreeing that AI-generated code was readable and useful. However, 76% of participants reported that AICAs struggled with complex algorithmic problems, and 82% found them inadequate in handling edge cases. In live competitive programming contests, Amazon Q ranked in the top 10% for some structured problem-solving scenarios but did not perform well in abstract reasoning tasks, while Tabnine exhibited static response patterns. Epistemic curiosity assessments revealed no strong correlation between AI tool usage and increased curiosity in programming, with 55% of participants expressing mixed motivation toward learning with AI. This study highlights the importance of Human-AI interaction in learning and the need to enhance AI-assisted environments to improve problem-solving skills and student engagement.

Keywords: AI Coding Assistants · Competitive Programming · Human-AI interaction

A. I. Cristea et al. (Eds.): AIED 2025, CCIS 2590, pp. 181–194, 2025.
https://doi.org/10.1007/978-3-031-99261-2_17

1 Introduction

The increasing integration of artificial intelligence (AI) into education has transformed how students learn programming [1]. AI-powered coding assistants, such as GitHub Copilot[1] and Amazon CodeWhisperer/Amazon Q[2], offer real-time code suggestions and syntactic assistance, promising to enhance coding efficiency [7,26]. However, their role as educational tools in competitive programming environments remains an open research question. While AI assistants streamline development by automating repetitive tasks, their potential to support deeper learning, problem-solving skills, and curiosity-driven exploration in programming education is not well understood.

Competitive programming provides an ideal setting to investigate the educational potential of AI coding assistants. It challenges learners to solve algorithmic problems under time constraints, reinforcing computational thinking and software engineering best practices. However, the effectiveness of AI tools in such contexts is uncertain. Can these tools genuinely support learning, or do they primarily serve as efficiency boosters? Do they enhance students' problem-solving abilities, or do they create dependencies that hinder deep learning? Moreover, what impact do they have on students' epistemic curiosity [8,13], the drive to seek new knowledge and understand complex programming constructs?

This proposed study addresses these questions by evaluating multiple AI coding assistants (AICAs) across different competitive programming environments through a group of competitive programming practitioners in India. We analyze their effectiveness using a rubric-based performance assessment, participant surveys, and observational data from coding sessions. Our findings reveal that while AI tools assist with syntax and basic problem-solving, they struggle with complex algorithmic reasoning, requiring human intervention. Additionally, we explore how programmers perceive these tools and whether they influence motivation and curiosity in learning programming. By examining the interplay between AI assistance and programming education, this study contributes to the broader discourse on integrating AI into learning environments. The insights gained will inform the development of AI-assisted educational frameworks that strike a balance between automation and active problem-solving.

2 Related Work

2.1 AI-Assisted Learning in Programming

AI-powered coding assistants are increasingly being integrated into software development and educational settings. Prior research has demonstrated their usefulness in reducing syntax errors and automating repetitive tasks [22]. However, several studies indicate that AI-generated code, while syntactically correct, often lacks logical depth and contextual adaptability [5,6]. Our findings align

[1] https://github.com/features/copilot.

[2] https://docs.aws.amazon.com/toolkit-for-vscode/latest/userguide/amazonq.html.

with this research, showing that AICAs perform well on structured problems but struggle with open-ended challenges, especially in a context and constraint (sample codes and test cases) given competitive programming environment.

2.2 Human-AI Collaboration in Software Engineering Education

The role of AI in software engineering education has been explored extensively. The work by [27] examined developer experiences with AI assistants, finding that AI enhances productivity but does not replace human problem-solving abilities. Similarly, [15] reported that AI tools improve efficiency in debugging but fail in conceptual reasoning. Our study builds on this by investigating how competitive programmers engage with AI tools under time constraints and whether such interactions foster learning or reliance.

2.3 Competitive Programming as a Learning Tool

Competitive programming is widely regarded as an effective way to develop algorithmic problem-solving skills [24]. However, the integration of AI assistants into this domain remains underexplored. In a recent study, [2] conducted a performance study of LLM-generated code on LeetCode, showing that while AI excels at syntax-based problems, it fails in strategic problem-solving. Similarly, [28] analyzed GPT-4's performance on LeetCode, noting its struggles with complex, multi-step reasoning. Our study confirms these findings by demonstrating that AICAs show a steep drop-off in performance when moving from structured to abstract problems, reinforcing the need for adaptive AI tutoring systems in programming education.

2.4 Comparison with Prior Research on AI Coding Assistants

Several prior studies have analyzed the strengths and limitations of AICAs. For example, the work of [3] evaluated AI-based code assistants for method generation tasks, finding that AI tools work best when problem constraints are well-defined. Our study extends this by showing how AI performs under high-pressure, time-sensitive scenarios. Another aspect of AICAs were discussed by [4], focusing on desirable characteristics for AI teaching assistants in programming education. While they emphasized the need for adaptive guidance, our study highlights that existing AI tools lack such adaptability, often requiring human intervention. The work of [21] studied code search behavior in AI assistants, noting that programmers often struggle to phrase queries effectively. Our study aligns with this, showing that AI assistants require iterative prompting to generate useful solutions in competitive programming.

2.5 Comparative Effectiveness of Prompt Engineering Techniques in Code Generation

Recent studies have explored various prompt engineering techniques to enhance code generation by large language models (LLMs). Structured Chain-of-Thought

(SCoT) prompting, which incorporates programming structures, outperformed traditional Chain-of-Thought (CoT) prompting by up to 13.79% in Pass@1 on benchmarks like HumanEval [14]. Similarly, Semantic Chain-of-Thought (SeCoT) prompting, which integrates code semantic information, achieved state-of-the-art performance on multiple benchmarks [23]. The DemoCraft system, utilizing in-context learning and demonstration selection with latent concept learning, showed a 2x increase in pass@k metrics compared to baselines on MBPP and HumanEval datasets [10]. Additionally, the CodePromptEval dataset was introduced to evaluate five prompt techniques, revealing that combining multiple techniques doesn't necessarily improve outcomes and highlighting a trade-off between correctness and quality in LLM-generated code [11].

Research Gap. By synthesizing these prior studies with our findings, we highlight the gap between AI-generated assistance and true learning impact. Unlike prior work that evaluates AI tools in general software engineering tasks, our study specifically addresses their role in competitive programming education, offering new empirical evidence on their limitations and potential enhancements.

3 Methodology

To assess the effectiveness of AICAs in competitive programming education, we conducted two empirical studies that evaluated multiple AICAs across different programming environments. This section details participant selection, AI tool choices, task design, data collection, and ethical considerations.

3.1 Participant Selection and Study Design

The study participants (Initially 45, out of which 18 were able to complete all the tasks) included undergraduate and postgraduate students from STEM backgrounds. An open call was announced, and participants were selected based on their familiarity with data structures and algorithms (DSA). No financial incentives were provided to ensure unbiased engagement. The study was divided into two phases:

1. Phase 1 (Controlled Kotlin-Based Study): Participants used six AICAs (GitHub Copilot, Amazon CodeWhisperer/Amazon Q, Codeium[3], Cody (Sourcegraph)[4], NinjaTech AI[5], and Tabnine[6]) to solve kotlin programming problems in a live competitive platform. Kotlin was selected because the participants were previously unfamiliar with the programming language. The goal was to determine if a new programming language could be mastered

[3] https://codeium.com/.
[4] https://sourcegraph.com/cody.
[5] https://www.ninjatech.ai/.
[6] https://www.tabnine.com/.

through AICAs. We also wanted to see whether the use of AICAs has any effect on the programmers' curiosity to learn the new language. Each coding tool was evaluated by three participants. Each participant interacted with the assigned AI tool for a fixed set of problems, ensuring a balanced dataset for comparative analysis. For this study, we used live competitive programming problems using Kotlin, a modern programming language gaining traction in both industry and competitive coding circles[7]. The choice of Kotlin was strategic, aligning with current trends in software engineering and offering unique advantages for our research. Kotlin, a statically typed language that runs on the Java Virtual Machine (JVM) [25], offers several features particularly beneficial for competitive programming which are: (i) Concise syntax, reducing boilerplate code compared to Java[8], (ii) Strong type inference capabilities, (iii) Functional programming support, including higher-order functions and lambda expressions[9], (iv) Seamless interoperability with Java libraries, including the Java Collections Framework. The programmers were provided resources prior to the study to familiarize themselves with Kotlin language syntax.

2. Phase 2 (Live Competitive Programming Contests): Two top performing AICAs (Amazon Q and Tabnine) based on previous study were evaluated in real-world programming contests on LeetCode, CodeChef, and Codeforces over two months.

3.2 Competition Environment Selection

Our choice of platforms - LeetCode, CodeChef, and Codeforces - emerged from their complementary nature in testing different aspects of programming capability: (i) LeetCode's structured environment with well-defined test cases, (ii) CodeChef's emphasis on mathematical and logical reasoning, and (iii) Codeforces' inclusion of visual and abstract problem descriptions. Participating in nine competitions (three per platform) over two months (Nov-Dec '24) allowed us to observe patterns and gather deep insights without being overwhelmed by data. This timeframe proved sufficient to witness both the consistent behaviors and edge cases that inform our reflections.

3.3 Evaluation Framework

Rather than implementing rigid metrics, we adopted a holistic observation approach focusing on:

---

[7] https://medium.com/@oliviachris5567/unlocking-the-power-of-kotlin-why-businesses-are-flocking-to-hire-kotlin-developers-in-2024-2cde9d494897.

[8] https://blog.kotlin-academy.com/kotlin-for-competitive-programming-803ef03e8683?gi=2a7411632061.

[9] https://www.reddit.com/r/Kotlin/comments/163kimo/is_kotlin_a_good_choice_for_competitive/.

1. Questions Solved (Test cases passed): Beyond mere numbers, we observed the nature of problems where AICAs succeeded or struggled.
2. Solution Quality and Error Patterns: We examined not just correctness, but the journey to reaching solutions. We reflected on recurring challenges and their implications for AICA development.
3. Adaptability: We noted how AICAs responded to iterative guidance and feedback.
4. Ratings: The relative performance of the AICAs compared to human participants. Table 1 gives the overall ratings obtained by these two AICAs. The table presents the individual ratings of each AI Coding Assistant (AICA) for competitions on their respective competitive coding platforms. Each platform includes three competitions, with the first row displaying the performance ratings of the AI Assistant for each competition. The second row provides the cumulative average of all the ratings for each AICA on the specific platform.

3.4 AI Tool Selection and Usage Constraints

The selection of AICAs was based on availability, industry adoption, and relevance to competitive programming. While GitHub Copilot is widely used in software engineering, Amazon Q and Tabnine were considered in phase 2 due to their distinct capabilities in structured code completion and adaptive learning as seen from the phase 1 of the study.

To comply with competitive programming platform policies, we imposed the following usage constraints:

1. AI tools were restricted to syntax assistance and minor code completion.
2. AI-generated solutions for full problem logic were prohibited.
3. Error detection and debugging assistance were allowed, but AI-generated answers were manually verified by participants before submission.

These constraints ensured the study focused on AI's role as a learning aid rather than an automated solver.

3.5 Task Design: Competitive Programming Challenges

Participants were assigned a range of competitive programming tasks, structured as follows:

1. Phase 1 (Controlled Study): 1 online event consisting of 9 Kotlin-based problems from Codeforces, covering sorting, recursion, dynamic programming, and graph traversal.
2. Phase 2 (Live Contests): 9 real-world problems over 2 months, spanning algorithmic problem-solving, optimization, and mathematical reasoning.

Each task was classified into low, medium, and high-complexity levels to analyze AI performance at different difficulty levels.

3.6 Data Collection and Evaluation Framework

A mixed-methods approach was employed to assess AI effectiveness:

Quantitative Analysis:

1. AI performance scores from contest ratings[10].
2. 15-point rubric-based evaluation[11], developed by authors and validated by SMEs, covering problem-solving efficiency, adaptability, error handling, and code readability.
3. Epistemic curiosity survey, adapted from Litman's curiosity model, using a 10-question Likert scale [12,17–19].

Qualitative Analysis:

1. Screen recordings to track AI-generated suggestions and user modifications.
2. Participant logbooks documenting AI interactions and problem-solving challenges.
3. Post-event structured interviews capturing user reflections on AI-assisted learning.

Ethical Considerations and Compliance with Platform Policies. Since many competitive programming platforms prohibit full AI-assisted problem-solving, we ensured compliance by:

1. Restricting AI usage to debugging and syntax support only.
2. Manually verifying AI-generated code before submission.
3. Following platform guidelines, including Codeforces' and LeetCode's restrictions on AI-assisted competition participation.
4. The study was conducted under an approved institutional review board (IRB) protocol[12], ensuring ethical compliance regarding participant consent, data privacy, and AI tool usage limitations.

4 Results and Analysis

4.1 AI Performance in Problem-Solving Across Different Competitive Programming Platforms

The study revealed that AICAs were effective in handling simple programming tasks but exhibited significant challenges with complex algorithmic problems. In the controlled Kotlin-based study, AICAs successfully generated correct solutions for low-complexity problems in 85% of cases but struggled with medium and high-complexity problems, achieving only 43% accuracy in these scenarios. In the live competitive programming contests, Amazon Q ranked in the top 10%

[10] How to Interpret Contest Ratings - https://codeforces.com/blog/entry/68288.
[11] Form Link.
[12] IRB no. IITB-IRB2023037.

for structured problem-solving tasks but failed when encountering abstract reasoning challenges. Tabnine, in contrast, exhibited static response patterns, often repeating similar approaches even when faced with different problem constraints. This highlights the "complexity cliff" (discussed in details in Sect. 5.3) in AI performance, where AI tools perform well in structured environments but fail when adaptability and deep problem comprehension are required.

Table 1. Ratings obtained by AICAs

AICAs	Competitive Programming Platforms								
	CodeChef			LeetCode			CodeForces		
Amazon Q	1232	1674	1586	1709	1852	1770	377	600	859
	1497.33			1777.00			612.00		
Tabnine	1371	1201	1211	1548	1481	1464	355	580	821
	1261.00			1497.66			585.33		

4.2 Student Perceptions of AI Assistance

Survey responses indicated mixed perceptions of AI tools in competitive programming. 59% of participants agreed or strongly agreed that AI-generated code was clean and readable, while 35% found AICAs user-friendly. However, 53% disagreed or strongly disagreed that AI-generated suggestions were contextually accurate, indicating limitations in AI's ability to adapt to specific problem requirements. Further, 41% of participants believed AI tools could assist with diverse programming tasks, but only 35% trusted AI-generated suggestions for complex competitive programming constructs like recursion and dynamic programming. These results suggest that while AICAs improve coding efficiency, they do not yet match human problem-solving intuition.

4.3 AI's Impact on Curiosity and Learning Behaviors

A key research question was whether AICAs influence epistemic curiosity in learning programming. The study found no strong correlation between AI usage and increased curiosity. 55% of participants indicated that they would work hard to solve a problem using AI assistance, but only 33% expressed enjoyment in exploring new programming concepts with AI tools. Furthermore, 44% indicated that they sometimes remained concerned about unresolved issues despite AI assistance. Participants who engaged with AI tools frequently were more likely to use them as debugging aids rather than as learning companions. This aligns with prior research suggesting that AICAs currently function more as efficiency boosters than educational scaffolds, reinforcing the need for AI-enhanced learning environments that encourage active problem-solving rather than passive code completion.

4.4 Comparative Analysis of AI Tools in Structured vs. Open-Ended Problem-Solving Contexts

The study's comparative analysis across structured and open-ended programming contexts reveals key differences in AI effectiveness:

1. Structured Problems (LeetCode, Kotlin-based study): AICAs performed well in solving syntax-based and pattern-recognition tasks, providing correct solutions in over 70% of structured problems.
2. Open-Ended Problems (Codeforces, CodeChef): AICAs struggled with dynamic problem-solving, optimization, and real-world constraints, with success rates dropping to 30% in contests requiring novel algorithmic solutions.
3. Edge Case Handling: 82% of participants reported that AICAs were inadequate at handling edge cases, often failing on boundary conditions that required logical reasoning beyond simple pattern-matching.

5 Discussion

The findings from this study provide valuable insights into the capabilities and limitations of AICAs in competitive programming education. This section discusses the strengths and weaknesses of AI-assisted learning, the role of AI as a tutor versus an assistant, the observed "complexity cliff", ethical considerations, and implications for designing AI-powered learning environments.

5.1 Strengths and Limitations of AI in Programming Education

AICAs demonstrate significant strengths in syntactic assistance, debugging, and code readability. In our study, 59% of participants found AI-generated code readable and useful, reinforcing prior findings that AI tools reduce cognitive load by automating routine coding tasks [20]. Additionally, AI-assisted solutions in structured programming challenges achieved over 70% correctness rates, highlighting their effectiveness in well-defined problem contexts. However, these tools exhibit major limitations when it comes to conceptual reasoning and problem adaptation. 76% of participants reported that AI assistants struggled with complex algorithmic problems, particularly in recursion, dynamic programming, and optimization tasks. Furthermore, 82% found AI tools inadequate for handling edge cases, aligning with previous studies that have shown AI's limited generalization abilities [2]. These results suggest that while AI tools can enhance productivity, they do not yet possess the adaptability required for high-level programming education.

5.2 Role of AICAs as a Tutor vs. Assistant

One of the central questions in AI-assisted education is whether AI tools function better as tutors (providing structured guidance) or assistants (offering support but requiring human oversight). Our findings suggest that current AICAs operate more effectively as assistants rather than tutors. While they excel in providing quick syntactic suggestions and debugging help, they fail in delivering adaptive, step-by-step learning pathways, a key characteristic of intelligent tutoring systems. A major gap is the lack of context-aware feedback. Unlike traditional intelligent tutoring systems, which offer scaffolded learning experiences, AICAs provide static, one-size-fits-all responses. This limitation highlights the need for more personalized AI-driven tutoring approaches, incorporating real-time adaptive feedback mechanisms to better support programming learners.

5.3 The "Complexity Cliff": Where AI Struggles in Advanced Problem-Solving

A critical finding from this study is the "complexity cliff", the observed sharp decline in AI performance as problem difficulty increases. While AICAs performed well on structured problems, their success rates dropped from 70% on simple syntax-based problems to 30% on open-ended, abstract reasoning tasks. This drop-off aligns with prior research indicating that AI tools rely heavily on pattern recognition and code completion rather than true problem-solving [28]. Our findings further reveal that AI-generated solutions often require extensive human intervention, as participants had to iteratively prompt the AI to generate more contextually accurate responses. The "complexity cliff" suggests that future AI models need to integrate multimodal learning capabilities, incorporating visual problem representations, logical reasoning enhancements, and interactive debugging suggestions, to bridge the gap between pattern-based assistance and true problem-solving adaptability. However, current AI applications in complex problem-solving are unevenly distributed across cognitive, metacognitive, social, and affective dimensions, with most progress in the cognitive domain [9].

5.4 Ethical Concerns and Platform Compliance

One major challenge in studying AI-assisted competitive programming is the ethical implications of AI tool usage. Many programming competition platforms prohibit AI-generated solutions, restricting usage to syntax suggestions and debugging assistance. Our study adhered to these guidelines by manually verifying AI-generated code and limiting AI intervention to permitted tasks. However, the broader issue of AI reliance and academic integrity remains. The increasing adoption of AICAs raises concerns about cognitive offloading, where students rely on AI to complete tasks rather than developing independent problem-solving skills. Prior research has highlighted the risk of over-reliance on AI tools, particularly in educational settings [15]. Addressing this challenge requires a balanced approach that integrates AI while reinforcing fundamental coding skills. To mitigate ethical risks, future AI-assisted learning environments should implement transparency mechanisms, such as AI-generated solution flags, plagiarism detection tools, and structured AI-assistance guidelines for programming competitions.

5.5 Implications for Designing AI-Powered Learning Environments

The findings from this study have significant implications for the design of AI-powered programming education tools. Firstly, it is essential to balance AI automation with human oversight, ensuring that AI tools act as collaborative learning partners rather than replacements for human problem-solving. Implementing hybrid learning models that combine AI-generated guidance with instructor-led reinforcement can maintain active student engagement. Additionally, the focus should be on encouraging active learning instead of mere autocompletion, prompting students to refine their problem-solving approaches rather than simply generating answers. Progressive hinting systems, where AI offers incremental guidance rather than full solutions, could enhance conceptual understanding. Furthermore, future AI tutors should adopt multimodal learning strategies by integrating interactive problem visualizations, logic flow diagrams, and contextual explanations to bolster learning outcomes. This includes moving beyond text-based code suggestions to feature voice-guided explanations, real-time debugging assistance, and adaptive challenge recommendations. Finally, ethical integration of AI in competitive programming environments is crucial; AI-powered tools should incorporate built-in compliance checks to prevent unauthorized AI usage during competitions, while educational institutions must provide clear policies to guide students on ethical AI engagement in learning settings.

5.6 AI's Impact on Curiosity in Learning

One of the key findings of this study is that AICAs have an ambiguous effect on epistemic curiosity. While some participants used AI suggestions as a starting point for deeper exploration, others relied on AI-generated code without questioning its correctness or efficiency. While 55% of participants indicated that AI-assisted problem-solving encouraged them to work harder to find a solution, only 33% reported an increase in curiosity about alternative algorithms or deeper problem understanding. This aligns with previous research indicating that AI tools often provide quick solutions without necessarily encouraging iterative exploration [16]. Our findings suggest that AI tools need to be redesigned to actively prompt students to consider multiple solutions, explore variations, and question their problem-solving approach. Our results indicate that AI tools must strike a balance between providing efficiency and fostering intellectual engagement.

6 Future Work and Conclusion

The findings of this study underscore the opportunities and challenges associated with integrating AICAs into competitive programming education. While these tools offer valuable support in syntax, debugging, and structured problem-solving, they fall short in adaptive reasoning and conceptual learning. This section outlines key areas for future research and summarizes the broader implications of our findings.

1. Multi-Institutional and Longitudinal Studies - Expanding this study to multiple universities and educational settings would provide a more diverse dataset and improve generalizability. Conducting a longitudinal study tracking students' progress over several months would help determine the long-term impact of AICAs on learning outcomes and problem-solving skills.
2. Enhancing AI Tools for Adaptive Learning - Current AI assistants lack adaptability in complex problem-solving scenarios. Future research should focus on integrating reinforcement learning models to improve AI's ability to learn from user interactions and adapt responses accordingly. AI tools should be designed to provide personalized feedback mechanisms rather than one-size-fits-all solutions. Real-time hints and progressive scaffolding techniques could be integrated into AI tutors to help students gradually arrive at solutions instead of receiving direct answers.
3. Balancing AI Automation with Human Problem-Solving - A key concern raised in this study is the risk of cognitive offloading, where students rely on AI-generated solutions instead of actively engaging in problem-solving. Future research should investigate human-AI collaboration models, where AI tools suggest problem-solving strategies instead of complete code implementations. Developing "explainable AI" (XAI) features, where AI-generated code includes step-by-step reasoning and contextual explanations, could help mitigate over-reliance.

This study provides empirical insights into the role of AICAs in competitive programming education. While these tools significantly improve syntax correction, debugging efficiency, and structured problem-solving, their limitations in complex reasoning, adaptability, and conceptual learning highlight the need for further improvements. AICAs should be positioned as supportive tools rather than substitutes for problem-solving. The findings reinforce the importance of hybrid learning models, where AI assists learners without diminishing critical thinking skills. The observed "complexity cliff" suggests that AI tools must evolve to support higher-order problem-solving and adaptive learning techniques. Future AI tutors should focus on interactive feedback, multimodal explanations, and scaffolding mechanisms to foster deeper engagement. AICAs should move beyond being static code generators to dynamic, context-aware learning partners. The next generation of AI-assisted learning environments should emphasize conceptual understanding over efficiency gains, ensuring that students develop strong algorithmic reasoning and problem-solving skills. By addressing these challenges, AI can transition from a passive coding assistant to an active learning facilitator, ultimately reshaping the way students learn programming. Future research must continue to explore how AI can be ethically and effectively integrated into programming education, ensuring that these tools enhance learning rather than replace fundamental cognitive processes.

References

1. Becker, B.A., Denny, P., Finnie-Ansley, J., Luxton-Reilly, A., Prather, J., Santos, E.A.: Programming is hard-or at least it used to be: educational opportunities and challenges of AI code generation. In: Proceedings of the 54th ACM Technical Symposium on Computer Science Education, vol. 1, pp. 500–506 (2023)
2. Coignion, T., Quinton, C., Rouvoy, R.: A performance study of LLM-generated code on leetcode. In: Proceedings of the 28th International Conference on Evaluation and Assessment in Software Engineering, pp. 79–89 (2024)
3. Corso, V., Mariani, L., Micucci, D., Riganelli, O.: Generating java methods: an empirical assessment of four AI-based code assistants. arXiv preprint arXiv:2402.08431 (2024)
4. Denny, P., MacNeil, S., Savelka, J., Porter, L., Luxton-Reilly, A.: Desirable characteristics for AI teaching assistants in programming education. In: Proceedings of the 2024 on Innovation and Technology in Computer Science Education, vol. 1, pp. 408–414 (2024)
5. Haindl, P., Weinberger, G.: Does ChatGPT help novice programmers write better code? Results from static code analysis. IEEE Access (2024)
6. Haindl, P., Weinberger, G.: Students' experiences of using ChatGPT in an undergraduate programming course. IEEE Access **12**, 43519–43529 (2024)
7. Haque, E.A., Brown, C., LaToza, T.D., Johnson, B.: Towards decoding developer cognition in the age of AI assistants. arXiv preprint arXiv:2501.02684 (2025)
8. Hüsing, S., Schulte, C., Sparmann, S., Bolte, M.: Using worked examples for engaging in epistemic programming projects. In: Proceedings of the 55th ACM Technical Symposium on Computer Science Education, vol. 1, pp. 443–449 (2024)
9. Joksimovic, S., Ifenthaler, D., Marrone, R., De Laat, M., Siemens, G.: Opportunities of artificial intelligence for supporting complex problem-solving: findings from a scoping review. Comput. Educ.: Artif. Intell. **4**, 100138 (2023)
10. Kapu, N.J., Sreejith, M.: DemoCraft: using in-context learning to improve code generation in large language models. arXiv preprint arXiv:2411.00865 (2024)
11. Khojah, R., Neto, F.G.D.O., Mohamad, M., Leitner, P.: The impact of prompt programming on function-level code generation. arXiv preprint arXiv:2412.20545 (2024)
12. Lee, S., Hsu, Y.T., Cheng, K.H.: Do curious students learn more science in an immersive virtual reality environment? Exploring the impact of advance organizers and epistemic curiosity. Comput. Educ. **182**, 104456 (2022)
13. Lee, S., Liang, J.C., Hsu, C.Y., Tsai, M.J.: Students' beliefs about computer programming predict their computational thinking and computer programming self-efficacy. Interact. Learn. Environ. **32**(8), 4088–4108 (2024)
14. Li, J., Li, G., Li, Y., Jin, Z.: Structured chain-of-thought prompting for code generation. ACM Trans. Softw. Eng. Methodol. **34**(2), 1–23 (2025)
15. Liang, J.T., Yang, C., Myers, B.A.: A large-scale survey on the usability of AI programming assistants: successes and challenges. In: Proceedings of the 46th IEEE/ACM International Conference on Software Engineering, pp. 1–13 (2024)
16. Lin, Z.: Why and how to embrace AI such as chatGPT in your academic life. Roy. Soc. Open Sci. **10**(8), 230658 (2023)
17. Litman, J., Hutchins, T., Russon, R.: Epistemic curiosity, feeling-of-knowing, and exploratory behaviour. Cogn. Emot. **19**(4), 559–582 (2005)
18. Litman, J.A.: Interest and deprivation factors of epistemic curiosity. Pers. Individ. Differ. **44**(7), 1585–1595 (2008)

19. Litman, J.A., Crowson, H.M., Kolinski, K.: Validity of the interest-and deprivation-type epistemic curiosity distinction in non-students. Pers. Individ. Differ. **49**(5), 531–536 (2010)
20. Liu, C., Zhang, X., Zhang, H., Wan, Z., Huang, Z., Yan, M.: An empirical study of code search in intelligent coding assistant: perceptions, expectations, and directions. In: Companion Proceedings of the 32nd ACM International Conference on the Foundations of Software Engineering, pp. 283–293 (2024)
21. Liu, J., Li, S.: Toward artificial intelligence-human paired programming: a review of the educational applications and research on artificial intelligence code-generation tools. J. Educ. Comput. Res. **62**(5), 1385–1415 (2024)
22. Liu, J., Xia, C.S., Wang, Y., Zhang, L.: Is your code generated by chatGPT really correct? Rigorous evaluation of large language models for code generation. In: Advances in Neural Information Processing Systems, vol. 36 (2024)
23. Ma, Y., et al.: Bridging code semantic and LLMs: semantic chain-of-thought prompting for code generation. arXiv preprint arXiv:2310.10698 (2023)
24. Mirzayanov, M., et al.: Codeforces as an educational platform for learning programming in digitalization. Olympiads Inform. **14**(133-142), 14 (2020)
25. Nakamura, H., Sonoyama, A., Kamiyama, T., Oguchi, M., Yamaguchi, S.: Performance study of Kotlin and Java programs with bytecode analysis. J. Inf. Process. **32**, 380–395 (2024)
26. Pan, Z., Xie, Z., Liu, T., Xia, T.: Exploring the key factors influencing college students' willingness to use AI coding assistant tools: an expanded technology acceptance model. Systems **12**(5), 176 (2024)
27. Pinto, G., De Souza, C., Rocha, T., Steinmacher, I., Souza, A., Monteiro, E.: Developer experiences with a contextualized AI coding assistant: usability, expectations, and outcomes. In: Proceedings of the IEEE/ACM 3rd International Conference on AI Engineering-Software Engineering for AI, pp. 81–91 (2024)
28. Vishnu, S., Sahil, Garg, N.: Unveiling the role of GPT-4 in solving LeetCode programming problems. Comput. Appl. Eng. Educ. **33**(1), e22815 (2025)

Gamified Intelligent Communities of Practice: A Path to Acknowledge and Support Underprivileged Teachers

Paula T. Palomino[1](✉), Luiz Rodrigues[2], Carlos Portela[3], Grace Lis Porto[4], Armando Toda[5], Ig Bittencourt[6], Seiji Isotani[7], and Thiago Cordeiro[6]

[1] Sao Paulo State College of Technology (FATEC), Matão, Brazil
paula.palomino@fatec.sp.gov.br
[2] Technological Federal University of Paraná (UTFPR), Curitiba, Brazil
luizrodrigues@utfpr.edu.br
[3] Federal University of Pará (UFPA), Belém, Brazil
csp@ufpa.br
[4] Federal University of São Carlos (UFSCar), São Carlos, Brazil
grace@estudante.ufscar.br
[5] São Paulo, Brazil
[6] Federal University of Alagoas (UFAL), Penedo, Brazil
{ig.ibert,thiago}@ic.ufal.br
[7] University of São Paulo (USP), São Paulo, Brazil
sisotani@icmc.usp.br

Abstract. This study explores the support of Artificial Intelligence in Education (AIED) in empowering teachers within Brazil's educational context. Recognizing the often-neglected importance of teachers in the learning ecosystem, we introduce a Gamified Intelligent Community of Practice (GICoP) tailored to enhance teacher competence and foster collaboration. While AIED research has predominantly emphasized student-centered approaches, our research highlights the importance of addressing teachers' professional development needs in response to technological advancements. By interlinking the objectives of the Common National Curriculum Base (BNCC) with advanced AI tools, our proposed GICoP seeks to create an environment where knowledge dissemination is intuitive and aligned with national standards. Our research also contributes to AIED by offering valuable insights and outlining challenges in formulating public policies that intertwine AI with a teacher-centric focus.

Keywords: Education · Communities of Practice · Artificial Intelligence · Gamification

1 Introduction

The heavy workload is a common problem for Brazilian teachers due to inadequate income, poor working conditions, lack of professional development

Armando Toda—Independent Researcher.

A. I. Cristea et al. (Eds.): AIED 2025, CCIS 2590, pp. 195–207, 2025.
https://doi.org/10.1007/978-3-031-99261-2_18

opportunities, and limited resources [5]. These problems result in a lack of time for lesson planning and preparation, grading assignments, and attending to parents and students, leading to long working hours. Ultimately, these issues can lead to mental health issues, such as stress, exhaustion, low morale, and burnout, negatively affecting their well-being, teaching quality, and job satisfaction [14].

While several reasons might explain teachers' work overload and lack of recognition, one major factor is the limited resources and funding available for the education sector, which might lead to underfunded schools, overcrowded classrooms, and a lack of support for teachers [14]. Based on that context, the necessity arises for innovative, scalable, and accessible solutions that can mitigate teacher workload, improve their well-being, enhance their professional development, and cultivate societal respect for this noble profession. Advancements in educational technology [16], communities of practice [21], and gamification [17] offer promising pathways in this direction.

To achieve that goal, we introduce the Gamified Intelligent Community of Practice (GICoP), which can be a powerful tool for bringing together educators from diverse backgrounds in a supportive, engaging, and interactive environment.

This article aims to conceptualize and propose a novel framework for a GICoP specifically designed to engage teachers. It focuses on leveraging gamification strategies and intelligent system designs to mitigate workload, enhance well-being and professional development, and foster collaborative problem-solving. Through this study, we contribute to the theoretical discourse on teacher workload and professional development, providing actionable insights for educational policymakers, administrators, and ed-tech developers.

2 Background

Integrating Artificial Intelligence in Education (AIED) has revolutionized how students learn and teachers instruct. AI-powered tools and systems can personalize learning experiences, adapt to individual student needs, and offer real-time feedback [3]. Those applications might improve the student learning experience and empower teachers by providing information on points that need attention, automating administrative tasks, and maximizing the time they have to help students [6].

Recently, the prominence of Natural Language Processing (NLP) in education is mainly due to large language models, such as Generative Pre-trained Transformer (GPT) and similar architectures [8]. These models are trained on vast amounts of text data, allowing them to acquire language knowledge, context, and patterns. GPT, for instance, leverages a transformer architecture that excels in various language-related tasks, including text generation, translation, summarization, and question-answering [12].

However, the rise of large language models (LLMs) in education brings ethical concerns to the forefront. One pressing issue is the potential for these models to inadvertently generate biased or inappropriate content [20]. LLMs learn from the data on which they are trained. If that data contains biases or misinformation, the models may inadvertently perpetuate these issues in educational content [8], posing a significant challenge for educators and institutions that strive to provide equitable and accurate learning materials.

There is a growing emphasis on the importance of filtering content generated from LLMs in educational settings to address these ethical issues. Content moderation and review mechanisms are crucial to ensure that materials created by these models align with academic standards, are free from bias, and comply with the institution's ethical guidelines. Human oversight, combined with AI-powered content analysis tools, can help maintain the quality and integrity of educational materials, fostering trust and ethical awareness.

In parallel with advancements in AI and content generation, gamification has emerged as a complementary strategy for enhancing engagement and motivation in educational contexts. While AI technologies offer powerful tools for supporting instruction and resource creation, gamification provides design principles that can foster sustained participation and a sense of purpose among users. Together, these approaches offer synergistic opportunities to rethink how teachers are supported, motivated, and empowered in their professional practice.

Gamification has been used in several domains, including education and learning, health and exercise, and crowd-sourcing, aiming to improve engagement, foster desired behaviors, and increase motivation [10]. Evidence from several meta-analyses supports gamification's effectiveness [9,10]. Despite robust empirical evidence of its potential [9], researchers have cautioned against the use of gamification [1].

To ensure the benefits of gamification and prevent adverse effects, it is necessary to follow a design process. Palomino *et al.* [15] recently developed a framework for content gamification, which is especially grounded on components from Gamification Heuristics (e.g., Purpose and Meaning; Scarcity) [18] and the Hero's Journey (e.g., Call to Action; Transformation) [19], among others, aiming to provide an immersive experience primarily based on game elements such as Narrative and Storytelling. Although the framework focused on content gamification, it was designed for learning. Thus, a framework for content gamification remains needed regardless of its application domain. Upon reviewing available options, our investigation led us to select this particular framework for adaptation primarily because it stands out as one of the few viable choices that can meet the intricate demands of engaging teachers. This decision was informed by the framework's demonstrated flexibility, making it an optimal base for adaptation rather than adopting another one without modification or attempting to develop a new framework from scratch.

3 Methods

First, we conducted a descriptive-qualitative study with the main objective of identifying the needs and perspectives of primary education teachers regarding the use of digital technologies in teaching. The data were collected over two months through interviews with elementary school teachers recruited using convenience sampling. We contacted 27 possible participants from the Brazilian government's policy, "Brasil na Escola" (Brazil in School), whose objective is to promote strategies and innovations that ensure permanence, learning, and school progression with equity. Of these, six teachers agreed to participate in the interviews. All participants are public school teachers from Alagoas, Brazil.

All interviews were conducted remotely via Google Meet in a secure and isolated environment, with no interference or participation from third parties. They were conversational, following a semi-structured script. A qualitative approach was chosen because the study aimed to explore teachers' perceptions and experiences in depth, rather than testing predefined hypotheses. Given the complexity of teachers' daily challenges and the need to uncover nuanced insights into their professional contexts, semi-structured interviews provided the flexibility necessary to capture rich, detailed narratives that would not have been accessible through a purely quantitative or survey-based method.

4 Results

This section presents the main findings of the qualitative study conducted with primary education teachers in Alagoas, Brazil. The results are organized into two thematic categories that emerged from the interview analysis: teachers' experiences with their daily professional activities (School Routine) and their perspectives on digital collaboration and community building (Digital Community).

In the School Routine theme, all six participants reported investing significant time and resources in developing lesson plans, drawing upon books, online materials, and video content. Diagnostic assessments were widely recognized as crucial for tailoring instruction to students' needs but were also described as highly time-consuming, often requiring between two and eight hours of preparation.

In the Digital Community theme, teachers expressed a strong interest in collaboration and resource sharing, but noted that opportunities for peer-to-peer interaction were limited in physical school settings. They expressed a desire for an online platform that would enable them to share materials efficiently, engage with colleagues across different regions, and participate in video conference discussions. Participants also emphasized the importance of a simple, user-friendly interface designed specifically for teachers, which facilitates meaningful digital interactions while maintaining a sense of human connection.

These findings provide the empirical foundation for conceptualizing the proposed Gamified Intelligent Community of Practice (GICoP).

5 Gamified Intelligent Community of Practice (GICoP)

Building on the findings from the empirical study presented in Sect. 4, we propose a Gamified Intelligent Community of Practice (GICoP) tailored to the needs and aspirations of primary education teachers. The following subsections outline the theoretical foundations, core components, and technical architecture of the proposed intervention.

5.1 Motivation

Based on the opportunity to help optimize teachers' jobs, this might be accomplished by providing additional support for their daily tasks, such as grading assignments and preparing lessons. In that context, we believe that a Community of Practice[1] (CoP) might help to ease teachers' workload and optimize tasks such as planning and preparing lessons. Then, by reducing teachers' workload, this intervention might consequently help to reduce teacher burnout and improve job satisfaction, among other factors.

To elaborate on how this intervention might address teachers' problems, we rely on the Theory of Change, which is a step-by-step outline of an intervention's long-term goals and expected outcomes [2]. Accordingly, we believe that such a CoP would increase teacher collaboration and the sharing of resources, such as lesson plans and assignments. As a consequence, it would provide an environment in which teachers would be recognized for their contributions (i.e., collaborations and sharing), help them develop themselves based on their peers' contributions, and, ultimately, reduce the workload of designing everything from scratch, thereby maximizing job satisfaction following acknowledgments for their merits.

Furthermore, recent research shows that AI and gamification, or game-based learning, can offer significant value in educational settings, enhancing student engagement and learning outcomes [11]. These technologies can also address concerns about teachers' engagement with digital platforms and help them find valuable resources to benefit from the CoP. Thus, to be successful, such an intervention should implement a Gamified Intelligent Community of Practice (GICoP).

5.2 GICoP Concept

The empirical study's findings directly informed the GICoP Concept Design. Teachers' reports of excessive workload motivated the integration of AI-driven support mechanisms, particularly for lesson planning and resource creation. Their expressed need for broader collaboration opportunities guided the adoption of a digital community model as the core structure of the intervention. Finally, their desire for a motivating and supportive environment underpinned the incorporation of gamification elements designed to foster engagement and professional development. Each component of the GICoP responds explicitly to the challenges and aspirations articulated by the participating teachers.

Figure 1 details the proposed GICoP.

The main objective of the CoP's Gamification is to design a cycle that allows teachers to be recognized for their merits, skills, and contributions, among other elements. To achieve this, we will use the Gamification Journey framework [15] as the basis for designing the gamification, which will be adapted to meet the needs of teachers.

[1] People with similar interests, passion or concerns that interact with each other regularly to learn how to do something better together [22].

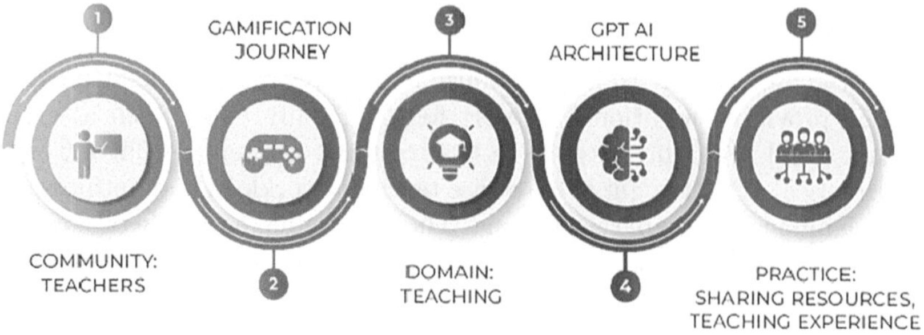

Fig. 1. Gamified Intelligent Community of Practice Framework.

5.3 Gamification Design

We adapted the Gamification Journey framework [15] to focus specifically on motivating and engaging teachers within the GICoP. While originally developed to foster student engagement, its structure—divided into four quadrants ("Call to Action", "Trials", "Transformation", and "Results")—offers a flexible foundation for designing immersive experiences for teacher professional development. Subsequently, each of these is described, and we demonstrate how to implement them. Specifically, we will explore Narrative and Storytelling, which are defined as follows:

- *Narrative*: The order of events occurring in a game through the user's experience. The user's implicit choices also influence this experience.
- *Storytelling*: How the environment's story (and context) is told and presented through text, voice, or even sensorially.

The Call to Action Quadrant. According to this quadrant, the system design should consider at least the following heuristics [15]:

- Purpose and Meaning: The presented content must be closely related to the teacher's daily life, their "common world". They should perceive a significant contribution and "gain" something from it, enriching their life experience.
- Immersion: The design should acknowledge that it brings someone from an "outside" world into a new world, effectively immersing the teacher in this new universe. The teacher must perceive that the journey they are about to embark on will be rewarding.
- Autonomy and Creativity: The system should convey a sense of autonomy to the teacher, allowing them to approach tasks as they see fit. It should foster their creativity and enable them to experience the journey in the best way possible.

The Trials Quadrant. This stage should present the pedagogical content in a way that takes the teacher out of their comfort zone, introduces them to the new, and challenges them to grow. To accomplish this, the system design should consider at least the following heuristics:

- Properties and Rewards: Utilize external motivational stimuli to compensate for stepping out of the 'comfort zone.' Examples include virtual goods, incentives, and a virtual economy.
- Scarcity: Employ external motivational stimuli related to exclusivity and status to provide confidence and satisfaction to the user, motivating them to continue the journey.
- Challenges and Competencies: Assuming the user is confident (thanks to the previous stage), the idea is to work with intrinsic motivation, integrate knowledge, and aim for the user to challenge themselves to test their competencies.

The Transformation Quadrant. In this quadrant, the system is ready to bring this transformation to light through ultimate challenges and experiences. Drawing a parallel to games, those could be seen as *boss fights*, for instance. For this, the Gamification journey framework suggests the following heuristics:

- Loss Avoidance: The design should apply pressure, using motivational stimuli that address urgency and the loss of rewards previously acquired.
- Change and Disruption: The system should propose innovations and explore stimuli that work on disruption control to motivate teachers to keep on their transformative arc.
- Unpredictability: The idea is to explore unpredictability, aiming to stimulate teachers' creativity and adaptive capabilities, which can be achieved through varied rewards and challenges that motivate them to create new resources based on what they have learned from others.

The Results Quadrant. The system should provide an environment in which teachers can express themselves and, based on that, follow the cycle, closing with the impression that something new will happen (i.e., a new journey). For this, the framework suggests the following three heuristics, as introduced next:

- Feedback: The system needs to provide clear insights that demonstrate to the teacher the process they have gone through up to this point. Notably, even though this heuristic should be part of one's whole journey, it is essential at this point.
- Relatedness: The system must ensure that teachers feel part of the community and understand that they are transforming agents of their journey. Accordingly, the system must inform them of these aspects and help them consolidate their feelings, which can be accomplished through social bonds.
- Completeness and Mastery: The system should demonstrate the results of the teacher's journey, highlighting where they started and where they are ending, which can be achieved through motivational stimuli that focus on progressive objectives and accomplishments.

5.4 AI Integration and Support

Our approach aims to empower teachers, and we strive to explore the benefits of AIED, such as increasing teaching competence and working efficiency [23]. Nevertheless, we understand that teachers may lack technological skills to, for instance, interact appropriately and benefit from AIED technologies [4]. Hence, our approach relies on the recent, increasing trend of Natural Language Processing (NLP) within AIED research [24].

Recent advancements in NLP have led to the development of large language models (LLMs) such as GPT-4, which power applications like ChatGPT [8]. Among the many applications of such technologies, these have been widely used for question-answering and text-completion tasks. For instance, one might ask a question and receive an answer from ChatGPT, and then ask it to generate a textual resource (e.g., an assignment for first-year elementary students). Hence, such solutions are valuable for our CoP's audience because they offer a communication interface similar to people's daily activities (i.e., using natural language). Similarly, these solutions hold great potential to empower teachers because they might generate a wide range of valuable resources, such as preliminary lesson plans and assignments, which can then be tailored to teachers' preferences, needs, and students' characteristics [23]. Therefore, we believe generative LLMs offer the functionalities our CoP's AI component needs.

Based on that context, the first step in implementing our AI component is selecting a generative large language model (LLM). At the time of writing, GPT architectures were probably the most advanced and popular alternatives. While ChatGPT, built on GPT architectures, plays a substantial role in its popularity, studies suggest that the performance efficiency of these architectures is noteworthy. However, the most recent models, such as GPT-3 and 4, are not open source and, from a product development perspective, can only be used through OpenAI's paid API. On the other hand, several generative models are openly available on platforms such as Huggingface[2], offering alternatives for developing products like the proposed CoP.

With the availability of generative AIs, ensuring that they generate meaningful resources for teachers [23] is paramount. Particularly, Brazil's primary education follows the Common National Curriculum Base (*Base Nacional Comum Curricular* - BNCC - in Portuguese): a normative document that defines an organic, progressive set of essential learning that all students must develop throughout the stages and modalities of primary education[3]. Therefore, to be of value to Brazilian teachers, the generated resources must be aligned with BNCC. Notably, the BNCC is openly available on the internet, and generative models like GPT-4 are often trained on content from the internet. Therefore, they likely possess the factual knowledge to create textual content aligned with BNCC. Nevertheless, achieving the expected results from such models depends on writing proper prompts. Thus, our AI component will internally use a *template prompt*

[2] https://huggingface.co/.
[3] http://basenacionalcomum.mec.gov.br/.

to complement teachers' prompts and ensure all requested resources are created in light of the BNCC, even when the teachers do not explicitly ask.

Although the approach above is expected to ensure resources are aligned with BNCC, we must acknowledge that the quality of an instructional resource depends on several criteria. Some examples are personal preferences, needs, goals, and evaluation criteria. Especially from a teacher's perspective, considering the student's characteristics and the extent to which they face difficulties is of utmost importance. Therefore, according to the collaborative nature of our CoP, the AI component will feature a reinforcement learning module. This module will iteratively learn to assess the quality of generated resources based on feedback gathered from teachers' interactions with the CoP. Then, it will be used to fine-tune the text generation model, improving its ability to create resources tailored to teachers' personal preferences and those of their students.

Lastly, one must note that text generation models often have no filter regarding specific characteristics of the content they create. To ensure our AI component creates BNCC-based materials, we use a template prompt that adds such requirements. Despite that, the text generation models are still susceptible to creating texts featuring abusive, offensive, discriminatory, and other unacceptable content [20]. Hence, our AI component will feature a classifier to ensure that no token or phrase within resources sent to teachers contains such content.

Figure 2 illustrates the AI component workflow. Users begin by entering a prompt into the Community of Practice (CoP), which forwards it to the AI component. The AI system complements the input with a template prompt to ensure alignment with the BNCC requirements before submitting it to the text generation model. The model generates a resource, which is then evaluated by a classifier. If the classifier detects unacceptable content, the model is prompted to regenerate the material; otherwise, the finalized resource is sent back to the CoP for teacher review. Feedback from teachers and interaction data are subsequently used to refine the model through a reinforcement learning module, progressively enhancing the quality and contextual relevance of future resources.

6 Discussion

This section reflects on the key contributions, opportunities, and challenges associated with the proposed Gamified Intelligent Community of Practice (GICoP).

The foray into developing and utilizing Artificial Intelligence in Education (AIED) to enhance teaching, rather than solely focusing on student-centric methodologies [4], represents an innovative shift in the pedagogical landscape. As we delve deeper into the intricacies of this proposed approach, several key facets stand out as potential pivot points for its success and scalability.

Traditionally, the primary beneficiaries of AIED have been students, who leverage AI's capabilities to tailor learning experiences individually [13]. We aim to fill a much-needed gap in the system by shifting the focus to teachers. Our approach's rationale to empower educators could lead to an amplified overall impact. By elevating teaching competence and enhancing work efficiency [23], we can cater to the multiplier effect of teachers on their many students.

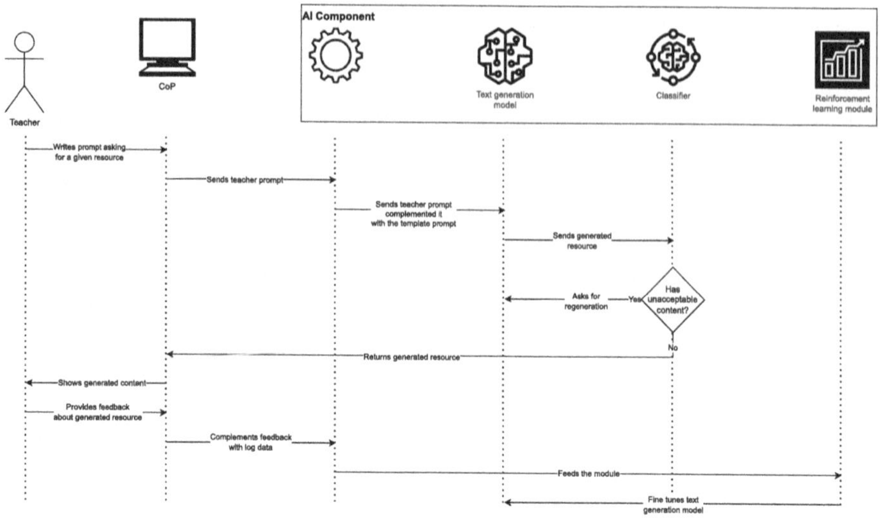

Fig. 2. Summary of the AI component of the proposed Community of Practice.

Recent literature shows that the increasing convergence of NLP and AIED presents a golden opportunity. The inception of large language models (LLMs) has revolutionized the AI realm. The ability of these systems to generate accurate and contextually relevant information through simple textual prompts is particularly significant, streamlining the AI-user interface and making it more intuitive and less intimidating, especially for educators who may not be inherently tech-savvy.

Ensuring the generated content aligns with Brazil's BNCC reflects a commitment to localized contextual relevance, as the AI-generated content is not just generic but tailored to specific curriculum mandates. While GPT models' broad training base likely covers the BNCC, using a "template prompt" adds an extra layer of assurance. This proactive approach mitigates potential misalignments that could compromise the content's relevance.

Drawing insights from teachers' feedback and fine-tuning the AI's outputs ensures the model's adaptability and evolution over time. This iterative learning approach mirrors the essence of pedagogy itself—a field where adaptability and continuous learning are paramount.

The emphasis on filtering and ensuring that the generated content is free from potentially harmful elements underscores the project's ethical foundations. Given the increasing concerns about AI perpetuating biases and potential harm, this content moderation layer is timely and essential.

The proposed AI component of the Community of Practice (CoP) blueprint, summarized succinctly in Fig. 2, provides a roadmap for potential extrapolation beyond the confines of the Brazilian educational framework. While the focus on BNCC alignment is pertinent for the current audience, the model's architecture can potentially cater to other curricula, making it a global tool. While the proposed approach offers significant promise, it is essential to acknowledge the critical challenges that must be addressed to ensure its effective implementation and long-term sustainability.

Despite teachers' crucial role in shaping societies, they often do not receive the recognition they deserve. There is a prevailing sentiment in many cultures that views teaching as a "fallback" profession or less prestigious than fields like medicine, law, or engineering. Integrating AI into teaching could exacerbate this issue.

Besides that, not all educators will welcome the introduction of AI in their classrooms. Veteran teachers, who have honed their teaching methods over decades, might view AIED as an unnecessary intrusion. Implementing a policy without considering the reservations and concerns of such educators could lead to friction and reduced efficacy.

AIED tools are often not user-friendly, requiring a certain level of digital literacy. There is a presumption that all teachers can seamlessly integrate and make the most out of these tools. However, with adequate training, many educators might be able to handle these AI tools. Developing any system in this context must address this limitation to ensure teachers understand and can innovatively use AIED in their classrooms.

From a policy perspective, integrating AIED into classrooms is about more than just developing the software. It requires hardware upgrades, infrastructure changes, and potentially revamped classroom designs to facilitate optimal AI integration. All these necessitate substantial financial investments. In regions with limited educational budgets, such as Alagoas, Brazil, prioritizing AI could mean sidelining other crucial aspects like classroom repairs, textbook upgrades, or even basic amenities. How can we overcome this challenge? Could Unplugged AI [7] be a path for that?

The implications of this study are multifaceted and potentially far-reaching. Firstly, for the teachers themselves, successfully implementing this GICoP could alleviate workload pressures, foster their professional growth, and enhance their job satisfaction and morale, indirectly contributing to improved teaching quality and student learning outcomes. From a societal perspective, the GICoP could help elevate societal respect and recognition for teachers by promoting community engagement and public acknowledgment of their efforts. For policymakers and administrators, the insights from this work could guide the design and implementation of effective support systems for teachers. For educational researchers, this work enriches the theoretical discourse around technology-enabled teacher communities and professional development, providing fertile ground for future research. Lastly, this work offers a practical blueprint for ed-tech developers to create innovative, user-friendly, and impactful solutions that address real-world educational challenges.

7 Final Remarks

Integrating Artificial Intelligence in Education (AIED) has opened new horizons for enhancing teacher competence and providing personalized support.

This paper presented the conceptualization of a Gamified Intelligent Community of Practice (GICoP) designed to bridge the gap between teachers' professional development needs and the opportunities offered by technological innovations. Rooted in Brazil's educational context and aligned with the Common National Curriculum Base (BNCC), the proposed framework leverages AI and gamification to directly address the challenges articulated by primary education teachers, particularly excessive workload, limited collaboration opportunities, and the need for motivational environments.

Our emphasis on empowering teachers stemmed from a keen recognition of their pivotal role in the educational ecosystem, an aspect often overshadowed by student-centered AIED initiatives. By shifting the focus to supporting educators through intuitive, accessible, and contextually relevant tools, we aim to foster not only individual competence but also collective professional growth through knowledge sharing and peer collaboration. Nevertheless, the practical application of the GICoP framework entails critical challenges. Effective adoption will depend on addressing disparities in digital literacy, safeguarding content quality and ethics, and ensuring infrastructural readiness, particularly in resource-constrained environments. While the proposed design offers a promising blueprint, it remains preliminary. Future empirical validation, including pilot studies and longitudinal evaluations, is essential to assess its real-world effectiveness, its impact on teacher workload and well-being, and its contribution to broader educational outcomes.

As education continues to evolve, equipping and empowering teachers must remain a central priority. By grounding technological innovation in educators' lived realities, we can ensure that tools like the GICoP not only enhance professional development but also strengthen the foundations of equitable, resilient, and transformative education systems.

References

1. Bai, S., Hew, K.F., Huang, B.: Is gamification "bullshit"? Evidence from a meta-analysis and synthesis of qualitative data in educational contexts. Educ. Res. Rev. 100322 (2020)
2. Brest, P.: The power of theories of change (2010)
3. Chen, X., Zou, D., Xie, H., Cheng, G., Liu, C.: Two decades of artificial intelligence in education. Educ. Technol. Soc. **25**(1), 28–47 (2022)
4. Crompton, H., Jones, M.V., Burke, D.: Affordances and challenges of artificial intelligence in k-12 education: a systematic review. J. Res. Technol. Educ. 1–21 (2022)
5. Gallo-Belluzzo, S.R., Inacarato, G.M.F., Silva, D.C.D., Carvalho, V.D.S., Amaral, V.D.S., Aiello-Vaisberg, T.M.J.: O imaginário de professores sobre ser professor nos dias atuais: considerações preliminares. O procedimento de Desenhos-Estórias na clínica e na pesquisa: 45 anos de percurso (2017)
6. Holmes, W., Bialik, M., Fadel, C.: Artificial Intelligence in Education. Globethics Publications (2023)

7. Isotani, S., Bittencourt, I.I., Challco, G.C., Dermeval, D., Mello, R.F.: AIED unplugged: leapfrogging the digital divide to reach the underserved. In: International Conference on Artificial Intelligence in Education, pp. 772–779. Springer (2023)
8. Kasneci, E., et al.: ChatGPT for good? On opportunities and challenges of large language models for education. Learn. Individ. Differ. **103**, 102274 (2023)
9. Kim, J., Castelli, D.M.: Effects of gamification on behavioral change in education: a meta-analysis. Int. J. Environ. Res. Public Health **18**(7), 3550 (2021)
10. Liu, L., Zhao, Y.Y., Yang, C., Chan, H.: Gamification for promoting advance care planning: a mixed-method systematic review and meta-analysis. Palliat. Med. **35**(6), 1005–1019 (2021)
11. McLaren, B.M., Adams, D.M., Mayer, R.E., Forlizzi, J.: A computer-based game that promotes mathematics learning more than a conventional approach. Int. J. Game-Based Learn. (IJGBL) **7**(1), 36–56 (2017)
12. OpenAI: GPT-4 technical report (2023)
13. Ouyang, F., Jiao, P.: Artificial intelligence in education: the three paradigms. Comput. Educ.: Artif. Intell. **2**, 100020 (2021)
14. Paiva, R.S.: The basic education teacher's ilness process: an analysis in the light of work sciences. Braz. J. Dev. **7**(7), 74820–74840 (2021)
15. Palomino, P.T., Nacke, L., Isotani, S.: Gamification of virtual learning environments: a narrative and user experience approach. In: Proceedings of the XXII Brazilian Symposium on Human Factors in Computing Systems, pp. 1–10 (2023)
16. Rodrigues, L., Oliveira, W., Toda, A., Palomino, P., Isotani, S.: Thinking inside the box: how to tailor gamified educational systems based on learning activities types. In: Proceedings of the Brazilian Symposium of Computers on Education, pp. 823–832. SBC (2019). https://doi.org/10.5753/cbie.sbie.2019.823
17. Toda, A.M., et al.: Analysing gamification elements in educational environments using an existing gamification taxonomy. Smart Learn. Environ. **6**(1), 16 (2019). https://doi.org/10.1186/s40561-019-0106-1
18. Tondello, G.F., Kappen, D.L., Ganaba, M., Nacke, L.E.: Gameful design heuristics: a gamification inspection tool. In: International Conference on Human-Computer Interaction, pp. 224–244. Springer (2019)
19. Vogler, C.: The Writer's Journey. Michael Wiese Productions, Studio City (2007)
20. Weidinger, L., et al.: Taxonomy of risks posed by language models. In: Proceedings of the 2022 ACM Conference on Fairness, Accountability, and Transparency, pp. 214–229 (2022)
21. Wenger, E.: Communities of Practice: Learning, Meaning, and Identity. Cambridge University Press (1999)
22. Wenger-Trayner, E., Wenger-Trayner, B.: Introduction to communities of practice: a brief overview of the concept and its uses (2015)
23. Xia, Q., Chiu, T.K., Zhou, X., Chai, C.S., Cheng, M.: Systematic literature review on opportunities, challenges, and future research recommendations of artificial intelligence in education. Comput. Educ.: Artif. Intell. 100118 (2022)
24. Xieling, C., Di, Z., Haoran, X., Cheng, G., Caixia, L.: Two decades of artificial intelligence in education: contributors, collaborations, research topics, challenges, and future directions. Educ. Technol. Soc. **25**(1), 28–47 (2022)

Artificial Intelligence and Public Education Policy: Improving Textbook Evaluation in Brazil

Sidarta Tenório[1], Álvaro Sobrinho[2(✉)], Daniel Rosa[3], Ranilson Paiva[1], Leonardo Marques[1,4], Diego Dermeval[1,4], Alan Pedro da Silva[1], Thiago Cordeiro[1], Seiji Isotani[4,5], and Ig Ibert Bittencourt[1,4]

[1] Federal University of Alagoas, Maceió, AL 57072-970, Brazil
sidarta.tenorio@sibi.ufal.br,
{ranilsonpaiva,alanpedro,thiago,ig.ibert}@ic.ufal.br,
leonardo.marques@cedu.ufal.br, diego.matos@famed.ufal.br

[2] Federal University of the Agreste of Pernambuco, Garanhuns, PE 55292-270, Brazil
alvaro.alvares@ufape.edu.br

[3] Federal Rural University of Pernambuco, Recife, PE 52171-900, Brazil

[4] Harvard Graduate School of Education, Cambridge, MA 02138, USA
sisotani@icmc.usp.br

[5] University of São Paulo, São Carlos, SP 13566-590, Brazil

Abstract. The quality of educational materials directly impacts students and educators, especially in underserved communities, where, for instance, textbooks often serve as the primary source for learning. This study examines how Artificial Intelligence (AI) can improve the evaluation process of textbooks distributed through the Brazilian Textbook Program. We developed an AI-powered system that relies on convolutional neural networks to classify textbook images into three categories: sharp, defocused-blurred, and motion-blurred. We experimented with frequency-domain preprocessing, including the Fourier and Haar transforms. Our findings indicate that models using Haar demonstrated greater consistency, with accuracy ranging from 69.08% to 87.14% and AUC between 90.27% and 93.95%. To assess the system's impact on the evaluation process, we conducted a randomized controlled trial with 76 textbook analysts. Results showed that analysts using the AI-assisted system reviewed more images than the control group while maintaining evaluation accuracy. Although pre-test and post-test score variations were modest, the system contributed to improving analysts' ability to assess image defects. This work demonstrates how AI can support large-scale educational policies and improve textbook quality assurance by streamlining and enhancing the evaluation of educational materials.

Keywords: CNNs · Textbooks · Image classification · Education policy

1 Introduction

Textbooks, though not a novel concept, continue to be crucial educational resources globally, serving as a strong foundation for learning and reaching vast

A. I. Cristea et al. (Eds.): AIED 2025, CCIS 2590, pp. 208–220, 2025.
https://doi.org/10.1007/978-3-031-99261-2_19

numbers of students [5,9]. In regions such as the Global South, textbooks represent the primary source of information, particularly for marginalized students. This underscores the importance of enhancing educational materials prior to distribution, a practice widely implemented in public initiatives like the Brazilian Textbook Program (called PNLD). AI can contribute to this process by improving the efficiency and precision of textbook assessments.

The PNLD undergoes a rigorous approval process implemented by a multidisciplinary team that performs comprehensive analyses to assess the editorial quality and pedagogical merits. The primary aim is to ensure resources offer adequate support for students and teachers in public education [11]. However, evidence indicates potential risks of decreased quality and increased time consumption during the PNLD triage process [16]. Developing AI-based systems can enhance effectiveness and efficiency in public education policies like the PNLD.

The triage stage is a critical component of the textbook evaluation process and represents the primary focus of this study. A triage team examines textbooks from diverse publishers, ensuring adherence to the quality standards outlined in a public call for textbooks. One aspect evaluated in this process is the quality of the textbook's visual elements. Enhancing the quality of textbook images can support the improvement of educational outcomes in public schools, as they are essential educational resources [13,18]. For instance, even a minor distortion in a textbook image can impact the comprehension of visually impaired students.

To implement our AI-based system, we conducted experiments using the Fourier and Haar transform methods [10]. We used a public dataset from Kaggle[1], which consists of 1,050 images with varying characteristics. Blurry images were generated by applying an algorithm to sharp ones. We defined three categories: sharp, defocused-blurred, and motion-blurred. Each class contained 350 images, ensuring a balanced dataset. We expanded the dataset from 1,050 to 8,100 and used the accuracy, precision, recall, and Area Under the Curve (AUC) metrics for validation. We applied 10-fold cross-validation to enhance reliability.

Our study aims to determine whether an AI-based system enhances the assessment process. We employed a parallel two arm controlled design with 1:1 randomization involving 76 analysts. The study compares analysts' performance in two scenarios: (1) without any system and (2) with the AI-based system. The primary contribution of this paper lies in its policy-relevant application rather than in the development of novel AI techniques.

2 Related Works

The study by Wang et al. [19] employs support vector machines for image classification. Their approach demonstrated superior performance compared to other blur classification methods. Wang et al. [20] introduced a Convolutional Neural Network (CNN) model combined with ensemble learning, proving its effectiveness in identifying blurred images. The study presents a classifier to distinguish

[1] https://www.kaggle.com/datasets/kwentar/blur-dataset.

four different types of blur. Silva et al. [15] emphasize that, despite the wide range of CNN architectures available, selecting the optimal layer structure and parameters depends on the specific requirements of the study design. Tiwari [17] explores the use of CNNs for gesture recognition considering blurred images. Our AI-based system draws inspiration from the findings of Tiwari [17].

Incorporating AI into educational processes can improve productivity and operational efficiency. Dwivedi et al. [7] explore the role of AI in optimizing assessment, highlighting how automation can reduce time and costs associated with repetitive tasks. Recent studies support this perspective by examining the impact of digital assistants in education, showing productivity gains [12].

3 Convolutional Neural Networks Architecture

CNNs comprise sequences of multiple layers of neurons, which perform a nonlinear operation on a linear transformation (previous layer's outputs) [14]. CNNs can include convolutional, pooling, batch normalization, or fully connected layers. Convolutional and fully connected layers have trainable weights, while pooling layers rely on fixed functions. The convolutional layer operates by moving a fixed-size window across the input feature map, extracting neighboring feature tiles, combining tiles with a learned weight matrix through a tensor product operation, and rearranging the results to form a new tensor [4].

The activation function plays a significant role. For instance, the ReLU layer is crucial in training the network and generating results for the subsequent pooling layer [8]. The pooling layer consists of a grid of units that summarize the activations of the neurons to which they are connected. This process enhances the machine's decision-making ability, ultimately leading to image classification in the output layer. In this study, we used a different architecture than that of Tiwari [17]. The max-pooling layers use a 2 $\times$ 2 window size. The first convolution layer is 32-3 $\times$ 3 convolution filters. The second convolution layer has 32-3 $\times$ 3 convolution filters. Followed by a max-pooling layer. The third layer is a convolution layer that has 64-3 $\times$ 3 convolution filters, followed by a max-pooling layer. The final convolution layer has 128-3 $\times$ 3 convolution filters, followed by a max-pooling layer. After the convolutional layers, there are two Linear layers composed by 1024 and 512 neurons, respectively. Each layer is followed by a Dropout and Batch-Normalization layer. At the end, there is one more Linear layer with output size 3 that represents each class of our problem.

In contrast to the prior study, which used a 32 $\times$ 32 image as input, we opted for a larger input size of 120 $\times$ 120. Our experiments revealed that using a 32 $\times$ 32 image did not improve the model's performance when applied to our dataset.

4 AI-Based System

The images underwent a preprocessing step to enhance their blur effects. We conducted three different experiments using the Fourier Transform and Haar Transform. We selected these two methods to evaluate which frequency-based approach performs better in a multiclass classification task, considering the three categories in the Kaggle blur dataset: Defocused Blur, Motion Blur, and Sharp.

Using the Fourier Transform, the workflow closely follows the methodology described by Tiwari [17] for handling blur. First, the images were captured, converted to grayscale, and processed with the Hann Window. Then, the Fourier Spectrum was computed for each image and labeled according to its corresponding original image.

In contrast, the Haar Transform approach involved extracting high-frequency coefficient matrices. These matrices, representing vertical and horizontal transformations, were combined into a single matrix. The resulting matrices were then labeled according to their original images and used for training.

Thus, we divided our dataset into training, testing, and validation sets, allocating 72.25%, 15.00%, and 12.75% of the data, respectively. In this initial experiment, using the Fourier transform for preprocessing resulted in an accuracy of 76.58%.

We proceeded with a 10-fold cross-validation training to enhance confidence in our results. Through this approach, we attained an average accuracy of 75.44%, an average precision of 78.38%, and an average recall of 76.05%.

We applied another image frequency transform to our dataset to enrich our experiments and compare them with the Fourier transform results. For this purpose, we selected the Haar wavelet transform to preprocess the entire dataset and started another training process.

In the first experiment using the Haar wavelet transform, we employed the same subset split proportions as for Fourier. The results revealed an 80.38% accuracy, surpassing the initial accuracy obtained with Fourier's transform. We conducted another 10-fold cross-validation session to increase the confidence in the Haar wavelet training. The results showcased an accuracy of 69.08%, with an average precision of 67.15% and an average recall of 68.11%. This suggests that our initial model trained using the Haar wavelet transform outperformed the results achieved in the cross-validation process.

We attained an average AUC of 90.27%. The average results obtained with the Fourier Transform outperformed those of models trained with the Haar transformation. The variability observed in the Haar models elucidates why the initially trained model exhibited higher values than the cross-validation results. This suggests that our models can differentiate well among the three classes.

Although we have performed a cross-validation process to validate the generalization of our models, we chose to carry out a new set of experiments by adding more images to our training. We decided to apply a blur filter to another set of images outside Kaggle's blurry dataset to increase the size of the blur dataset. To do this, we took 1000 new sharp images from a widely used COCO dataset[2] and separately applied motion and defocused filters to them, adding 3000 new images to the entire dataset. To ensure consistency with the blurry effects presented in the blur dataset, we used the same techniques used on Kaggle's blurry dataset in our data generation process. It is important to ensure that the new

[2] https://cocodataset.org/#home.

images present the same effects as the old images and do not bias the metrics, as we have increased the number of images by almost four times.

After constructing this new version of the blur dataset, we conducted two additional single-classification experiments using the Fourier and Haar transforms. While both models demonstrated comparable performance, the Haar wavelet model achieved higher accuracy than the Fourier model. By running these experiments and applying the same blur filter from the original dataset to generate more blurred images, we confirm that, as expected, increasing the number of training samples leads to improved model performance.

We performed another cross-validation process to increase confidence in our results using the dataset's new version. The ROC curves of the Haar transform followed a more stable pattern than the Fourier curves.

As the last experiment, we ran a new training set by doubling the images in our dataset. It was enlarged and had its images transformed to analyze its impact on performance, focusing our discussion on using the Fourier or Haar approaches. We included 1350 more images to each class in our expanded dataset, adding 4050 new images to reach 8100. This time, we could see that the models evaluated using the Fourier transform became even more unstable than we have seen previously. Meanwhile, the Haar wavelet models maintained more stable ROC curves. The average accuracy, using Fourier, achieved across all folds was 72.44%, with precision and recall values of 74.52% and 72.69%, respectively. The AUC averaged 84.93%. For Haar, the average accuracy achieved across all folds was 87.14%, with precision and recall values of 87.80% and 87.25%, respectively. The AUC averaged 93.95%, indicating this was the best-performing model.

It was evident how the results from the Fourier transform performed worse than in the first data generation experiment. However, the Haar wavelet transform achieved its best results since the beginning of the experiments. This means that the generation of new data continues to increase our classifiers' performance. All files and scripts used in the experiments are accessible in our repository[3]. We developed a system using the best-performing IA model and web technologies. Our system recommends whether a failure exists for each textbook image, which the PNLD's triage process staff then decides.

5 Impact Analysis

To analyze the impact of the IA-based system, we conducted a parallel 2-arm Randomized Controlled Trial (RCT) with a 1:1 randomization. The evaluation takes place concurrently with the PNLD's actual triage process. Analysts remotely assess textbooks submitted for a specific public call for textbooks. During the triage process, we recruited analysts to be allocated into either the intervention or control groups. We rely on the CONSORT Statement and other relevant guidelines to formulate and define the study protocol [2].

[3] https://github.com/Danielgol/Blur-Recomendation-System.

The primary Research Question (RQ) relies on the population, intervention, comparison intervention, outcome, and duration components. Thus, we formulated the primary RQ: *Does the AI-based system enhance the level of understanding among analysts in assessing failures in textbooks' images, compared to the current manual approach?*

We defined secondary RQs: (RQ1) Does using the system increase the productivity of analysts in the assessment? (RQ2) Does using the system increase the quality of the assessment of analysts? The secondary RQs guided the specification of hypotheses: productivity is not increased (H0-1), productivity is increased (HA-1), quality is not increased (H0-2), and quality is increased (HA-2).

We apply hypothesis tests, such as the student's t-test, to examine validity. All analysts in the intervention and control groups conducted pre-test and post-test evaluations for the primary RQ. We defined and used image failure assessment tests for both pre-test and post-test assessments. We designed the tests to have similar complexity levels to ensure comparability. We assessed the primary outcome, focusing on improving understanding of image failures, using Analysis of Covariance (ANCOVA) [6]. ANCOVA analysis is warranted due to our study design, which involves pre-test and post-test evaluations.

Regarding the secondary RQs, we employed metrics to evaluate the hypotheses: (1) the number of images to measure productivity, considering a window of 30 min of evaluation; and (2) the quality of validations, computing accuracy, precision, and recall. This study involves two dependent variables: the productivity of analysts and the quality of their validations. The independent variable is the AI-based system. To ensure comprehensive and accurate outcomes reporting, we adhere to the CONSORT-outcome 2022 extension [3].

At the time of this study, the PNLD involved 140 analysts actively engaged in the triage process of textbooks. To calculate the sample size, we randomly selected 76 analysts with an 80% confidence level and a 5% margin of error. The control group used spreadsheets and the native image viewer application provided by his/her Operating System (OS). The treatment group used a web-based application designed to display the images (similar to the OS viewer), present AI-generated recommendations, and record the analysts' final decisions. Analysts in both groups similarly accessed the images, one by one, and recorded their final decisions. Based on our experience in textbook assessment, we carefully selected images from the dataset that were not included in the AI model's training set. The AI model evaluated them, and the results were presented to the treatment group by a graphical interface alongside each corresponding image.

Analysts received the baseline test and were randomly allocated into almost equal-sized control and intervention groups. We provided a second test to the participants after the baseline assessment. To create randomized arm allocation sequences, we employed the Maximally-Tolerated-Imbalance (MTI) method [1]. The use of the MTI method is intended to minimize selection biases and enhance the reliability of the RCT. We conducted the RCT from June 26 to June 29, 2024. Brazil's Federal University of Alagoas's ethics committee approved the protocol under the number 6.545.809 in 2023.

We applied two distinct statistical tests to investigate the differences between the control and treatment groups based on the data's normality properties, as determined by the Shapiro-Wilk test: the student's t-test for total responses and the Mann-Whitney U test for correct response percentage. We used the student's t-test to compare the total number of responses between the groups, as the data followed a normal distribution indicated by the Shapiro-Wilk test (p-value $=$ 0.1846 for the control group and p-value $=$ 0.2994 for the treatment group). We applied the Mann-Whitney U test for the correct response percentage, as this variable did not follow a normal distribution in either group (p-value $=$ 0.0015 for the control group and $p - value < 0.0001$ for the treatment group).

The statistical analysis revealed a student's t-value of -13.9190 regarding total responses, indicating a substantial difference between the groups analyzed. The associated p-value was less than 0.0001, suggesting this difference is not due to chance and is statistically significant (HA-1).

Table 1 presents the descriptive statistics for the control and treatment groups, providing mean values, standard deviations, medians, and ranges of total responses, which formed the basis for applying the student's t-test. The difference between the means (96.08 for the control group and 452.51 for the treatment group) and the range of standard deviations underscore this analysis's importance, highlighting the intervention's impact on the response volume.

Table 1. Descriptive statistics of total responses.

Statistic	Control	Treatment
Participants	36	39
Mean	96.08	452.51
Standard Deviation	33.78	150.10
Minimum	28.00	150.00
1st Quartile (25%)	72.75	349.00
Median (50%)	92.50	431.00
3rd Quartile (75%)	112.25	547.50
Maximum	183.00	775.00

Figure 1 illustrates the distributions of response values for both groups. The boxplot highlights the asymmetry and range of responses in the treatment group compared to the control group. This provides a view of the data variability and the presence of outliers. Most responses in the control group concentrate in lower ranges, while the treatment group spans a broader range of responses.

For correct response percentages, the Mann-Whitney U statistic analysis yielded a value of 446.0000, with a p-value of 0.0067. This p-value, smaller than the standard significance level (i.e., 0.05), provides statistical evidence to support rejecting the null hypothesis. The investigated factor positively influenced

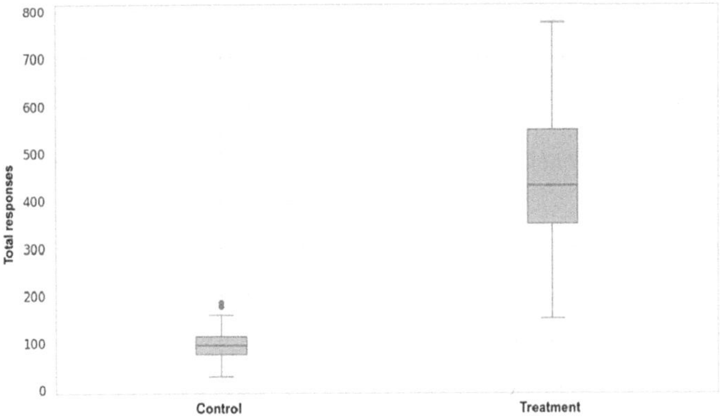

Fig. 1. Total responses between the control and treatment groups.

the performance of the treatment group in terms of correct response percentage (HA-2). This finding aligns with the hypothesis that using AI can improve the precision of responses. The analysis provided by the Mann-Whitney U test complements the performance evaluation by distinguishing the groups based on variability in correct response percentages rather than focusing on means.

Figure 2 shows the distribution of correct response percentages for each participant in both groups, highlighting the range and dispersion of values. This illustrates how individual percentages vary within each group, avoiding a superficial interpretation based solely on the means (86.05% for the control group and 88.76% for the treatment group). The treatment group's mean is only slightly higher and the data dispersion suggests differences in distributions, justifying the Mann-Whitney U test for a more detailed analysis.

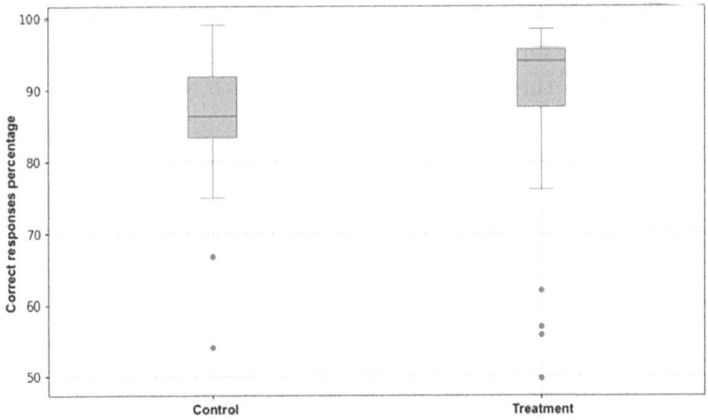

Fig. 2. Correct responses between the control and treatment groups.

Moreover, Table 2 presents the means and standard errors of accuracy, precision, and recall metrics for the control and treatment groups. We calculated such metrics to provide a deeper discussion on quality. To evaluate whether the differences between the group metric means were statistically significant, we applied the student's t-test to each metric (accuracy, precision, and recall).

Table 2. Performance results for control and treatment groups.

Performance Metric	Treatment Group		Control Group	
	Mean	Standard Error	Mean	Standard Error
Accuracy	0.8903	0.0184	0.8378	0.0272
Precision	0.9221	0.0102	0.9165	0.0257
Recall	0.8903	0.0184	0.8378	0.0272

Accuracy represents the proportion of correct classifications relative to the total classifications, reflecting the overall ability of analysts to make accurate decisions. Unlike the correct response percentage, this metric considers the number of true positive and true negative responses divided by the total number of instances. We calculated the accuracy for each participant and the mean accuracy per group. The mean accuracy in the treatment group was 0.8903, while in the control group, it was 0.8378 (Table 2). This difference suggests that participants in the treatment group who used the AI tool demonstrated superior performance.

The t-test value was 1.6101, with a p-value of 0.1116. Since the p-value exceeds the standard significance level of 0.05, there is insufficient evidence to reject the null hypothesis. Although the treatment group achieved slightly higher accuracy, this difference may be attributed to chance, indicating that using the AI-based system may not significantly impact accuracy. However, these analyses do not account for the number of images evaluated by each group, which was considerably higher for the treatment group.

The mean precision for the treatment group was 0.9221, while the control group's was 0.9165 (Table 2). This metric is essential for evaluating confidence in positive classifications. With a t-value of 0.2053 and a p-value of 0.8379, the test indicated that the difference in precision between the groups was also not statistically significant. The AI-based system may not significantly improve precision, suggesting that the minor observed variations may be due to chance.

We calculated the recall, with a mean of 0.8903 for the treatment group and 0.8378 for the control group (Table 2). The test revealed that the difference between the groups is also not statistically significant, with a t-value of 1.6101 and a p-value of 0.1116. Thus, although the treatment group showed slightly higher recall, the observed variation is insufficient to demonstrate impact.

Therefore, the accuracy, precision, and recall provide a detailed analysis of the quality of the participants' classifications. Although the positive results regarding the correct response percentage, the statistical results for the accuracy, precision, and recall metrics indicate that the observed differences between the control and treatment groups were not statistically significant, suggesting that using the AI-based system did not consistently improve classification quality (HA-2).

The lack of statistical significance in the performance metrics indicates that, although the AI tool may have increased the number of responses, it did not significantly affect the quality of classifications. These results partially support the null hypothesis regarding quality. This analysis contributes to understanding the RQs by highlighting that, despite a potential productivity gain, the use of AI did not provide a clear advantage in the quality of evaluations. However, given that the treatment group evaluated a considerably larger number of images, the similarity in metric values compared to the control group can be seen as positive, as it indicates that at least the same level of quality was maintained.

We conducted the ANCOVA to evaluate post-test performance (*OkPos*) between the control and treatment groups, considering pre-test performance (*OkPre*) as a covariate. The term "Corrected Model" represents the variance explained after adjusting for the influence of the covariate *Pre-Test Scores* (*OkPre*), allowing for a more precise analysis of the variation in post-test performance (*OkPos*). With an F-value of 4.824 and a p-value of 0.011, the corrected model achieved statistical significance, indicating that initial performance and the experimental condition (control group vs. treatment group) are relevant components for participants' outcomes. In addition to the experimental condition, participants' initial performance has a considerable influence, reinforcing the necessity of controlling this factor when evaluating the impact of the intervention.

The model's intercept, with an F-value of 37.187 and a p-value of less than 0.001, highlights a consistent difference in post-test performance, regardless of the group or pre-test performance. The overall mean of post-test scores is significantly different from zero, demonstrating that, in general, participants achieved correct answers after the intervention, whether in the control or treatment group. Participants gained a substantial number of correct responses after the experiment, indicating consistency in correct answers beyond chance.

When considering pre-test performance (*OkPre*) as a covariate, we observe a statistically significant influence on post-test outcomes ($F = 6.750$, $p = 0.011$). This underscores the necessity of accounting for participants' prior performance when interpreting the impact of the intervention, as those with higher initial performance tend to maintain superior performance in the post-test. This control is essential to ensure a precise and unbiased analysis of the AI-based system's impact, isolating the effect of initial performance.

Additionally, the factor *Group*, which distinguishes the control group from the treatment group, also showed statistical significance ($F = 4.952$, $p = 0.029$). Even after adjusting for participants' initial performance, the treatment group using the AI-based system achieved significantly better performance than the control group. This evidence supports the hypothesis that the AI intervention improves participant understanding of class images (primary RQ).

The R^2 and adjusted R^2 values of 0.117 and 0.093 indicate that approximately 9.3% of the variance in post-test accuracy results can be attributed to the model, even after adjusting for initial performance. From the 9.3%, 6.37% related to the treatment group. While the adjusted R^2 value is relatively modest, the statistical significance of the group and covariate effects demonstrates that the model provides relevant insights for the analysis. This result suggests that, although other factors influence participants' performance, the AI intervention and prior performance still play significant roles in the variability, supporting the value of this experimental approach. The results suggest that, despite the relatively modest variance, the AI-based system can assist in enhancing human analysts' understanding when evaluating different categories of image defects.

6 Conclusions

This study represents an advancement toward implementing and deploying an AI-based system to aid decisions regarding textbook image quality within the scope of a substantial educational public policy in Brazil. Our findings underscore the effectiveness of using CNN, trained with images preprocessed using the Haar transform method, for image quality assessment as a multi-class problem. Focusing solely on image clarity introduces pedagogical limitations and poses a threat to the validity of our study. However, clarity can have a meaningful impact on the quality of educational materials, demanding proper assessments.

Moreover, this study represented a relevant effort to evaluate the impact of the AI-based system on the textbook triage process within the world's most significant textbook distribution program, which operates in a developing country. The experiment showed that the system increased the group's productivity, with the treatment group evaluating a more significant number of images compared to the control group. The system successfully boosts productivity without compromising evaluation quality.

Acknowledgements. This research was funded by FNDE and the Brazilian Ministry of Education, Brazil, Decentralized Execution Term (Termo de Execução Descentralizado—TED) 10320—Inteligência Aumentada para Validação e Análise de Atributos do PNLD.

References

1. Berger, V.W., Odia, I.: Characterizing permuted block randomization as a big stick procedure. Contemp. Clin. Trials Commun. **2**, 80–84 (2016). https://doi.org/10.1016/j.conctc.2016.01.001
2. Bhide, A., Shah, P.S., Acharya, G.: A simplified guide to randomized controlled trials. Acta Obstet. Gynecol. Scand. **97**(4), 380–387 (2018). https://doi.org/10.1111/aogs.13309
3. Butcher, N.J., et al.: Guidelines for reporting outcomes in trial reports: the CONSORT-outcomes 2022 extension. JAMA **328**(22), 2252–2264 (2022). https://doi.org/10.1001/jama.2022.21022

4. Cong, S., Zhou, Y.: A review of convolutional neural network architectures and their optimizations. Artif. Intell. Rev. **56**(3), 1905–1969 (2023). https://doi.org/10.1007/s10462-022-10213-5
5. Delgado, P., Vargas, C., Ackerman, R., Salmerón, L.: Don't throw away your printed books: a meta-analysis on the effects of reading media on reading comprehension. Educ. Res. Rev. **25**, 23–38 (2018). https://doi.org/10.1016/j.edurev.2018.09.003
6. Dugard, P., Todman, J.: Analysis of pre-test-post-test control group designs in educational research. Educ. Psychol. **15**(2), 181–198 (1995). https://doi.org/10.1080/0144341950150207
7. Dwivedi, Y.K., et al.: Artificial intelligence (AI): multidisciplinary perspectives on emerging challenges, opportunities, and agenda for research, practice and policy. Int. J. Inf. Manag. **57**, 101994 (2021). https://doi.org/10.1016/j.ijinfomgt.2019.08.002
8. Gu, J., et al.: Recent advances in convolutional neural networks. Pattern Recogn. **77**, 354–377 (2018). https://doi.org/10.1016/j.patcog.2017.10.013
9. Heyneman, S.P., Farrell, J., Sepulveda-Stuardo, M.: Textbooks and Achievement: What We Know. No. 298 in World Bank Staff Working Paper, World Bank, Washington, DC, October 1978. Available in English, French, and Spanish
10. Li, M., Ma, Z., Wang, Y.G., Zhuang, X.: Fast Haar transforms for graph neural networks. Neural Netw. **128**, 188–198 (2020)
11. Manoel, C.C., Silva, M., Valero, P.: Happy and healthy families! financial mathematics and the making of the homus oeconomicus. In: International Mathematics Education and Society Conference (2019)
12. Nazari, N., Shabbir, M.S., Setiawan, R.: Application of artificial intelligence powered digital writing assistant in higher education: randomized controlled trial. Heliyon **7**(5) (2021). https://doi.org/10.1016/j.heliyon.2021.e07014
13. Purwaningtyas, T.: Didactic symbol of visual images in EFL textbook: multi-modal critical discourse analysis. Pedagogy: J. Engl. Lang. Teach. **8**(1), 51–63 (2020). https://doi.org/10.32332/pedagogy.v8i1.1959
14. Silva, C.A.C.F., et al.: A multi-objective grammatical evolution framework to generate convolutional neural network architectures. In: IEEE Congress on Evolutionary Computation (CEC) (2021)
15. da Silva, C.A., et al.: A multi-objective grammatical evolution framework to generate convolutional neural network architectures. In: 2021 IEEE Congress on Evolutionary Computation (CEC), pp. 2187–2194. IEEE (2021). https://doi.org/10.1109/CEC45853.2021.9504822
16. Sobrinho, A., et al.: Towards digital transformation of the validation and triage process of textbooks in the Brazilian educational policy. Sustainability **15**(7) (2023). https://doi.org/10.3390/su15075861
17. Tiwari, S.: A blur classification approach using deep convolution neural network. Int. J. Inf. Syst. Model. Des. (IJISMD) **11**(1), 93–111 (2020). https://doi.org/10.4018/IJISMD.2020010106
18. Wang, C., Fu, W., Cheng, L., Wang, Y., Duan, S.: Teaching with picture books on deaf and hard-of-hearing students' creativity. J. Deaf Stud. Deaf Educ. **26**(2), 278–295 (2021). https://doi.org/10.1093/deafed/enaa041

19. Wang, R., Li, W., Li, R., Zhang, L.: Automatic blur type classification via ensemble SVM. Signal Process.: Image Commun. **71**, 24–35 (2019). https://doi.org/10.1016/j.image.2018.08.003
20. Wang, R., Li, W., Zhang, L.: Blur image identification with ensemble convolution neural networks. Signal Process. **155**, 73–82 (2019). https://doi.org/10.1016/j.sigpro.2018.09.027

Prediction of Online Mathematics Test Efficiency Based on Stacked Integrated Models: A Case Study of NAEP Data

Chengliang Wang, Haoming Wang(✉), Yutong Lai, Zihan Xiao, and Xianlong Xu

East China Normal University, Shanghai 200062, China
wfrank0222@gmail.com

Abstract. This study based on data from the National Assessment of Educational Progress (NAEP) in the United States, developed an online mathematics test efficiency prediction model using stacked ensemble learning. The research compared the performance of two machine learning algorithms—Random Forest and Gradient Boosting—and combined them to construct a stacked ensemble model. Through ten-fold cross-validation, the model achieved an average F1 score of 74.92% and a recall rate of 77.45%. Feature importance analysis based on SHAP (SHapley Additive exPlanations) indicated that question type played a dominant role in predicting student test efficiency, while learning behavior features had a relatively small impact. This study innovatively integrated stacked ensemble learning with explainable artificial intelligence techniques, enhancing prediction accuracy while providing reliable support for educational decision-making.

Keywords: Stacked Integrated Model · Online Mathematics Test · Student Efficiency Prediction · SHAP · Explainable Artificial Intelligence

1 Introduction

With the rapid development of educational informatization, online testing has become a key tool for assessing student learning outcomes [19]. Online testing can collect real-time response data from students and provide personalized learning feedback and suggestions [12, 20]. This feature is particularly important in the field of mathematics, as the multidimensional data generated during online math tests (such as response time and types of errors) provides rich information for a deeper understanding of students' learning processes and thought patterns. However, the current educational assessment system still primarily relies on standardized transcripts. This approach fails to fully utilize the vast amount of data generated during online testing and struggles to capture the ability traits exhibited by students during problem-solving [1]. Therefore, accurately predicting students' online test performance becomes crucial. By deeply analyzing online test data, we can more accurately assess students' actual levels and provide data support for educational decision-making [10].

A. I. Cristea et al. (Eds.): AIED 2025, CCIS 2590, pp. 221–229, 2025.
https://doi.org/10.1007/978-3-031-99261-2_20

This study uses data from the National Assessment of Educational Progress (NAEP) as the research foundation [14]. Since its launch in 1969, NAEP has systematically evaluated the academic performance of U.S. students in subjects such as mathematics, reading, and science through regularly administered standardized tests. Based on this, the study aims to build an accurate stacked model for predicting students' online math test efficiency. By comparing the predictive performance of different machine learning algorithms and employing evaluation methods such as ten-fold cross-validation, Receiver operating characteristic (ROC) curve analysis, and Kappa coefficient, the study will systematically assess the model's prediction effectiveness. Additionally, feature importance analysis and Shapley additive explanations (SHAP) value interpretation techniques will be used to identify the key factors influencing test efficiency, providing empirical evidence for optimizing online testing systems and improving teaching strategies. Therefore, this study poses the following two research questions:

RQ1: To what extent can machine learning algorithms predict students' online test efficiency?
RQ2: Which feature variables have the greatest impact on predicting students' online test efficiency? How do these key features influence the prediction results?

2 Related Work

2.1 Research Status on Online Test Efficiency

Research on online test efficiency primarily focuses on two aspects: evaluation methods and influencing factors. Traditional methods such as Item Response Theory can quantify students' accuracy and time investment, but they have limitations when handling complex online education data, making it difficult to dynamically capture students' behavioral differences online [2]. Regarding influencing factors, previous studies have identified various potential variables such as subject knowledge, learning strategies, and online behaviors, finding that response time distribution and interactive behaviors are important indicators. However, these studies face challenges such as limited prediction accuracy [22]. Additionally, even with high prediction accuracy, educators often struggle to intuitively understand which factors truly impact test efficiency [18]. In recent years, large-scale assessment data, such as those from NAEP, have provided rich behavioral and background information for research. However, existing studies rarely conduct systematic analysis of online math test efficiency and have not fully integrated the advantages of machine learning and educational psychology [8]. Therefore, how to make accurate efficiency predictions based on these data and enhance the model's interpretability has become an important research direction in this field.

2.2 Application of Model Stacking Combined with Explainable Artificial Intelligence

With the widespread application of machine learning in the field of educational data mining, model stacking has become a research focus. Stacking integrates multiple base models and conducts secondary learning, demonstrating excellent performance in tasks such as student performance prediction [21, 24]. This technique effectively integrates multi-modal data, such as time series and interaction logs, enhancing the model's adaptability [5]. At the same time, to meet educators' demand for model interpretability, researchers have started to focus on explainable artificial intelligence (XAI) technologies, such as SHAP value analysis. These methods quantify the contribution of features to prediction outcomes, helping to identify key factors [17]. For example, studies have found that factors such as mathematical self-efficacy and course competency often show high SHAP values [7]. The research approach that combines model stacking with explainable artificial intelligence meets the requirement of "accurate prediction" while also providing the capability for "explainable decision-making," offering a new perspective for understanding online math test efficiency.

3 Method

3.1 Data Source

The NAEP dataset contains 864 students and a total of 29,708 online learning records. Each record includes multiple dimensions such as response time, answering behavior, and question completion status, with a total of 65 feature variables. These features comprehensively document the learners' behavioral performance during the online testing process.

3.2 Feature Processing and Model Construction

In the data preprocessing phase, the study uses the median and the most frequent value to handle missing values in numerical and categorical features, respectively [3], and employs the SMOTE technique to address the data imbalance issue [16]. During the feature engineering stage, the study sequentially applies near-zero variance analysis, Variance Inflation Factor (VIF), and Recursive Feature Elimination (RFE) algorithms to select key features. In terms of model construction, the study chooses Random Forest, known for its excellent noise resistance [23], and Gradient Boosting, which is good at handling difficult-to-classify samples [6], as base models. Logistic Regression is used as the meta-learner to construct a stacked ensemble model to improve overall prediction performance.

3.3 Evaluation Methods

This study adopts a multidimensional evaluation framework, including accuracy, precision, recall, and F1 score, to quantitatively assess model performance. It also introduces ROC curves, Area Under Curve (AUC) values, and Kappa coefficients to evaluate the model's classification discrimination ability and consistency level [4]. To deeply analyze the model's decision-making mechanism, the study uses the SHAP method, which precisely quantifies feature contributions and reveals how they influence prediction outcomes in different scenarios [11]. In experimental validation phase, 10-fold cross-validation is applied, and the dataset is split into training and test sets with an 8:2 ratio to reduce model evaluation bias.

4 Results

4.1 Base Model Classification Performance Analysis

This study evaluates the classification performance of two base models: Random Forest and Gradient Boosting. Based on 10-fold cross-validation, the performance metrics of the two models are shown in Table 1. In comparison, the Gradient Boosting model showed slightly lower performance across all metrics. However, the Kappa value of the Gradient Boosting model (0.4029) was slightly higher than that of Random Forest (0.3822), indicating a slight advantage in classification consistency. The ROC curves shown in Fig. 1 further validate that both models exhibit reliable classification capabilities.

Table 1. Comparison of performance of random forest and gradient boosting models.

k-fold	Random Forest					Gradient Boosting				
	Accuracy	ROC_AUC	Precision	Recall	F1	Accuracy	ROC_AUC	Precision	Recall	F1
1	63.77	67.99	69.77	71.43	70.59	60.87	68.17	67.44	69.05	68.24
2	79.71	85.8	83.33	83.33	83.33	73.91	79.54	78.57	78.57	78.57
3	72.46	70.99	72.55	88.1	79.57	73.91	66.84	75.00	85.71	80.00
4	66.67	67.81	73.17	71.43	72.29	60.87	61.38	68.29	66.67	67.47
5	76.81	74.25	78.26	85.71	81.82	75.36	74.34	79.07	80.95	80.00
6	63.77	66.75	70.73	69.05	69.88	60.87	62.43	69.23	64.29	66.67
7	65.22	69.58	70.45	73.81	72.09	63.77	70.81	71.79	66.67	69.14
8	75.36	82.4	81.58	75.61	78.48	68.12	78.66	75.68	68.29	71.79
9	60.87	61.15	64.58	75.61	69.66	59.42	58.45	63.27	75.61	68.89
10	68.12	74.65	70.21	80.49	75.00	63.77	72.74	69.05	70.73	69.88
Ave	69.28	72.14	73.46	77.46	75.27	66.09	69.33	71.74	72.65	72.07

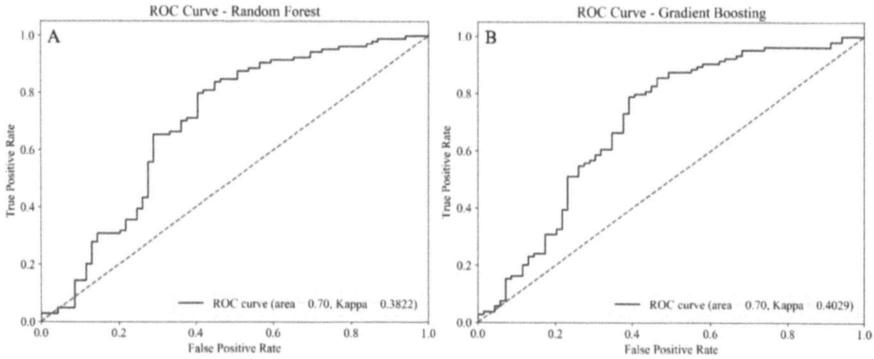

Fig. 1. ROC curve analysis of base models. (A) ROC curve of the random forest model. (B) ROC curve of the gradient boosting model.

4.2 Stacked Model Performance Evaluation

To further improve classification performance, we constructed a stacked model based on the two base models. Through 10-fold cross-validation, as shown in Fig. 2A, the stacked model demonstrated robust performance: the average accuracy reached 68.70%, ROC_AUC was 71.66%, precision was 72.81%, recall was 77.45%, and the F1 score reached 74.92%. As shown in Fig. 2B, the model's ROC curve displayed good classification performance, with an AUC value of 0.70. Additionally, the Kappa coefficient of 0.4230 reflects a high degree of consistency between the predicted results and the actual labels. The stacked model exhibited stable and satisfactory performance across all key metrics, confirming the effectiveness of this approach.

A Stacked model classification performance (%)

k-fold	Accuracy	ROC_AUC	Precision	Recall	F1
1	63.77	68.25	69.77	71.43	70.59
2	78.26	84.74	81.4	83.33	82.35
3	75.36	70.11	74.51	90.48	81.72
4	68.12	65.52	72.73	76.19	74.42
5	75.36	74.07	77.78	83.33	80.46
6	62.32	66.23	70	66.67	68.29
7	65.22	69.93	70.45	73.81	72.09
8	73.91	81.97	79.49	75.61	77.5
9	57.97	61.15	61.54	78.05	68.82
10	66.67	74.65	70.45	75.61	72.94
Average	68.70	71.66	72.81	77.45	74.92

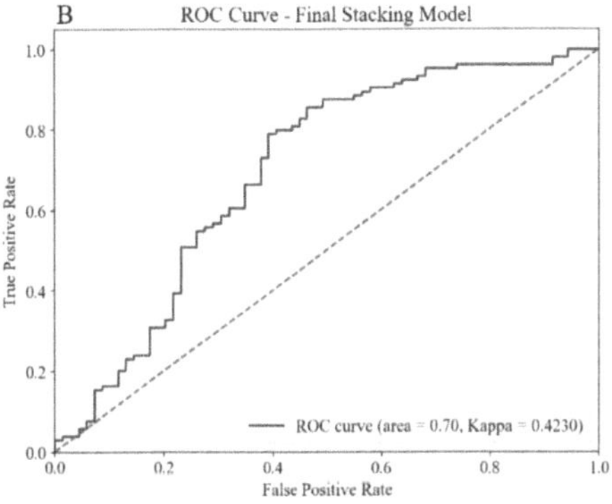

Fig. 2. Classification performance evaluation of the stacked model. (A) distribution of key performance metrics of the stacked model. (B) ROC curve analysis of the stacked model.

4.3 Feature Optimization and Selection

This study used 10-fold cross-validation to evaluate the F1 scores of 64 features. As shown in Fig. 3, when the number of features reached 29, the F1 score achieved its optimal value of 0.7513. Further increasing the number of features resulted in only slight fluctuations in the F1 score, with no significant improvement observed. Therefore, we selected the top 29 features to construct the final feature subset, successfully reducing the feature dimension from 64 to 29. This not only reduced computational complexity but also maintained the model's optimal performance. The experimental results indicate that these 29 features are sufficient to effectively capture the key discriminative information.

4.4 Feature Importance Analysis

Through SHAP value analysis, we explored the impact of each feature in the optimal feature subset on online learning efficiency. The analysis was conducted from two dimensions: question type and learning behavior, as shown in Fig. 4. In terms of question type, Single Choice 13 (0.15 to 0.20) had a significant positive impact on the prediction, while Single Choice 1 (−0.25 to −0.20) and Single Fill Blank 1 (−0.15 to −0.20)

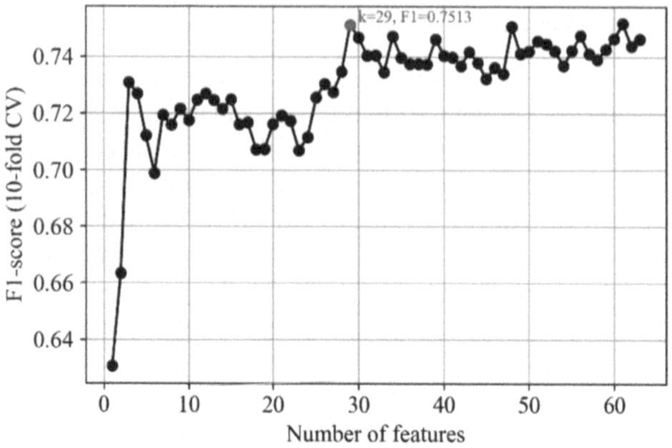

Fig. 3. Analysis of the relationship between feature count and F1 score.

showed strong negative impacts. This suggests that these questions have a high degree of discrimination. Regarding learning behaviors, most behavioral features had relatively small SHAP values. However, it is noteworthy that certain learning behaviors, such as viewing exam instructions (−0.10 to −0.15), drawing (−0.10 to −0.15), and changing topics (−0.10 to −0.15), exhibited slight negative effects. This finding suggests that some non-core learning behaviors may distract students' attention. In contrast, time-management-related behaviors, such as clicking on progress navigation (0.05), showed a slight positive effect.

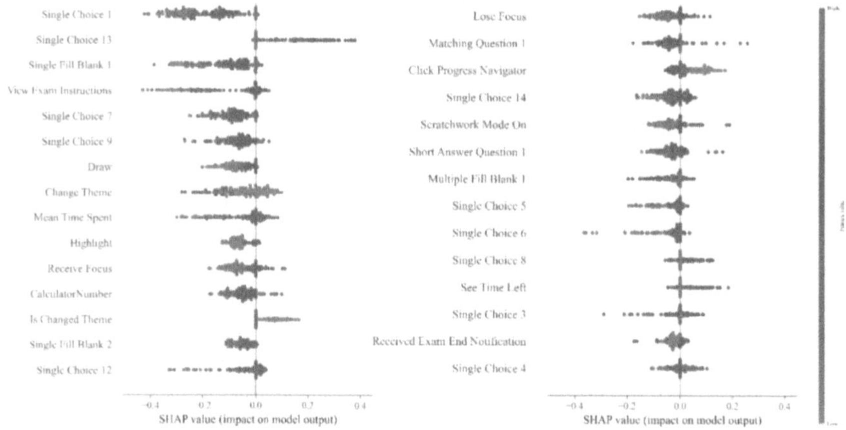

Fig. 4. Feature importance analysis based on SHAP values.

5 Discussion

Regarding ***RQ1***, this study found that machine learning algorithms demonstrate moderate predictive ability in forecasting students' online test efficiency. The Random Forest model achieved an accuracy of 69.28%, indicating that while machine learning methods have some predictive potential, their performance still has clear limitations. These limitations mainly stem from two factors: First, in educational settings, student behavior is dynamically influenced by multiple factors, and this complexity makes it difficult for static predictive models to fully capture. Second, students' behavior patterns during testing have significant temporal dependencies, posing additional challenges for traditional machine learning models. As noted by Lan et al. [9], the difficulty in educational prediction tasks largely arises from the time-varying nature of learner behaviors.

In terms of algorithm performance comparison, significant differences were observed between different algorithms. The Random Forest model outperformed in most evaluation metrics, reflecting its unique advantage in handling nonlinear relationships in educational data. It is worth noting that the Gradient Boosting model showed a relative advantage in terms of the Kappa value (0.4029 vs. 0.3822), suggesting it may be more advantageous when dealing with imbalanced datasets. In this study, the stacked model, by integrating the strengths of different algorithms, achieved stable performance, demonstrating the value of hybrid models in educational prediction tasks. These findings highlight that the choice of algorithm should be based on the specific characteristics of the application scenario, rather than simply optimizing a single evaluation metric.

As for ***RQ2***, the results of the study indicate that question type features have a dominant influence on predicting students' online test efficiency. Specifically, the polarization effect of Single Choice questions reveals that the order of questions may significantly affect students' test performance. This finding suggests that a well-designed progression of question difficulty plays an important role in helping students build confidence and maintain a steady pace during the test [13]. In contrast, the influence of most learning behavior features is relatively limited, which may be due to the nature of standardized testing environments: students' attention is primarily focused on solving the problems, and their behavioral choices are more restricted, reducing the discriminative power of behavioral features.

By optimizing features, this study reduced the 64 features to 29, achieving optimal predictive performance. This finding suggests that, despite the complexity and variability of the educational process, its core influencing factors are relatively concentrated. Feature importance analysis shows that, although time management-related behaviors have a small impact, they exert a positive effect, while some auxiliary behaviors (such as viewing instructions and drawing) show negative effects. The differential impact patterns of these features provide important insights for the design of online testing platforms: while ensuring basic functionality, the focus should be on optimizing the core answering process and minimizing unnecessary features that could distract students. This approach not only improves test efficiency but also helps more accurately assess students' actual ability levels.

6 Conclusion and Implication

In summary, this study focuses on the prediction of online math test efficiency, constructing a predictive model based on stacked ensemble learning and conducting empirical research using data from the NAEP. The results show that by effectively integrating the strengths of multiple base models, the stacked model demonstrates outstanding predictive performance. Feature importance analysis reveals that question type features play a dominant role in predicting test efficiency, particularly the difficulty level and presentation order of multiple-choice questions, which have a significant impact on student performance. Although learning behavior features contribute relatively little to the prediction results, they provide a unique perspective for understanding student behavior patterns.

This study innovatively combines stacked ensemble learning with XAI techniques, not only significantly improving the accuracy of online test efficiency predictions but also providing a solid empirical foundation for personalized teaching strategies and intervention plans in educational practice. Future research directions may focus on exploring the temporal characteristics of learner behaviors and their underlying mechanisms, and consider incorporating more advanced deep learning methods to further enhance the predictive capability and interpretability of the model.

Disclosure of Interests. The authors declare no competing interests.

References

1. Bi, H., et al.: Quality meets diversity: a model-agnostic framework for computerized adaptive testing. In: 2020 IEEE International Conference on Data Mining (ICDM), pp. 42–51. IEEE, New York (2020). https://doi.org/10.1109/ICDM50108.2020.00013
2. Chen, Y., Li, X., Liu, J., Ying, Z.: Item response theory—a statistical framework for educational and psychological measurement. arXiv preprint arXiv:2108.08604 (2021)
3. Deshkar, P.A., et al.: Data pre-processing solution using statistical and data mining techniques. In: Garg, L., et al. (eds.) AI Technologies for Information Systems and Management Science. ISMS 2023. Lecture Notes in Networks and Systems, vol. 1136, pp. 109–120. Springer, Cham (2024)
4. Dol, S.M., Jawandhiya, P.M.: Classification technique and its combination with clustering and association rule mining in educational data mining—a survey. Eng. Appl. Artif. Intell. **122**, 106071 (2023). https://doi.org/10.1016/j.engappai.2023.106071
5. Fang, T., Huang, S., Zhou, Y., Zhang, H.: Multi-model stacking ensemble learning for student achievement prediction. In: 2021 12th International Symposium on Parallel Architectures, Algorithms and Programming (PAAP), pp. 136–140. IEEE, New York (2021)
6. Fernandes, E., Holanda, M., Victorino, M., Borges, V., Carvalho, R., Van Erven, G.: Educational data mining: Predictive analysis of academic performance of public school students in the capital of Brazil. J. Bus. Res. **94**, 335–343 (2019). https://doi.org/10.1016/j.jbusres.2018.02.012
7. Huang, Y., Zhou, Y., Chen, J., Wu, D.: Applying machine learning and SHAP method to identify key influences on middle-school students' mathematics literacy performance. J. Intell. **12**(10), Article no. 10 (2024). https://doi.org/10.3390/jintelligence12100093

8. Ji, C.S., Yee, D.S.-W., Rahman, T.: Mapping state proficiency standards onto the NAEP scales: results from the 2019 NAEP reading and mathematics assessments. NCES 2021–036. National Center for Education Statistics, Washington, DC (2021)
9. Lan, A.S., Studer, C., Baraniuk, R.G.: Time-varying learning and content analytics via sparse factor analysis. In: Proceedings of the 20th ACM SIGKDD International Conference on Knowledge Discovery and Data Mining (KDD), pp. 452–461. ACM, New York (2014). https://doi.org/10.1145/2623330.2623631
10. Liu, Y., Huang, Z., Wang, G.: Student learning performance prediction based on online behavior: an empirical study during the COVID-19 pandemic. PeerJ Comput. Sci. **9**, e1699 (2023). https://doi.org/10.7717/peerj-cs.1699
11. Lundberg, S.: A unified approach to interpreting model predictions. arXiv preprint arXiv:1705.07874 (2017)
12. Dai, J., Zhang, X., Wang, C.L.: A meta-analysis of learners' continuance intention toward online education platforms. Educ. Inf. Technol. **29**, 21833–21868 (2024). https://doi.org/10.1007/s10639-024-12654-7
13. Moneva, J., Tribunalo, S.M.: Students' level of self-confidence and performance tasks. Asia Pac. J. Acad. Res. Soc. Sci. **5**(1), 42–48 (2020)
14. National Center for Education Statistics: NAEP Nations Report Card - National Assessment of Educational Progress - NAEP. National Center for Education Statistics, Washington, DC (2019)
15. Ning, R., Waters, A.E., Studer, C., Baraniuk, R.G.: SPRITE: a response model for multiple choice testing. arXiv preprint arXiv:1501.02844 (2015)
16. Sahlaoui, H., Alaoui, E.A.A., Agoujil, S., Nayyar, A.: An empirical assessment of SMOTE variants techniques and interpretation methods in improving the accuracy and the interpretability of student performance models. Educ. Inf. Technol. **29**(5), 5447–5483 (2024). https://doi.org/10.1007/s10639-023-12007-w
17. Salih, A.M., et al.: A perspective on explainable artificial intelligence methods: SHAP and LIME. Adv. Intell. Syst. **2024**, 2400304 (2024). https://doi.org/10.1002/aisy.202400304
18. Saqr, M., López-Pernas, S.: Why explainable AI may not be enough: predictions and mispredictions in decision making in education. Smart Learn. Environ. **11**, 52 (2024). https://doi.org/10.1186/s40561-024-00343-4
19. Wang, C.L., Chen, X.J., Yu, T., Liu, Y.D., Jing, Y.H.: Education reform and change driven by digital technology: a bibliometric study from a global perspective. Humanit. Soc. Sci. Commun. **11**, 256 (2024). https://doi.org/10.1057/s41599-024-02717-y
20. Wang, C.L., Dai, J., Zhu, K.K., Yu, T., Gu, X.Q.: Understanding the continuance intention of college students toward new e-learning spaces based on an integrated model of the TAM and TTF. Int. J. Hum.-Comput. Interact. **40**(24), 8419–8432 (2023). https://doi.org/10.1080/10447318.2023.2291609
21. Wang, C.L., Chen, X.J., Hu, Z.B., Jin, S., Gu, X.Q.: Deconstructing university learners' adoption intention towards AIGC technology: a mixed-methods study using ChatGPT as an example. J. Comput. Assist. Learn. **41**, e13117 (2025). https://doi.org/10.1111/jcal.13117
22. Xiao, W., Hu, J.: Analyzing effective factors of online learning performance by interpreting machine learning models. IEEE Access **11**, 132435–132447 (2023). https://doi.org/10.1109/ACCESS.2023.3334915
23. Zhou, X., Ding, P.L.K., Li, B.: Improving robustness of random forest under label noise. In: 2019 IEEE Winter Conference on Applications of Computer Vision (WACV), pp. 950–958. IEEE, New York (2019). https://doi.org/10.1109/WACV.2019.00106
24. Öz, E., Bulut, O., Cellat, Z.F., Yürekli, H.: Stacking: an ensemble learning approach to predict student performance in PISA 2022. Educ. Inf. Technol. **30**, 7753–7779 (2025). https://doi.org/10.1007/s10639-024-13110-2

Structural and Semantic Analysis Techniques for Translation Evaluation of Educational Materials

Leonardo Brandão Marques[1,5](✉), Seiji Isotani[2,5], Bruno Pimentel[1,5], João Vitor L. B. do Nascimento[3,5], Mina Isotani[4,5], Ig I. Bittencourt[1,5], and Catherine E. Snow[1,5]

[1] Federal University of Alagoas (UFAL), Penedo, Brazil
leonardo.marques@cedu.ufal.br
[2] University of Sao Paulo, São Paulo, Brazil
[3] Estácio College at Alagoas, Penedo, Brazil
[4] Sunrise School, São Paulo, Brazil
[5] Harvard University, Cambridge, USA
https://www.nees.ufal.br/conselho-administrativo/

Abstract. Cross-lingual semantic similarity analysis can be a valuable tool for evaluating the quality of educational material adaptation and cultural fit across cultures and languages. This study assesses semantic and cultural quality from an Adolescent Literacy program (Word Generation, HARVARD & SERP) adapted into Brazilian Portuguese. Using Sentence-BERT MiniLM, and XLM-RoBERTa, we evaluate semantic similarity in two cases: English text vs. Portuguese version and English text vs. back-translated English. Back-translation yields higher similarity due to literal translation, while cultural adaptation introduces linguistic shifts with slightly lower, yet high, similarity scores. Multilingual BERT offers stable alignment; Sentence-BERT MiniLM finds sentence-level similarities; XLM-RoBERTa captures detailed token-level shifts. Findings highlight the translation fidelity and cultural relevance trade-off and emphasize the need for automated and expert validation. The study suggests an AI-assisted validation framework for educational material adaptation, advocating for integrating qualitative and automated measures.

Keywords: Cross-cultural · educational material · cosine similarity · BERT · mBERT · XLM-RoBERTa

1 Introduction

Cross-lingual semantic similarity is a valuable strategy for assessing educational text adaptation across cultures. It helps countries with developing economies overcome educational barriers, enhancing their teaching while retaining evidence-based strategies and resources of the original material's context.

A. I. Cristea et al. (Eds.): AIED 2025, CCIS 2590, pp. 230–241, 2025.
https://doi.org/10.1007/978-3-031-99261-2_21

While automatic evaluation of translated educational materials using semantic and structural analysis offers numerous benefits, it is essential to acknowledge these techniques' inherent challenges and limitations [22]. A key challenge is the complexity of human language [9]. Educational materials often include idioms, cultural references, and specialized terms that automatic systems struggle to interpret accurately, leading to literal translations that are grammatically correct but semantically incorrect. Cultural context, often beyond current automatic tool capabilities, is also necessary. Another challenge is language ambiguity, where different translations can convey the same meaning using varied words or structures. Automatic metrics relying on a single "gold standard" might unfairly penalize valid translations that differ from this reference, particularly in educational contexts where diverse linguistic choices can enhance clarity and pedagogy [8]. This is especially true in educational scenarios, where polysemous vocabulary is frequent and necessary for academic learning [21].

Developing automatic evaluation metrics usually uses human evaluations as quality benchmarks [11]. However, due to language subjectivity, it is common that studies show low agreement among evaluators [12]. Current metrics, despite improvements, struggle to assess nuanced translation quality, especially in literary texts or materials needing specific styles. Educational translations require suitable reading levels and pedagogical effectiveness, which are challenging to evaluate solely with automated tools.

This study compares original English educational material handout texts with Portuguese-Brazilian adaptations and their back-translated version. The primary objective is to quantify the degree of semantic preservation during the adaptation process and the extent to which the cultural nuances and linguistic variations were respected. By utilizing a multi-tier approach that is initiated by automatic evaluations based on bidirectional encoder representations from transformers (BERT), we aim to comprehensively assess semantic preservation during the adaptation process.

1.1 Ensuring Evidence-Based Education for All

The rapid expansion of digital technologies has compelled traditional education to transform, requiring teachers to prioritize critical thinking skills, information literacy, and, above all, reading comprehension [5,10,20]. This effort supports the United Nations' SDG 4, which ensures inclusive, quality education, fosters lifelong learning, and reduces inequalities for vulnerable groups [23].

However, structural challenges, such as the difficulty in managing tasks that require reading beyond the basic level (UNESCO, 1978) and the low academic performance evidenced in national and international assessments [15,16], reinforce the need for innovative proposals that promote qualitative advancements in education.

In this context, innovative programs like WordGen Weekly, developed by Catherine Snow and Suzanne Donovan, stand out as they expand students' academic vocabulary and stimulate critical argumentation through interdisciplinary activities involving subjects such as Language, Mathematics, Science, and Social Studies [5,6,20,21].

In Brazil, like many countries in the Global South, ensuring quality educational materials that reflect local nuances, rather than just offering literal translations, is crucial for equitable, high-quality education. To preserve *WordGen's* evidence-based integrity while enabling large-scale roll-out, we combined traditional expert review with an AI-assisted evaluation layer that scores each adapted handout for semantic fidelity and cultural relevance. Although automatic translation resources exist, linguistics and educational specialists must review the adaptations to ensure the semantic quality of educational texts [1,18]. Thus, this work proposes an automated validation process to assist specialists in educational material cultural adaptation. This hybrid workflow safeguards the program's proven instructional design, directs resources toward iterations that matter most, and offers a replicable pathway. This proposal can contribute to creating a replicable model for future educational projects and promoting inclusive education aligned with the principles of SDG 4.

The contributions of this work include: 1) the use of AI to evaluate the semantic similarity of the translated material; 2) the analysis of choices made by specialists in cultural translation; and 3) the measurement of semantic similarity in the context of cross-cultural validation.

2 Background

2.1 Materials Support to Development of Debating and Argumentative Skills

The *WordGen Weekly* was conceived as an innovative initiative to expand academic vocabulary and strengthen essential skills, such as critical language and the ability to adopt different perspectives, especially in school contexts with high socioeconomic vulnerability [5]. Developed by researchers and specialists in academic literacy, the program is based on studies highlighting the challenges students face in high-poverty environments, promoting debate and argumentation in the classroom to reduce educational inequalities [19].

Empirical studies indicate that *WordGen Weekly* positively impacts vocabulary acquisition and reading comprehension, especially when consistently implemented in low-income high-risk area schools, including students for whom English is a second language [6]. The approach values critical discussion and reinforces academic language development, enriching students' linguistic repertoire and significantly improving their ability to analyze and understand content [5,21].

A key feature of these programs is their interdisciplinary and transversal design, achieved through a well-organized set of educational materials, such as handouts for students and teachers, that guide and give cohesion to the implementation of the activities [13]. These resources integrate a common theme across Language, mathematics, science, and social studies, ensuring alignment throughout the curriculum. This approach supports teachers in presenting academic vocabulary pertinent to the week's discussions. It helps students see the connections between concepts across different subjects, ultimately deepening their understanding and broadening their academic perspective [10,11].

The research on reading motivation, especially for adolescent students, guided the dilemma-based discussions proposed on WordGen. Fostering authentic engagement by presenting complex issues without clear answers, motivating students to read and explore texts rich in target vocabulary. Since WordGen's classroom implementation heavily relies on debate and argumentation; selecting engaging and appropriately controversial topics is essential. Preserving this characteristic is crucial when adapting these activities to another cultural context, as the relevance of dilemmas depends on local norms, values, and discourse styles. Effective cultural adaptation ensures that students remain engaged with meaningful debates while maintaining the academic and linguistic benefits of the original design. This underscores the importance of thoughtful cultural adaptation, ensuring that the way the topics are presented resonates with students in diverse contexts while maintaining their relevance and educational value [13,20].

2.2 Cross-Cultural Validation

Cross-cultural validation ensures that assessments and informative texts are practical in different cultural contexts. This process goes beyond simple translation, requiring adaptation to guarantee conceptual, semantic, idiomatic, and experiential equivalence, as minor cultural variations can alter respondents' perceptions and compromise data comparability [2,24].

Wild et al. [24] emphasize that inconsistent translation methodologies can undermine data validity and hinder the aggregation of global datasets. Therefore, translations must be carried out by experienced natives, accompanied by rigorous review to harmonize versions and identify discrepancies, strengthening the integrity of results, minimizing selection bias, and promoting inclusion in multicultural contexts.

3 Related Works

3.1 Natural Language Processing (NLP) for Semantic and Structural Text Analysis

Automated evaluation methods can provide scores reflecting the semantic quality of translations, which have become increasingly crucial in machine translation research. These metrics offers numerous benefits of various translation systems, eliminating substantial expenses and complexities associated with direct human translation, and complement semantic embedding techniques like those previously described. Rather than focusing on the translation process, this approach emphasizes evaluating the similarity between an original text and its translated version after the translation has already occurred.

Based on the calculation and comparisons of insertions, deletions, and substitutions, the simplest techniques must turn to text equivalents. This can compare words and phrases similarly to the Levenshtein distance calculation [14]. The issue of word order is a clear drawback in this metric and other lexical measurements. Because of linguistic diversity, translations may recreate the reference sentence differently while preserving the same meaning, which would unfairly punish a competent translation.

An alternative is to use more complex techniques, such as those based on text embedding. This allows for calculating the textual similarity (sentence similarity) [11]. Word embeddings are dense vector representations of words or sentences in continuous vector spaces. These embeddings are learned from large text corpora and capture semantic relationships between words, placing semantically similar words closer together in the vector space. Popular embedding models include Word2Vec, GloVe, BERT, and Sentence-BERT. For cross-language translation evaluation (e.g., English and Portuguese), multilingual embedding models (e.g., multilingual Sentence-BERT or XLM-Roberta) are commonly used to align semantic representations across different languages into a shared embedding space. Source and translated texts were converted into multilingual vector representations, and semantic similarity was quantified using cosine similarity [22]. A higher cosine similarity indicates greater semantic similarity between texts, allowing for flexible evaluation beyond exact word matching and effectively capturing paraphrases and meaning variations [25].

The application of textual similarity as a metric for translation quality evaluation has shown promising results. Recent analyses demonstrate that textual similarity correlates strongly with human evaluation scores across various language pairs, consistently outperforming traditional metrics like Human Translation Edit Rate (HTER) and model confidence scores (ML_{eval}). Specifically, textual similarity exhibits superior predictive capabilities when assessed using Generalized Additive Mixed Models (GAMM), highlighting its reliability and robustness. These findings underline the value of integrating textual similarity into translation quality estimation frameworks, suggesting it can effectively complement or even surpass traditional evaluation methods [4].

Understanding nuanced semantic relationships through multilingual embeddings is pivotal for assessing the quality of translated educational materials. Such embeddings yield insights into how languages encode and convey meaning, providing a flexible framework to evaluate translation accuracy beyond exact word matching and effectively capturing paraphrases and variations in wording.

Related works highlight BERT (Bidirectional Encoder Representation from Transformers) as a revolutionary model in natural language processing, utilizing a bidirectional deep learning approach to capture the complete context of words in a text [3]. Its applications span tasks such as classification, prediction, translation, and semantic analysis, underscoring its relevance in overcoming the limitations of traditional methods. Recent research has explored adaptations and enhancements of BERT for specific domains, increasing its effectiveness and accuracy in varied contexts [7]. Furthermore, the development of SBERT (Sentence-BERT) has demonstrated how BERT's architecture can be fine-tuned to generate optimized sentence embeddings, enabling precise evaluation of semantic similarity between texts [17].

4 Method

This study utilizes a mixed-methods approach, integrating both quantitative and qualitative methods to assess the role of artificial intelligence in cross-cultural validation of educational materials. It compares expert linguistic translations with machine translations generated by ChatGPT, focusing on semantic similarity through AI tools [17] and the selection of main ideas in the texts. Grounded in established practices of cross-cultural validation [2,24], the methodology evaluates whether AI can account for cultural and pedagogical nuances, aiding in the inclusive and universal adaptation of content.

The research further employs a comparative strategy, blending quantitative semantic similarity metrics [17] with qualitative analyses of cultural adaptation adequacy by experts. Its ultimate aim is to lay the groundwork for an automated cultural validation model that can be replicated in future educational initiatives while adhering to quality and inclusive education principles. By demonstrating the potential of AI to improve the translation and adaptation of educational materials, the study aspires to advance education tailored to the cultural and pedagogical needs of diverse contexts [1,2,18,24].

4.1 Procedures, Metrics and Data Analysis

Step 1: Selection, Adaptation of Teaching Materials, and Evaluation of Relevant Content by Specialists. The set of 18 teaching texts was extracted from the Harvard Graduate School of Education (HGSE) and SERP WordGen Weekly handouts[1] program, available on the Strategic Education Research Partnership (SERP) portal [19]. Its content was adapted to the Brazilian context by a linguistics specialist holding a doctoral degree, following the criteria for cross-cultural validation [2,24]. This material, translated by the specialist, was subsequently evaluated by a team of experts to determine which parts of the text were most relevant for understanding the teaching material. This expert evaluation was later compared to an evaluation conducted by ChatGPT, which had the same objective. To ensure adequate translation of educational materials, the expert needs to be aware of:

1. **Structural aspects:** ensure that the phrases construction follow the original intended argumentation flow. Adjust the text style, aware of the peculiarities of Portuguese;
2. **Factual and Data Adjustment:** normative and legislation (i.e., the minimum driving age in US is 16, in Brazil is 18), cited data (i.e., estimates of school dropouts in each country), and historic and recent events (i.e., the book being banned from public schools in US);

[1] https://www.serpinstitute.org/wordgen-weekly.

3. **Cultural nuances:** some topics can be seen as too controversial for the school setting (i.e., sex education or school dress coding that is not an issue in Brazil);
4. **Curricular adjustments:** the adjustments on math and science activities that the handouts present to exemplify the use of the weekly academic vocabulary across disciplines.

Step 2: Back Translation by the Chat Bot and Content Evaluation by the AI Tool. We utilized the AI tool ChatGPT 4.0 to re-translate the material into its original language, using a PROMPT developed explicitly for this purpose. The versions produced were meticulously documented, along with the lexical and cultural choices made by the AI. These were subsequently evaluated and compared to the translation completed by the specialist and the evaluations conducted by the experts.

In turn, the evaluation of the most relevant parts of the teaching texts was carried out using ChatGPT 4.0, employing the following PROMPT: The prompt was: "*Translate the following Portuguese from Brazil (targeting Middle School students from the 2024 year) text to English. Be aware of the factual and normative data presented into the original text. The result should make the cultural adaptations and change data and examples to turn relevant to American Middle School students from 2024:*". The results of this framework are presented in the following section.

Step 3: Automatic Comparisons of Structure, Lexical Richness, and Choices in Cultural Adaptation. Comparisons of lexical richness between the texts were done using the SBERT tool, a modification of the pre-trained BERT network that employs siamese and triplet network structures to derive semantically meaningful sentence embeddings [17].

5 Results

5.1 Automatic Semantic Analysis

This analysis takes a closer look at how standard BERT and multilingual BERT (mBERT) models perform when preserving meaning between English originals and their Portuguese-Brazilian adaptations in educational materials. The observed statistical differences help us better understand how these models handle multiple languages and how reliable they are for comparing meaning across languages (Table 1).

Both models show relatively high average similarity scores (0.774 for BERT vs. 0.790 for mBERT), suggesting that the Portuguese adaptations maintain substantial semantic alignment with the original English texts overall. The slightly higher mean and median values for mBERT indicate that it may detect semantic relationships that monolingual BERT misses when comparing languages. The small difference between mean and median in standard BERT (0.774 vs.

0.768) suggests a fairly symmetrical distribution of similarity scores. In contrast, the larger difference in mBERT (0.790 vs. 0.800) indicates a slight negative skew, with more scores falling above the mean, potentially reflecting mBERT's enhanced ability to recognize cross-lingual semantic equivalence.

Table 1. Summary statistics of similarity scores

Statistic	Multilingual BERT	mBERT	Paraphrased mBERT
Count	18	18	18
Mean	0.7748	0.7714	0.7898
Std. Dev.	0.2145	0.1537	0.1395
Min	0.1509	0.3751	0.3947
25%	0.7473	0.6733	0.7265
50% (Median)	0.8555	0.8227	0.7997
75%	0.9099	0.8816	0.9110
Max	0.9510	0.9630	0.9576

The most striking difference is in standard deviation: 0.037 for BERT versus 0.136 for mBERT—nearly four times higher variability in mBERT scores. This substantial difference reveals fundamental differences in how these models operate in cross-lingual contexts:

Consistency vs. Sensitivity: Standard BERT produces remarkably consistent scores across all document pairs (low SD = 0.037), suggesting it may be applying a more uniform judgment that potentially overlooks language-specific nuances. By contrast, mBERT's higher variability (SD = 0.136) indicates greater sensitivity to different translations and adaptations.

Cross-lingual Feature Detection: mBERT likely detects more features specific to cross-lingual comparison, identifying strong preservation and significant divergences that standard BERT might miss. This explains why some document pairs receive much higher or lower scores in mBERT.

Cultural Adaptation Sensitivity: The higher standard deviation in mBERT suggests it may be more responsive to cultural adaptations, recognizing when meaning is preserved despite substantial cultural localization—or conversely, when cultural adaptations have significantly altered the original semantic content.

Cosine Similarity and BERTScore: Cosine similarity is a straightforward way to quantify semantic similarity among the original handout English, its adapted/translated Portuguese version, and the back-translated Portuguese-to-English version. The mBERT, RoBERTA, and LSA multilingual models encoded the texts into vectors, then computed cosine similarity. A high cosine similarity (close to 1.0) indicates that the adaptation preserved the meaning well. This works because multilingual transformers learn a shared semantic space, allowing direct comparison of English–Portuguese pair meaning, and the direct original English to the back-translated English version. BERTScore analyzes two token-embedded texts using a pre-trained BERT model. It matches candidate text tokens to reference text tokens based on the cosine similarity of BERT embeddings to calculate Precision, Recall, and F1. BERTScore is suitable for translation/adaptation evaluation since it captures semantic overlap without exact word matching. Additionally, BERTScore for mBERT and XLM-R permits cross-lingual comparison. The new BERTScore outperforms BLEU/METEOR in rewarding good paraphrases or adaptations, even if phrasing differs [25]. (BLEU chose a lexically similar but semantically off translation, while BERT-based grading properly preferred the real meaning match. In cultural adaptation, BERTScore's Recall statistic shows how much of the original content's meaning is kept, while Precision shows if the adaptation contributed to the content. A high F1 indicates that the altered text accurately conveyed the original.

Table 2. Detailed semantic-similarity results for MiniLM and XLM-RoBERTa

Model/Metric	Precision	Recall	F1 or Cosine
Sentence-BERT MiniLM			
Semantic similarity (ENG $\leftrightarrow$ PT)	–	–	0.8648
Semantic similarity (ENG $\leftrightarrow$ back-ENG)	–	–	0.8947
XLM-RoBERTa			
BERTScore	0.8792	0.8884	0.8837

Sentence-BERT MiniLM Model. Similarity between the original English text and back-translated English (0.8947) is significantly higher than that of the adapted Portuguese (0.8648) (Table 2). The back-translation kept much of the original phrasing and structure, possibly due to its literal translation into English. The somewhat lower similarity score for the original English vs the adapted Portuguese shows that cultural adjustments changed meaning or language. While not a direct translation, these alterations maintained the original text's semantic intent. The 0.8648 similarity score shows that the altered text retains semantic equivalence, allowing naturalization into Brazilian Portuguese while retaining the original content.

The evaluation of XLM-RoBERTa's performance in semantic similarity reveals key metrics: a precision of 0.8792 confirms alignment with the original text, avoiding extraneous information; a recall of 0.8884 highlights the retention of core content in the Portuguese version; and an F1-score of 0.8837 demonstrates a balanced preservation of meaning with appropriate adjustments.

The comparison between XLM-RoBERTa and Sentence-BERT MiniLM highlights their shared capacity for strong semantic retention while showcasing distinct strengths. XLM-RoBERTa's BERTScore F1 (0.8837) closely matches Sentence-BERT MiniLM's semantic similarity score (0.8648), validating the reliability of both models. XLM-RoBERTa excels in precision and recall, more effectively capturing detailed token-level semantic relations. However, its slower processing speed (0.09 sentences per second) limits its scalability, making Sentence-BERT MiniLM the preferred choice for larger datasets due to its efficiency.

The findings emphasize the nuanced relationship between semantic preservation and cultural adaptation in translation quality, aligning with the study's objectives. Back-translation yields higher similarity scores (0.8947) than adaptation (0.8648), reflecting its literal nature. In contrast, cultural adaptation introduces meaningful changes to enhance clarity and engagement for Brazilian students, preserving core meanings while addressing local needs. The comparison underscores Sentence-BERT MiniLM's efficiency in sentence-level evaluations and XLM-RoBERTa's sensitivity to adaptation subtleties. Importantly, lower similarity scores in adaptations signify successful cultural and linguistic adjustments rather than a lack of translation fidelity. Future research should integrate qualitative analysis and human evaluations to complement automated scoring, further refining frameworks for cross-cultural educational translation.

6 Discussion and Conclusion

The experiment underscores a tradeoff between semantic accuracy and cultural relevance in cross-linguistic communication adaptation. While machine back-translation resembles the source text more closely, it leans towards literal equivalence at the expense of natural fluency. In contrast, the adapted Portuguese version's lower similarity score reflects deliberate adjustments to align with linguistic and cultural norms. Effective adaptation prioritizes clarity and readability over strict literal fidelity, as excessively high similarity scores may indicate unnatural or rigid translations in the target language.

The comparison between Sentence-BERT MiniLM and XLM-RoBERTa demonstrates how model selection influences semantic similarity evaluation. Sentence-BERT MiniLM, with its efficiency and broad sentence-level similarity scoring, is well-suited for large-scale assessments. Meanwhile, XLM-RoBERTa offers a more precise token-level analysis, capturing detailed semantic nuances but demanding greater computational resources. This precision makes XLM-RoBERTa particularly valuable for examining translation fidelity and adaptation efficacy, revealing the balance between preserving original meaning and accommodating cultural and contextual adjustments.

A key finding is that similarity scores alone cannot assess adaptation quality. Lower similarity scores between the original English and adapted Portuguese paragraph may indicate good cultural adaptation, not a failure. Translation involves conveying meaning and ensuring the content resonates with the audience. Adapting materials for Brazilian middle schoolers is crucial to ensure engagement and accessibility. Contextual shifts may make culturally particular references, idioms, or sentence patterns in the source material unavailable, making some loss in accuracy acceptable and even required for appropriate adaptation.

More accurate evaluation of cultural adaptation quality requires more than just numerical similarity measurements, which sometimes lack face validity. While BERT-based models effectively measure semantic recall, they lack readability, tone, and cultural relevance. A data-driven assessment must balance automated similarity analyses with human review to preserve the input text's message, appeal, and educational value. Hybrid evaluation frameworks combining AI-connected metrics and human language evaluations may provide deeper insights into effective adaptation in future studies.

References

1. Canen, A.: Child education and literacy learning for multicultural societies: the case of the Brazilian national curricular references for child education (NCRs). Compare: J. Comp. Int. Educ. **33**(2), 251–264 (2003). https://doi.org/10.1080/0305792032000070011
2. Beaton, D.E., Bombardier, C., Guillemin, F., Ferraz, M.B.: Guidelines for the process of cross-cultural adaptation of self-report measures. Spine **25**(24), 3186–3191 (2000). https://doi.org/10.1097/00007632-200012150-00014
3. Devlin, J., Chang, M.W., Lee, K., Toutanova, K.: BERT: pre-training of deep bidirectional transformers for language understanding. In: Proceedings of the 2019 Conference of the North American Chapter of the Association for Computational Linguistics: Human Language Technologies (Volume 1: Long and Short Papers), pp. 4171–4186 (2019)
4. Fomicheva, M., et al.: MLQE-PE: a multilingual quality estimation and post-editing dataset. In: Calzolari, N., et al. (eds.) Proceedings of the Thirteenth Language Resources and Evaluation Conference. pp. 4963–4974. European Language Resources Association, Marseille, France (2022). https://aclanthology.org/2022.lrec-1.530/
5. Jones, S.M., et al.: Experimental effects of word generation on vocabulary, academic language, perspective taking, and reading comprehension in high-poverty schools. J. Res. Educ. Effect. **12**(3), 448–483 (2019)
6. Kim, H.Y., Hsin, L.B., Snow, C.E.: Reducing academic inequalities for English language learners: variation in experimental effects of word generation in high-poverty schools. Int. J. Biling. Educ. Biling. **24**(7), 1024–1042 (2018). https://doi.org/10.1080/13670050.2018.1535574
7. Koroteev, M.V.: BERT: a review of applications in natural language processing and understanding. arXiv preprint arXiv:2103.11943 (2021)
8. Lake, B.M., Murphy, G.L.: Word meaning in minds and machines. Psychol. Rev. **130**(2), 401 (2023)
9. Landauer, T.K.: LSA as a theory of meaning. In: Handbook of Latent Semantic Analysis, pp. 15–46. Psychology Press (2007)

10. Lawrence, J.F., Crosson, A.C., Paré-Blagoev, E.J., Snow, C.E.: Word generation randomized trial: discussion mediates the impact of program treatment on academic word learning. Am. Educ. Res. J. **52**(4), 750–786 (2015). https://doi.org/10.3102/0002831215579485
11. Lee, S., et al.: A survey on evaluation metrics for machine translation. Mathematics **11**(4) (2023). https://doi.org/10.3390/math11041006. https://www.mdpi.com/2227-7390/11/4/1006
12. Li, C., Wang, X., Qian, L.: Exploring syntactic complexity and text readability in an ELT textbook series for Chinese English majors. SAGE Open **15**(1) (2025). https://doi.org/10.1177/21582440251323619. Original work published 2025
13. Lin, A.R., Lawrence, J.F., Snow, C.E., Taylor, K.S.: Assessing adolescents' communicative self-efficacy to discuss controversial issues: findings from a randomized study of the word generation program. Theory Res. Soc. Educ. **44**(3), 316–343 (2016)
14. Navarro, G.: A guided tour to approximate string matching. ACM Comput. Surv. **33**(1), 31–88 (2001). https://doi.org/10.1145/375360.375365
15. OECD: Education in Brazil: An international perspective. OECD Publishing, Paris (2021). https://doi.org/10.1787/60a667f7-en
16. OECD: Pisa 2022 results (volume i): The state of learning and equity in education. OECD Publishing, Paris (2023). https://doi.org/10.1787/53f23881-en.
17. Reimers, N., Gurevych, I.: Sentence-BERT: sentence embeddings using Siamese BERT-networks. arXiv preprint arXiv:1908.10084 (2019)
18. Remillard, J., Van Steenbrugge, H., Bergqvist, T.: A cross-cultural analysis of the voice of curriculum materials. In: Proceedings of ICMT-2014, The International Conference on Mathematics Textbook Research and Development, pp. 395–400. University of Southampton (2014)
19. SERP Institute: Wordgen weekly (n.d.). https://www.serpinstitute.org/wordgen-weekly. Accessed 10 Mar 2025
20. Snow, C.E.: Motivation and engagement in language and literacy development. In: Segers, E., van den Broek, P. (eds.) Developmental Perspectives in Written Language and Literacy. John Benjamins, Amsterdam (2017)
21. Snow, C.E., Uccelli, P.: The challenge of academic language. In: The Cambridge Handbook of Literacy, pp. 112–133. Cambridge University Press (2009)
22. Sun, K., Wang, R.: Textual similarity as a key metric in machine translation quality estimation (2024). https://arxiv.org/abs/2406.07440
23. United Nations: Transforming our world: The 2030 agenda for sustainable development. United Nations (2015)
24. Wild, D., et al.: Principles of good practice for the translation and cultural adaptation process for patient-reported outcomes (PRO) measures: report of the ISPOR task force for translation and cultural adaptation. Value Health **8**(2), 94–104 (2005). https://doi.org/10.1111/j.1524-4733.2005.04054.x
25. Yousuf, O., Tang, G., Jin, Z.: Improving bertscore for machine translation evaluation through contrastive learning. IEEE Access (2024)

Advancing MOOCs Personalization: The Role of Generative AI in Adaptive Learning Environments

Ridouane Oubagine[1](✉), Loubna Laaouina[1], Adil Jeghal[1], and Hamid Tairi[2]

[1] LSATE Laboratory, National School of Applied Science, Sidi Mohamed Ben Abdellah University, 30000 Fez, Morocco
{ridouane.oubagine,loubna.laaouina,adil.jeghal}@usmba.ac.ma
[2] LISAC Laboratory, Faculty of Sciences Dhar El Mehraz, Sidi Mohamed Ben Abdellah University, 30000 Fez, Morocco

Abstract. The integration of Generative Artificial Intelligence (GAI) into Massive Open Online Courses (MOOCs) represents a transformative shift in virtual schooling, notably enhancing personalization, engagement, and studying outcomes. This comprehensive literature review examines the position of GAI in optimizing MOOCs, emphasizing its packages in adaptive studying pathways, real-time personalized feedback, and intelligent evaluation systems. By leveraging large language fashions (LLMs) and statistics-driven insights, GAI allows dynamic content technology, customized quizzes, and automatic educational help, tailoring instructional experiences to individual rookies' wishes. Despite its capability, the adoption of GAI in MOOCs increases crucial challenges, including moral worries, privacy dangers, algorithmic biases, and the integrity of AI-generated exams. Additionally, the shortage of clean regulatory frameworks affords limitations to ensuring transparency, equity, and responsible AI deployment in education. This overview identifies key study gaps and future guidelines, focusing on GAI's sustainable and equitable integration in MOOCs to foster inclusive and robust virtual learning environments. By addressing these challenges, GAI can redefine online schooling, making MOOCs more excellent, interactive, private-sized, and accessible for numerous international novices. The findings presented in this look at function as a roadmap for researchers, educators, and policymakers seeking to harness the strength of GAI in the continuous evolution of era-more advantageous mastering environments.

Keywords: Generative Artificial Intelligence (GAI) · MOOCs · Adaptive Learning · Personalized Education · Intelligent Tutoring Systems (ITS) · Automated Feedback

1 Introduction

The continued rise of artificial intelligence (AI) is showing greater influence on various industries, education being one of them. Of recent technological developments, perhaps one of the most transformational is the potential of AI to create personalized learning

A. I. Cristea et al. (Eds.): AIED 2025, CCIS 2590, pp. 242–254, 2025.
https://doi.org/10.1007/978-3-031-99261-2_22

where learners engaged in the learning process become better at absorbing knowledge, thereby enhancing access to education. With the evolution of AI, adaptive learning systems can now modify content, pace, and style in real time and do not need large cohorts of learners to work [1].

Massive Open Online Courses (MOOCs) have been at the center of global digital education because of their expansive scalability and accessibility. However, in spite of the broad reach and low barrier of entry for the integration of MOOCs, dropout rates, motivation stagnation for learners, and failure to accommodate individual learning preferences still plague the fledgling medium [2]. Personalization is not only a catalyst but also a functional component as a homemade answer to these matters, in view of the fact that it allows learners to overcome barriers to course completion and achievement while distributing course material to be in line with the need and goal of the learner [3]. These have simulated operational effectiveness in the traditional AI applications envisaged by MOOCs (course recommendations, adaptive pathways in learning, automated grading, etc.). They are also often not interactive and have a top content-generated approach, making them less surely the personalized [4] content generation system. A Personalized Digital Education AI (GAI): Powered by the latest LLMs, ADI takes personalized digital education to the next level. Within the context of MOOCs, GAI can provide support to offer real-time content creation, instant feedback, and adaptive assessment features that allow for more flexible and personalized learning [5]. When learners interact with this content, courses can be adapted in real-time to their individual needs, helping mitigate dropout and boost success. But the use of GAI in MOOCs also invokes some significant questions. In current discussions, ethical concerns such as data privacy, algorithmic fairness, and academic integrity are still paramount [6, 7]. Moreover, AI-generated answers and feedback can lack pedagogical or curricular viability without systematic evaluation, leading to contextually niche or academically questionable content [1].

By systematic literature review, this paper examines the changing role of GAI in the context of MOOCs. Abstract: The purpose of this paper is to address the technological affordances, pedagogical advantages, and moral dilemmas presented by large-scale generative AI in the context of large-scale MOOCs, as well as to highlight gaps in the existing literature and suggest avenues for responsible and equitable incorporation of these technologies.

2 State of the Art

Recent research emphasizes the heterogeneity of learners in Massive Open Online Courses (MOOCs), and it is crucial to note that every student has different needs, preferences, and learning styles. This starkly contrasts with typical MOOCs, where students typically follow the same learning path, which may not be the most effective way to learn. The absence of personalization often results in disengagement and a higher rate of dropouts, with learners struggling to be connected with the course and teaching methods [2]. To address these challenges, personalized learning has become a cornerstone strategy for increasing the effectiveness of MOOCs, leading to increased engagement and improved learning outcomes [5]. MOOCs have benefited significantly from Artificial Intelligence (AI), facilitating personalization through recommendation systems, adaptive learning algorithms, and personalized feedback mechanisms. Real-time adaptive learning systems [3]. Recommendation systems are beneficial in customizing MOOCs by monitoring student behavior, preferences, and previous experience to recommend appropriate collaborative learning resources to learners. These systems filter course materials according to engagement patterns [4] guiding the learners through content suitable to their needs and interests.

The former utilizes data on a real-time basis to develop a granular capture of the knowledge level of each learner by constantly evaluating learners through well-structured assessments and optimizing instructional delivery to be adaptive to their real-time dynamic learning needs. The latter use adaptive learning algorithms to adjust the difficulty and pace of the instructional content to ensure that each learner progresses based on their unique knowledge level and learning speed [4]. One significant step forward in the growth of AI-powered personalization has been the inclusion of affective computing, which identifies the emotional status of learners and provides them with a need-based engagement experience. Emotions heavily influence learning, whether helpful (making them more engaged and motivated) or distressful (making it harder to focus and self-regulate their behavior). Systems of this kind utilize data on facial expressions, voice tone, and interaction patterns, which enables constant adjustments to course materials [1]. Bringing this into practice matters because it helps alleviate some of the frustration for learners and gives them space to focus on learning more productively.

While kickingly AI in MOOCs has excellent potential, widespread use in MOOCs has challenges. There are increasing concerns about general algorithm bias because the input data that feeds AI models are mainly collected from historical records, reinforcing existing inequalities and creating unjust conditions for students who engage in learning [2]. Data privacy is also a key issue, with AI-driven personalization dependent on the gathering and processing a vast amount of user data, which can create ethical problems regarding informed consent and the security of data [6]. Another concern is scalability, as AI-powered personalization requires considerable computational resources and infrastructure, which may not be possible for educational environments constrained by resource limitations [7]. Besides those ethical questions, the academic integrity of AI-based content is still under discussion: Generative AI (GAI) can produce individualized and customized evaluations and comments, but this also leads to challenges concerning the authenticity of content, plagiarism, and misinformation [8]. Without regulation and monitoring, AI-generated responses may lack accuracy and pedagogical quality, resulting in wrong learning [9].

Researchers and educators must address these challenges by establishing ethical guidelines and frameworks for AI-driven personalization in MOOCs. That includes establishing transparent AI governance, algorithmic fairness, and stronger learner data protection policies [10]. Perplexity-based adaptive learning approaches will likely benefit from multimodal data analysis that leads to cognitive, behavioral, and emotional insights yielding truly personalized and ethical learning through hybrid AI models that leverage multimodal data [5].

3 Methodology

Following the objective outlined in the introduction, this study is based on a systematic literature review that adheres to the PRISMA guidelines. The purpose is to ensure transparency, reproducibility, and rigor in identifying and selecting relevant research on the integration of Generative Artificial Intelligence (GAI) in MOOCs.

Search Strategy and Sources

The search was conducted across two major databases: **IEEE Xplore** and **ScienceDirect,** covering peer-reviewed articles published between 2022 and 2024. Boolean operators and a combination of keywords were used, such as:

("MOOCs" OR "massive open online courses") AND ("generative AI" OR "artificial intelligence" OR "machine learning") AND ("adaptive learning") AND ("educational technology") AND ("personalization" OR "e-learning").

Selection Process

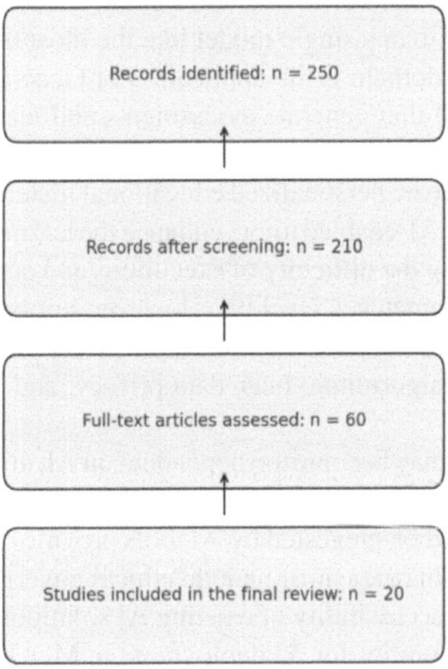

Fig. 1. Flow Diagram showing study selection process

The figure below Fig. 1 illustrates the selection process using a PRISMA flow diagram. It shows that out of 250 initially identified records, 210 remained after duplicate removal and initial screening, and 60 full-text articles were thoroughly assessed. **20 studies** met all criteria and were included in the final synthesis.

4 Key Concepts in AI-Powered MOOCs

Artificial Intelligence (AI) enables the personalization of learning in MOOCs, such as the use of AI-based recommendation systems, adaptive learning models, and personalized feedback mechanisms in order to enhance automatic engagement and retention of learners [2, 3]. However, despite those advancements, challenges in scalability, algorithmic bias, and data privacy remain, with questions about the fairness and reliability of AI-produced content within education [1]. Adaptive solutions evolved into more sophisticated responsive muscle solutions with the advent of Generative AI (GAI), able to dynamically curate content and personalize user experiences in real time. However, the long-term sustenance and ethics of GAI are still hotly debatable. The section delves into state-of-the-art machine learning techniques that power personalization solutions in MOOCs. It evaluates their efficacy, challenges, and prospects to facilitate fair and practical learning.

4.1 AI-Driven Personalization in MOOCs

Artificial Intelligence (AI) has emerged as a powerful tool in enriching MOOCs' personalization, offering personalized content recommendations, feedback on assignments and tests, and personalized learning tracks. The machine-learning approach empowers data according to the interaction between the learner and the model, which is an instructive model requirement, rather than a single model like the most traditional way [2, 3]. One such breakthrough in this domain is the application of Generative AI (GAI) and Large Language Models (LLMs) that generate assessments and learning resources based on resource suggestions and enrich learner inputs as the user progresses through the learning experience using GAI to create personalized educational material [4, 5]. Also, intelligent recommender systems and AI-enabled tutors enhance the learning experience from one to one if it comes to correcting the difficulty of executions and contextualized explanations relative to the learner's competence level [9]. However, notwithstanding its potential to transform teaching and learning practices, the deployment of AI in education brings up significant issues such as algorithmic bias, data privacy, and a lack of transparency in recommendations [6, 7].

Additionally, students may become too dependent on AI, affecting their critical thinking skills, thus necessitating hybrid models of AI-human positions wherein educators ensure that what is explored or suggested by AI tools provides a logical and fair learning path [11, 12]. A focus on this area must include ethical governance strategies, inclusive algorithms, and increased accessibility of existing AI solutions for low-resource institutions to enmesh the opportunities for AI deployment in MOOCs and the natural risks it can bring [10]. Therefore, although Generative AI has the potential to revolutionize the process of personalization in online learning at a radical level, its incorporation requires supervision before finally helping build an educational future worthy of society.

This Fig. 2 illustrates the various elements of integrating Artificial Intelligence (AI) into MOOCs, highlighting its role in personalizing and optimizing learning reviews. It represents an interconnected technique wherein AI contributes via advanced content version mechanisms, raising moral issues and requiring powerful collaboration among era and educational stakeholders.

Fig. 2. AI's Role in MOOC Personalization

4.2 Intelligent Recommender Systems in MOOCs

Intelligent recommender systems are widely used to personalize MOOCs as they analyze learners' behaviors, preferences, and past interactions for customized learning content and resources. Collaborative and content-based filtering to improve engagement and accessibility to learning materials has long been applied. However, traditional approaches show great difficulties in dealing with the complexity of individual learning styles and tend to generate less relevant recommendations [2]. The Application of a deep learning-based model in recommendation systems (NLP) integrated dynamic interaction between learners and how their knowledge progresses over time [3].

Generative AI (GAI) is a groundbreaking approach that can result in real-time adaptation of learning pathways, automated resource generation, and more personalized recommendations tailored to the upcoming needs of every student [4]. Moreover, effective recommender systems consider the emotional and cognitive data such as facial expression, level of engagement with content, and behavioral response to provide a more holistic recommendation by customizing learning prompts based on learners' emotional states and motivation levels [1]. Nonetheless, as various advances in AI models have been witnessed, issues such as algorithmic bias (whereby AI datasets result in biased recommendations) and the problem of scalability (real-time recommendation engines that are employed in large-scale massive open online courses still require extensive computational resources) [6] assists in improving the effectiveness and fairness of these systems and optimal reduction of algorithmic bias; involved improvements in technological infrastructures; and efforts to build regulatory frameworks that guarantee responsible and inclusive use of AI in educational recommendations [7]. So, although Generative AI redefines the MOOC and makes its use of recommendations more accurate and personalized, its inclusion must be adequately regulated to ensure equitable and adaptive education for all student profiles.

4.3 Adaptive Learning and Personalized Feedback in MOOCs

Powered by Artificial Intelligence (AI), adaptive learning has transformed the online education landscape by allowing Massive Open Online Courses (MOOCs) to automatically tailor course material, difficulty level, and pace to each student according to engagement and performance [4]. In contrast to traditional pedagogical approaches that prescribe a one-size-fits-all learning trajectory to every student, AI-powered adaptive systems monitor the strengths and weaknesses of individual learners as they interact with content, providing real-time adjustment of their paths to accelerate progression and improve knowledge retention in time [8].

Adding to this development are advances based on Generative AI (GAI) and Large Language Models (LLMs) that allow for the immediate generation of individualized learning materials, interactive assessments, and personalized feedback that increase motivation and pedagogical suitability [5]. It relies on real-time feedback to promote personalized learning and create (shattering) automated assessment systems offering more than the mere automated correction of errors; it also seeks to explain them and provide recommendations for progress and additional resources according to students' errors [12]. Nevertheless, the impact of these technologies significantly improves the MOOC experience. However, they also present crucial challenges, such as assessment accuracy, algorithmic bias, and dependence on AI-produced content, which can also be inaccurate or need pedagogical contextualization [6]. Moreover, the lack of human supervision in some adaptive platforms may lead to questions about the quality of recommendations provided by AI systems and their alignment with broader learning goals, emphasizing the necessity for blended approaches that integrate AI and human-mediated learning to ensure tailored and relevant educational experiences [9]. Therefore, in order to fully realize the potential of adaptive learning in massive open online courses (MOOCs), future research efforts should prioritize the creation of more transparent AI models, the enhancement of the trustworthiness of automated feedback, and the establishment of hybrid pedagogies that synergize AI tools with human wisdom, leading to personalized, equitable, and effectiveness-oriented learning experiences for all [10].

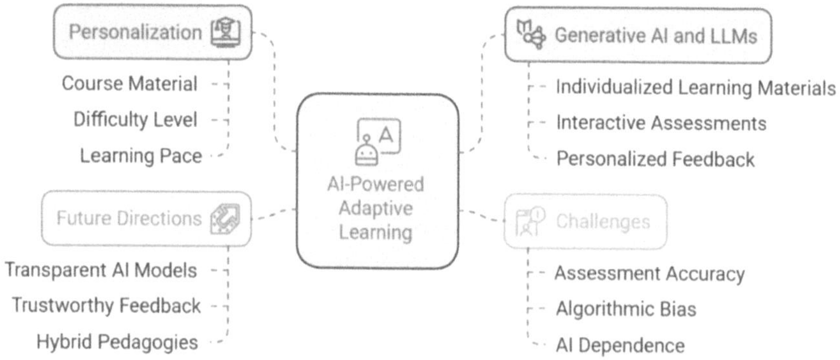

Fig. 3. AI-Driven Adaptive Learning in MOOCs

This Fig. 3 illustrates the key components of AI-powered adaptive mastering, emphasizing its position in improving personalization, leveraging generative AI, advert-dressing demanding situations, and guiding future instructions. It highlights how AI-pushed personalization adjusts direction substances, problem degrees, and mastering tempo to meet male or female pupil wishes. Additionally, the combination of generative AI and large language models (LLMs) allows the introduction of individualized learning materials, interactive assessments, and personalized remarks.

4.4 Impact of Generative AI on Learner Engagement and Outcomes in MOOCs

Incorporating Generative Artificial Intelligence (GAI) into MOOCs dramatically restructuring learner engagement approaches and learning outcomes. MOOCs, however, struggle with high dropout rates and relatively low levels of interactivity [5]. A significant promise of GAI is its provision of virtual tutors and conversational agents that can deliver instant feedback, contextually relevant explanations, and customized guidance aligned with the pace and individual requirements of each learner [4, 9]. It becomes easier to create interactive content like gamified quizzes, simulations, and narrative-based assessments that can be personalized, thus encouraging active participation and deeper involvement in the learning process [3, 13].

For example, Affective AI utilizes facial expressions and attention levels to alter content to ensure a conducive environment using engagement and emotion levels, enabling a rich, responsive, adaptive learning environment [1, 14]. On the other hand, over-dependence on automated tools might hinder critical thinking and problem-solving skills, as learners may rely on the solutions provided by AI instead of building reasoning [6]. Additionally, the influence of culture and differences in access to AI technologies may impact the success of engagement strategies, highlighting the importance of more flexible and adaptive AI models [7, 15]. It will be important to integrate GAI in a pedagogical manner that promotes collaboration between groups and individuals, benefitting from the advantages of hybrid AI-human approaches that place emphasis on using AI as an assistive tool rather than a replacement, which will better guarantee engaging, ethical and pedagogical learning experiences [9, 16].

4.5 Technical and Ethical Challenges in AI-Powered MOOCs

Responsible and Equitable Adoption of AI in MOOCs: Technical and Ethical Implications Data privacy and security are significant concerns. AI systems need adequate data on student learning behavior for analysis. This data collection poses challenges to consent, transparency, and protection of personal information, mainly as proprietary models are used to analyze learner performance [6, 7]. Addressing those risks, others propose to incorporate privacy-preserving tailored AI models that use cryptographic and decentralized approaches like blockchain to protect interactions and guarantee ethical processing of the data [8, 9].

Another critical challenge is biased algorithms that can affect the fairness and accessibility of MOOCs. Using AI models trained on historical data risks the continuity of those inequalities through the favoritism towards particular profiles, recommending biased content to specific learners, which, in consequence, unfairly conditions learning experiences [2, 10]. In the context of education, this calls for the necessity of inclusive and adaptable recommendation systems so that all students have equal access to resources and opportunities regardless of their socioeconomic status or prior knowledge [8]. Finally, the quality and reliability of AI-generated content are major academic issues. Though a great boon of adaptability and personalization, automated assessments and instant feedback also face threats regarding incorrect responses, out-of-context content generation, and lack of pedagogical oversight [12, 17]. Trained on data until October 2023, human monitors of AI validation loops will be critical to preventing these technologies from compromising pedagogical integrity and producing inaccurate learning outcomes [9].

The use of AI for MOOCs raises a key technical challenge regarding the accessibility and scalability of such courses. Using more advanced algorithms and large-scale personalization models requires significant computing resources that could further widen the digital divide between well-resourced institutions and those with relatively little access to AI [6, 7]. In improving the relationship with these limitations, the advancement of nimble and open-source AI solutions personalized for low-asset conditions is most important, ensuring democratized and inclusive access to AI-empowered advancements in MOOCs [15].

Clear regulatory frameworks and evidence-based ethical policies must be implemented to promote responsible, transparent, and equitable uses of AI in education in light of these challenges. To realize the full potential of AI-driven innovation, AI-powered MOOCs must align with digital ethics, data protection standards, and pedagogical fairness [8, 9].

This Fig. 4 contrasts the blessings and drawbacks of synthetic intelligence (AI) in learning, particularly in MOOCs. Among its blessings, AI permits personalized getting-to-know paths, immediate comments, elevated scalability, adaptability to beginners' desires, and more pedagogical performance. However, it also provides demanding situations: statistics privacy risks, algorithmic bias, content material reliability problems, excessive aid consumption, and ethical worries. Thus, while AI complements getting to know, its integration must be carefully regulated to ensure accountable and equitable use.

5 Analysis and Discussion

Artificial Intelligence (AI) and Generative AI (GAI) are integrated into MOOCs, which is a new breakthrough to further personalize learning pathways, engagement, and optimize learning outcomes in online education. Furthermore, the package of AI systems especially in the form of intelligent recommender systems, adaptive learning models, and personalized feedback mechanisms contribute to student experience and dropout rate reduction [4, 5]. Yet, such technological development does not come without challenges related to algorithmic fairness, transparency, and data privacy [6, 7].

Possibly one of the greatest contributions AI can make to MOOCs is personalizing learning experiences through content tailored to the individual learner. Online courses can adjust the complexity of its exercises, the teaching styles, and the pace of learning based on student interactions with the course materials and performance as measured

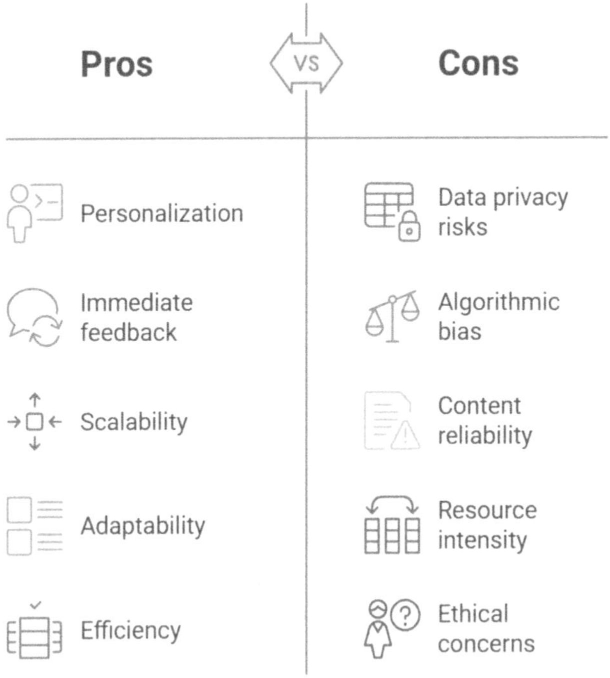

Fig. 4. AI in MOOCs

through well-defined metrics in the form of adaptive learning models and intelligent recommender systems [2, 3]. The incorporation of such individualized learning methods improves student engagement, especially when merged with immersive technologies, like gamification and interactive simulations [18, 19]. In addition, AI-based virtual tutors and conversational agents offer real-time support, enhancing student engagement and comprehension [4, 9]. While these tools both provide personalized educational support, they might also cause an overdependence on AI, which can interfere with learning critical thinking skills [6].

Another significant enhancement of AI-powered MOOCs has been the automation of feedback and evaluations. While standard assessments are frequently static and impersonal, AI-driven platforms can provide instant, comprehensive feedback, enabling students to spot errors without delay [12]. Adaptive AI assesses the difficulty level and type of feedback based on each student's understanding of the subject, thus allowing learning to be interactive and engaging [8, 17]. Contracting with AI-powered organizations to evaluate assessments is an established practice in many sectors, including health care and education, due to benefits such as cost savings through automation and increased accuracy over human evaluators. However, concerns about quality, reliability, algorithmic bias, and inaccuracies in produced content must also be considered [7]. One approach that can be used in solving this challenge is to use AI in combination with human supervision, where educators validate AI-supported recommendations regarding their pedagogical relevance [9].

However, using AI in MOOCs brings significant technical and ethical difficulties. The first concern is data privacy protection, given that AI algorithms require the analysis of big datasets of learners to improve further course personalization [6, 7]. There are no coherent regulations on data usage in MOOCs, leaving student surveillance and privacy at risk [10]. On the other hand, a more critical point of view regarding personalized adaptation is related to the probabilistic nature of the underlying learning algorithms, which could favor some learner groups and exclude others from the personalized recommendations or even lead to inequitable learning experiences based on biases in training datasets [2, 8]. You should take measures to ensure the implementation of bias-correction mechanisms [5]. And diversify the training datasets. Moreover, scalability is another critical technical challenge that keeps its access slim because advanced AI models used for teaching and detecting require much computational power, rendering it inaccessible for most low-resource institutions we discussed above [15]. Open-source, low-weight AI solutions for limited infrastructures can be developed for a broader use of technologies [11].

Contributions and Recommendations for AI Integration in MOOCs
Here are a few recommendations to promote the responsible and sustainable adoption of AI in MOOCs:

1. **Towards inclusive and transparent AI models:** integrating bias detection and correction mechanisms for equitable personalization of learning pathways [2, 5]. This encompasses training on broad datasets and regulating the algorithms for other ethical practices to combat discrimination in content recommendations.
2. **Develop regulatory frameworks and ethical standards:** To improve user confidence, clear and well-communicated policies around the management of student data need to be put in place along with the development of privacy-protecting technologies (e.g., blockchain) [7, 9].
3. **Promote a hybrid AI-human learning model:** Indeed, rather than replace educators, AI should be used to assist instructors by automating routine tasks as human pedagogical expertise supervises and reviews AI-generated content [8, 17].
4. **Could enable wider and broader aspects of AI technologies:** Academic institutions and tech companies can collaborate; AI-driven education can be made available for low-resource institutions [11]. Cloud-based and open-source AI solutions have the potential to facilitate more widespread use of AI resources in MOOCs [15].
5. **Support active student engagement by using interactive methods:** The use of AI tools allows for the introduction of novel pedagogical mechanisms: gamification, interactive simulations, and collaborative learning [3, 19]. Data Mining and Predictive Analytics (DMPA) in MOOCs improve learning and reduce dropout rates.

6 Conclusion

The wide adoption of Artificial Intelligence (AI) and Generative AI (GAI) in MOOCs represents a paradigm shift in digital education that promises personalized learning, adaptive instruction, and increased engagement while at the same time tackling long-standing issues such as high dropout and disengaged learners [4, 5]. AI-based recommendation systems, adaptive assessment, and intelligent learning tutors provide personalized learning experiences but raise serious concerns about data privacy, algorithmic bias, and the reliability of AI-generated content [6, 7].

Future work should then focus on bias attacking in AI models, building the AI ethics governance framework, AI-human hybrid instruction, and resource-efficient and scalable approaches that are effective in low-resource settings [2, 8, 15] in developing practical and responsible AI. Furthermore, exploring innovative AI-based pedagogies, such as gamification, immersive learning, and collaborative AI-assisted education, should be expanded to maximize engagement and learning outcomes [3, 19]. However, by tackling these challenges with a focus on responsible AI deployment and interdisciplinary collaboration, we can reform MOOCs to become even more inclusive, efficient, and personalized systems for education that will serve the interests of learners everywhere.

References

1. Harnadi, B., Widiantoro, A.D., Prasetya, F.H.: Investigating the behavioral differences in the acceptance of MOOCs and E-learning technology. Comput. Hum. Behav. Rep. **14**, 100403 (2024)
2. Rodrigo, C., Iniesto, F., Garcia-Serrano, A.: Applying andragogy for integrating a MOOC into a formal online learning experience in computer engineering. Heliyon **10**(1) (2024)
3. Zhu, J., Liao, Y., Lu, D.: Design and optimization of personalized education system based on intelligent algorithms. Procedia Comput. Sci. **243**, 514–522 (2024)
4. Maity, S., Deroy, A.: Generative AI and its impact on personalized intelligent tutoring systems (2024). https://doi.org/10.35542/osf.io/kawr5
5. Noviandy, T., et al.: Embrace, don't avoid: reimagining higher education with generative artificial intelligence. J. Educ. Manag. Learn. **2**(2), 81–90 (2024). https://doi.org/10.60084/jeml.v2i2.233
6. Kanont, K., et al.: Generative-AI, a learning assistant? Factors influencing higher-ed students' technology acceptance. Electron. J. E-Learn. **22**(6), 18–33 (2024). https://doi.org/10.34190/ejel.22.6.3196
7. Bozkurt, A.: Generative AI, synthetic contents, open educational resources (OER), and open educational practices (OEP): a new front in the openness landscape. Open Praxis **15**(3), 178–184 (2023). https://doi.org/10.55982/openpraxis.15.3.579
8. Arslan, B., et al.: Opportunities and challenges of using generative AI to personalize educational assessment. Front. Artif. Intell. **7**, 1460651 (2024). https://doi.org/10.3389/frai.2024.1460651
9. AlAli, R., Wardat, Y.: Opportunities and challenges of integrating generative artificial intelligence in education. Int. J. Relig. **5**(7), 784–793 (2024). https://doi.org/10.61707/8y29gv34
10. Wirzal, M., Nordin, N., Halim, N., Bustam, M.: Generative AI in science education: a learning revolution or a threat to academic integrity? A bibliometric analysis. Jurnal Penelitian Dan Pengkajian Ilmu Pendidikan E-Saintika **8**(3), 319–351 (2024). https://doi.org/10.36312/e-saintika.v8i3.2127
11. Kadaruddin, K.: Empowering education through generative AI: innovative instructional strategies for tomorrow's learners. Int. J. Bus. Law Educ. **4**(2), 618–625 (2023). https://doi.org/10.56442/ijble.v4i2.215
12. Page, E., Meyers, G., Billings, E.: Theory to practice: an assessment framework for generative AI. Int. J. Educ. Technol. **5**(4) (2024). https://doi.org/10.61669/001c.124250

13. Slamet, J., Basthomi, Y., Ivone, F., Eliyanah, E.: Utilizing an SDL approach in designing a gamification-based MOOC to enhance autonomous learning. J. Inf. Technol. Educ. Res. **23**, 010 (2024). https://doi.org/10.28945/5278
14. Badali, M., Hatami, J., Banihashem, S., Rahimi, E., Noroozi, O., Eslami, Z.: The role of motivation in MOOCs' retention rates: a systematic literature review. Res. Pract. Technol. Enhanced Learn. **17**(1) (2022). https://doi.org/10.1186/s41039-022-00181-3
15. Xu, J., Li, J.: Effects of AI affordances on student engagement in EFL classrooms: a structural equation modeling and latent profile analysis. Eur. J. Educ. **59**(4) (2024). https://doi.org/10.1111/ejed.12808
16. Morales, M., Amado-Salvatierra, H., Medina, J., Barchino, R., Hernández-Rizzardini, R., Teixeira, A.: Personalized feedback in massive open online courses: harnessing the power of LangChain and OpenAI API. Electronics **13**(10), 1960 (2024). https://doi.org/10.3390/electronics13101960
17. Tenakwah, E., et al.: Generative AI and higher education assessments: a competency-based analysis. Front. Artif. Intell. **7** (2023). https://doi.org/10.21203/rs.3.rs-2968456/v1
18. Jayavardhini, P.: AI study partner: development of an LLM and Gen AI-enhanced study assistant tool. Int. J. Sci. Res. Eng. Manag. **08**(03), 1–5 (2024). https://doi.org/10.55041/ijsrem29626
19. Smaili, E., Khoudda, C., Sraidi, S., Charaf, M.: An innovative approach to prevent learners' dropout from MOOCs using optimal personalized learning paths: an online learning case study. Stat. Optim. Inf. Comput. **10**(1), 45–58 (2022). https://doi.org/10.19139/soic-2310-5070-1206

Does the Early Bird Get the Worms? K-12 Teachers' Perceptions on the Use of Generative Artificial Intelligence in Chinese Classrooms

Dong Yang, Yi Zhang, Tore Hoel, and Li Yuan(✉)

College of Education for the Future, Beijing Normal University, Zhuhai 519087, China
l.yuan@bnu.edu.cn

Abstract. China is recognized as the second leading nation in the global advancement of artificial intelligence (AI), and integrating AI into educational practices has become a national priority. To develop an effective professional development program, it is essential to understand teachers' perspectives, as this knowledge informs decision-making and promotes a balanced approach that maximizes AI's benefits in education while upholding the principles of effective and inclusive learning. However, there is a notable gap in understanding the perceptions of teachers, particularly early adopters of Generative AI (GenAI) within the K-12 context in China. This qualitative study conducted in-depth interviews with eight subject teachers identified as "early adopters" of GenAI in K-12 classrooms. The research explored the roles, affordances, and challenges teachers face when leveraging AI to enhance student learning. The findings revealed a diverse range of experiences among teachers. Some regarded GenAI as a valuable supplemental tool, while others expressed concerns about its accuracy and associated risks. Notably, the study suggested that if GenAI is not implemented thoughtfully, it could hinder critical thinking skills. Additionally, the suitability and impact of AI varied across different subject areas. These insights are crucial for designing tailored professional development programs that effectively harness the potential of GenAI in K-12 education.

Keywords: Generative artificial intelligence · China · K12 education · teacher perceptions

1 Introduction

Artificial Intelligence (AI) has quickly become a significant transformative influence across various global sectors, as evidenced by evolving policy initiatives highlighting its growing importance. In education, AI has the potential to transform both learning and teaching practices fundamentally [1]. Countries such as China, Korea, Canada, the USA, and members of the European Union have strived to claim a place in the field of AI Education, which sees artificial intelligence teaching as a strategic priority rather than just a curriculum, have issued several incentive policies to facilitate AI education and cultivate AI talents [2].

A. I. Cristea et al. (Eds.): AIED 2025, CCIS 2590, pp. 255–268, 2025.
https://doi.org/10.1007/978-3-031-99261-2_23

Teachers are the leading actors in effectively integrating AI into compulsory education. While there is consensus among researchers and relevant stakeholders about the importance of AI curriculum within K-12 education [3, 4] and several initiatives to further popularize the subject in schools [5], teachers' experience and perceptions regarding using AI tools (e.g., GenAI) are still largely unknown in the academic world. Understanding teachers' perspectives is crucial for informed decision-making and fostering a balanced approach that optimizes AIED's contributions [6]. Compared to other fields of educational technology, such as gamification and blended learning, AIEd research is more scattered and less organized. As a result, the impact of AI on education remains unclear [7], and more research is needed to understand whether and how these emerging technologies benefit education [8]. Using a semi-structured interview, we interviewed eight K12 teachers of various subjects on the perceived role of GenAI in shaping their pedagogical practices, affordances of GenAI, and challenges and limitations encountered when leveraging AI to enhance student learning. Borrowing the idea of diffusion of innovation theory, they were considered the "early adopters" of GenAI in their school subject teaching.

2 Literature Review

2.1 Artificial Intelligence Education (AIED) Development in China

In the global AI competition, China has been considered a world leader in applying artificial intelligence in education [9, 10]. Early in 2017, the Chinese government published the State Council's National Strategy for AI Development [11]. This policy document has been frequently cited in the media as a pivotal moment in China's AI development and fostered the notion of a comprehensive national strategy for achieving global AI dominance [12, 13]. Moreover, integrating AI into educational practices has emerged as a national strategy within China, with AI education increasingly recognized as essential at the K–12 level. For instance, in February 2024, China's Ministry of Education announced 184 AI education pilot bases in primary and secondary schools to explore further new concepts, new models, and new schemes of AI education [14]. Such emphasis has expanded from national to municipal levels, with cities like Beijing, Shanghai, and Shenzhen announcing that AI will be compulsory in K -12 education [15]. Meanwhile, as a newly-emerged discipline, artificial intelligence education in China faces the dilemma of inadequate theoretical guidance and limited practical experience [2]. For instance, there is a lack of overall planning, which draws insights from AI theories to guide artificial intelligence education in the fundamental education stage. Against such a background, exploring the status quo of AI development in K-12 is crucial, especially the affordance, role, and challenges of GenAI as perceived by teachers.

2.2 Teachers as Early Adopters of GenAI in Education

In any technological context, there are individuals who actively engage with the technology, those who are passive observers, and those who express skepticism or outright refuse to engage [16]. Similarly, GenAI has had a powerful ripple in the field of education because it has the potential to reshape the way teachers teach and students learn.

Still, educators have mixed responses to its ubiquitous use. Some teachers are eager to embrace technology in the classroom, see it as a way to personalize learning and prepare their students for life and careers in the future [17]. Others are uncertain and hesitant to embrace it due to their lack of knowledge and experience with GenAI [18]. This phenomenon can be analyzed through various theoretical frameworks, with the diffusion of innovation theory being particularly significant. According to the diffusion of innovation theory, individuals with high levels of individual innovativeness are more likely to adapt to and use new applications or technologies than others [19]. One distinct group of individuals are called early adopters, who serve as influential role models within their communities, often earning respect for their willingness to embrace innovation [15].

AI-powered tools, such as GenAI, can be considered new technology. Teachers in this study are regarded as early adopters based on the argument that they are influential role models within their schools. However, existing literature often treats these factors in isolation, overlooking how they collectively influence teachers' ability to integrate GenAI into their pedagogical practices effectively. Notably, there is a scarcity of studies focusing on the perceptions of GenAI among Chinese teachers, particularly early adopters in their subject areas. This paper addresses this gap by proposing a comprehensive framework that investigates GenAI's roles, affordances, and challenges in K-12 classrooms, based on in-depth interviews with eight early adopter teachers across China. Specifically, we pose the following research questions:

a) How do Chinese teachers (as early adopters) perceive the role of GenAI in shaping their pedagogical practices within the K-12 classroom?
b) What are the perceived affordances of GenAI in Chinese K-12 education? and
c) What challenges do teachers encounter when leveraging AI to enhance student learning?

3 Method

3.1 Study Setup

For our research purposes, we conducted a series of in-depth, semi-structured interviews among Chinese K-12 teachers. The interview was hosted online using the Tencent Meeting software the Beijing Normal University provided. Each teacher was interviewed separately. To recruit participants, we used a convenience sampling method, a form of non-probability sampling, to reach potential teachers from our professional networks. It was chosen due to its cost-effective and simplistic nature, and it helps generate a potential hypothesis or study objective [20]. In addition, we followed Stratton's methods to improve the dependability of convenience sampling [20] and the credibility of our qualitative study. As a result, eight teachers were recruited in our study.

3.2 Sampling

This study selected eight Chinese K-12 teachers using a convenience sampling technique. The rationale for choosing early adopters as a sampling strategy for this research lies in their unique role in the diffusion of innovation. These individuals are typically more experienced with the technology, enabling them to provide valuable insights, detailed use cases, and reflective feedback on its affordances and limitations. There were three males and three females in the sample. Their age ranged from 26 to 57 (M age = 36). They teach science (2), math (2), computer science (1), English (1), Chinese (1), and innovations (1). On average, the participants had a teaching experience of 12.4 years and 9.8 years of experience with AI. The demographics of the sample are listed in Table 1.

Table 1. Participant demographics

Coding	Gender	Age	Subject	Teaching experience	Experience with AI
JZ	Female	37	Chinese	13	3
DL	Male	38	Innovation	15	20
WJ	Female	33	English	9	2
XW	Male	57	Science	39	35
ZH	Female	36	Science	10	2
YL	Male	30	Math	3	7
HL	Male	26	Math	1	2
SU	Male	34	Computer science	9	7

3.3 Data Collection and Validation

The research team developed a semi-structured interview protocol to investigate the roles, affordances, and challenges associated with teachers' utilization of generative AI technologies. This approach facilitates access to respondents' perceptions and experiences in an engaging manner [21]. The interviews were conducted in Chinese and included three main questions following a demographic section. Utilizing Tencent Meeting software, the online interviews encouraged participants to express their views in detail. The friendly and interactive nature of the discussions allowed participants to elaborate on their perspectives freely. On average, each interview lasted about 60 min, resulting in 9 h of audio data, which were later transcribed using OpenAI's Whisper. All audio transcriptions were then double-checked by two of the group members for accuracy and matching. Lastly, as explained below, the researchers organized the audio files for later transcription and analysis.

3.4 Data Analysis

The transcripts were translated into English and analyzed using NVivo 12 for coding and thematic identification, following Braun and Clarke's [22] framework. Initially, researchers familiarized themselves with the data through repeated readings to gain comprehensive insights. Next, they developed initial codes aligned with the research objectives, focusing on significant phrases and concepts related to generative AI's roles, affordances, and challenges. These codes were then interconnected and synthesized into broader themes, ensuring alignment with study objectives. The extracted themes were rigorously reviewed and labeled to represent the identified constructs accurately. Consistent with Braun and Clarke [22], themes underwent validation through collective author review, ensuring coherence and clarity. The consistency rate between validated themes and finalized coding was 86%.

4 Result

4.1 How do Chinese Teachers (as Early Adopters) Perceive GenAI's Role in Shaping Their Pedagogical Practices in the K-12 Classroom?

The first core question concerns the perceived role of GenAI in K-12 teachers' pedagogical practices. Our analysis identified three distinct roles: GenAI as a content provider, teaching assistant, and facilitator of student-centered learning. We present them separately, along with the relevant quotes and examples.

GenAI as a Content Provider
Generative AI can create teaching materials like lesson plans, exercises, and assessments. This frees up teachers' time for other tasks like planning engaging activities or providing individualized support. For example, English teacher WJ mentioned how he lets students use AI tools on iPads to seek information.

> *"In class, since kids have limited life experience, especially with hard-to-follow knowledge, I can let them use Doubao AI to ask about the potential concepts or statements. In such ways, I let them use tools to look for information and ask for help."*

GenAI as a Teaching Assistant
AI can help provide personalized feedback and support to students, particularly in large classes with limited individual attention. When asked about the role of GenAI in his classroom, teacher YL gave us his perspective:

> *"It serves as a teaching assistant. It can handle real-time feedback... Even in traditional classrooms, it is important since many students do not dare ask questions; they do not want to feel stupid; they have peer pressure, right? Moreover, they care how teachers see them. As a result, they do not ask questions at all. Therefore, it would be really nice if AI could fill this gap."*

Moreover, teacher also briefly mentioned:

> *"Em...It takes the role of similar to a tutor or assignment assistant. For instance, when teaching programming, quite often, when a student starts to learn, he makes many mistakes. He cannot see them with his own eyes. For example, some punctuation marks are written in English, but in Chinese, for example, he missed a semicolon in this sentence. For example, he misspelled one in this sentence, but*

his computer will report an error when it is compiled...That is a bit of a challenge for him. If the lecturer is helping to check, that may be a huge workload. In this situation, I may go directly to let him use AI for help".

GenAI as a Facilitator of Student-Centered Learning
GenAI can empower students to take ownership of their learning. Students can explore ideas, conduct research, and receive feedback by interacting with AI, fostering critical thinking and problem-solving skills. For example, teacher WJ expressed his perspectives on this:

"Actually, with AI, my classroom comes alive. (The classroom structure) changed from teacher-led to student-centered in a way that they support and interact with each other. It contributed to the classroom atmosphere to some extent."

4.2 What are the Perceived Affordances of GenAI in Chinese K-12 Education?

Regarding GenAI's perceived affordance in K-12 classrooms, teachers in our study believe that applying GenAI tools in the classroom contributed to their teaching efficiency and productivity, enabled personalized learning, and, importantly, expanded creative boundaries among students. In the following section, we present the three themes and quotes when necessary.

Enhanced Efficiency and Productivity
Generative AI can automate tasks like generating teaching materials, assessing student work, and providing feedback, saving teachers time and effort.

"Indeed, AI has greatly improved our teaching efficiency and resource volume in lesson preparation. It has significantly increased our constructive efficiency. Before GenAI, searching for resources from the Internet or textbooks was time-consuming. However, now, with highly accurate GenAI tools such as ChatGPT, it efficiently generates good-quality teacher resources."

Personalized Learning
AI can cater to students' needs and learning styles by providing differentiated materials and feedback. AI offers graded materials that reduce teachers' workload.

"For example, our first-year high school children may be at B1 and B2 level, but if we want to give our senior high school children English schematics, it may be more complicated. This generation is much better than the Oxford Dictionary because there is only one edition of the Oxford Dictionary, and the difficulty level is higher than the comprehension ability of our children in grades 9 and 10. Then, after the vocabulary and phrases are shown, we can make AI generate, for example, multiple-choice fill-in-the-blank exercises.

Expanding Creative Boundaries
Generative AI can help students explore new ideas and push creative boundaries by providing novel prompts, generating diverse content, and offering unique perspectives. Teachers expressed how GenAI can be harnessed to achieve this. For instance, one teacher mentioned:

> *"We asked students to do a future city design project. This year's theme is different, and the theme is "floating city". (In this project), The student's first step is to let the AI help them imagine what it will be like and see if there is any difference between what he makes and what they expect. And then it actually helps people to open their minds, help them to find the inspiration for this kind of creativity."*

4.3 What Specific Challenges do Teachers Encounter When Leveraging AI to Enhance Student Learning?

Under this research question, teachers in the investigation expressed the concerns, difficulties, and challenges they encountered when leveraging GenAI to enhance student learning and teaching efficiency. Five distinct challenges emerged: access and equity, uncertainty and lack of control, over-reliance and critical thinking, ethical concerns and data privacy, and lack of teacher training and support.

Access and Equity
Not all teachers and students have equal access to high-quality AI tools due to factors like cost, VPN requirements, and account restrictions. This creates a digital divide that needs to be addressed. Teachers mentioned one challenge: Not all teachers can access high-quality AI like ChatGPT. They may need to find smaller AIs to solve current problems. This limits teachers' and students' access to AI.

> *"The thing that we do not think we can handle at the moment is that the AI is not entirely open. For teachers and students, if we want to use (Chat) GPT, we need to have a VPN, and of course, the account needs to be verified, and it needs a foreign mobile phone or email address, so not all teachers have access to high-quality AI like ChatGPT.*

Uncertainty and Lack of Control
The unpredictable nature of AI-generated content can challenge teachers accustomed to structured lesson plans and predetermined outcomes. Many teachers are hesitant to use AI because of the uncertainty. Many are unprepared for the new viewpoints AI might generate in class. For example, one teacher expressed his concerns:

> *"Many teachers are afraid to introduce AI in the classroom because it brings too much uncertainty. He brings a lot of new ideas that teachers are not prepared for. Because you do not know what new ideas the AI will give you in the classroom."*

Over-Reliance and Critical Thinking
There is a concern that students may become overly reliant on AI and neglect to develop

their critical thinking and problem-solving skills. We worry about students over-relying on AI. Some students submit assignments generated by AI without any real learning. For instance, teacher YL expressed his concern on this:

"The biggest problem that students have with it is that after they get the access, they do not stay on top of the class objectives... Because it is so much fun, they get through the objectives so quickly, and then they start playing with other things, and that is the biggest problem, is that they do not want to put the computer back."

Similar situations were mentioned by teacher WJ:

"But because the Doubao AI is installed on the kids' iPads, we are worried that they might use the AI to do homework, that might be cheating, and they might use the AI to engage in some less learning-related or entertaining conversations, but of course, there is no harm in letting kids explore the boundaries of what it can do.... Of course, the most serious thing we are worried about is that he/she will not be able to complete his /her homework independently when he /she uses AI... He turns in his homework under false pretences, but the whole time, he is just typing in a few commands, and then he turns in this homework, but he is not acquiring it."

Ethical Concerns and Data Privacy

Concerns about the ethical implications of using AI in education, including data privacy, algorithmic bias, and the potential misuse of AI-generated content, exist. The biggest risk is whether the resources we submit will be used to train the AI or collected by third-party companies, which raises concerns about data breaches and privacy.

"In fact, one of the most significant risks that generative AI presents to us is that we have to determine whether the resources that we have submitted to us will be used to train its models or whether our third-party companies will collect these resources and data, and that is one of the biggest problems. The second one is actually, like I just said, we are in the process of finding a balance (on) products and content of GenAI. I know we have seen an ethical framework so far, but in China at least, there is not a clear document that protects the ethical use of AI and the code of ethics put in place."

Lack of Teacher Training and Support

Many teachers lack the necessary training and support to integrate AI into their teaching practices effectively. Providing comprehensive professional development opportunities, clear guidelines, and ongoing support is crucial for successful AI implementation in education. For instance, one participant mentioned:

"In addition to the mindset, we can do more and better in our workshops.

Because of time constraints, teachers receive very general training. What is lacking is hands-on, lesson-specific training. For example, a workshop for a single subject.

If we had subject-based workshops, I am sure they would be much better off if they had that, whatever it is."

Teacher ZH and JZ also expressed their concern about inappropriate teacher development programs. For instance:

"Our school used to invite Prof. X from Y University to give us training; he just talked about how AI can be used (for example) to generate PPT, teach design, and then help to assist in the lesson preparation. These are more outside the classroom, but I do not recall any training on using it in the classroom. There is less training on using it to interact with students or design lessons in the teaching process."

5 Discussion

The integration of generative AI (GenAI) in K-12 education calls for a critical examination of its benefits, contextual contradictions, and systemic implications. While this study highlights the perspectives of Chinese early-adopter teachers, it also reveals tensions that challenge the narrative of AI as an unequivocal educational advantage. Below, we analyze the roles, affordances, and challenges of GenAI through a reflective lens, situating our findings within broader socio-political and cultural contexts.

5.1 Re-examining GenAI's Roles: Efficiency vs. Pedagogical Agency

The integration of GenAI in K-12 education has sparked interest among educators regarding its roles in shaping pedagogical practices. By framing GenAI as a content provider, teaching assistant, and facilitator of student-centered learning, we risk obscuring deeper pedagogical trade-offs. First, GenAI's role as a content provider emphasizes its ability to create tailored educational materials. A recent meta-analysis indicates that AI can significantly outperform humans in content creation [23]. GenAI generates dynamic, interactive content that adapts to individual learning styles, enhancing engagement and comprehension. However, this efficiency may lead to homogenized content, privileging algorithmic outputs over teacher creativity. In China's exam-centric system, reliance on AI for lesson plans could standardize curricula and restrict teaching diversity [2]. This reflects critiques of "automation bias," where efficiency undermines educators' agency [23]. Second, GenAI is recognized for its potential to support classroom dynamics and enhance instructional effectiveness. AI tools can offer insights into student performance, allowing educators to tailor their pedagogical approaches. This predictive capability fosters a responsive educational environment, aligning with studies emphasizing AI's role in strengthening teacher-student relationships [24]. Nevertheless, positioning GenAI as a teaching assistant raises concerns about commodifying teacher labor and reframing pedagogy as a technical rather than relational endeavor [25].

Last but not the least, GenAI facilitates a significant shift toward student-centered learning. AI technologies promote active engagement and collaboration, fostering ownership of learning. This aligns with findings that AI applications enhance academic support services [26] and improve student motivation [27]. However, aligning AI's exploratory ethos with China's collectivist educational values presents challenges, as Confucian traditions emphasizing authority may conflict with AI-promoted exploration.

5.2 Affordances: The Myth of Neutrality in Personalization and Creativity

Teachers identified three primary affordances of GenAI: enhanced efficiency, personalized learning, and expanded creative boundaries. The most significant affordance is the enhancement of efficiency and productivity. Bankins et al. [27] highlight that productivity benefits arise from effective human-AI collaboration. Tools like ChatGPT can create tailored lesson plans, significantly easing educators' workloads. However, effective use requires a holistic evaluation of GenAI tools, as meta-studies show that human-AI combinations may perform worse than their human-only counterparts [23]. Personalized learning is another critical area where GenAI impacts education. Its adaptability allows for tailoring content to individual student needs, accommodating diverse learning styles. However, the emphasis on efficiency often obscures the labor involved in "taming" AI, as teachers spend time refining prompts and verifying outputs, masking this effort in narratives of time-saving efficiency [28].

GenAI can potentially expand creative boundaries in education, enhancing innovation in design processes [29]. This creative potential stimulates collaboration and innovative thinking among students, which is vital for future challenges. However, while GenAI expands possibilities, it risks prioritizing algorithmic novelty over critical inquiry, potentially reflecting superficial engagements with complex issues. In conclusion, GenAI's affordances encompass enhanced efficiency, personalized learning, and expanded creativity. As educators recognize GenAI's potential, developing strategies for its effective integration is crucial for creating engaging and innovative learning environments.

5.3 Challenges: Structural Inequities and the Illusion of Agency

Despite its potential, teachers face significant challenges when integrating GenAI. Interviews revealed five key themes: access and equity, uncertainty and lack of control, over-reliance and critical thinking, ethical concerns, and lack of training. First, access to GenAI technologies is critical, particularly in China, where disparities can exacerbate inequalities, since unequal access limits potential benefits, leading to a digital divide that affects marginalized students. Policymakers must ensure equitable access to GenAI regardless of socioeconomic status to democratize educational opportunities. Second, introducing GenAI can create uncertainty and a perceived lack of control among educators. In general, uncertainty may hinder engagement with GenAI, as teachers worry about reliability and bias. This feeling can overwhelm educators, stifling their willingness to adopt these tools. Third, over-reliance on GenAI challenges developing students' critical thinking skills. Excessive dependence can impair analytical skills and diminish creativity [29]. Therefore, Educators must create instructional strategies that promote critical thinking alongside GenAI use. Fourth, ethical concerns related to data privacy are paramount. This reflected an urgent need for robust frameworks to protect user privacy, as data misuse can erode trust in educational settings. Educators must navigate these ethical considerations, particularly regarding fairness and bias in AI algorithms. Fifth, the integration of GenAI is often hindered by insufficient teacher training and support. According to our findings, some feel unprepared to use GenAI effectively, leading to underutilization and highlighting the urgency of designing specific professional development programs in AI integration for pedagogical contexts. Without substantial training, teachers remain mere operators of tools rather than decision-makers in the classroom.

Furthermore, systemic inequities underpin these challenges. China's urban-rural divide, exacerbated by VPN restrictions, creates a two-tiered educational system, mirroring global disparities in EdTech adoption [30]. This raises questions about whether national AI policies genuinely democratize access or privilege urban centers. Continued research and dialogue are essential to harness GenAI's potential in Chinese K-12 education responsibly, fostering equitable learning environments for all students.

6 Limitations and Future Work

Almost every study has its limitations, and our study is no exception. Given the limitations of this study, readers should be cautious when interpreting our results. Firstly, the specific sample size and selection criteria may limit the generalizability of our findings. Utilizing a convenience sampling method, we recruited eight participants identified as early adopters of GenAI in educational settings. Stratton [20] noted that convenience sampling in qualitative research often reflects the participants' motivation. Consequently, varying levels of motivation may introduce bias. Our research selected teachers based on their interest in the topic and willingness to contribute their perspectives, likely mitigating some of this motivational bias. Furthermore, drawing upon categorization of innovation adopters [19], our limited sample restricts our ability to explore the diverse profiles of teachers. This did not align with our study's primary objectives but presents intriguing avenues for future research. While understanding K-12 teachers' general perceptions of GenAI is vital, it's equally important to investigate the differences in their preferences, teaching styles, and usage patterns. For example, as Zhai [31] indicates, early adopters may engage with GenAI differently, revealing potential discrepancies in perceptions and usage across contexts. Future studies should also examine the alignment between teachers and students in acceptable GenAI usage, providing evidence-based guidelines for its integration into educational practices. A deeper understanding of how teachers evaluate the capabilities and limitations of GenAI is crucial for its effective incorporation into teaching. Additionally, our reliance on interviews as the sole data collection method limits our findings. Incorporating surveys, fieldwork, and classroom observations could strengthen our results through a mixed-methods approach. This could offer a more comprehensive view of teachers' attitudes and practices regarding new technologies like GenAI, highlighting significant directions for future research.

7 Conclusion

Teachers are the leading actors in effectively integrating AI into compulsory education. Understanding teachers' perspectives is paramount to developing an effective professional development program. However, existing literature frequently examines these factors in isolation, failing to recognize their collective influence on teachers' ability to integrate Generative AI (GenAI) into their pedagogical practices. This qualitative study involved in-depth interviews with eight subject teachers identified as "early adopters" of GenAI in K-12 classrooms. The findings revealed diverse experiences; while some teachers viewed GenAI as a valuable supplemental tool, others raised concerns regarding its accuracy and associated risks. Notably, the study indicated that hasty or unthoughtful implementation of GenAI could impede the development of critical thinking skills. Furthermore, the appropriateness and impact of AI varied across different subject areas. These insights are essential for developing tailored professional development programs that effectively leverage the potential of GenAI in K-12 education.

Acknowledgments. This study is supported by the 2024 Guangdong Education Science Planning Project (Higher Education) (Grant No. 2024GXJK690) and the Guangdong Higher Education Society 14th Five-Year Plan 2024 Higher Education Project (Grant No. 24GYB210).

Disclosure of Interests. The authors have no competing interests to declare relevant to this article's content.

References

1. Cheng, E.C.K., Wang, T.: Leading digital transformation and eliminating barriers for teachers to incorporate artificial intelligence in basic education in Hong Kong. Comput. Educ. Artif. Intell. **5**, 100171 (2023). https://doi.org/10.1016/j.caeai.2023.100171
2. Huang, X.: Aims for cultivating students' key competencies based on artificial intelligence education in China. Educ. Inf. Technol. **26**, 5127–5147 (2021). https://doi.org/10.1007/s10639-021-10530-2
3. Long, D., Magerko, B.: What is AI literacy? Competencies and design considerations. In: Proceedings of the 2020 CHI Conference on Human Factors in Computing Systems, pp. 1–16. Association for Computing Machinery, New York (2020). https://doi.org/10.1145/3313831.3376727
4. Touretzky, D., Gardner-McCune, C., Martin, F., Seehorn, D.: Envisioning AI for K-12: what should every child know about AI? In: Proceedings of the AAAI Conference on Artificial Intelligence, vol. 33, pp. 9795–9799 (2019). https://doi.org/10.1609/aaai.v33i01.33019795
5. Jatileni, C.N., Sanusi, I.T., Olaleye, S.A., Ayanwale, M.A., Agbo, F.J., Oyelere, P.B.: Artificial intelligence in compulsory level of education: perspectives from Namibian in-service teachers. Educ. Inf. Technol. **29**, 12569–12596 (2024). https://doi.org/10.1007/s10639-023-12341-z
6. Alwaqdani, M.: Investigating teachers' perceptions of artificial intelligence tools in education: potential and difficulties. Educ. Inf. Technol. (2024). https://doi.org/10.1007/s10639-024-12903-9
7. Holmes, W., Bidarra, J., Køhler Simonsen, H.: Artificial intelligence in higher education: a roadmap and future perspectives. In: Køhler Simonsen, H. (ed.) Forsøg med Uddannelses-digitalisering og Hybride Formater, pp. 121–138. SmartLearning, Copenhagen, Denmark (2021)
8. Chiu, T.K.F., Xia, Q., Zhou, X., Chai, C.S., Cheng, M.: Systematic literature review on opportunities, challenges, and future research recommendations of artificial intelligence in education. Comput. Educ. Artif. Intell. **4**, 100118 (2023). https://doi.org/10.1016/j.caeai.2022.100118
9. Savage, N.: The race to the top among the world's leaders in artificial intelligence. Nature **588**, S102–S102 (2020)
10. Yuan, L.: Where does AI-driven education, in the Chinese context and beyond, go next? Int. J. Artif. Intell. Educ. **34**, 31–41 (2024). https://doi.org/10.1007/s40593-023-00341-6
11. New Generation Artificial Intelligence Development Plan. http://www.gov.cn/zhengce/content/2017-07/20/content_5211996.htm (open in a new window). Accessed 16 Feb 2025

12. "China's AI Awakening." MIT Technology Review. https://www.technologyreview.com/2017/10/10/148284/chinas-ai-awakening/. Accessed 16 Feb 2025
13. China's plan to 'lead' in AI: purpose, prospects, and problems. http://newamerica.org/cybersecurity-initiative/blog/chinas-plan-lead-ai-purpose-prospects-and-problems/. Accessed 16 Feb 2025
14. Chinese schools start new semester with more AI-empowered facilities. https://english.www.gov.cn/news/202409/02/content_WS66d55340c6d0868f4e8ea738.html. Accessed 16 Feb 2025
15. Hine, E., Floridi, L.: Artificial intelligence with American values and Chinese characteristics: a comparative analysis of American and Chinese governmental AI policies. AI Soc. **39**, 257–278 (2024). https://doi.org/10.1007/s00146-022-01499-8
16. Cross, R., Parker, A., Sasson, L.: Networks in the Knowledge Economy. Oxford University Press (2003)
17. Channa, F.R., Sarhandi, P.S.A., Bugti, F., Pathan, H.: Harnessing artificial intelligence in education for preparing learners for the 21st century. Elem. Educ. Online. **20**, 3186 (2021)
18. Ottenbreit-Leftwich, A., et al.: Lessons learned for AI education with elementary students and teachers. Int. J. Artif. Intell. Educ. **33**, 267–289 (2023). https://doi.org/10.1007/s40593-022-00304-3
19. Pinho, C., Franco, M., Mendes, L.: Application of innovation diffusion theory to the E-learning process: higher education context. Educ. Inf. Technol. **26**, 421–440 (2021). https://doi.org/10.1007/s10639-020-10269-2
20. Stratton, S.J.: Population research: convenience sampling strategies. Prehospital Disaster Med. **36**, 373–374 (2021). https://doi.org/10.1017/S1049023X21000649
21. Shen, Y., Guo, H.: "I feel AI is neither too good nor too bad": unveiling Chinese EFL teachers' perceived emotions in generative AI-mediated L2 classes. Comput. Hum. Behav. **161**, 108429 (2024). https://doi.org/10.1016/j.chb.2024.108429
22. Braun, V., Clarke, V.: Using thematic analysis in psychology. Qual. Res. Psychol. (2006)
23. Vaccaro, M., Almaatouq, A., Malone, T.: When combinations of humans and AI are useful: a systematic review and meta-analysis. Nat. Hum. Behav. **8**, 2293–2303 (2024). https://doi.org/10.1038/s41562-024-02024-1
24. Renz, A., Hilbig, R.: Prerequisites for artificial intelligence in further education: identification of drivers, barriers, and business models of educational technology companies. Int. J. Educ. Technol. High. Educ. **17**, 14 (2020). https://doi.org/10.1186/s41239-020-00193-3
25. Selwyn, N.: The future of AI and education: some cautionary notes. Eur. J. Educ. **56**(3), 420–433 (2021). https://doi.org/10.1111/ejed.12532
26. Zawacki-Richter, O., Marín, V.I., Bond, M., Gouverneur, F.: Systematic review of research on artificial intelligence applications in higher education – where are the educators? Int. J. Educ. Technol. High. Educ. **16**, 39 (2019). https://doi.org/10.1186/s41239-019-0171-0
27. Bankins, S., Ocampo, A.C., Marrone, M., Restubog, S.L.D., Woo, S.E.: A multilevel review of artificial intelligence in organizations: implications for organizational behavior research and practice. J. Organ. Behav. **45**, 159–182 (2024). https://doi.org/10.1002/job.2735
28. Crawford, K.: The atlas of AI: power, politics, and the planetary costs of artificial intelligence (2021)
29. Casteleiro-Pitrez, J.: Generative artificial intelligence image tools among future designers: a usability, user experience, and emotional analysis. Digital **4**, 316–332 (2024). https://doi.org/10.3390/digital4020016

30. Howell, P., Wu, C., Aryal, A.: Creating a sustainable IS career ecosystem: examining the cross-section of the diversity management approach and the social cognitive career theory: research in progress. In: Proceedings of the 2023 Computers and People Research Conference, pp. 1–5. Association for Computing Machinery, New York (2024). https://doi.org/10.1145/3579168.3632725
31. Zhai, X.: Transforming teachers' roles and agencies in the era of generative AI: perceptions, acceptance, knowledge, and practices. J. Sci. Educ. Technol. 1–11 (2024). https://doi.org/10.1007/s10956-024-10174-0

MAS-CPS Assessor: A System for Evaluating Collaborative Problem-Solving Skills in Multi-agent Environments

Yu Chen[1(✉)], Jiaqi Bao[2], Yutong He[1], Bian Wu[1], and Yiling Hu[1]

[1] Department of Education Information Technology, Faculty of Education, East China Normal University, Shanghai 200062, China
51264108009@stu.ecnu.edu.cn

[2] School of Journalism and Communication, Jinan University, Guangzhou 510630, China

Abstract. In the context of intelligent technologies empowering learning environments, cultivating students' collaborative problem-solving (CPS) skills through effective teamwork to address complex tasks has become an integral component of the 21st-century core competencies framework. However, the diagnosis of CPS skills, as the primary step in personalized development, faces numerous challenges. Traditional "human-to-human" assessment methods lack sample independence, while "human-to-computer" assessment methods, although overcoming this limitation, suffer from issues of unrealistic scenario construction. To address this challenge, this study focuses on the assessment of CPS skills in a human-to-computer collaborative setting, leveraging the superior performance of generative AI in text understanding and reasoning. Based on a multi-agent technology framework and using the PISA 2015 assessment approach as a foundation, this study constructs an agent collaboration environment powered by generative AI technology to simulate highly realistic human-to-human collaborative scenarios. This method aims to accurately evaluate students' actual levels of collaborative problem-solving skills. From the perspectives of the construction and implementation of multi-agent collaborative scenarios and the measurement of CPS skills, this study provides new ideas and implementation pathways for educational assessment empowered by generative AI.

Keywords: Collaborative Problem-Solving · Multi-Agent · Educational Assessment · Large Language Model

1 Introduction

In the knowledge-intensive and digital society, real-world problems involve complex factors from multiple domains [3, 8]. Learners must integrate professional knowledge in collaborative scenarios to solve problems [16]. Cultivating Collaborative Problem Solving (CPS) ability and enabling efficient teamwork to solve complex tasks have become important cultivation directions in the field of education [26]. Assessing CPS skills is crucial for evaluating learners' CPS development and the first step in personalized CPS

A. I. Cristea et al. (Eds.): AIED 2025, CCIS 2590, pp. 269–282, 2025.
https://doi.org/10.1007/978-3-031-99261-2_24

training. Creating task environments mirroring real-world scenarios and establishing accurate assessment tools can precisely reflect learners' CPS abilities, helping teachers adjust strategies and laying the foundation for a data-driven CPS training framework.

The transformative iteration of generative artificial intelligence has enabled machines to simulate human neural network structures and logical reasoning capabilities, approximating human problem understanding, proactive thinking, and the proposal of rational action plans [1]. Through appropriate agent settings and task design, agents can simulate real human-human collaboration. How to measure CPS competency levels by creating human-computer interaction scenarios that mimic human-human collaborative problem-solving based on individual student characteristics, and to provide appropriate suggestions, has become a key issue [18]. However, single-agent human-computer collaborative interaction methods struggle to meet the complex and diverse teamwork demands of real problem-solving processes, thereby limiting their application scope and practical effectiveness in CPS competency assessment and development. In contrast, multi-agent technology, leveraging the powerful language understanding and generation capabilities of large language models, can more accurately simulate the collaborative behaviors and thinking patterns of real human teams. The situational awareness, imitative learning, guided learning, and collaborative learning capabilities demonstrated in task-solving processes have the potential to overcome many challenges faced by single-agent systems in real human collaborative problem-solving scenarios. This, in turn, offers new possibilities and more effective solutions for constructing CPS competency assessment environments that reflect real-world scenarios [25].

To this end, this study focuses on the assessment scenario of collaborative problem-solving skills and designs and develops a CPS assessment system based on large language models and multi-agent technology, supporting open-ended dialogues in the collaborative process. The system aims to simulate human-human collaborative problem-solving modes in real-world scenarios to assess learners' CPS skills. This study not only explores the design and implementation of multi-agent systems supported by large language models from a technical perspective but also further investigates their application in educational assessment, providing a new pathway for technology-enhanced educational evaluation. Specifically, this study addresses the following two questions:

(1) How to Build a Multi-Agent Collaborative Environment Based on a LLM to Measure Learners' CPS Competence?
(2) How does MAS-CPS Assessor perform on CPS assessments?

2 Related Work

CPS capability refers to an individual's ability to effectively integrate information and resources, collaborate with others to share understanding, and ultimately reach consensus and find solutions in a collective problem-solving process [2, 5]. Currently, computer-based CPS assessment methods worldwide can be categorized into two approaches: Human-Agent (H-A) collaboration and Human-Human (H-H) collaboration. The former, exemplified by PISA (2015), requires students to collaborate with simulated partners driven by computer applications. The latter, represented by ATC21S, involves students collaborating with other participants to complete assessment tasks [4, 17]. In the H-A method, student-agent interactions and dialogue paths are based on pre-designed standardized items [15]. Students must select language expressions or interaction behaviors from fixed options, which are considered indicators of a specific dimension of CPS ability [23]. In contrast, the H-H method involves real-time dialogue among group members through collaborative software.

Comparing the two assessment methods, the H-A approach ensures student independence during the assessment process and can integrate a competency framework to construct tasks that deeply assess each element of problem-solving, supporting horizontal comparisons at the individual level [9]. However, due to predefined procedural rules, agent intelligence struggles to match students' personalized expression needs, and pre-set options cannot fully represent the complex real interactions among team members [22]. The H-H method, on the other hand, offers a more authentic dialogue scenario, minimizing the differences between real-world task collaboration and experimental collaborative methods [10]. However, the casual nature of communication can lead to interdependence among group members, potentially affecting the validity of individual CPS ability assessment [11, 24].

Despite criticism for its lack of authenticity in human collaboration, the H-A assessment method has significant advantages in experimental control and standardization. Research has focused on enhancing task richness, assessment precision, and natural language interaction in human-computer contexts. For instance, Rojas et al. [20] developed a CPS assessment system for elementary students, conducting three rounds of iterative experiments to validate the system from both activity design and psychometric tool perspectives. They found that students' performance in human-computer and human-human collaborative problem-solving was similar, but human-computer interaction offered advantages in data management and analysis. Additionally, Kuo et al. [13] embedded a chat collaboration window in the problem-solving process, triggering collaborative dialogues based on students' behavioral operations. Their results showed that the human-computer collaboration method demonstrated reliability and validity in ability assessment but required natural language technology integration to enhance dialogue authenticity and create open-ended collaborative tasks.

With the advancement of large language models (LLMs), multi-agent technology can introduce multiple agent structures with different identity cognitions, defined roles, and coordinated cooperation mechanisms to simulate complex collaborative modes in real problem-solving scenarios [1]. Compared to single-agent systems, multi-agent systems offer stronger social simulation capabilities, greater flexibility, and broader adaptability. Current applications of LLM-based multi-agent systems in social behavior simulation, debate, and information retrieval have shown significant potential. For example, Xi et al. [25] found that multi-agent systems can engage in cooperative or competitive interactions, acquiring social knowledge and experience through imitation, guidance, and collaborative learning. Li et al. [14] assessed the interaction capabilities of multi-agent systems in team tasks, revealing emergent collaborative behaviors and high-order theory of mind abilities among agents, which allowed real-time task progress recognition and adjustment of team collaboration strategies. These findings highlight the substantial educational application potential of LLM-based multi-agent collaborative modes in simulating complex teaching scenarios and enhancing learner engagement in collaborative processes. However, research in educational contexts remains in the exploratory phase of constructing and implementing application models. For instance, Zhang et al. [27] used the AutoGEN framework to develop a novel multi-agent classroom environment based on LLMs, demonstrating significant effects on enhancing student learning experiences through dynamic simulation. Yet, further optimization is needed in areas such as long-term memory storage, real-time responsiveness, and authenticity of scenario construction.

Therefore, integrating LLM-based multi-agent technology into H-A assessment methods, through effective collaborative scenario design and construction of H-A and Agent-Agent interactions, has the potential to overcome the current bottleneck of interaction authenticity in H-A assessments. By simulating real human communication through open-ended dialogue mechanisms, the reliability and authenticity of H-A assessment methods can be significantly improved.

3 System Design and Architecture

3.1 Design and Implementation of Collaborative Environments

To ensure the reliability and feasibility of the collaborative environment and CPS skill assessment, this study referenced the "Visit" task from PISA 2015 [17], in which students are required to collaborate with three intelligent agents—George, Rachel, and Brad—to develop a study plan. As shown in Fig. 1, in the original assessment system, students receive predefined messages and select responses from given options to advance the collaborative process. These options, based on the PISA 2015 CPS assessment framework, represent different levels of collaboration. The sequence of students' option selections during the collaboration serves as the basis for evaluating their CPS skills. Therefore, to create a CPS assessment environment that closely resembles real-world collaborative scenarios, this study integrates a multi-agent system framework based on large language models. The design and implementation are approached from three dimensions: Partner Agentization, Task Deconstruction, and Session Controller.

Collaboration Partner Agentization

The CPS process encompasses challenges such as group communication, conflict management, consensus building, and solution development, requiring students to engage in multi-round, task-oriented interactions within an information-rich environment. However, simply exchanging responses from large language models (LLMs) inevitably leads to issues such as repetitive instructions, false responses, and ambiguous role positioning.To address these challenges, this study draws on the agent settings from the "Visit" task of PISA 2015 and defines three types of agents: Assistant, Classmate, and Real Student. Each agent ai $\in$ A is based on an LLM and incorporates a system-specific prompt Pa, a role description Da tailored to the collaborative context, and distinct functional modules ρ (e.g., RAG retrieval, memory mechanisms). Therefore, the agent set can be represented as A = ρ (LLM, Pa, Da). The following sections will elaborate on the settings for each type of agent.

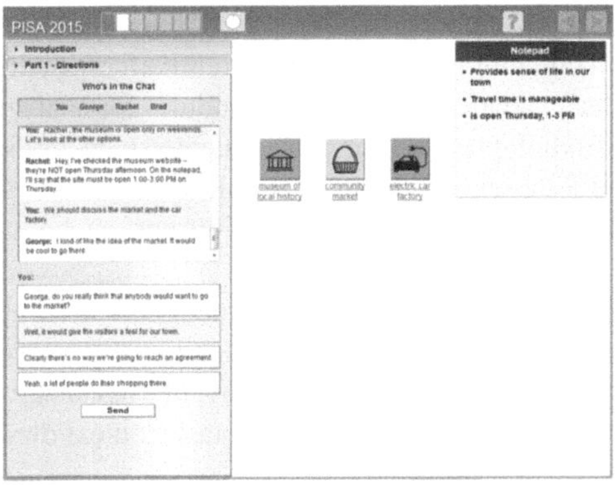

Fig. 1. Illustration of the PISA 2015 CPS task the Visit as retrieved from the official OECD report on released field trial cognitive items.

(1) Assistant Agent

The Assistant agent is responsible for controlling the collaborative process, with its core responsibilities focusing on topic intervention from three dimensions: relevance of the current topic to the collaborative task, progress toward the task objectives, and achievement of assessment criteria. On one hand, it requires robust capabilities in content understanding and reasoning to distill the core content of the current topic. On the other hand, it must provide suggestive or fallback dialogues based on the task objectives and corresponding assessment criteria, thereby guiding the topic's direction.

To balance responsiveness with the text comprehension and reasoning abilities of large language models, the Assistant agent is integrated with the GPT-4 Turbo interface. Additionally, the agent's functionality includes modules for analyzing CPS assessment dimensions in long dialogues, Retrieval-Augmented Generation (RAG) knowledge base retrieval, and long-term memory retrieval. These modules enable the agent to access the collaborative task context and key historical dialogue information stored in the knowledge base, and to integrate the CPS ability dimensions measured in past dialogues.

(2) Partner Agents

The Partner agent aims to simulate the interactive peer roles found in authentic collaborative teams. By configuring the cognitive styles and CPS ability levels of different agents within the system prompt, the Partner agent can exhibit diverse collaborative behaviors and communication patterns. This differentiated setting helps to simulate the individual variability within a real collaborative team, thereby supporting the full mobilization of students' engagement in the collaborative process and the comprehensive assessment of their CPS abilities across multiple dimensions.

Specifically, we have designed two Partner agents—Deep Thinker and Class Clown—to collaborate with students. The Deep Thinker has a higher level of CPS ability, focusing on the depth and breadth of problem-solving solutions, while the Class Clown has a lower CPS ability but higher social interactivity, guiding students in topic discussions and thereby helping them to demonstrate their true problem-solving abilities in collaborative exchanges. Both Partner agents are integrated with the GPT-4o interface and are equipped with modules for analyzing CPS assessment dimensions in short dialogues, Retrieval-Augmented Generation (RAG) knowledge base retrieval, and long-term memory retrieval. These modules assist the agents in understanding the collaborative task context of the current dialogue and the expected CPS ability dimensions to be assessed.

Class Clown - This agent is designed as an experienced atmosphere enlivener and discussion initiator, skilled in context comprehension to propose new topics for discussion. With a moderate level of CPS ability, this agent actively initiates discussions with collaborative partners based on the task context during the collaborative process.

Deep Thinker - This agent is designed as a seasoned problem analyst, proficient in deconstructing problems and engaging in deep thinking. With a high level of CPS ability, this agent demonstrates a profound understanding of the collaborative task and focuses on enhancing the feasibility and perspective of proposed solutions for the current topic. During the collaborative process, this agent efficiently integrates the diverse knowledge and skills of team members, identifies and addresses key nodes in complex problems, and engages in in-depth discussions with collaborative partners to ensure the achievement of team goals.

(3) Real Student Agent

This agent represents a real student and is used solely to obtain input from the participant student. It standardizes the format of text input through functional definitions.

Collaborative Task Deconstruction

To develop a CPS assessment environment that supports open-ended dialogic collaboration with agents, this study deconstructed the PISA 2015 "Visit" task, which is traditionally based on predefined message exchanges. Participants are required to complete three core tasks sequentially: "Deciding on the Study Tour Destination", "Allocating Preparation Tasks", and "Handling Unexpected Situations". As shown in the Fig. 2, the collaborative environments for each task are independent of one another. Participants collaborate with agent teams that have task-specific Role Descriptions and System Prompts, oriented by CPS assessment dimensions and task objectives. Interactions between tasks are connected through the CPS Identification and Information Extraction APIs, which extract information from each segment of the interaction. On one hand, these APIs analyze the CPS dimensions measured in the current task. On the other hand, they summarize and persist key information from the dialogue (e.g., study tour theme, destination) into a vector database to provide memory storage for subsequent tasks. Through this approach, the interaction records and shared information link each task phase, forming a complete collaborative task sequence.

Specifically, in Task 1, students need to collaborate with three agent partners to select an appropriate study tour destination. To reach a decision, the team must share and discuss their preferences, resolve any misunderstandings, and make a final choice. In Task 2, based on the chosen destination, students discuss the preparation plan and task allocation for the study tour with the three agent partners. In Task 3, we introduce an unexpected situation: "A student falls ill during the study tour and needs to be taken to the hospital and their parents notified". Students must communicate with their three partners to develop a response plan and summarize the approach to handling the unexpected issue, ensuring the smooth continuation of the study tour.

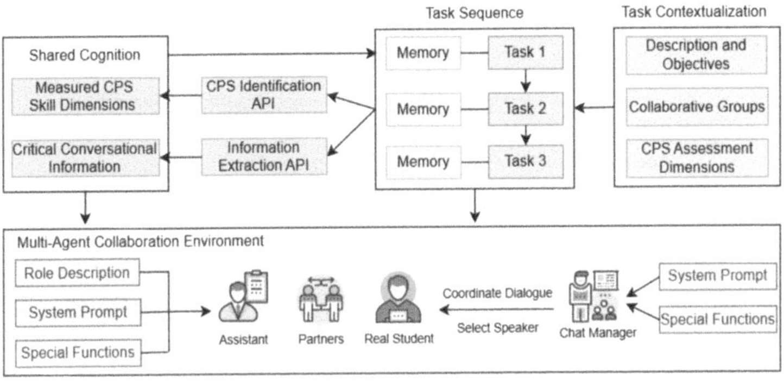

Fig. 2. Framework of Task Deconstruction and Collaborative Patterns.

Collaboration Session Controller

Unlike multi-agent systems with standardized operating procedures [12, 19], collaborative problem-solving is a dynamic group chat without a strict workflow. The interactions among agents are not dependent on predefined rules; therefore, agents need to dynamically determine appropriate times to speak in order to manage the collaborative process. Consequently, as shown in Fig. 3, we designed and implemented a session controller that observes, decides, and directs agent behavior based on the current collaborative state. It comprises three modules:

Collaborative State Reception: The collaborative state is composed of three core pieces of information. The "Dialogue History" serves as the primary source for identifying the progress of the current task dialogue, obtained by reading the multi-agent conversation. The "Role Description" identifies the agent composition for the current collaborative task. The "Cognition Memory" links the current task with key information from previous task collaboration dialogues, retrieved from the vector database through similarity calculations.

Collaborative Process Advancement: This module focuses on the selection of collaborative agents and the updating of dialogue memory. Based on the SelectorGroupChat class provided by the AutoGEN multi-agent framework, we set a selector prompt. Additionally, we defined an Agent Selection Function through the addition of Functions, stipulating that the interval between two consecutive utterances by the Real Student should not exceed two rounds of dialogue to ensure their engagement in the collaborative process. We also added methods for extracting and updating dialogue memory, extracting key information from the received messages, and asynchronously invoking the

CPS assessment interface. The results are then persisted into the "Cognition Memory" and "Task Memory" vector databases, respectively.

Agent Group Management: To prevent agents from entering a blocked state due to the execution of special functions (e.g., persisting CPS assessment information), which could hinder the collaborative process, this module synchronizes agent states. It identifies currently available agents and synchronizes them with the "Role Description" information.

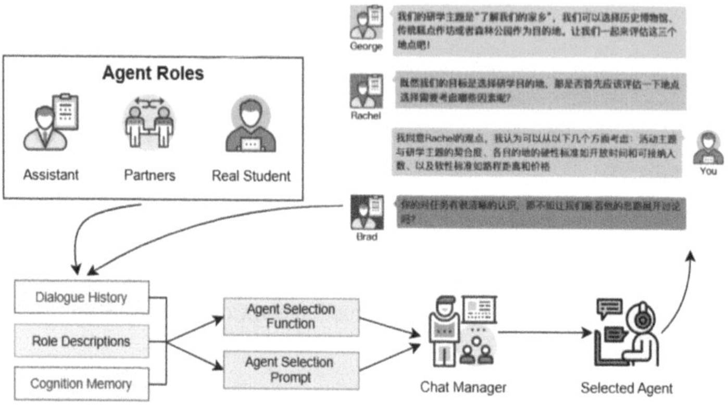

Fig. 3. Mechanism of Collaboration Session Controller.

3.2 Design and Implementation of the Interactive Interface

To optimize the user experience in collaborative learning through rational functional zoning, as shown in Fig. 4, this study drew on the interaction interface design of PISA 2015 and divided the collaborative interface into four core areas: "Collaborative Task Selection and Introduction" (red section), "Collaborative Member Management" (yellow section), "Dialogue Collaboration" (green section), and "Notebook" (blue section).

Collaborative Task Selection and Introduction: This area serves as the navigation entry point of the collaborative interface, functioning as the "primary navigation" in the team collaboration interface. It provides an overall introduction to and selection of collaborative tasks. The design references the assessment framework of PISA 2015, using clear task categorization and guidance to help learners quickly locate their target tasks. Notably, the three assessment tasks are sequential; only after completing the preceding task can the next one be unlocked.

Collaborative Member Management: This section displays the members of the collaborative team for the current task. Clicking on a specific agent reveals its description.

Dialogue Collaboration: This area is the core interactive zone of the collaborative interface, supporting real-time interaction between students and the agent team. It allows

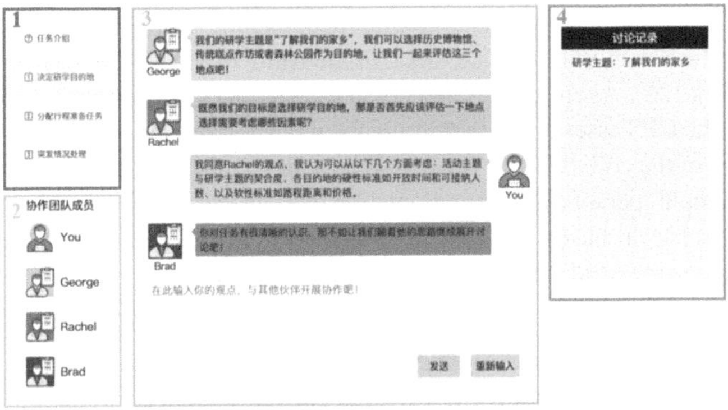

Fig. 4. Implementation of the Interactive Interface (Functional Areas Highlighted in Different Colors) (Color figure online)

learners to view dialogues among agents and access historical conversation information, providing a dynamic and visual collaborative window for interaction.

Notebook: This area serves as a space for recording and reflecting on the collaborative process. It logs key information from each collaborative task, such as the study tour theme, study plan, and task allocation.

3.3 System Architecture

Based on the design and implementation of the collaborative environment described above, the Fig. 5 illustrates the system architecture of the MAS-CPS Assessor from a system development perspective.

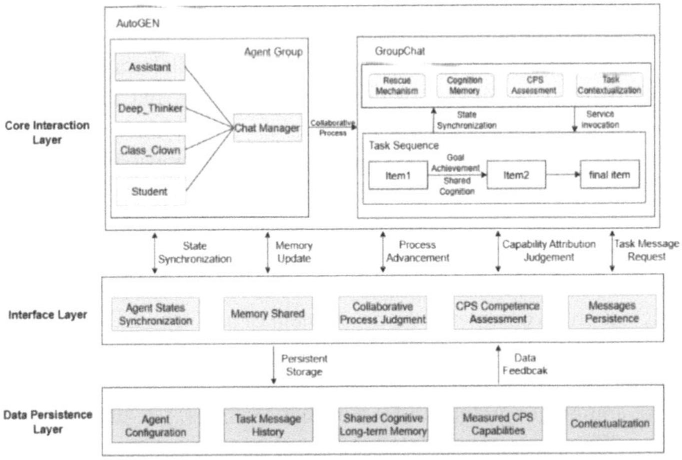

Fig. 5. System Architecture Diagram.

Core Interaction Layer: This layer is crucial for collaborative environment construction. It's based on the SelectorGroupChat mode of the AutoGEN 0.4.5 multi-agent collaboration framework [26]. Agent groups and collaborative tasks are designed and partitioned according to the CPS assessment requirements of the collaborative scenario. To improve the coherence of the collaborative process, interaction authenticity, and the focus of CPS ability assessment, personalized functions like dialogue rescue mechanisms and cognitive memory retrieval based on collaborative session history are implemented through agent prompt settings and function definitions.

Interface Layer: It offers essential interface support for agent collaboration and CPS ability assessment. For example, to establish the shared cognition mechanism, operations on memory are carried out within the "Memory" abstract class. By integrating the Mem0 long-term memory framework, the information extraction interface is first called to summarize key info from a single task and store it in a remote vector database. When later tasks make dialogue requests, similarity matching is done to return relevant key conversation information, thus enabling memory retrieval.

Data Persistence Layer: This layer manages the scenario information needed for collaboration and stores messages from the collaborative process. On one side, it has to process and persist data from different sources and types, such as scenario info and message dialogues. On the other side, it must be compatible with data request interfaces during collaboration and quickly return preprocessed, formatted data.

4 Results

To test the effectiveness of the MAS-CPS Assessor system in CPS ability assessment and H-A collaborative processes, we invited four eighth-grade students from a middle school in Shanghai. Among them, two students had high levels of CPS ability, while the others had lower levels. As shown in Table 1, each student was required to independently collaborate with a team of three agents to complete the collaborative task, with the goal of assessing six dimensions of CPS ability through the collaboration process. During this process, we focused particularly on the system's coverage of CPS ability assessment indicators and the fluency of collaborative dialogue.

To this end, three researchers coded the CPS ability dimensions associated with the experimental sessions based on the CPS skill assessment criteria by Rojas et al. [20]. As shown in Table 2, we statistically analyzed key indicators of the experimental process, including the number of dialogues, dialogue role distribution, CPS assessment dimensions involved, and experimental duration. It was found that in the predefined subtasks, each experimental group completed the assessment of all six CPS dimensions. The proportion of participant dialogues was above 30%, and the number of dialogue messages in each group was within the predefined maximum (35). This indicates that each group passed the task completion judgment by the Chat Manager, i.e., determining the study tour destination. Thus, it can be concluded that all participants successfully completed the collaborative task, met the requirements of the CPS assessment dimensions, and demonstrated high engagement, thereby showing the system's usability in assessing CPS ability through open-ended H-A dialogues.

Table 1. Experimental Collaborative Scenario Setup Information.

Task Objectives	Task Description	CPS Assessment Dimensions	Agent Group
Determination of Study Tour Destination	Determination of the Final Study Tour Destination through Criteria Setting and Analysis of Advantages and Disadvantages	B2 - Identifying and describing tasks to be completed C1 – Communicating with team members about the actions to be performed C2 – Enacting plans C3 – Following the rules of engagement D1 – Monitoring and repairing the shared understanding D2 - Monitoring action results and evaluating the success in solving the problem	George (Assistant) Rachel (Class Clown) Brad (Deep Thinker) You (Real Student)

Focusing on the performance of participants with different CPS abilities in the assessment system, the experimental results preliminarily revealed the following two characteristics:

(1) The high-CPS group completed the task in a shorter time with more dialogue turns, indicating that they could collaborate more effectively with the agent team and complete the task more quickly. This reflects that high CPS ability enables participants to adapt more rapidly to the collaborative environment and collaboratively develop solutions.
(2) In terms of message distribution, the high-CPS group's dialogues were more focused on interactions between the Real Student (S) and Deep Thinker (D), while the low-CPS group had a higher proportion of dialogues involving the Assistant (A) and Class Clown (C). This suggests that participants with high CPS ability could more actively express their opinions and engage in in-depth discussions with the collaborative team on solutions. In contrast, those with low CPS ability required more guidance from their peers to adapt to the collaborative environment.

Table 2. Experimental Session Analysis Results (S: Real Student; A: Assistant; C: Class Clown; D: Deep Thinker).

Assessment ID	Total Messages	Message Distribution	CPS Dimensions	Time
High CPS-1	31	S: 13; A: 4; C: 5; D: 9	6	5:12
High CPS-2	32	S: 14; A: 5; C: 5; D: 8	6	5:33
Low CPS-1	27	S: 9; A: 7; C: 6; D: 5	6	7:20
Low CPS-2	26	S: 9; A: 7; C: 6; D: 4	6	6:52

5 Conclusion, Discussion and Future Work

The MAS-CPS Assessor is a CPS assessment system using large language models and multi-agent technology. Students have open-ended dialogues with agent teams on collaborative tasks. The agent team builds dialogue contexts according to CPS assessment goals and analyzes students' responses to measure their ability levels. To address the long-standing criticism of the H-A assessment system's reliance on predefined options for interactive dialogue, we propose a universal design and implementation strategy from three perspectives: agent team construction, deconstruction and progress management of collaborative tasks, and control of collaborative conversations. First, the construction of the agent team must fully consider the complexity of the collaborative task and the situational nature of the assessment objectives. By combining the roles of the System Prompt agents, we clarify the positioning and capabilities of each agent in both the assessment and collaborative scenarios, thereby forming an efficient collaborative team. On this basis, we deconstruct the collaborative task from the perspective of model memory capacity, establishing a shared cognitive mechanism that connects each collaborative task. This mechanism is derived from key information extracted from each collaborative task, ensuring that each task focuses on its specific assessment dimension. Focusing on the progress of the collaborative process and the interaction patterns among agents, we design a conversation controller that synchronizes the current multi-agent collaborative state (including agent and collaborative conversation history) at the granularity of each dialogue request, coordinating the sequence of agent conversations. This approach achieves a collaborative process that combines CPS competency assessment objectives with smooth and natural interactive dialogues. The above-mentioned strategy aims to optimize the interaction between agents and students to enhance the collaborative authenticity of the H-A assessment method. It is also applicable to educational scenarios represented by multi-agent-facilitated group learning. By analyzing specific application scenarios, our study hopes to propose a new way to promote human-agent collaboration [7, 27].

The study constructs an assessment scenario based on the "Visit" task from PISA 2015. Results from a small-scale assessment experiment show that the MAS-CPS Assessor system can measure the expected CPS competency dimensions through dialogue guidance while maintaining student engagement. This demonstrates its potential for real-world CPS competency assessment. Innovatively, it breaks through the limitations of H-A interaction authenticity by creating open-ended interaction methods based on large language models and multi-agent technology. By simulating a real collaborative interaction environment through agent team construction, it provides a new practical solution for empowering CPS assessment in educational scenarios [14]. Additionally, by analyzing individual performance in the assessment, we found that high-CPS-level groups were more proactive in collaborating with the multi-agent team and engaged in deeper discussions during the collaborative planning process, consistent with the findings of Rojas et al. [21] and Chang et al. [6].

However, this study is subject to several limitations. First, the construction of the multi-agent collaborative environment is based on the AutoGEN framework. Although it is expandable in terms of agent function definition and dialogue agent selection rules, it struggles to adapt to more complex logic, such as asynchronous dialogues between students and agents, within the existing architecture. Second, the system's effectiveness has only been validated through case analysis in a small sample, and further expansion of the sample size is needed. Finally, the construction of realistic collaborative scenarios and the refinement of the CPS assessment framework require long-term research and improvement.

In future research, several promising directions for optimization are worth exploring: (1) Conducting large-scale system testing to validate the system's effectiveness in CPS competency assessment and collecting feedback from participants on the interaction process to optimize agent interaction patterns and collaborative environment construction. (2) Expanding the application scenarios of the system by varying the scenarios and defining agent function rules to explore the application of large language model-based multi-agent technology in personalized learning scenarios, such as assisting students in group cooperative learning in educational settings.

Funding. This work was supported by the National Natural Science Foundation of China (grant number 72304099).

References

1. Aher, G.V., Arriaga, R.I., Kalai, A.T.. Using large language models to simulate multiple humans and replicate human subject studies. In: 2023 40th International Conference on Machine Learning (ICML), pp. 337–371. PMLR, Hawaii, USA (2023)
2. Andrews-Todd, J., Forsyth, C.M.: Exploring social and cognitive dimensions of collaborative problem solving in an open online simulation-based task. Comput. Hum. Behav. **104**, 105759 (2020)
3. Camacho-Morles, J., Slemp, G.R., Oades, L.G., Morrish, L., Scoular, C.: The role of achievement emotions in the collaborative problem-solving performance of adolescents. Learn. Individ. Differ. **70**, 169–181 (2019)
4. Care, E., Griffin, P.: An approach to assessment of collaborative problem solving. Res. Pract. Technol. Enhanc. Learn. **9**(3), 367–388 (2014)
5. Chai, H., Hu, T., Wu, L.: Computer-based assessment of collaborative problem solving skills: a systematic review of empirical research. Educ. Res. Rev. **43**, 100591 (2024)
6. Chang, C.J., et al.: An analysis of student collaborative problem solving activities mediated by collaborative simulations. Comput. Educ. **114**, 222–235 (2017)
7. Cheng, Y., et al.: Cooper: coordinating specialized agents towards a complex dialogue goal. In: 2024 39th Annual AAAI Conference on Artificial Intelligence, vol.38, pp. 17853-17861. AAAI Press, Pennsylvania, USA (2024)

8. De Boeck, P., Scalise, K.: Collaborative problem solving: processing actions, time, and performance. Front. Psychol. **10**, 1280 (2019)
9. Graesser, A., Kuo, B.C., Liao, C.H.: Complex problem solving in assessments of collaborative problem solving. J. Intell. **5**(2), 10 (2017)
10. Harding, S.M.E., Griffin, P.E., Awwal, N., Alom, B.M., Scoular, C.: Measuring collaborative problem solving using mathematics-based tasks. AERA Open **3**(3), 1–19 (2017)
11. Herborn, K., Stadler, M., Mustafić, M., Greiff, S.: The assessment of collaborative problem solving in PISA 2015: can computer agents replace humans? Comput. Hum. Behav. **104**, 105624 (2020)
12. Hong, S., et al.: MetaGPT: meta programming for multi-agent collaborative framework. arXiv preprint arXiv:2308.00352 (2023)
13. Kuo, B.C., Liao, C.H., Pai, K.C., Shih, S.C., Li, C.H., Mok, M.M.C.: Computer-based collaborative problem-solving assessment in Taiwan. Educ. Psychol. **40**(9), 1164–1185 (2020)
14. Li, G., Hammoud, H., Itani, H., Khizbullin, D., Ghanem, B.: CAMEL: communicative agents for “mind” exploration of large language model society. Adv. Neural. Inf. Process. Syst. **36**, 51991–52008 (2023)
15. Li, M., Liu, H., Cai, M., Yuan, J.: Estimation of individuals’ collaborative problem solving ability in computer-based assessment. Educ. Inf. Technol. **29**(1), 483–515 (2024)
16. Li, S., Pöysä-Tarhonen, J., Häkkinen, P.: Patterns of action transitions in online collaborative problem solving: a network analysis approach. Int. J. Comput.-Support. Collab. Learn. **17**(2), 191–223 (2022)
17. OECD.: PISA 2015 Results: Collaborative Problem Solving (2017)
18. Ouyang, F., Xu, W., Cukurova, M.: An artificial intelligence-driven learning analytics method to examine the collaborative problem-solving process from the complex adaptive systems perspective. Int. J. Comput.-Support. Collab. Learn. **18**(1), 39–66 (2023)
19. Qian, C., et al.: ChatDev: communicative agents for software development. arXiv preprint arXiv:2307.07924 (2024)
20. Rojas, M., et al.: Assessing collaborative problem-solving skills among elementary school students. Comput. Educ. **175**, 104313 (2021)
21. Rojas, M., Nussbaum, M., Moreno, C.: Validating a collaborative problem-solving assessment tool across educational stages. Comput. Educ. **227**, 105228 (2025)
22. Stadler, M., Herborn, K., Mustafić, M., Greiff, S.: The assessment of collaborative problem solving in PISA 2015: an investigation of the validity of the PISA 2015 CPS tasks. Comput. Educ. **157**, 103964 (2020)
23. Sun, C., Shute, V.J., Stewart, A., Yonehiro, J., Duran, N., D’Mello, S.: Towards a generalized competency model of collaborative problem solving. Comput. Educ. **143**, 103672 (2020)
24. Tang, P., Liu, H., Wen, H.: Factors predicting collaborative problem solving: Based on the data from PISA 2015. Front. Educ. **6**, 619450 (2021)
25. Xi, Z., et al.: The rise and potential of large language model based agents: a survey. arXiv preprint arXiv:2309.07864 (2023)
26. Zhang, S., et al.: Understanding student teachers’ collaborative problem solving: insights from an epistemic network analysis (ENA). Comput. Educ. **183**, 104485 (2022)
27. Zhang, Z., et al.: Simulating classroom education with LLM-empowered agents. arXiv preprint arXiv:2406.19226 (2024)

Doctoral Consortium

Evaluating the Impact of LLM-Generated Assignment Report Summaries in Intelligent Tutoring Systems

Wen-Chiang Ivan Lim(✉) and Neil T. Heffernan

Worcester Polytechnic Institute, Worcester, MA 01609, USA
{wlim,nth}@wpi.edu

Abstract. Large language models (LLMs) have the potential to support teachers by addressing challenges related to time constraints and data literacy, helping them engage more effectively with student learning analytics. While intelligent tutoring systems (ITS) provide valuable data, many teachers find it challenging to extract important insights for instructional decision-making. This study explores how LLM-generated summaries based on assignment reports can provide concise, actionable insights that improve teachers' ability to understand reports and respond to student performance. Using a randomized design, we aim to evaluate teachers' perceived usefulness of summaries, their effect on engagement with reports, and the impact on student participation and outcomes.

Keywords: Human-centered computing · large language models · learning analytics · data-informed teaching practices

1 Motivation for Research

Intelligent tutoring systems (ITS) have gained popularity in classrooms as teachers become more confident in using these technologies. Ideally, all teachers would have ample time and data literacy skills to analyze learning analytics found in ITS to inform instruction. Yet, unsurprisingly, a large-scale survey of over 40,000 teachers identified limited time as the major obstacle to data use, and insufficient technical capacity as the secondary obstacle [5].

Lack of Time. While learning analytics guide instruction, they are only part of teachers' responsibilities. Beyond teaching, they manage classrooms and administrative tasks. In U.S., teachers work an average of 46.2 h per week—the fifth highest among TALIS systems [6]. Of this, 28.1 h are spent teaching (second highest), but only 7.2 on individual planning (ranked 19th). This imbalance suggests limited time to review learning analytics and inform instruction.

Insufficient Technical Capacity. Many teachers lack training in data literacy [4], often finding data overwhelming or unclear to apply. While professional development can help, the time and effort required to interpret data remain barriers. For teachers already pressed for time, these obstacles hinder insight extraction, even with simple dashboards or tables.

A. I. Cristea et al. (Eds.): AIED 2025, CCIS 2590, pp. 285–290, 2025.
https://doi.org/10.1007/978-3-031-99261-2_25

2 Proposed Solution and Practical Concerns

Simplifying teachers' access to insights from learning analytics reports can help in two key ways: (a) making it easier to inform instruction and (b) addressing challenges related to time and training.

Proposed Solution. Providing a concise textual summary of reports would allow teachers to quickly grasp actionable insights about student learning, rather than sift through complex data tables or visualizations to interpret key findings.

This approach is not new—healthcare research shows that summaries can be as effective, if not more, than visualizations for supporting decision-making [8]. Notably, it suggested that for novice medical staff, textual summaries enhanced decision-making while reducing visual complexity and cognitive load.

Although education may not be as high-stakes as healthcare, the need for quick insights in classrooms remains. Whether reviewing homework from last night or adapting instruction on-the-fly, teachers must respond in real time. Well-structured summaries could help them focus on teaching rather than analyzing.

Practical Concerns. While using LLMs to generate summaries is promising, application and adoption depends on teacher perceptions and acceptance. Some may question their accuracy or dismiss them as superficial gimmicks. Understanding and addressing these perceptions is critical to evaluating feasibility.

There are also risks. Overreliance on LLM-generated summaries may hinder data literacy development. Since summaries present only a limited snapshot to avoid overload, teachers might rely on them without engaging with full data. This could lead to missed opportunities for deeper analysis. Implementation must therefore complement, not replace, teachers' engagement with learning analytics.

3 Research Questions

In this study, we refer to the learning analytics report as the assignment report and evaluate whether LLM-generated summaries support teachers in using it. To assess their effectiveness and address the concerns outlined above, we pose the following research questions and subquestions:

- **RQ1:** Do teachers find these LLM-generated assignment summaries useful?
  - Do teachers perceive summaries as adding value to assignment report?
  - When are summaries useful for instructional decisions?
  - Will having a summary change their existing teaching and ITS practices?
- **RQ2:** How do LLM-generated summaries impact teacher behavior?
  - Do teachers with access to summaries engage with more assignment reports than those without access?
  - Do teachers with access exhibit different engagement patterns?
- **RQ3:** How do LLM-generated summaries impact student outcomes?
  - Do teachers with access have higher student participation rates?
  - Are they more likely to have students with higher performance on assignments, compared to students working on the same problem sets?

4 Background

4.1 ASSISTments Online Learning Platform to Learning Analytics

ASSISTments [1] is an online math platform that allows teachers to create assignments using K-12 mathematics problems from open educational resources such as Illustrative Mathematics and Engage New York. Teachers can then track students' progress in real time through the assignment report, accessing various types of learning analytics (see Fig. 1). These include: (a) overall student progress (e.g., students who completed or did not start the assignment); and (b) color-coded student scores at assignment and problem levels to facilitate interpretation. Teachers can also review detailed student reports (Fig. 2) to view all attempts across the assignment for deeper insights into student performance.

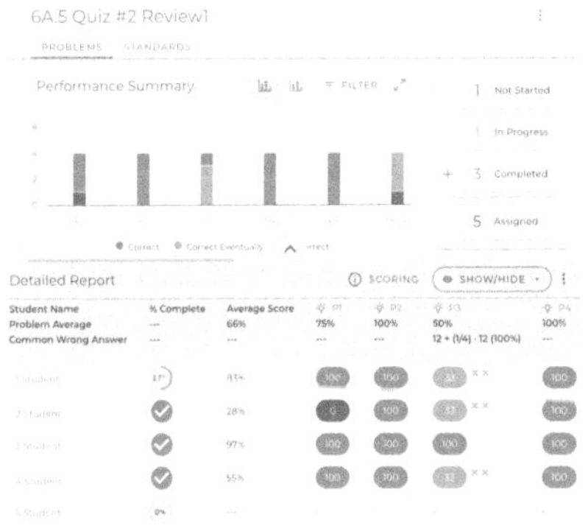

Fig. 1. Assignment report for problem set

Fig. 2. Student details report

4.2 Preprocessing Data for LLM-Generated Summaries

Prior work in [3] examined two key preprocessing steps to support LLM-generated assignment report summaries:

Structuring Learning Data. Data from ASSISTments were organized into four datasets—covering assignment-level metrics, student performance, problem statistics, and student-problem interactions. These included relevant aggregates (e.g., average scores, prior comparisons, common wrong answers) that help surface instructional patterns such as student difficulties and challenging problems.

Tabular Data Processing. The study compared direct text conversion (plain text and JSON), rule-based aggregation, and sentence serialization. Sentence serialization emerged as most effective, yielding higher-quality insights, fewer errors, and relatively lower token usage. Though evaluation was limited in scale, its findings guide our use of sentence serialization for data processing here.

5 Current Progress

This study builds on [3], which showed that LLMs can generate assignment report summaries from ASSISTments data, with sentence serialization proving most effective. We extend this work by developing a scalable AWS-based pipeline that nightly extracts, preprocesses, and serializes tabular learning analytics, then generates summaries via LLM with minimal manual effort. These summaries are integrated into teacher-facing reports on ASSISTments via the Userflow API. This automation is critical to enable a large-scale randomized field experiment, where summaries can be delivered consistently and reliably over time.

By enabling summary generation at scale for real-world classrooms, this work examines whether LLM-generated summaries increases teacher engagement with learning analytics and improve student outcomes. Before launching the experiment, it is important to establish that teachers perceive these summaries as useful rather than superficial or distracting. The following subsection summarizes early user feedback to assess these perceptions.

5.1 Obtaining Initial Feedback on Assignment Report Summaries

In November 2024, ASSISTments surveyed teachers who had viewed at least three assignment reports. Among the 31 respondents (about 5% of those eligible), 87% found summaries "useful" or "very useful", and 77% were satisfied or very satisfied with the current report interface.

When shown sample summary statements, teachers highlighted two as especially helpful: (1) "Students struggled with problems 3 and 5. Over 40% of the class needed hints or multiple tries." and (2) "Many students incorrectly answered '9' in Q1. You may need to reteach scale drawing and proportional relationships." These examples—which we will use for few-shot prompting—were seen as clear, actionable, and pedagogically relevant.

While the low response rate limits generalizability, it is similar to typical Userflow survey rates. The feedback suggests that even experienced users value well-crafted summaries. Newer teachers may benefit even more, gaining quicker insight into reports and developing deeper engagement over time. These findings, alongside the technical groundwork laid in earlier work, motivate continued improvements to the summarization process.

5.2 Ongoing Work

While [3] compared summarization methods, it did not examine prompt engineering or LLM constraints. In this work, we refine the system prompt and sentence serialization templates to improve summary quality, with new variants tailored to different assignment types and student outcomes. We also evaluate different LLMs, as their effective context windows vary despite similar reported limits [2]. Our goal is to improve accuracy and efficiency through better preprocessing and by leveraging models with longer effective context windows. To support teacher decision-making, each summary now also includes a direct link to the full overtime class report with student performance trends (see Fig. 3).

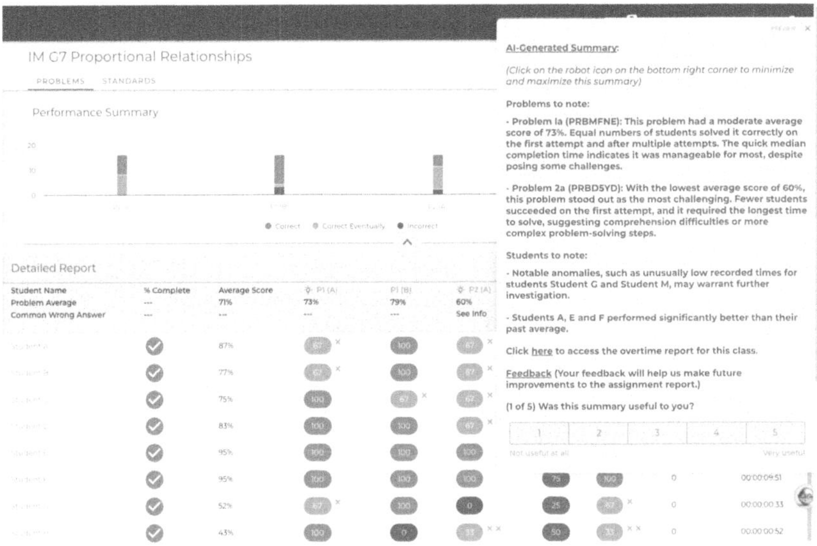

Fig. 3. Example of AI generated summary for assignment report

6 Proposed Evaluation Methodology

ASSISTments uses Userflow [7] for onboarding and guided tours, enabling random A/B assignment of teachers to a control group or a summary group. Control teachers use the platform as usual, viewing reports without summaries. In the summary group, LLM summaries are created nightly for due assignments and are viewable at teacher's discretion, hidden by default to protect student privacy.

Experiment Timing and Causal Goals. The experiment will launch at the start of next academic year, when teacher activity peaks and sample size is highest. Comparing the two groups will estimate the causal effect of summary access on teacher engagement and student outcomes. Specifically, we assess whether summaries (a) increase report usage, (b) lead to more assignments, (c) promote deeper report engagement, and (d) improve student participation and performance. Teacher surveys evaluate perceived usefulness and instructional value.

Limitations and Future Extensions. This study evaluates pre-generated summaries only. While on-demand generation or teacher customization is not yet supported, these are promising future directions to better align with classroom practices and increase adoption of LLM-based tools.

7 Expected Contributions

This study contributes to learning analytics, educational technology, and human-centered AI by investigating if LLMs can increase teachers' engagement with the assignment reports. Using a randomized controlled trial, it establishes causal links between access to LLM-generated summaries, teacher engagement, and student outcomes. This enables examination of both the benefits and unintended consequences of such summaries. The findings can inform the design of AI-driven educational tools that enhance teachers' use of learning analytics.

Acknowledgments. We would like to thank our many past and current funders including NSF (2118725, 2118904, 1950683, 1917808, 1931523, 1940236, 1917713, 1903304, 1822830, 1759229, 1724889, 1636782, 1535428, 2215842, 2341948, & 2153481); IES (R305N210049, R305D210031, R305A170137, R305A170243, R305A180401, R305A120125, R305R220012, & R305T240029); GAANN (P200A120238, P200A180088, & P200A150306); DOE (U411B190024 S411B210024, & S411B220024); ONR (N00014-18-1-2768); NIH (via a SBIR R44GM146483), Schmidt Futures, BMGF, CZI, Arnold, Hewlett, the Jaffe Foundation, and anonymous donors. None of the opinions expressed here are those of the funders.

References

1. Heffernan, N.T., Heffernan, C.L.: The ASSISTments ecosystem: building a platform that brings scientists and teachers together for minimally invasive research on human learning and teaching. Int. J. Artif. Intell. Educ. **24**(4), 470–497 (2014). https://doi.org/10.1007/s40593-014-0024-x
2. Hsieh, C.P., et al.: RULER: what's the real context size of your long-context language models? In: First Conference on Language Modeling (2024). https://openreview.net/forum?id=kIoBbc76Sy
3. Lim, W.C., Heffernan, N., Eroshenko, I., Khumwang, W., Chan, P.C.: Leveraging LLMs for assignment report summaries to support teacher insights in intelligent tutoring systems. In: Proceedings of the 12th ACM Conference on Learning @ Scale (L@S '25) (2025). accepted Work in Progress paper. To appear
4. Michos, K., Schmitz, M.L., Petko, D.: Teachers' data literacy for learning analytics: a central predictor for digital data use in upper secondary schools. Educ. Inf. Technol. **28**(11), 14453–14471 (2023). https://doi.org/10.1007/s10639-023-11772-y
5. Moore, R., Croft, M.: Reducing Barriers to Educator Data Use. Technical report R1662, ACT Inc (January 2018). https://www.act.org/content/dam/act/unsecured/documents/R1662-data-use-barriers-2018-01.pdf
6. U.S. Department of Education, Institute of Education Sciences, National Center for Education Statistics: TALIS 2018 U.S. Highlights Web Report (NCES 2019-132 and NCES 2020-069). Technical report, National Center for Education Statistics (2019). https://nces.ed.gov/pubsearch/pubsinfo.asp?pubid=2019132, Available online
7. Userflow: Userflow Documentation (August 2024). userflow Documentation. https://docs.userflow.com/docs. Accessed 30 Nov 2024
8. Van Der Meulen, M., Logie, R.H., Freer, Y., Sykes, C., McIntosh, N., Hunter, J.: When a graph is poorer than 100 words: a comparison of computerised natural language generation, human generated descriptions and graphical displays in neonatal intensive care. Appl. Cogn. Psychol. **24**(1), 77–89 (2010). https://doi.org/10.1002/acp.1545

Fair for Whom? Investigating School Identity, Algorithmic Fairness, and Educational Technologies

Clara Belitz(✉)

University of Illinois Urbana–Champaign, Champaign, IL 61820, USA
cbelitz2@illinois.edu

Abstract. Algorithmically driven systems are now frequently found in educational spaces in the form of adaptive tutoring systems and online learning environments. Previous research has shown that well-integrated technologies can augment good human teaching practices. How such integration is measured, however, is an evolving debate within the field of artificial intelligence for education. Research that seeks to measure equity in these systems often assumes the social categories that are used for comparison, typically by using already existing demographic data such as race or sex. While practical, these assumptions may limit the transferability of subsequent conclusions. This work explores dimensions of algorithmic justice questions through interviews with students. This research engages middle and high school students directly, as the least powerful but arguably most impacted individuals in the algorithmic tutoring ecosystem, about their own identities, their relationship to these technologies, and the way their data are used.

Keywords: Equity · Social Identity · Qualitative Methods · Intelligent Tutoring Systems

1 Introduction

As algorithmically driven adaptive technologies have become more commonplace, it has become possible to research both the fits and misfits of these systems (e.g., [19]). Research has been done using the massive amounts of data created by users of these systems [1], but students themselves are often not included in the research directly [19]. Youth are a frequently missing critical voice in research [11]; "student voice" research specifically engages youth in school settings, identifying students as "consequential stakeholders" who "bear the burden of decisions most often made by others" [4]. Researchers may miss important concerns, weaknesses, or ethical issues that students are able to identify.

As such, this research will involve asking the users of these technologies, middle and high school students, about their experiences with intelligent tutoring systems (ITSs). ITSs seek to support learners on an individualized basis, often

A. I. Cristea et al. (Eds.): AIED 2025, CCIS 2590, pp. 291–296, 2025.
https://doi.org/10.1007/978-3-031-99261-2_26

using machine learning models to serve students questions based on predictions of content mastery or knowledge gain [20]. Individual predictions for each student are built from information extracted from records of students' interactions with the software. The underlying models that feed these predictions need to be able to serve large, diverse populations of students to address a wide variety of learning needs [12]. While the benefits of these systems have been studied, questions remain about whether all types of students experience these benefits equally [16]. This work proposes qualitative research to further understand the relationship between social identities, educational technologies, and algorithmic equity. In this context, social identities refer to any constructed category, such as race, gender, or student identity [3]. Which social identity is chosen for comparison will therefore impact how equity is defined. This doctoral research will ask the following questions:

- RQ1. Are student perceptions of educational technologies related to their social identities, including that of their learning identity? If so, in what ways?
- RQ2. What types of information do students think are acceptable for use in predictive systems for educational outcomes?

Using semi-structured interviews and a card sorting activity, this work will explore student identities, their perceptions of educational technology, and what information they see as appropriate for use in predictive systems.

2 Background and Related Work

The field of education has a rich history of engaging with questions of equity, which has directly impacted research being done today [13,17]. Learning sciences and computer science have inherited this tradition, investing in increasing gender diversity in computing [15], asking how AI helps and hurts [18], and exploring the impact of specific predictive algorithms [14,22]. Educational technology researchers have also argued that "better data beat big data" [23], demonstrating the value of effective and accurate data. In addition, there has been a push for research to be done *with* young people, as opposed to *on* them [21].

Social identity is an important facet of power and equity, including for educational technologies. For example, math identity influences learning outcomes and is impacted by both internal (e.g., self belief) and external (e.g., teacher enthusiasm) factors [6,9]. Math identity is not, however, a protected demographic class in the way sex or race may be in some places, such as the United States. Though it is not legally protected, it is still a valuable social identity to understand in the context of ITSs and education more broadly. In addition, protected demographic categories have been critiqued for failing to capture nuance or otherwise missing information [10]. Questions of equity therefore require careful selection of the social categories across which comparisons will be made. This research specifically builds on my prior work, which has argued for the expansion of which categories are used for comparison [3] as well as demonstrated the importance of learning identity for student outcomes with an ITS [2]. As such, a section of the interview focuses on the participant's relationship to school and learning.

3 Method

Fifteen students will participate in one-hour sessions comprising a semi-structured interview and a card sorting activity. The number of participants was chosen because this is an exploratory study, and it does not seek to be representative of all young people in the United States. This sample size reflects choices made in similar studies, including a large-scale study of middle-school students' perceptions of the concept of AI which used 12 students in their initial pilot [8] and preliminary research on STEAM Teachers' perceptions of AI in education [7].

The interview questions will cover student identity, perception about technology in the classroom through an intelligent tutoring system, and opinions about teaching and learning. For example, Question 14 of the interview guide, "Are there things you like or dislike about learning math?," will help elicit students feelings towards mathematics and how they view their own math identity. The follow-up question asks whether the ITS they use makes these things better or worse. The goal of the interviews will be to expand upon findings from prior social identity surveys [3], deepen findings about the importance of student identity [2], and further investigate student opinions about educational technology. Interviews will be audio-recorded and the transcriptions will be analyzed using thematic analysis [5], allowing for both previously documented and emergent topics to be explored.

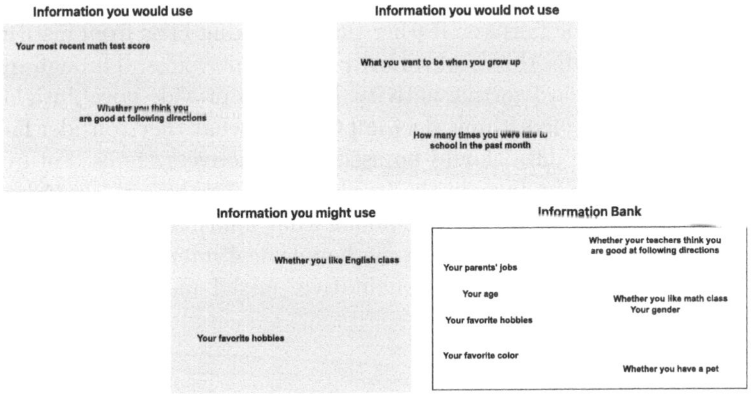

Fig. 1. A digital prototype of the card sorting activity with a subset of the features students will see. Some of the features remain in the "feature bank" while others have been distributed.

The card sorting activity will ask students about what they feel is appropriate usage of their data in educational spaces. The activity will then be repeated with a new prompt, where students will be asked to move cards when presented with new information about the predictive value of the features that they deemed unacceptable. The activity is analog, to facilitate engagement and make the experience more youth friendly. In addition, the paper and ability to move items around will help students think through the scenarios more easily.

This activity uses features that fall into one of three categories: demographic information, proxies for protected information, and proxies for behavioral or technological usage data 1. There will also be two feature of extraneous information (their favorite color and favorite animal), which can be used as a baseline for comparison. The activity will be audio recorded, allowing for thematic analysis of the transcript. At the end of each scenario, students will be asked to explain a few of their choices, with possible prompting by the interviewer. These results will be analyzed for commonalities, trends, and differences across participants.

The activity is phrased as "you" for each piece of information with the goal of prompting students to talk about themselves. Students are the experts in their own perceptions more than those of their peers, teachers, or teachers. The goal of this phrasing is to encourage students to share their opinions, rather than give answers that they think I want to hear or that they believe are generally true. The students will be reminded that there are no right or wrong answers.

4 Expected Contributions

Recruitment for this project is underway, and the first participant session has been conducted. In the card sort, the first participant indicated comfort with personalization as long as it was not "invasive" (e.g., family income) or "irrelevant" (e.g., favorite animal). The first contribution of this work is to center student perspectives on algorithmic fairness. Rather than analyzing ITSs from institutional or technical standpoints, this work prioritizes student voices. Through qualitative interviews and a card sorting activity, I hope to provide novel insights into how students perceive algorithmic decision making, what they consider fair, and how they believe their data should be used in predictive systems. Future work may be able to use the findings in the (re)design of ITSs by considering student responses to the card sorting activity about what information is acceptable for use. In addition, this work presents a methodological intervention, encouraging the field to recognize the value of qualitative, mixed-methods, and student-centered approaches.

5 Conclusion

Researchers, policymakers, and institutional leaders are actively engaged in lively debate about the role of technology in a variety of domains, including in United States' educational spaces. But those who are most directly impacted by these decisions are frequently left out of the conversation. This work seeks to contribute to the field by centering student perspectives on intelligent tutoring systems, offering insights into how social identity and personal experiences shape their perceptions. By using qualitative methods to explore students' views on fairness, equity, and data usage, this research contributes to ongoing discussions about ethical AI in education. The results may help inform future work on responsible integration of adaptive learning technologies and provide a foundation for more student-inclusive approaches to the use of AI for education.

Acknowledgments. Thank you to Drs. Nigel Bosch and Tobias Ley for feedback on earlier versions of this manuscript.

References

1. Ambarish, J., Fancsali, S.E., Ritter, S., Nixon, T., Berman, S.R.: Generalizing and extending a predictive model for standardized test scores based on Cognitive Tutor interactions. In: Proceedings of the 7th International Conference on Educational Data Mining, pp. 369–370. International Educational Data Mining Society (IEDMS), Online (2014)
2. Belitz, C., et al.: Exploring student identity in adaptive learning systems through qualitative data. In: Artificial Intelligence in Education: 25th International Conference, AIED 2025. Springer-Verlag (2025)
3. Belitz, C., Ocumpaugh, J., Ritter, S., Baker, R.S., Fancsali, S.E., Bosch, N.: Constructing categories: moving beyond protected classes in algorithmic fairness. J. Am. Soc. Inf. Sci. **74**(6), 663–668 (2022). https://doi.org/10.1002/asi.24643
4. Brasof, M., Levitan, J.: Student Voice Research: Theory, Methods, and Innovations from the Field. Teachers College Press, New York, NY (2022)
5. Braun, V., Clarke, V.: Using thematic analysis in psychology. Qual. Res. Psychol. **3**(2), 77–101 (2006). https://doi.org/10.1191/1478088706qp063oa
6. Cribbs, J.D., Hazari, Z., Sonnert, G., Sadler, P.M.: Establishing an explanatory model for mathematics identity. Child Dev. **86**(4), 1048–1062 (2015). https://doi.org/10.1111/cdev.12363
7. Cruz, S., Duque, D., Carvalho, V.: STEAM teachers' perceptions of artificial intelligence in education: preliminary research. In: Proceedings of the 16th International Conference on Computer Supported Education, pp. 278–285. SCITEPRESS - Science and Technology Publications, Angers, France (2024). https://doi.org/10.5220/0012713700003693
8. Demir, K., Güraksın, G.E.: Determining middle school students' perceptions of the concept of artificial intelligence: a metaphor analysis. Participatory Educ. Res. **9**(2), 297–312 (2022). https://doi.org/10.17275/per.22.41.9.2
9. Frenzel, A.C., Goetz, T., Pekrun, R., Watt, H.: Development of mathematics interest in adolescence: influences of gender, family, and school context. J. Res. Adolesc. **20**(2), 507–537 (2010). https://doi.org/10.1111/j.1532-7795.2010.00645.x
10. Hanna, A., Denton, E., Smart, A., Smith-Loud, J.: Towards a critical race methodology in algorithmic fairness. In: Proceedings of the 2020 Conference on Fairness, Accountability, and Transparency, pp. 501–512. ACM, Barcelona, Spain (2020). https://doi.org/10.1145/3351095.3372826
11. Hautea, S., Dasgupta, S., Hill, B.M.: Youth perspectives on critical data literacies. In: Proceedings of the 2017 CHI Conference on Human Factors in Computing Systems, pp. 919–930. ACM, Denver Colorado USA (2017). https://doi.org/10.1145/3025453.3025823, https://dl.acm.org/doi/10.1145/3025453.3025823

12. Heffernan, N.T., Heffernan, C.L.: The ASSISTments ecosystem: building a platform that brings scientists and teachers together for minimally invasive research on human learning and teaching. Int. J. Artif. Intell. Educ. **24**(4), 470–497 (2014). https://doi.org/10.1007/s40593-014-0024-x
13. Hutchinson, B., Mitchell, M.: 50 years of test (un)fairness: lessons for machine learning. In: Proceedings of the Conference on Fairness, Accountability, and Transparency, pp. 49–58. FAT* '19, Association for Computing Machinery, New York, NY, USA (2019). https://doi.org/10.1145/3287560.3287600
14. Jiang, W., Pardos, Z.A.: Towards equity and algorithmic fairness in student grade prediction. In: Proceedings of the 2021 AAAI/ACM Conference on AI, Ethics, and Society, pp. 608–617. AIES '21, Association for Computing Machinery, New York, NY, USA (2021). https://doi.org/10.1145/3461702.3462623
15. Kim, K.A., Fann, A.J., Misa-Escalante, K.O.: Engaging women in computer science and engineering: promising practices for promoting gender equity in undergraduate research experiences. ACM Trans. Comput. Educ. **11**(2), 1–19 (2011). https://doi.org/10.1145/1993069.1993072
16. Kizilcec, R.F., Piech, C., Schneider, E.: Deconstructing disengagement: Analyzing learner subpopulations in massive open online courses. In: Proceedings of the Third International Conference on Learning Analytics and Knowledge, pp. 170–179. LAK '13, Association for Computing Machinery, New York, NY, USA (2013). https://doi.org/10.1145/2460296.2460330
17. Ladson-Billings, G., Tate, W.F.: Toward a critical race theory of education. Teach. Coll. Rec. **97**(1), 47–68 (1995). https://doi.org/10.1177/016146819509700104, publisher: SAGE Publications
18. Madaio, M., Blodgett, S.L., Mayfield, E., Dixon-Román, E.: Beyond fairness: structural (in)justice lenses on AI for education. In: Holmesand, W., Porayska-Pomsta, K. (eds.) The Ethics of Artificial Intelligence in Education: Current Challenges, Practices and Debates, p. 37. Routledge, England, UK (2022)
19. Martin, A.C., Ying, K.M., Rodríguez, F.J., Kahn, C.S., Boyer, K.E.: Intelligent support for all? A literature review of the (in)equitable design & evaluation of adaptive pedagogical systems for CS education. In: Proceedings of the 53rd ACM Technical Symposium on Computer Science Education - Volume 1. SIGCSE 2022, vol. 1, pp. 996–1002. Association for Computing Machinery, New York, NY, USA (2022).https://doi.org/10.1145/3478431.3499418
20. Nkambou, R., Bourdeau, J., Mizoguchi, R., Kacprzyk, J. (eds.): Advances in Intelligent Tutoring Systems, Studies in Computational Intelligence, vol. 308. Springer, Berlin, Heidelberg (2010). http://link.springer.com/10.1007/978-3-642-14363-2
21. Winters, N., Eynon, R., Geniets, A., Robson, J., Kahn, K.: Can we avoid digital structural violence in future learning systems? Learn. Media Technol. **45**(1), 17–30 (2020). https://doi.org/10.1080/17439884.2020.1708099
22. Wise, A.F., Sarmiento, J.P., Boothe Jr., M.: Subversive learning analytics. In: LAK21: 11th International Learning Analytics and Knowledge Conference, pp. 639–645. LAK21, Association for Computing Machinery, New York, NY, USA (2021). https://doi.org/10.1145/3448139.3448210
23. Yudelson, M.V., Fancsali, S.E., Ritter, S., Berman, S.R., Nixon, T., Joshi, A.: Better data beat big data. In: Proceedings of the 7th International Conference on Educational Data Mining, pp. 205–208. International Educational Data Mining Society, London, United Kingdom (2014)

Fine-Tuning Large Language Models for Knowledge Tracing Harnessing Insights from Explainable AI

Deliang Wang[1], Gaowei Chen[1], and Yu Lu[2,3](✉)

[1] Faculty of Education, The University of Hong Kong, Pokfulam, Hong Kong, China
[2] School of Educational Technology, Beijing Normal University, Beijing, China
[3] Advanced Innovation Center for Future Education, Beijing Normal University, Beijing, China
luyu@bnu.edu.cn

Abstract. Deep learning based knowledge tracing (DLKT) models involve using deep learning techniques and learners' exercise-answer data to model their knowledge state and predict their performance on future exercises. Explainable artificial intelligence (xAI) reveals that DLKT models rely on learners' correctness, skills, and recency information from previously answered records. Inspired by the explanations, this study fine-tuned two large language models (LLMs) to perform the KT task by incorporating these key factors into the prompt. Experimental results demonstrate that Llama-3.2-3B achieves higher accuracy and precision scores compared to both Deep Knowledge Tracing (DKT) and Mistral-7B-v0.3. However, DKT outperforms both LLMs in terms of AUC, F1, and recall scores. Our approach, as a pilot study, highlights a promising direction for future research on the application of LLMs in KT.

Keywords: Knowledge tracing · large language model · explainable AI

1 Context and Problem Statement

Knowledge tracing (KT) involves the application of machine learning techniques to learners' sequential exercise-answer data, aiming to model their knowledge state and predict their performance on future exercises. KT models can be incorporated into online learning platforms for tracking students' learning progress and providing individualized services [7]. Early research in this field utilized Markov chains [1] and logistic regression [9] to model knowledge states. While these models offered a basic analysis, their performance was limited.

Since 2015, researchers have increasingly employed deep neural networks, such as recurrent neural networks [10], to represent students' knowledge states and predict their performance. Deep learning-based KT (DLKT) models demonstrate superior performance. However, unlike the interpretable parameters and

A. I. Cristea et al. (Eds.): AIED 2025, CCIS 2590, pp. 297–302, 2025.
https://doi.org/10.1007/978-3-031-99261-2_27

structures of Markov chain-based and logistic regression-based KT models, DLKT models functions as "black box" models, predicting learners' knowledge mastery without explanations. To address this limitation, researchers have explored explainable artificial intelligence (xAI) techniques to elucidate DLKT models [6,7,11], yielding valuable insights. For instance, it has been shown that when predicting learner performance on a specific skill, DLKT models prioritize past responses related to that skill over irrelevant skills [6,7]. Recently, the potential of large language models (LLMs) for KT tasks has been investigated, leveraging their robust capabilities in language understanding, reasoning, and generation (e.g., [5]). While these studies either adopted LLM structures or created customized prompts for KT, they often overlooked the insightful explanations of the underlying mechanisms of DLKT models.

To address this gap, this study instructs LLMs to utilize the explanations from xAI for performing the KT task and investigates whether they can achieve outstanding performance.

2 Method

2.1 Dataset

This study utilized the ASSISTment 2009 dataset [2] to fine-tune LLMs for the KT task. Specifically, the "skill builder" version of the dataset was selected. All duplicated exercise sequences and those without labeled skills were eliminated. To ensure that the LLMs could effectively utilize information on correctness, skill, and recency to predict learners' knowledge mastery, we further filtered the dataset to include only exercise sequences longer than five interactions. Consequently, the refined dataset comprises 323,687 exercise records spanning 110 skills from 3,510 learners. We randomly selected 80% of the sequences for training, reserving the remaining 20% for testing purposes.

2.2 Fine-Tuning LLMs

Explanations from Explainable AI. Given a sequence of previous exercise records of a student, a DLKT model aims to predict the student's performance on future exercises, specifically the probability of correctly answering the exercise. To elucidate how these predictions are made, prior research leverages xAI methods, such as LRP and Deep SHAP, to determine the contributions of each past exercise record to the prediction. By analyzing numerous explanations of individual predictions, common patterns in DLKT models can be identified. Firstly, correct exercise records typically have positive contributions to the prediction, whereas incorrect exercise records have negative contributions. Secondly, exercise records that pertain to the same skill as the predicted exercise exhibit significantly larger contributions than those related to other skills. Thirdly, recent exercise records have significantly larger contributions compared to older exercise records. Detailed explanations of these findings are available in previous research [6–8].

Instruction with context	The following sequence represents a learner's previous exercise records in a mathematics course. Each exercise record is described by a tuple consisting of three elements, for example, (1, 'Box and Whisker', 'answered wrongly'). In this tuple, the number '1' indicates the question order, 'Box and Whisker' signifies the skill tested by the question, and 'answered wrongly' denotes that the student responded incorrectly. Since the beginning, the learner has answered 16 exercises, where 12 exercises were correctly answered and the overall correct rate is 0.75. During his recent 8 exercises, he has correctly answered 6 exercises and the recent correct rate is 0.75. Among all exercises, the learner has answered 1 exercise(s) on Percent Of, where 1 exercises were correctly answered and the correct rate on the target skill is 1.0. Based on the historical exercise records, such as the overall correctness rate, the learner's performance on recent exercises and exercises on relevant skills, please predict the learner's correctness in answering the exercise on Percent Of.
Input	Past exercise records: [(1, 'Circle Graph', 'answered correctly'), (2, 'Circle Graph', 'answered correctly'), (3, 'Circle Graph', 'answered correctly'), (4, 'Ordering Positive Decimals', 'answered wrongly'), (5, 'Ordering Positive Decimals' 'answered correctly'), (6, 'Ordering Positive Decimals', 'answered correctly'), (7, 'Ordering Positive Decimals 'answered wrongly'), (8, 'Ordering Positive Decimals', 'answered correctly'), (9, 'Ordering Positive Decimals' 'answered correctly'), (10, 'Ordering Positive Decimals', 'answered correctly'), (11, 'Ordering Fractions', 'answered wrongly'), (12, 'Ordering Fractions', 'answered wrongly'), (13, 'Ordering Fractions', 'answered correctly'), (14, 'Ordering Fractions', 'answered correctly'), (15, 'Ordering Fractions', 'answered correctly'), (16, 'Percent Of', 'answered correctly')] Predicted exercise: 'Percent Of
Expected output	Answered correctly

Fig. 1. An example of the prompt set for LLMs

Prompt Inspired by Explanations. Inspired by these explanations, we meticulously designed a prompt to fine-tune the Llama-3.2-3B and Mistral-7B-v0.3 models, as illustrated in Fig. 1. The prompt consists of two primary components: the instruction with context and the input exercise records. In the instruction with context, we first elucidated the meaning of each element in an exercise record, such as $(1, Box\ and\ Whisker, answered\ wrongly)$, as demonstrated in the first paragraph of Fig. 1.

Secondly, we provided LLMs with several key indicators related to the explanations, as outlined in the second paragraph of Fig. 1. Specifically, positive contributions of a past exercise record indicate an increased probability that a learner will correctly answer a future exercise, whereas negative contributions suggest a decreased likelihood. Given that correct exercise records usually have positive contributions and incorrect exercise records have negative contributions, we infer that a higher number of correct exercise records in the sequence increases the probability of a correct response in a future exercise. Therefore, we incorporated the total number of exercise records, the number of correct exercise records, and the overall correct rate into the prompt. Considering the significant contributions of exercise records on the same skill as the predicted exercise, we included the total number of exercise records, the number of correctly answered records on the same skill, and their correct rate in the prompt. Similarly, given the significant contributions of recent exercise records compared to older records, we included the number of recently completed exercise records (i.e., the second half of a sequence), the number of correctly answered recent records, and their correct rate in the prompt.

Finally, we instructed the LLMs to predict the learner's response to an exercise on the target skill (i.e., the predicted skill) using the contextual information and historical exercise records provided. This included the learner's overall correct rate, the correct rate of the most recent half of the exercises, and the correct rate for exercises on the predicted skill in previous exercise records, as shown in the third paragraph of Fig. 1. The input of the prompt consisted of the learner's past exercise records, while the expected output was designated as either "answered correctly" or "answered wrongly" in this study.

Fine-Tuning. We utilized Low-Rank Adaptation (LoRA) [4] to fine-tune two LLMs in a parameter-efficient manner: Llama-3.2-3B and Mistral-7B-v0.3. The Llama-3.2-3B and Mistral-7B-v0.3 models were chosen for their open-source availability and superior performance. In contrast to full fine-tuning, which requires substantial computational resources to update all model weights and may result in performance degradation on other tasks, parameter-efficient fine-tuning (PEFT) updates a significantly smaller set of parameters. This approach is cost-effective and maintains the performance of LLMs across various tasks. LoRA is a well-established PEFT technique. Due to space constraints, detailed technical aspects of LoRA can be found in [4].

Despite using LoRA to expedite the fine-tuning process, completing one epoch of training still required more than one month. To further accelerate training, we decided to train and predict the learner's performance on only the last 20% of each sequence, rather than on every exercise. During fine-tuning, the rank, optimizer, batch size, and learning rate were set to 16, AdamW, 48, and 2e-4, respectively. After fine-tuning, the LoRA adapter can either function independently or be merged into the original LLM.

2.3 Baseline

We also trained the DKT model in a similar manner as a baseline. Specifically, we computed and backpropagated the loss for the last 20% of each sequence in the training set. We adhered to the guidelines outlined in previous works [3, 10] for the implementation and hyperparameter tuning of DKT. The optimizer was configured as Adam with a learning rate of 0.001. The batch size was set to 100, and the dropout probability was selected from $\{0, 0.25, 0.5\}$. We also employed early stopping. The hidden state dimension was chosen from $\{50, 100, 200\}$, and the number of recurrent layers was selected from $\{1, 2\}$.

3 Preliminary Results

Figure 2 illustrates the performance metrics of DKT, Llama-3.2-3B, and Mistral-7B-v0.3 in the ASSISTment 2009 dataset, encompassing accuracy, AUC, F1, precision, and recall scores. Considering that KT is a binary classification task, our analysis primarily focuses on the accuracy and AUC scores. In terms of accuracy, DKT achieves a score of 0.728, while Llama-3.2-3B attains 0.740 in the LoRA version and 0.742 in the merged version. Mistral-7B-v0.3 achieves an accuracy of 0.729 in the LoRA version and 0.715 in the merged version. Consequently, Llama-3.2-3B demonstrates superior accuracy compared to both DKT and Mistral-7B-v0.3. Regarding AUC scores, DKT achieves 0.754, while Llama-3.2-3B scores 0.666 in both the LoRA and merged versions. Mistral-7B-v0.3 achieves AUC scores of 0.634 in the LoRA version and 0.610 in the merged version. Therefore, DKT outperforms both LLMs in terms of AUC. For the other metrics, DKT surpasses the LLMs in both F1 and recall scores, whereas the LLMs exhibit higher precision scores compared to DKT.

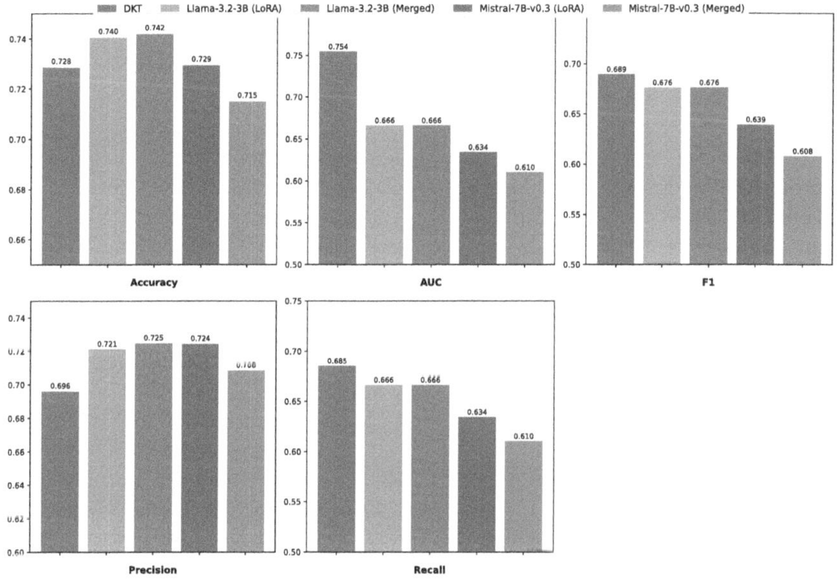

Fig. 2. Performance of DKT and LLMs

4 Conclusions and Future Directions

This study investigates the potential of fine-tuning LLMs for knowledge tracing by integrating explanations derived from DLKT models into prompts. Prior research has demonstrated that DLKT models rely on correctness, skill, and recency of previous exercise records to predict learners' future performance. We, therefore, incorporate factors associated with these indicators into the prompt, such as the learner's overall correct rate, the correct rate of the most recent half of the exercises, and the correct rate for exercises on the predicted skill in previous records. We then fine-tune Llama-3.2-3B and Mistral-7B-v0.3 using LoRA and evaluate their performance against DKT on ASSISTments 2009.

Experimental results reveal that Llama-3.2-3B outperforms both DKT and Mistral-7B-v0.3 in terms of accuracy and precision scores. However, DKT achieves superior performance in AUC, F1, and recall scores compared to both Llama-3.2-3B and Mistral-7B-v0.3. While our approach does not achieve state-of-the-art performance across all metrics, it highlights the promising potential of LLMs in KT tasks. We expect to explore extending the prompts by incorporating additional meaningful information, applying the approach to multiple LLMs, and evaluating its efficacy across diverse KT datasets in future research.

Acknowledgment. This research is supported by the National Natural Science Foundation of China (No. 62477003).

References

1. Corbett, A.T., Anderson, J.R.: Knowledge tracing: modeling the acquisition of procedural knowledge. User Model. User-Adap. Inter. **4**, 253–278 (1994)
2. Feng, M., Heffernan, N., Koedinger, K.: Addressing the assessment challenge with an online system that tutors as it assesses. User Model. User-Adap. Inter. **19**, 243–266 (2009)
3. Gervet, T., Koedinger, K., Schneider, J., Mitchell, T., et al.: When is deep learning the best approach to knowledge tracing? J. Educ. Data Min. **12**(3), 31–54 (2020)
4. Hu, E.J., et al.: Lora: low-rank adaptation of large language models. arXiv preprint arXiv:2106.09685 (2021)
5. Li, H., et al.: Explainable few-shot knowledge tracing. arXiv preprint arXiv:2405.14391 (2024)
6. Lu, Y., Wang, D., Chen, P., Meng, Q., Yu, S.: Interpreting deep learning models for knowledge tracing. Int. J. Artif. Intell. Educ. **33**(3), 519–542 (2023)
7. Lu, Y., Wang, D., Chen, P., Zhang, Z.: Design and evaluation of trustworthy knowledge tracing model for intelligent tutoring system. IEEE Transactions on Learning Technologies (2024)
8. Lu, Yu., Wang, D., Meng, Q., Chen, P.: Towards interpretable deep learning models for knowledge tracing. In: Bittencourt, I.I., Cukurova, M., Muldner, K., Luckin, R., Millán, E. (eds.) AIED 2020. LNCS (LNAI), vol. 12164, pp. 185–190. Springer, Cham (2020). https://doi.org/10.1007/978-3-030-52240-7_34
9. Pavlik, P.I., Cen, H., Koedinger, K.R.: Performance factors analysis–a new alternative to knowledge tracing. In: Artificial Intelligence in Education, pp. 531–538. Ios Press (2009)
10. Piech, C., et al.: Deep knowledge tracing. Adv. Neural Inf. Process. Syst. **28** (2015)
11. Wang, D., Lu, Y., Zhang, Z., Chen, P.: A generic interpreting method for knowledge tracing models. In: International Conference on Artificial Intelligence in Education, pp. 573–580. Springer (2022)

Assessing the Effectiveness of GenAI Tutoring for Tertiary Academic Probation Students: A Repeated Measures Study Using ChatGPT

Mario Adkins(✉)

Cornerstone University, Grand Rapids, MI 49525, USA
mario.adkins@cornerstone.edu

Abstract. Generative artificial intelligence (GenAI) has precipitately emerged as a transformative tool in higher education, particularly in student learning. This study investigates the efficiency of ChatGPT as a GenAI-assisted tutoring tool for academically at-risk students enrolled in IDS-097 Academic Accountability, a 1-credit intervention course designed to support students on academic probation. Using a quantitative repeated measures design, 50 students completed two equivalent exams: Exam #1 (without GenAI support) and Exam #2 (after utilizing ChatGPT as a studying and tutor aid). A paired-sample t-test revealed a statistically significant improvement in student performance ($t(49) = -8.21$, $p < 0.001$), with mean scores increasing from Exam #1: 72.1% (C) to Exam #2: 85.4% (B+) and demonstrating a large effect size (Cohen's $d = 1.13$). A Pearson correlation analysis ($r = 0.74$, $p < 0.001$) and pre- and post-surveys analyzed through Cronbach's Alpha ($\alpha = 0.93$) suggest a positive relationship between ChatGPT usage and learning gains. Findings align with the Technology Acceptance Model and Cognitive Load Theory, inferring that GenAI-assisted tutoring supports knowledge acquisition. Future research should investigate long-term memory retention for students, faculty-led GenAI support initiatives, and GenAI-incorporated tertiary studies utilizing broader student types for increased generalization.

Keywords: Generative Artificial Intelligence · ChatGPT · Tutoring · Studying

1 The Problem Addressed, State of the Art in AIED and its Relevance to Education

1.1 The Problem Addressed as Unstructured GenAI Utilization

As the Director of Academic Success at Cornerstone University, a private Christian liberal arts institution located in Grand Rapids, Michigan serving undergraduate and graduate students in traditional, adult, and online learning, the unstructured use of generative artificial intelligence (GenAI) by students in their academic coursework is increasingly becoming normalized. This occurrence is currently being assessed via two major initiatives. The first initiative is creating short-term guidance for both faculty and students. The Director of Academic Success has created the institution-wide policy "Generative

A. I. Cristea et al. (Eds.): AIED 2025, CCIS 2590, pp. 303–308, 2025.
https://doi.org/10.1007/978-3-031-99261-2_28

AI and Academics", which provides directives on how GenAI can be used in coursework, assessments, and academic integrity compliance [1]. This policy is included in all syllabi at the institution and within the university's academic catalog. The second initiative is leveraging institutional leadership expertise and emerging GenAI technologies to design and implement structured GenAI usage procedures and integrated Gen-AI learning frameworks for higher education institutions.

1.2 State of the Art Regarding GenAI and AIED

The rapid proliferation of free, publicly accessible GenAI platforms such as the Chat Generative Pre-Trained Transformer (ChatGPT) has catalyzed a significant shift in the field of Artificial Intelligence in Education (AIED). While foundational advancements in large language models, natural language processing, and intelligent tutoring systems predate this technological surge, recent GenAI popularity and accessibility has amplified interest in their educational applications. Researchers Gligorea and Cioca claim that the intersection of GenAI and learning technologies enable higher education to balance innovation with institutional integrity, particularly through adaptive content generation and student engagement features [2]. Yan et al., in a controlled experimental study revealed that GenAI agents augmented with scaffolding mechanisms significantly improved students' comprehension of ophthalmic learning outcomes, highlighting the potential for GenAI within complex educational learning environments [3]. Despite this momentum, substantial pedagogical and ethical concerns persist. Al-Mamary et al. identified that while GenAI adoption is increasing among faculty and students, the absence of institutional clarity on academic integrity and coursework authorship has produced inconsistent usage policies and perceptions [4]. Zhang et al. provide a scoping review that outlines the evolving impact of GenAI on educational assessments, stressing the tension between automation and pedagogical authenticity [5]. These concerns are further researched in McDonald et al.'s analysis of institutional policy documents across U.S. universities, where the authors found that while many institutions acknowledge GenAI's instructional capabilities, few have developed actionable, detailed guidelines to support equitable and ethical adoption for students and faculty [6].

1.3 GenAI Impact on Higher Education Curricula and Learning

The popularity of GenAI and its availability to matriculated students has caused disruptions for both academic administrators and course instructors. When faculty design a course, it includes learning outcomes students should fully understand and grasp at the conclusion of the course. When students use GenAI to generate entire essays or answer prompts without engaging with course content, it bypasses the intended cognitive effort required by the assignment. For instance, instead of analyzing a case study to develop critical thinking skills, students could copy a ChatGPT-generated response to said case study, resulting in superficial understanding and reduced learning engagement, which may undermine intended learning outcomes in courses where original student synthesis and critical analysis are primary learning objectives.

2 Theoretical Framework and Research Design Approach

2.1 Theoretical Framework for GenAI Curricula Support

This study utilized an integrated theoretical framework combining four established models to evaluate the academic impact of GenAI support for students on academic probation. Each model contributes a complementary lens: Bloom's Revised Taxonomy (BRT) provides a cognitive scaffolding to categorize and interpret the complexity of student responses in distinguishing between lower-order (e.g., recall) and higher-order (e.g., synthesis, evaluation) learning [7]. While not originally designed for GenAI, BRT serves as a heuristic framework to analyze student outputs across varying cognitive levels. Vygotsky's Zone of Proximal Development (ZPD) is adopted conceptually to frame the notion of "instructional scaffolding"—a role now mirrored in GenAI-assisted learning where GenAI provides dynamic prompts that adapt to students' inputs [8]. It is acknowledged that ZPD predates GenAI and does not natively support machine-mediated instruction, yet its foundational emphasis on learner support within an optimal developmental zone remains relevant. To assess the adoption and engagement with GenAI software, the Technology Acceptance Model (TAM) is employed to evaluate two key constructs: perceived usefulness and ease of use. These variables help explain students' willingness to incorporate GenAI into their study practices and the extent to which it influences their academic motivation [9]. Finally, Cognitive Load Theory (CLT) supports an analysis of how GenAI may or may not minimize extraneous cognitive load, which refers to unnecessary mental effort often caused by unclear instructional delivery or cognitively inefficient learning materials, while enhancing germane load, which supports schema construction and intentional learning [10].

2.2 Methodology: Repeated Measures Design and Exam Structure

This study employed a repeated measures design involving 50 students on academic probation enrolled in a required academic accountability course. All 50 students completed both assessments: Exam 1 without any GenAI support and Exam 2, one-week later, after structured tutoring engagement with ChatGPT as a study aid. The exams covered identical neuroscience content related to learning mechanisms such as neuron structure, synaptic transmission, neuroplasticity, and the role of the hippocampus in memory, but used different question formats to minimize memorization effects. Specifically, Exam 1 employed multiple-choice questions (MCQs) for foundational recall and short-answer questions (SAQs) for applied reasoning; in Exam 2, the formats were reversed (SAQs for recall, MCQs for application), ensuring equivalent cognitive demands while preventing direct repetition. This "MCQ-SAQ flip" stratagem was implemented to reduce test familiarity bias and emphasize conceptual understanding [11]. Additionally, all students were given a PDF study guide with specific instructions on prompt engineered questions [12] to ask ChatGPT to aid in studying exam content (Figs. 1 and 2).

2.3 Additional Research Study Factors and Future Considerations

Implementing GenAI within a higher education context requires more than access to software interfaces; it requires structured leadership, thoughtful pedagogy, and what scholars

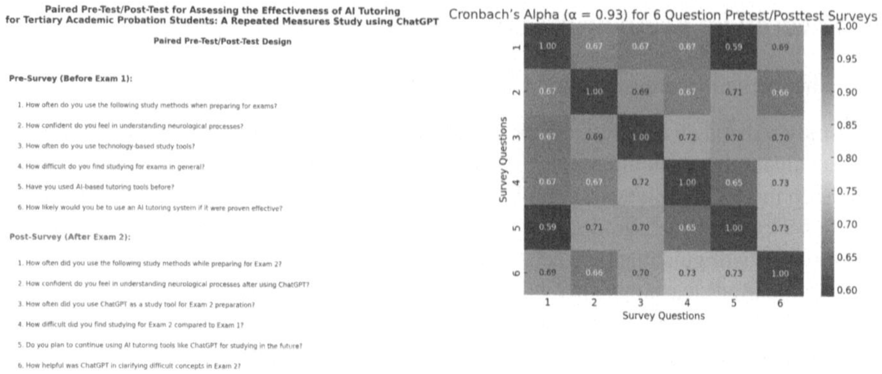

Fig. 1. A display of the paired pre- and post-survey questions used to assess changes in study behaviors, academic confidence, and perceptions of GenAI. Inter-item correlations shown in the heatmap indicate strong internal consistency, with a Cronbach's Alpha of $\alpha = 0.93$, validating the reliability of the instrument in evaluating the effects of ChatGPT tutoring.

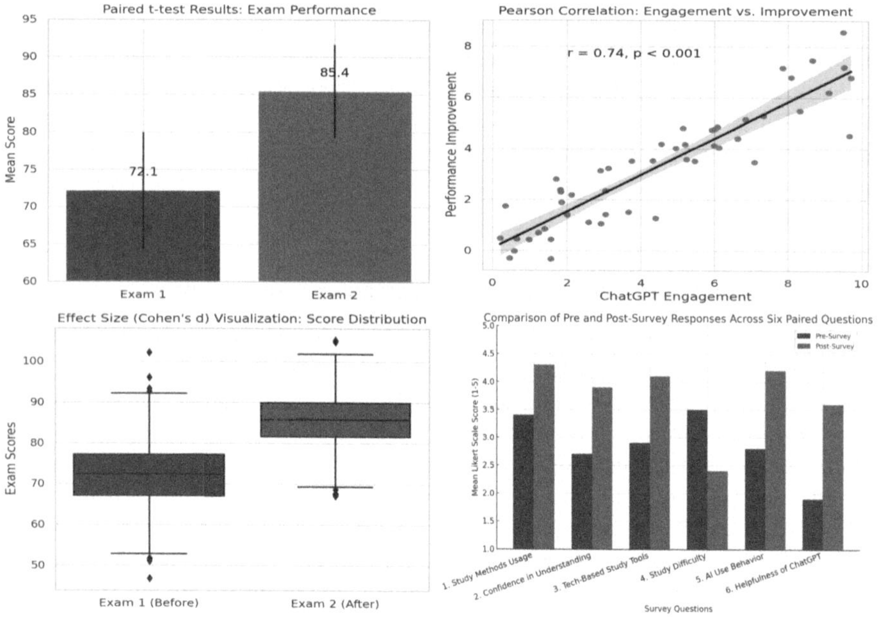

Fig. 2. Exam average scores improved from 72.1 (C) to 85.4 (B+): ($t(49) = -8.21$, $p < 0.001$), with a strong correlation between ChatGPT usage and performance gains ($r = 0.74$, $p < 0.001$). A large effect size (Cohen's $d = 1.13$) and Likert-scale paired pre- and post-survey results verified post-intervention gains in study behavior, confidence, and GenAI expediency.

increasingly refer to as AI literacy—a working understanding of how AI systems function, their ethical constraints, and their responsible application in academic settings. In this study, the Director of Academic Success, serving also as the instructor for IDS-097

Academic Accountability, facilitated the intervention by designing a five-page neuroscience study guide aligned with course content and demonstrating how to integrate it into ChatGPT. Students were shown prompt engineering techniques [12] to generate custom practice questions and were instructed to verify GenAI outputs against course textbooks and materials to reinforce content accuracy. Further dissertation research must expand on this study by including a randomized control trial structure, longitudinal retention tracking, validated instruments for cognitive and behavioral measures, and broader student sampling across disciplines and academic standings.

3 Contributions and Impact of Research to AIED

3.1 Learning Sciences and Future AIED Research Direction for GenAI

The learning science domain encompasses studies on how individuals learn and how instructional methods, technologies, and environments can be designed to enhance learning outcomes by integrating research from cognitive science, educational psychology, artificial intelligence technology, and social learning theories [13]. Learning sciences also examines theories of learning such as constructivism, TAM, and CLT—the latter two cited in this piloted research study. Many empirical research studies on GenAI's intersection within education neither focus on integrative incorporations nor are there a substantial amount of longitudinal research studies available. Throughout the Ed.D. program, the Director of Academic Success is committed to assisting fill this identified research gap by designing intentional research studies and collaborating with researchers, educators, instructional designers, data scientists, and key GenAI stakeholders throughout his dissertation process. GenAI and computer science are interconnected entities, from machine learning and deep learning using neural network architectures such as transformers, generative adversarial networks, and diffusion models [14], to more technical specs such as software engineering and programming to automate code generation, refactoring, and debugging [15]. A contribution to the computer science domain would be focused research on the human-centric nature of GenAI as a supplementary tool, not a primary replacement for human-driven work [16]. Additionally, checks and balances must be imposed within GenAI frameworks to ensure that machine learning databases are filled with equitable data that is culturally sensitive to diverse users, is fully accessible to persons with disabilities, expand language options to increase the populations of users and researchers, ethically and securely stores data in cloud and physical servers that are protected from data breaches, and prioritize GenAI-related support to at-risk populations [17] such as intentional GenAI support instruments for students facing adverse academic difficulties.

References

1. Cornerstone University: "Generative AI." Academic Policies & Procedures. https://catalog.cornerstone.edu/academics/policies-procedures/generative-ai/. Accessed 11 May 2025
2. Gligorea, L., Cioca, L.: Generative AI in higher education: Balancing innovation and integrity. Br. J. Biomed. Sci. (2024). https://www.frontierspartnerships.org/journals/british-journal-of-biomedical-science/articles/10.3389/bjbs.2024.14048/full

3. Yan, L., et al.: The effects of generative AI agents and scaffolding on enhancing students' comprehension of visual learning analytics. arXiv preprint arXiv:2409.11645 (2024). https://arxiv.org/abs/2409.11645
4. Al-Mamary, Y.H., Alfalah, A.A., Shamsuddin, A., Abubakar, A.A.: Generative AI and the future of higher education: a threat to academic integrity? Int. J. Educ. Technol. High. Educ. **21**, 1–15 (2024). https://educationaltechnologyjournal.springeropen.com/articles/10.1186/s41239-024-00453-6
5. Zhang, Y., Li, X., Wang, H.: A scoping review on how generative artificial intelligence transforms assessment in higher education. Int. J. Educ. Technol. High. Educ. **21**, 1–20 (2024). https://educationaltechnologyjournal.springeropen.com/articles/10.1186/s41239-024-00468-z
6. McDonald, N., Johri, A., Ali, A., Hingle, A.: Generative artificial intelligence in higher education: evidence from an analysis of institutional policies and guidelines. arXiv preprint arXiv:2402.01659 (2024). https://arxiv.org/abs/2402.01659
7. Anderson, L.W., Krathwohl, D.R.: A taxonomy for learning, teaching, and assessing: a revision of Bloom's taxonomy of educational objectives. Longman (2001)
8. Vygotsky, L.S.: Mind in Society: The Development of Higher Psychological Processes. Harvard University Press (1978)
9. Davis, F.D.: Perceived usefulness, perceived ease of use, and user acceptance of information technology. MIS Q. **13**(3), 319–340 (1989)
10. Sweller, J., Ayres, P., Kalyuga, S.: Cognitive load theory. Springer (2011). https://doi.org/10.1007/978-1-4419-8126-4
11. Roediger, H.L., Karpicke, J.D.: Test-enhanced learning: taking memory tests improves long-term retention. Psychol. Sci. **17**, 249–255 (2006)
12. Korzyński, P., Mazurek, G., Krzypkowska, P., Kurasiński, A.: Artificial intelligence prompt engineering as a new digital competence: analysis of generative AI technologies such as ChatGPT. Entrep. Bus. Econ. Rev. **11**(3), 25–38 (2023)
13. Bates, A.W.: Managing Technological Change: Strategies for College and University Leaders. Jossey-Bass (1999)
14. Greeno, J.G., Engeström, Y.: Learning in activity. In: Sawyer, R.K. (ed.) The Cambridge Handbook of the Learning Sciences, pp. 128–147. Cambridge University Press (2014)
15. Vaswani, A., Shazeer, N., Parmar, N., et al.: Attention is all you need. In: Advances Neural Information Processing System, vol. 30 (2017)
16. Riedl, M.: Human-centered artificial intelligence and the future of generative AI. AI Soc. **37**, 435–449 (2023)
17. Zawacki-Richter, O., Jung, I., Bond, M., et al.: Practical and ethical challenges of large language models in education: a review of ChatGPT in teaching and learning. Br. J. Educ. Technol. **54**(6), 1245–1264 (2023). https://doi.org/10.1111/bjet.13370

AI-Powered Classification of Medical Students' Professionalism Profiles

Chang Cai[1], Minyang Chow[1,3], Ruth Choe[1], and Xiuyi Fan[1,2](✉)

[1] Lee Kong Chian School of Medicine, Nanyang Technological University, Singapore, Singapore
xyfan@ntu.edu.sg

[2] College of Computing and Data Science, Nanyang Technological University, Singapore, Singapore

[3] Group Clinical Education, National Healthcare Group, Singapore, Singapore

Abstract. Medical professionalism is fundamental to ethical and competent healthcare, with early lapses during medical school linked to future disciplinary actions in clinical practice. It is essential to identify and analyze the factors contributing to professionalism issues among medical students. However, existing assessment methods rely on subjective evaluations, making them susceptible to bias, time-consuming, and retrospective rather than proactive. To address these limitations, this study develops an AI-driven method to identify medical professionalism profiles and determine key factors of professionalism issues. First, we build a classifier to identify professionalism lapses. Subsequently, the XAI method is applied to extract the main features. Then, we use a clustering algorithm to identify professionalism profiles. Finally, expert interviews with medical educators and hospital administrators can guide the development of tailored interventions based on AI-generated professionalism profiles. This study introduces a transparent AI-driven framework for professionalism assessment, enabling tailored interventions to support the development of ethical and competent physicians.

Keywords: Medical Professionalism · AI Predictions · Explainable AI

1 Introduction

Medical professionalism is essential for ensuring high-quality clinical care, maintaining patient trust, improving physician-patient relationships, and promoting career satisfaction among healthcare professionals [3,6,10]. According to Cruess and Cruess's (2008) paper [5, p. 579],

> ***professionalism*** *serves as the basis of medicine's relationship with society, and many have termed this relationship a social contract.*

Unprofessional behaviour among medical students and practitioners undermines healthcare quality and patient safety, while also imposing significant societal costs through resource waste and workforce attrition [9,11].

A. I. Cristea et al. (Eds.): AIED 2025, CCIS 2590, pp. 309–315, 2025.
https://doi.org/10.1007/978-3-031-99261-2_29

Identifying unprofessional behaviour in medical students becomes an increasing focus within medical education. Research demonstrates that early professionalism lapses in medical school are correlated with future professionalism issues in clinical practice [8,9]. For example, students cited for nonprofessional behaviour, such as academic dishonesty and interpersonal conflicts, are more likely to face professionalism issues during residency [9]. Mak-van der Vossen et al. (2017) summarize four medical students' unprofessional behaviours: "*failure to engage, dishonesty, disrespectful behaviour, and poor self-awareness*" [12, p. 4], all of which should be remediated to prevent future issues. Therefore, early detection of professionalism lapses in medical students is critical to mitigating long-term consequences and enhancing professionalism training.

Despite its importance, the assessment of medical professionalism remains insufficient. Current methods predominantly rely on subjective assessments, such as faculty evaluations, peer reviews, direct observations, and self-reported surveys [10]. These approaches are prone to bias and are often time-consuming. Moreover, few studies address non-academic factors or focus on the personalised intervention of professionalism issues. As a result, broader patterns that define professionalism profiles, such as demographic variables, pre-admission performance and in-school performance, are not explored sufficiently.

Advances in Artificial Intelligence (AI) present promising opportunities to improve the assessment of medical professionalism. In contrast to time-consuming and labour-intensive human assessments, AI enhances efficiency. AI can process complex datasets to identify at-risk students, facilitating timely prediction and early interventions [2,4]. Nevertheless, AI models frequently operate as "black boxes," making it difficult for users to comprehend the reasoning behind specific classifications [1]. Explainable AI (XAI) mitigates this challenge by clarifying the features influencing AI classifications [7], which may inform targeted interventions based on the extracted features. However, AI-driven tools in medical professionalism remain underexplored.

This study aims to use AI to detect professionalism issues, cluster students into different profiles, and interpret key features. Specifically, the study seeks to answer the following research questions:

- RQ1: What are the key features that identify professionalism issues?
- RQ2: What distinct medical professionalism profiles can be identified, and what key factors distinguish these profiles?

The contributions of this study are threefold. First, this study develops an AI-based method to identify distinct professionalism profiles among medical students. Second, the integration of XAI may ensure transparent and interpretable results by highlighting key factors of professionalism risk. Third, this research may offer data-driven recommendations for educators and administrators. These recommendations aim to enable personalized interventions based on students' profiles and extracted features.

2 Method

2.1 Study Design

We developed a five-stage AI-driven process (see Fig. 1). First, data were collected and preprocessed through data cleaning, missing value processing, feature construction, data formatting, and data coding. Subsequently, statistical analysis is performed. Next, a classifier is implemented to identify professionalism issues. XAI, particularly Shapley Additive Explanations (SHAP), is used to identify the most influential factors. Finally, a clustering algorithm is used to categorize students into distinct profiles.

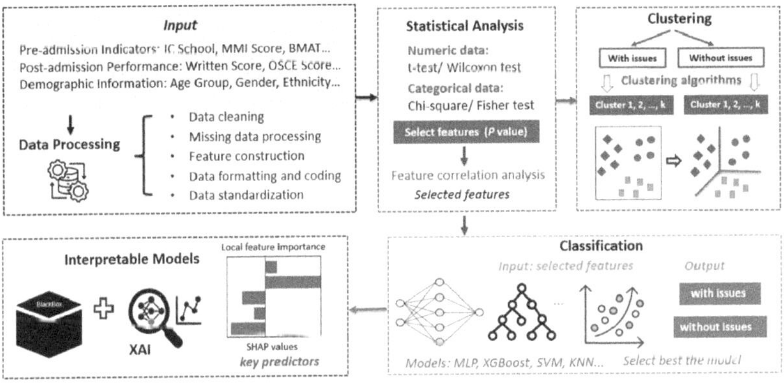

Fig. 1. AI-Driven Method for Identifying Medical Professionalism Profiles.

2.2 Data Collection and Preprocessing

Data was collected by researchers in this study from 2016 to 2024. The dataset comprises records from 1,048 medical students, encompassing all individuals enrolled in a Singapore medical school since 2016. Personally identifiable information is removed to protect privacy. As shown in Table 1, the dataset contained 46 features, comprising demographic factors, academic performance, clinical performance, pre-admission performance, and professionalism issues.

Observable Professionalism Metrics (CHERI Comments): CHERI Comments serve as our primary observable metric for professionalism issues. These comments are recorded by faculty using institution-wide criteria and document specific unprofessional behaviours, such as failing to mark attendance or marking attendance but not attending. Professionalism issues were coded as follows: any student with at least one documented CHERI Comment in any academic year was assigned a value of 1 (with issue), and all others were assigned 0 (no issue).

The data preprocessing involved the following steps:

Table 1. Overview of Data Types and Variable examples.

Type	Variable Examples
Demographic	Gender, Age, Ethnicity, ..., Mather's Occupation
Academic Performance	Year1_Written, ..., Year5_Written
Clinical Performance	Year2_Objective Structured Clinical Examination (OSCE), ..., Year5_OSCE
Pre-admission Performance	Knowledge in Biology, Biomedical Admission Test (BMAT) Score, ..., Multiple Mini Interview (MMI), University Admission Score (UAS)
Professionalism issues	Year 1 CHERI Comments, ..., Year 5 CHERI Comments

- *Data cleaning:* Samples with excessive missing values were excluded. Specifically, students missing all OSCE scores across academic years were removed.
- *Missing data processing:* Variables with more than 35% missing values were excluded. Numerical missing values were imputed using means or rounded means. Missing values in the categorical variable Dormitory_B were imputed using corresponding Dormitory_A entries.
- *Feature construction:* Derived variables were created from OSCE and written scores. The earliest score was the initial score, and the average across years was the final score. For both, max, min, and ranges were computed. Differences between written and OSCE scores were also derived.
- *Data formatting, encoding, and scaling:* School types were reclassified into three tiers based on academic reputation, and UAS was recoded into three levels based on its original values across different grading scales. Categorical variables were encoded using one-hot encoding. Numerical features were standardized using centering and scaling before classification.

2.3 Statistical Analysis (To Answer RQ1)

Statistical tests were selected based on variable types. For unordered categorical variables, the Chisquare test was used; Fisher's exact test was applied if expected counts were <5. For numerical variables, normality was assessed using the ShapiroWilk test. If normal, Levene's test determined variance equality, followed by a t-test or Welch's t-test. If non-normal, the Wilcoxon test was used. Bonferroni correction was applied to control for multiple comparisons, with adjusted $p < 0.05$ considered significant. Variables with $p < 0.1$ were selected for further analysis.

2.4 Classification and XAI (To Answer RQ1)

To identify student professionalism issues, we will train AI models using the selected variables as predictors. Specifically, we will use Logistic Regression, XGBoost, Random Forest, SVM, MLP, and KNN. The performance of each model was evaluated and compared using metrics such as accuracy, precision, recall, and F1-score. Then, SHAP values will be used to evaluate feature contributions. The top features were extracted to create a feature importance summary.

2.5 Clustering (To Answer RQ2)

Clustering will be performed separately for students with and without professionalism issues to identify latent subgroups within each group. First, students will be stratified by professionalism status; then, hierarchical clustering will be applied within each stratum. The optimal number of clusters (k) is selected using the Silhouette Coefficient. To interpret the clusters, intracluster means of selected variables will be compared against overall group means. Statistical analysis will be conduct between subgroups to identify distinguishing features.

3 Preliminary Results

After data processing, the final dataset comprised 897 samples, including 356 (39.7%) students with professionalism issues and 541 (60.3%) without. 44 features were used for this study. A series of Wilcoxon rank-sum tests (with Bonferroni correction) identified five features that differed significantly between groups (See Table 2). Figure 2 shows the distribution of these five features ($p\_adjusted < 0.05$) between students with and without professionalism issues. Students flagged for professionalism issues showed lower median scores across all four written-exam metrics (Final Written, Max_Written, Min_Written, and Year2_Written). In contrast, pre-admission academic performance (MMI_S6 scores) is shifted upward among those with professionalism issues.

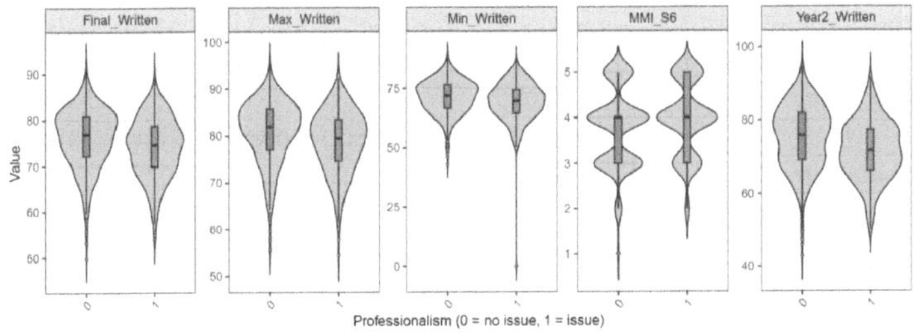

Fig. 2. Distribution of Selected Features by Professionalism Group (Violin + Boxplot)

Table 2. Statistical Test Results.

Variable	Type	Test	p-value	p-adjusted
Year2_Written	Continuous	Wilcoxon test	3.32×10^{-11}	1.49×10^{-9}
Max_Written	Continuous	Wilcoxon test	2.17×10^{-7}	9.75×10^{-6}
Final_Written	Continuous	Wilcoxon test	6.41×10^{-7}	2.88×10^{-5}
Min_Written	Continuous	Wilcoxon test	3.04×10^{-5}	1.37×10^{-3}
MMI_S6	Continuous	Wilcoxon test	7.37×10^{-4}	3.32×10^{-2}

4 Future Work

In the next phase, we will use AI to detect professionalism issues and identify key factors through XAI. We will then apply clustering to define distinct professionalism profiles. A panel of experts will independently review these profiles and provide qualitative feedback. This mixed-method approach combines data-driven insights with expert judgment to enhance trust and acceptance among educators and administrators.

Acknowledgments. This research is supported by the Ministry of Education, Singapore (Grant IDs: LKCMedicine Start up Grant).

References

1. Adadi, A., Berrada, M.: Peeking inside the black-box: a survey on explainable artificial intelligence (XAI). IEEE Access **6**, 52138–52160 (2018)
2. Almalawi, A., Soh, B., Li, A., Samra, H.: Predictive models for educational purposes: a systematic review. Big Data Cognit. Comput. **8**(12), 187 (2024)
3. Bhardwaj, A.: Medical professionalism in the provision of clinical care in healthcare organizations. J. Healthc. Leadership, 183–189 (2022)
4. Cai, C., Zhu, G., Ma, M.: A systemic review of AI for interdisciplinary learning: Application contexts, roles, and influences. Educ. Inf. Technol., 1–47 (2024). https://doi.org/10.1007/s10639-024-13193-x
5. Cruess, R.L., Cruess, S.R.: Expectations and obligations: professionalism and medicine's social contract with society. Perspect. Biol. Med. **51**(4), 579–598 (2008)
6. Desai, M.K., Kapadia, J.D.: Medical professionalism and ethics. J. Pharmacol. Pharmacother. **13**(2), 113–118 (2022)
7. Dwivedi, R., et al.: Explainable AI (XAI): core ideas, techniques, and solutions. ACM Comput. Surv. **55**(9), 1–33 (2023)
8. Haymaker, C.M., VanDerKolk, K., Rhodes, M., Young, J., Graves, L.: Screening for conscientiousness at start of residency: early detection of professionalism concerns. PRiMER: Peer-Reviewed Rep. Med. Educ. Res. **8**, 28 (2024)
9. Krupat, E., et al.: Do professionalism lapses in medical school predict problems in residency and clinical practice? Acad. Med. **95**(6), 888–895 (2020)
10. Li, H., Ding, N., Zhang, Y., Liu, Y., Wen, D.: Assessing medical professionalism: a systematic review of instruments and their measurement properties. PLoS ONE **12**(5), e0177321 (2017)

11. Macneill, P., Joseph, R., Lysaght, T., Samarasekera, D.D., Hooi, S.C.: A professionalism program in medical education and training-from broad values to specific applications: Yll school of medicine, Singapore. Med. Teach. **42**(5), 561–571 (2020)
12. Mak-Van Der Vossen, M., et al.: Descriptors for unprofessional behaviours of medical students: a systematic review and categorisation. BMC Med. Educ. **17**, 1–12 (2017)

Identifying and Fostering Self-regulated Learning Among Computer Programmers Using Artificial Intelligence Systems

Aditya Rajmane[1](✉), Ramkumar Rajendran[1], and Kshitij Sharma[2]

[1] Centre for Educational Technology, Indian Institute of Technology Bombay, Mumbai, India
{aditya.rajmane,ramkumar.rajendran}@iitb.ac.in

[2] Department of Computer Science, Norwegian University of Science and Technology, Trondheim, Norway
kshitij.sharma@ntnu.no

Abstract. Self-Regulated Learning (SRL) strategies are essential for learners to effectively plan, monitor, and reflect on their learning progress, ultimately enhancing problem-solving skills. This research focuses on real-time detection of SRL strategies within a Computer-Based Learning Environment (CBLE) by analyzing click-stream and log data. Using process mining and machine learning, we aim to identify key programming actions that indicate SRL strategies and develop adaptive interventions that provide personalized guidance. The study employs a mixed-methods approach, combining qualitative and quantitative techniques such as surveys, interviews, and statistical modelling to gain a comprehensive understanding of SRL in programming contexts. The goal is to design intelligent learning systems that deliver real-time feedback and scaffold learners' SRL strategies, enabling them to become more independent and efficient problem solvers. This research contributes to AI in Education, Learning Sciences, and Computer Science by advancing SRL detection and fostering meaningful, data-driven educational interventions.

Keywords: Self Regulated Learning · Artificial Intelligence in Education · Computer Based Learning Environments · Learning Analytics · Programming Education

1 Introduction

Self-Regulated Learning (SRL) is a self-directed and cyclic process involving three phases: forethought, performance, and self-reflection. [20]. The cyclic nature provides the learner with the freedom to continue the cycle until the learning objective is achieved. According to 2018 guidance report by The Education Endowment Foundation (EEF) [4], self-regulated learners have an awareness of their strengths and weaknesses - Learners can motivate themselves to engage

A. I. Cristea et al. (Eds.): AIED 2025, CCIS 2590, pp. 316–321, 2025.
https://doi.org/10.1007/978-3-031-99261-2_30

in, and improve, their learning. Developing learners' metacognitive knowledge of how they learn, and their self-awareness about their knowledge as learners, is an effective way to improve their learning outcomes [4].

Self-Regulated learning is studied in multiple contexts like reading, writing comprehension, mathematics, science learning, problem-solving tasks, and collaborative learning environments [10,11,14]. One particular challenging context for SRL is Open-Ended Learning Environments (OELE) as it involves unstructured tasks with minimal guidance, difficulty in providing equitable outcomes and scaffolds [12]. Without effective self-regulation, students may struggle with cognitive overload, lack of direction, and inefficient problem-solving approaches [15]. Hence, it is important to identify and foster SRL in these challenging conditions. Computer programming requires having complex cognitive processes, and supporting self-regulated learning skills to help achieve higher positive outcomes in problem-solving for learners in computer programming courses. [13]. Hence it is important to understand how learners regulate during programming to provide augmented SRL scaffolds.

In the context of computer programming [1] SRL is studied from multiple angles such as the relationship between SRL and academic success, interventions for specific SRL elements [18] and how CS (Computer Science) students at different levels showcase SRL. [5]. SRL in computer programming (SRCP) examines how students define a task, set goals, create plans, utilize resources, and employ strategies to navigate the programming space. [8]. CS students, especially novice learners, need help performing effective SRL. [9]. Current research trends in past decade have moved towards using click-stream and multimodal data to detect traces of SRL [2] [6] [17], there is also a trend in using Artificial Intelligence to detect SRL [19]. An effort is also made to use modelling the cyclic nature [7] and analyzing the temporal nature of SRL [11]. There is an effort to analyze the students' SRL beliefs towards Computer Science Programming [18], yet there is a gap in the literature when it comes to real time identification of SRL and analyzing them as a temporal process in the context of computer programming which is done using intelligent systems.

2 Research Proposal

2.1 Research Questions

Our overarching research goal is to identify Self-Regulated Learning (SRL) strategies that are used by computer programmers, develop targeted interventions that foster SRL, and evaluate the effectiveness of these adaptive systems in supporting and enhancing SRL beliefs in real time.

Following is a list of our proposed broad research questions:

- RQ1 - How can the SRL strategies of a learner be understood while they code?
- RQ2 - Can exploring the existing literature develop a method to identify SRL strategies of the learners using trace data?

- RQ3 - Is developing an adaptive learning system or scaffolds to support SRL among computer programmers effective?

3 Proposed Research

The proposed research is a part of the larger research project (INNOAIED). This project aims to study the impact of APS (AI-Powered Softwares) as well as look at a comparative study of how socio-cultural factors play a role in computer programming.

3.1 Computer Based Learning Environment

To collect the log data in the form of a click-stream and to achieve control over the learning environment we plan to augment an existing python programming learning environment. Pyguru is a python based learning environment [16], the environment effectively logs actions such as Logging in or out into the learning environment, Assessment (Program IDE) page is viewed, Executing the code on embedded Code Editor, Visiting the video Player, Highlighting features is used, In-video question is attempted, etc. The augmentation would be in making additional changes to the logging mechanism for logging behaviours that exhibit SRL beliefs in programming tasks, and further creating a CBLE that would foster SRL process in real-time.

3.2 Data Collection and Methodology

For the construct of SRL Beliefs in Computer Programming, the data that will be analyzed is click-stream and log data. The data will be temporal. Additionally, a questionnaire such as the Motivated Strategies for Learning Questionnaire (MSLQ) [3] will be administered to effectively triangulate the results provided by the system that measures SRL. Semi-structured interviews post-intervention will also be collected.

For Research Question 1 - The proposed study will employ a mixed methods approach. Quantitative Analysis will include process mining to examine the log data that contains click-stream, identifying the patterns of SRL behaviours. Statistical and ML models will be used to detect correlations between programming actions and SRL indicators. Qualitative Analysis will be used to understand learners' experience and will employ thematic coding. Additionally, analysis of A Self-Regulated Learning (SRL) questionnaire will provide triangulation, which will ensure valid intervention effectiveness.

For Research Question 2 - A systematic literature review will be conducted to analyze existing methods for identifying SRL through trace data. We will review existing methods for identifying SRL strategies using trace data in programming environments. This will provide an insight into developing a framework that applies process mining and machine learning to detect SRL occuSring instances from coding logs, such as debugging patterns and revision frequency. To ensure accuracy, findings will be validated through self-reports and qualitative analysis.

For Research Question 3 - We will design adaptive scaffolds and intelligent support systems to foster SRL in learners of computer programming. This includes real-time adaptive prompts and personalized feedback dashboards that guide learners based on their SRL strategies. The effectiveness of these interventions will be evaluated through experimental studies, measuring their impact on learners' self-regulation and coding performance.

A timeline for the study is shown in 1.

3.3 Expectations from the Study

The study aims to contribute to the identification and fostering of Self-Regulated Learning (SRL) among novice programmers. Expected outcomes include

1. The development of an SRL behaviour detection model that is comparable to traditional questionnaire-based assessments in identifying SRL actions in real-time programming tasks.
2. Insights into key programming actions that indicate SRL behaviours
3. The design and evaluation of adaptive learning systems that provide personalized feedback to foster SRL.
4. Empirical evidence on the effectiveness of technology-enhanced learning (TEL) systems in improving SRL skills which in turn would improve programming performance.

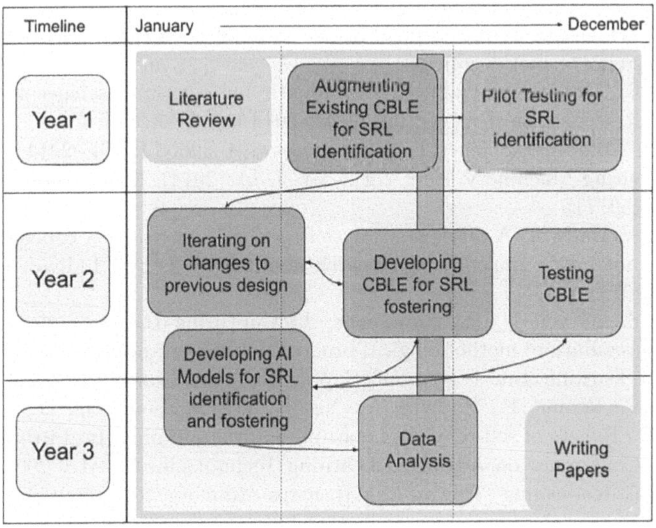

Fig. 1. Timeline

4 Contribution to AIED, Computer Science and Learning Science Communities

This research contributes to the AIED community by connecting Learning Sciences and Computer Science through interventions that foster Self-Regulated Learning (SRL) in programming education. From a Learning Sciences perspective, it advances understanding of SRL strategies by capturing real-time behaviours using process mining and learning analytics, differentiating itself from self-report methods or methods that do not account for real-time coding. It also informs the design of systems that foster self-regulation strategies among novice programmers. From a Computer Science standpoint, the study integrates artificial intelligence to detect, analyze, and support SRL strategies, contributing to the development of future intelligent systems and adaptive feedback mechanisms. This research hence would advance the understanding of learning constructs such as SRL in a granular and temporal way.

References

1. Arakawa, K., Hao, Q., Greer, T., Ding, L., Hundhausen, C., Peterson, A.: In situ identification of student self-regulated learning struggles in programming assignments, pp. 467–473 (2021)
2. Cristea, T., Snijders, C., Matzat, U., Kleingeld, A.: Unobtrusive measurement of self-regulated learning: a clickstream-based multi-dimensional scale. Educ. Inf. Technol. **29**(11), 13465–13494 (2024). https://doi.org/10.1007/s10639-023-12372-6
3. Duncan, T., pintrich, p., smith, d., Mckeachie, W.: Motivated strategies for learning questionnaire (MSLQ) manual (2015). https://doi.org/10.13140/RG.2.1.2547.6968
4. EEF (2021). https://educationendowmentfoundation.org.uk/education-evidence/guidance-reports/metacognition
5. Falkner, K., Vivian, R., Falkner, N.J.: Identifying computer science self-regulated learning strategies. In: Proceedings of the 2014 Conference on Innovation & Technology in Computer Science Education, pp. 291–296. ITiCSE 2014, Association for Computing Machinery, New York, NY, USA (2014). https://doi.org/10.1145/2591708.2591715
6. Järvelä, S., Hadwin, A.: Triggers for self-regulated learning: a conceptual framework for advancing multimodal research about SRL. Learn. Individ. Differ. **115**, 102526 (2024)
7. Järvelä, S., Järvenoja, H., Malmberg, J.: Capturing the dynamic and cyclical nature of regulation: methodological progress in understanding socially shared regulation in learning. Int. J. Comput. Support. Collab. Learn. **14**, 425–441 (2019)
8. Kumar, V., Winne, P., Hadwin, A., Nesbit, J., Jamieson-Noel, D., Calvert, T., Samin, B.: Effects of self-regulated learning in programming. In: Fifth IEEE International Conference on Advanced Learning Technologies (ICALT 2005), pp. 383–387. IEEE, Kaohsiung, Taiwan (2005). https://doi.org/10.1109/ICALT.2005.131, http://ieeexplore.ieee.org/document/1508706/
9. Loksa, D., Ko, A.J.: The role of self-regulation in programming problem solving process and success. In: Proceedings of the 2016 ACM Conference on International Computing Education Research, pp. 83–91. ACM, Melbourne VIC Australia (2016). https://doi.org/10.1145/2960310.2960334

10. Lämsä, J., et al.: Measuring secondary education students' self-regulated learning processes with digital trace data. Learn. Individual Differences **118** (2025). https://doi.org/10.1016/j.lindif.2024.102625, https://www.scopus.com/inward/record.uri?eid=2-s2.0-85213944634&doi=10.1016%2fj.lindif.2024.102625&partnerID=40&md5=c9b05960ec86d726824d550e8b94391c, cited by: 0
11. Nath, D., Gasevic, D., Fan, Y., Rajendran, R.: CTAM4SRL: a consolidated temporal analytic method for analysis of self-regulated learning. In: Proceedings of the 14th Learning Analytics and Knowledge Conference, pp. 645–655. LAK 2024, Association for Computing Machinery, New York, NY, USA (2024). https://doi.org/10.1145/3636555.3636926
12. Teo, T.W., Choy, B.H.: In: Tan, O.S., Low, E.L., Tay, E.G., Yan, Y.K. (eds.) Singapore Math and Science Education Innovation. ETLPPSIP, vol. 1, pp. 43–59. Springer, Singapore (2021). https://doi.org/10.1007/978-981-16-1357-9_3
13. Pedrosa, D., Cravino, J., Morgado, L., Barreira, C.: Self-regulated learning in computer programming: Strategies students adopted during an assignment. In: Allison, C., Morgado, L., Pirker, J., Beck, D., Richter, J., Gütl, C. (eds.) Immersive Learning Research Network. pp. 87–101. Springer International Publishing, Cham (2016). https://doi.org/10.1007/978-3-319-41769-1_7
14. Semensato, M.T., Pilatti, L.A., Silva, F.D.E., Pinheiro, N.A.M.: Revisão sistemática de estudos sobre a autorregulação da aprendizagem da matemática no ensino superior. Bolema: Boletim de Educação Matemática **37**(75), 218–249 (2023). https://doi.org/10.1590/1980-4415v37n75a11
15. Shin, Y., Song, D.: The effects of self-regulated learning support on learners' task performance and cognitive load in computer programing. J. Educ. Comput. Res. **60**(6), 1490–1513 (2022). https://doi.org/10.1177/07356331211052632
16. Singh, D., Subramaniam, H., Rajendran, R.: PyGuru: a programming environment to facilitate measurement of cognitive engagement. In: International Conference on Computers in Education (2022). https://library.apsce.net/index.php/ICCE/article/view/4507
17. Winne, P.H.: Construct and consequential validity for learning analytics based on trace data. Comput. Hum. Behav. **112**, 106457 (2020)
18. Wu, Y.: Self-regulated learning interventions in computer programming: challenge, model and direction. In: 2023 5th International Conference on Computer Science and Technologies in Education (CSTE), pp. 25–29. IEEE, Xi'an, China (2023). https://doi.org/10.1109/CSTE59648.2023.00012, https://ieeexplore.ieee.org/document/10261579/
19. Yildirim-Erbasli, S.N., Gorgun, G., Bulut, O.: Enhancing self-regulated learning with artificial intelligence-powered learning analytics, pp. 57–83. IGI Global (2023). DOIurlhttps://doi.org/10.4018/979-8-3693-0066-4.ch004, https://services.igi-global.com/resolvedoi/resolve.aspx?doi=10.4018/979-8-3693-0066-4.ch004
20. Zimmerman, B.J.: Self-regulated learning and academic achievement: an overview. Educ. Psychol. **25**(1), 3–17 (1990). https://doi.org/10.1207/s15326985ep2501_2

Personality-Aware Conversational Intelligent Tutoring System with GenAI: Studying the Effect on Learners in Introductory Programming

Amani Alrobai[1,2] and Alexandra I. Cristea[1(✉)]

[1] Durham University, Durham, UK
{amani.alrobai,alexandra.i.cristea}@durham.ac.uk
[2] Princess Nourah bint Abdulrahman University, Riyadh, Kingdom of Saudi Arabia
aalrabaie@pnu.edu.sa

Abstract. Recently, with growing public use of large language models (LLMs), researchers and educators have begun thinking about the role of LLMs in learning. Conversational Intelligent Tutoring Systems (CITSs) powered by Generative AI (GenAI) have added a new dimension to educational technology, characterised by enhanced contextual understanding, real-time content generation, and dynamic feedback elaboration. These systems are paving the way towards a broader adaptive education, surpassing the limitations of conventional educational approaches. Despite these capabilities, several important challenges remain unaddressed, such as understanding how learner personalisation can be systematically and quantifiably designed to scaffold learning supported by CITS and LLMs. Thus, the main purpose of this study is to design, implement, and evaluate *a framework for personalised CITS that leverages GenAI to accommodate learner variety in personality traits in learning.* The proposed CITS will be tested with learners, focusing on introductory programming and assessing the impact of personalisation on knowledge retention and learners' experience. Furthermore, the research aims to *formulate a set of performance metrics to evaluate the effectiveness of CITSs* in conversation quality and learner engagement.

Keywords: Generative AI · Intelligent Tutoring System · Learner Modelling · Personalised Learning · Natural Language Processing

1 Introduction

Large-scale classes with more than 100 learner, who have varying ability levels, unique preferences and personalities are becoming the default higher education scenario. Whilst educators and researchers have become more aware of the limitations of one-size-fits-all education paradigm, accommodating the learners' variety and providing individualised support in traditional teaching are both labour- and resource-intensive, and thus extremely hard to implement [3].

A. I. Cristea et al. (Eds.): AIED 2025, CCIS 2590, pp. 322–327, 2025.
https://doi.org/10.1007/978-3-031-99261-2_31

In this space towards effective personalised learning systems, AI conversational agents(CAs) have emerged as a promising solution. Growing evidence highlights their benefits in educational contexts, such as enhancing learning outcomes, motivation, self-efficacy [20].

Generative AI (GenAI) leverages LLMs models like GPT-4 and GPT-3.5. These models have revolutionised the capabilities of CAs in contextual understanding, real-time content generation, and dynamic feedback elaboration. Researchers and educators have begun to think about the use those LLMs as educational tools [10]. CITSs powered by GenAI can offer a more immersive and interactive learning experience in a human-like conversational manner. Despite these advancements, several critical challenges remain, such as understanding how learner personalisation can be systematically and quantifiably designed to scaffold learning supported by CITS and LLMs. These challenges must be addressed to unlock the full educational potential of GenAI and ensure its practical application across learning settings [10].

This research's long-term goal is to *structure a framework that harnesses GenAI capabilities within CITS to accommodate learner variety in personality traits*. We will also study the effect of personalised interaction on knowledge retention and learners' experience. Further, we will *formulate quantitative measures to evaluate the effectiveness of personalised CITS in conversation quality and learner engagement*. Within the scope of this research, we aim to explore the following main research question:

- **RQ:** How can CITS powered by GenAI be designed to accommodate learner variety (e.g. personality traits) effectively, in a systematic and quantifiable way? with emerging sub-questions:
  - **RQ1:** Can personality traits be inferred from learners' interactions with a chatbot during open-ended interview questions using NLP and machine learning?
  - **RQ2:** How can the inferred dominant personality trait be algorithmically mapped to scaffolding strategies and implemented through prompt engineering in an LLM-powered CITS within simulated tutoring scenarios?
  - **RQ3:** How does personalisation within CITS powered by GenAI, as a conversational educational application, impact on learner knowledge retention and experience across different learner profiles?
  - **RQ4:** How can quantitative measures be formulated to comprehensively evaluate the effectiveness of personalised CITS for diverse learners?

2 Related Work

LLMs have been extensively studied in computing education [15]. There has been much research focusing on the LLMs capabilityties' in solving programming exercises [8], generating code explanations [4], and answering question [16]. A recent study [12] has introduced Code-Help, which is an LLM-powered assistant tool designed for programming courses. Students have expressed a strong preference for scaffolding features rather than direct solutions. LLMs excel at zero- and few-shot reasoning, driving interest in prompt engineering to enhance interactions. This advancement supports the development of CITSs, enabling them to simulate individual learners' variety for both cognitive and non-cognitive aspects [13]. Additionally, it allows CITSs to accommodate learners of different age, cultures, and learning levels [1], as well as adapt to learning styles and confidence levels [14].Researchers consider personality as crucial in software engineering education, aiming to tailor teaching to specific student personalities [7]. While there has been much research on the adoption of pedagogical CAs, few researchers have taken adaptation to Big Five personality traits [17].

To the best of our knowledge, no prior research has leveraged CITS powered by LLMs with systematical and quantifiable personalisation scaffolding, as targeted by this research. Specifically, this study explores the potential of LLMs in tailoring scaffolding support to diverse learner personalities within programming scenarios, advancing CITS to enhance innovation and adaptability in learning.

3 Proposed Methodology

The research intends to test theory-based hypotheses and follows an experimental approach. It is based on Big Five personality traits [5], and the framework for designing adaptive tutoring in an intelligent conversational agent system [11]. The research is divided into four phases: (1) personality trait inference from open-ended interview-style conducted by a chatbot, which addresses (RQ1) (Sect. 4.1); (2) a simulation study, to optimising personality-aware prompt templates for tutoring personalisation, to answer (RQ2) (Sect. 4.2); (3) development of a framework for personalised CITS powered by LLMs to better fit the needs of individual learners (Sect. 4.3); (4) systematic evaluation of the CITS with real users and the development of quantitative measures, to address (RQ3)(RQ4) (Sect. 4.4).

4 Future Research Directions and Expected Contributions

4.1 Personality Trait Inference from Chatbot Interactions

This phase addresses RQ1, as well as to the cold start problem of CITS to adapt tutoring strategies without prior learner data. The learner personality predictor is created by building a *chatbot that conducts open-ended interview-style conversations*[1]. The learner responses will be collected and then *analysed using Natural Language Processing (NLP)* techniques, such as *linguistic feature extraction.* A *machine learning model* predicts Big Five personality traits from the extracted features [9]. The results are validated against self-reported assessments like the Big Five Inventory (BFI-44) to ensure accuracy.

[1] Open-ended questions for interviewer chatbot.

4.2 Simulation for Educational Conversation

We will build a *multiagent communication environment* following [13] using GPT-4, which effectively distinguishes personality traits through generated language, and interactions. One agent for *the teacher role*, which will be configured for personality-adaptive dialogue within *a retrieval-augmented generation (RAG) framework*, using role-play prompt templates for the five personality based on recommended activities from [2]. The other agent for *learner role*, Where we construct profiles of learners with dominant personality traits. A structured knowledge base will be built with domain-specific content tailored to personality profiles. The model will run iteratively in multi-agent simulations, generating personalised responses based on agent traits and context. Post-simulation analysis will evaluate simulated learner personality, agent adaptability, and retrieval accuracy to refine adaptive prompting strategies – thus answering RQ2. To evaluate realism and accuracy, a pilot study will be conducted in which real learners replace synthetic agents.

4.3 Development of Personality-Aware CITS

The proposed CITS will be designed in a modular structure, which allows the tutoring domain to be changeable, by keeping the knowledge of the system separate from the functionality. The learning system consists of 6 main modules (see Fig. 1). *The knowledge base* consists of the introduction of programming materials, explanations and exercises. *The personality predictor* agent infers learner personality, ensuring personalised tutoring from the first interaction (Sect. 4.1). *The learner profile* is where all the learner data, such as learner personality traits and the chat history with the AI agents, are stored for further analysis. *The user interface* is responsible for displaying the learning content and chat area. *The conversation manager* is the central manager of the system, responsible for communicating with all components and managing the learner interactions. The GPT-4 will be used to *train 5 different personality-adaptive AI agents.* We will utilise *chain-of-thought* and *role-playing* prompting techniques to enhance agents' ability in solving and explaining learning activities recommended for each personality type by [2].

4.4 Experimentation and Evaluation

In the final phase, a proposed CITS will be tested in an iterative way, by conducting a controlled experiment, to answer (RQ3). In the post-experiment, the survey will include *the User Experience Questionnaire (UEQ)* [18] and selected items from the *Unified Theory of Acceptance and Use of Technology* (UTAUT) [19]. *Mining the learner interactions* will help identify effective and ineffective teaching sessions, enabling the creation of *performance metrics* to evaluate the effectiveness of CITSs, as guided by [6], to answer (RQ4).

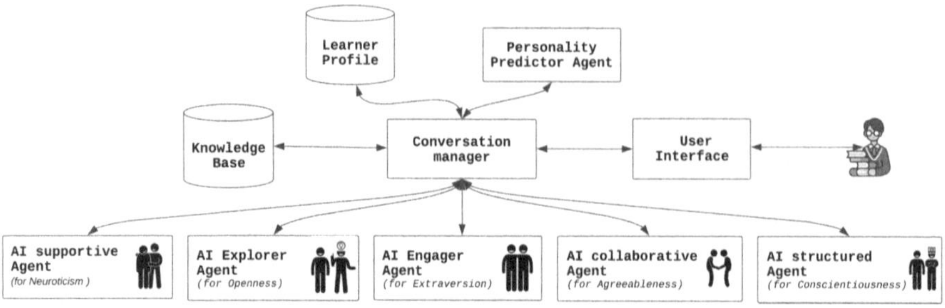

Fig. 1. Personality-Aware LLM-driven CITS architecture.

4.5 Expected Research Contributions

This research contributes to AIED by exploring systematical and quantifiable personalisation scaffolding with GenAI. From a learning sciences perspective, this research is noteworthy, because it aims to address the gap in understanding how GenAI can be used to support learners in dialogic scaffolding, to better learning opportunities. From a computer science perspective, this research is needed, as it attempts to explore the ability of GenAI technology to adapt to individual characteristics in conversations, towards more personalised interactions, and potentially enhancing learning efficiency.

References

1. Abolnejadian, M., Alipour, S., Taeb, K.: Leveraging ChatGPT for adaptive learning through personalized prompt-based instruction: a CS1 education case study. In: Extended Abstracts of the CHI Conference on Human Factors in Computing Systems, pp. 1–8. ACM, New York, NY, USA (2024). https://doi.org/10.1145/3613905.3637148
2. Baruth, O., Cohen, A.: Personality and satisfaction with online courses: the relation between the Big Five personality traits and satisfaction with online learning activities. Educ. Inf. Technol. **28**(1), 879–904 (2023)
3. Broadbent, J., Panadero, E., Boud, D.: Implementing summative assessment with a formative flavour: a case study in a large class. Assess. Eval. High. Educ. **43**(2), 307–322 (2018)
4. Chen, E., Lee, J.E., Lin, J., Koedinger, K.: GPTutor: great personalized tutor with large language models for personalized learning content generation. In: Proceedings of the Eleventh ACM Conference on Learning @ Scale, vol. 1831 CCIS, pp. 539–541. ACM, New York, NY, USA (2024)
5. Costa, P.T., McCrae, R.R.: A five-factor theory of personality. In: Handbook of personality: Theory and research, vol. 2, no. 01 (1999)
6. Cristea, A.I., et al.: The engage taxonomy: SDT-based measurable engagement indicators for MOOCs and their evaluation. User Model. User-Adapted Interact. **34**(2), 323–374 (2024). https://doi.org/10.1007/s11257-023-09374-x
7. Cruz, S., Da Silva, F.Q., Capretz, L.F.: Forty years of research on personality in software engineering: a mapping study. Comput. Hum. Behav. **46**, 94–113 (2015). https://doi.org/10.1016/J.CHB.2014.12.008

8. Denny, P., Kumar, V., Giacaman, N.: Conversing with copilot: exploring prompt engineering for solving CS1 problems using natural language. In: SIGCSE 2023 - Proceedings of the 54th ACM Technical Symposium on Computer Science Education, vol. 1, pp. 1136–1142 (2023). https://doi.org/10.1145/3545945.3569823
9. Jayaratne, M., Jayatilleke, B.: Predicting personality using answers to open-ended interview questions. IEEE Access **8**, 115345–115355 (2020)
10. Kasneci, E., Sessler, K., et al.: ChatGPT for good? On opportunities and challenges of large language models for education. Learning and Individual Differences **103**, 102274 (2023). https://doi.org/10.1016/j.lindif.2023.102274
11. Latham, A., Crockett, K., McLean, D., Edmonds, B.: Adaptive tutoring in an intelligent conversational agent system. In: Nguyen, N.-T. (ed.) Transactions on Computational Collective Intelligence VIII. LNCS, vol. 7430, pp. 148–167. Springer, Heidelberg (2012). https://doi.org/10.1007/978-3-642-34645-3_7
12. Liffiton, M., Sheese, B., Savelka, J., Denny, P.: CodeHelp: using large language models with guardrails for scalable support in programming classes. In: ACM International Conference Proceeding Series, p. 11 (2023)
13. Liu, Z., Yin, S.X., Lin, G., Chen, N.F.: Personality-aware student simulation for conversational intelligent tutoring systems. In: Proceedings of the 2024 Conference on Empirical Methods in Natural Language Processing, pp. 626–642. Association for Computational Linguistics, Stroudsburg, PA, USA (2024)
14. Park, M., Kim, S., et al.: Empowering personalized learning through a conversation-based tutoring system with student modeling. In: Extended Abstracts of the CHI Conference on Human Factors in Computing Systems, pp. 1–10. ACM, New York, NY, USA (2024)
15. Prather, J., Denny, P., et, a.: The robots are here: navigating the generative AI revolution in computing education. ITiCSE-WGR 2023 - Proceedings of the 2023 Working Group Reports on Innovation and Technology in Computer Science Education, pp. 108–159 (2023). https://doi.org/10.1145/3623762.3633499
16. Savelka, J., Agarwal, A., et al.: Thrilled by your progress! Large language models (GPT-4) no longer struggle to pass assessments in higher education programming courses. ICER 2023 - Proceedings of the 2023 ACM Conference on International Computing Education Research V.1, pp. 78–92 (8 2023)
17. Schlimbach, R., Rinn, H., Markgraf, D., Robra-Bissantz, S.: A literature review on pedagogical conversational agent adaptation. Tech. rep., PACIS 2022 Proceedings. 20 (2022)
18. Schrepp, M., Hinderks, A., Thomaschewski, J.: Design and evaluation of a short version of the user experience questionnaire (UEQ-S). Int. J. Interact. Multimedia Artif. Intell. **4**(6), 103 (2017)
19. Venkatesh, V., Morris, M.G., et al.: User acceptance of information technology: toward a unified view. MIS Q. Manage. Inf. Syst. **27**(3), 425–478 (2003)
20. Wu, R., Yu, Z.: Do AI chatbots improve students learning outcomes? Evidence from a meta-analysis. Br. J. Educ. Technol. **55**(1), 10–33 (2024)

Perceptions of Secondary School Students Solving Math Word Problems in a Foreign Language with Generative AI Support

Betsy Araujo Grando(✉), Jiyeong Yi, and Stephen B. Gilbert

Iowa State University, Ames, IA 50011, USA
betsya@iastate.edu

Abstract. Generative Artificial Intelligence (GAI) tools are creating new opportunities to support student learning, especially for those studying mathematics in a foreign language. The present study investigates how secondary students at a private school in Venezuela, whose primary language is Spanish, engage with a multilingual GAI agent, developed using Microsoft Copilot Studio, to understand and solve math word problems written in English. Motivated by the need to prepare students for international college entrance exams such as the Scholastic Aptitude Test (SAT), this research aims to analyze students' interactions with the GAI tutor and evaluate its effects on their persistence and confidence in mathematics. The Dynamic Strategic Math (DSM) framework and self-efficacy based on Social Cognitive Theory will be integrated to explore how GAI-supported scaffolding influences the students' abilities in solving math word problems. A multi-case study and convergent parallel mixed-methods design will enhance the study's rigor. By identifying the benefits and challenges of integrating AI tools into educational support systems, this study aims to provide valuable insights for developing and implementing effective, personalized AI tutors and promoting equitable access to educational resources for students from diverse backgrounds.

Keywords: Math Word Problems · Foreign Language · Generative AI

1 Context and Problem Statement

Generative AI (GAI) tools may offer new opportunities for inclusive educational support, particularly for students learning mathematics in a foreign language. Math word problem-solving has been a primary focus of research because it is one of the most challenging tasks for mathematics learners to master [1]. Research highlights the cognitive demands on students studying mathematics in a foreign or second language, indicating that lexical and syntactic complexities can hinder their problem-solving abilities compared to peers studying in their home language [2, 3]. While existing studies suggest that GAI can scaffold language learning [4] and enhance mathematics skills [5], gaps remain in understanding students' strategies when using GAI agents and their impact on student persistence and self-efficacy, specifically about solving math word problems in a foreign language. This study aims to get insights into how Spanish-speaking students

A. I. Cristea et al. (Eds.): AIED 2025, CCIS 2590, pp. 328–333, 2025.
https://doi.org/10.1007/978-3-031-99261-2_32

interact with a multi-lingual AI agent to understand and solve English-written math word problems and explore the effects of these interactions on the students' math self-efficacy, which is currently an underexplored topic.

Most of the past research about artificial intelligence in mathematics education has been done in the United States and China, followed by other countries in Asia and Europe [6, 7]. More research in developing countries is needed to understand the diverse needs of students worldwide. This study addresses this need by recruiting secondary school students enrolled in a private school in the Bolivarian Republic of Venezuela who may need to take international college admission tests in English to reach their educational aspirations in a foreign country.

The findings from this study will be valuable to students and educators striving to optimize the use of AI as an inclusive educational support tool, especially when human resources are scarce. Additionally, this study has the potential to contribute to the ongoing discourse on generative AI in education, particularly regarding its role in reducing the cognitive demands of learning mathematics in a foreign language. The findings could inform the development of effective instructional strategies and help promote more equitable access to educational resources, ultimately supporting all learners in achieving academic success in mathematics.

2 Research Questions

This research includes the following topics: Spanish-speaking secondary school students, Math Word Problems written in a Foreign Language (MWP-LX), Math Word Problem Self-Efficacy (MWP-SE), and Generative AI (GAI) as a support tool. In this study, LX represents the foreign language (English) to avoid language hierarchy implications [8].

The research questions are:

1. How do secondary school students use GAI to solve math word problems written in a foreign language (MWP-LX)?
   a. How do secondary school students use GAI to understand MWP-LX?
   b. What initial strategy do secondary school students apply to solve MWP-LX using GAI?
   c. How do secondary school students demonstrate persistence when using GAI to solve MWP-LX?
   d. How does the use of GAI influence the accuracy of solutions to MWP-LX among secondary school students?
2. How does the use of GAI influence secondary school students' MWP-SE regarding their ability to solve an MWP-LX?

3 Methodology

3.1 Theoretical Framework

This study integrates the Dynamic Strategic Math (DSM) framework [9] and Self-efficacy based on Social Cognitive Theory [10] to explore how GAI-supported scaffolding influences the students' abilities to understand and solve math word problems.

DSM is a comprehensive strategy designed specifically for English learners to enhance their comprehension of math word problems. This approach involves providing tailored feedback that aligns with students' language comprehension levels. The DSM framework incorporates four distinct levels of linguistic modification for math word problems and three systematic scaffolding procedures [9]. Key strategies within this framework include guiding students to identify pertinent questions, key terms, and numerical data. The ultimate objective is to empower students to assume leadership roles in their learning, thus actively practicing and applying the method. Given its collaborative nature, this strategy is particularly suited for integration with GAI in student interactions.

Self-efficacy, as defined by Bandura [10], refers to individuals' beliefs in their capabilities to achieve specific performance standards. According to social cognitive theory, self-efficacy can be influenced through four primary mechanisms: mastery experiences, social modeling, alteration of physical and emotional states, and verbal persuasion [10]. A study validating Bandura's self-efficacy hypothesis [10] specifically applied to high school students solving math word problems confirmed that self-efficacy is a robust predictor of math performance and persistence in problem-solving [11]. Research examining math word problem self-efficacy among language learners remains limited.

DSM's collaborative strategies, such as linguistic modifications and guided problem-solving, can be operationalized through GAI to provide personalized feedback, reducing language barriers and creating mastery experiences. These experiences, in turn, may enhance students' math self-efficacy. By bridging pedagogical scaffolding (DSM) and motivational psychology (self-efficacy), this framework provides a basis for the possibility that GAI could simultaneously address cognitive and affective barriers to math word problem-solving success in foreign language contexts.

3.2 Approach

To achieve an in-depth and comprehensive understanding of the interactions of secondary school students with a GAI agent for solving MWP-LX, the study will conduct an exploratory multi-case study research design [12] with 3 to 5 participants, instead of a quantitative research method alone. A convergent parallel mixed-methods approach will enhance the study's rigor and minimize biases.

A non-probability purposive sampling will ensure that the selected participants possess relevant characteristics, such as proficiency in English (intermediate/advanced level) and appropriate secondary school grade level (4th year in Venezuela, equivalent to 11th grade in the United States). Collaboration with school staff will facilitate recruitment, ensuring a well-defined participant pool that reflects the target population.

The self-efficacy questionnaire will be similar to the one used by Programme for International Student Assessment (PISA) [13, 14], which was administered to 15-year-old students worldwide. It will be adjusted to follow Bandura's guidelines [15]. The results of this questionnaire will be complemented by doing pre-and-post interviews with questions related to the four sources of self-efficacy.

The AI agent, nested in Microsoft Copilot Studio, will be designed and implemented. The description for the AI agent behavior will include DSM and social cognitive theory strategies and instructions to promote critical thinking instead of providing the solution to the MWP-LX. Participants will access it with individual accounts by using pseudonyms. Before the students can access the AI agent, an introduction about AI literacy will be provided to create awareness of ethical and privacy considerations when interacting with the AI agent.

A MWP rubric will be developed to classify the linguistic and mathematical demands. A bank of word problems with different difficulty levels will be drafted using GAI, asking it to follow the Mathematics Common Core State Standards (U.S. academic standards) for the student grade level, and reviewed by collaborators in the math education and linguistics fields. The participants can continue practicing with the GAI agent between sessions and ask the agent to provide further math word problems.

3.3 Data Collection and Analysis

The data collection will occur virtually via Zoom in at least seven sessions of around 1 hour each. More sessions can be included if the participant requires more time to solve the MWPs. The data collection includes the following:

1. MWP-SE Questionnaire: Administered before and after the intervention, this Likert scale and online instrument will assess students' attitudes and MWP-SE. (RQ2)
2. Pre- and Post-Tests: Participants will complete tests without using GAI, designed to measure their understanding of MWP-LX, consisting of items similar to the MWP-SE questionnaire. (RQ1, RQ2)
3. Individual Interviews: Conducted pre- and post-experience, these interviews will explore participants' confidence levels in solving MWP-LX aligned to the self-efficacy sources [10] and perceptions regarding the utility of the GAI tool. (RQ1, RQ2)
4. Think-Aloud Protocol: Participants will verbalize their thought processes while solving MWP-LX without and using the GAI agent. This will provide insights into their cognitive strategies and engagement with the task and tool. (RQ1)
5. Screen Recording, Observation, and Interaction Data: Sessions will be recorded to capture real-time interactions with the GAI tool, allowing for detailed analysis of user behavior and tool usage. Additionally, the interaction data will be retrieved from Microsoft Copilot Studio for further analysis. (RQ1)
6. Participant Artifacts: Photos of their handwritten work will be collected. (RQ1)

Data will be analyzed using a combination of qualitative and quantitative approaches tailored to specific research questions:

- An inductive thematic analysis and behavioral coding will be employed for data analysis from interviews, think-aloud protocols, and interaction with the GAI tool, allowing for the identification of recurring themes, patterns, and strategies across participants' experiences. A process description will be developed. A peer-debriefing will be conducted on the thematic analysis to ensure analysis rigor, and a second coder will complete the behavioral coding task for reliability, establishing interrater reliability.
- Quantitative data from the questionnaire and tests will be analyzed using descriptive statistics and a non-parametric inferential analysis, such as the Wilcoxon test, to compare the pre- and post-intervention results.

To strengthen the validity of the findings, multiple sources of evidence will be triangulated, and a replication logic will be employed within the multiple case studies. Replication logic involves treating each case as an individual experiment, allowing patterns to be confirmed or contrasted across cases. A comprehensive case study protocol, complete with a data management plan, will ensure the reliability of the study.

This multifaceted approach is expected to yield rich data that will contribute valuable insights into the effects of GAI tools on foreign language math problem-solving strategies among secondary school students.

4 Conclusions and Future Directions

A pilot study with one participant in the first half of 2025 will help develop a preliminary codebook and improve data collection instruments and protocols. In the third quarter of 2025, comprehensive data will be gathered from additional participants using these refined tools, incorporating feedback from the AIED 2025 Doctoral Consortium. Data analysis will occur in the final quarter of 2025, followed by dissertation writing and defense in the first half of 2026.

This study will provide valuable insights into the role of AI in diverse settings for solving math word problems. It will explore how students use AI tools in math learning when they need language support. The research could inform instructional strategies for using GAI to boost learners' math self-efficacy. Additionally, it may lead to more effective learning methods that leverage AI to reduce cognitive demands and enhance learning outcomes.

From a computer science standpoint, the study could offer insights into developing more effective GAI tools to support math learning for students who need additional language support. Furthermore, the study will provide valuable behavioral insights into how users engage with AI in educational settings, informing the development of more user-friendly and effective AI tools for learning.

This study will promote equitable access to educational resources by demonstrating how AI can support students with diverse needs. This aligns with the broader goals of the AIED community and the AIED 2025 conference theme, ensuring all learners have access to inclusive and personalized educational technologies. The study will also address the underexplored area of AI's impact on solving math word problems in a foreign language, especially with participants from a developing country. The findings could inform educational policy and practice, guiding the integration of AI tools in classrooms to support diverse student populations.

Disclosure of Interests. The authors have no competing interests to declare that are relevant to the content of this article.

References

1. Verschaffel, L., Schukajlow, S., Star, J., Van Dooren, W.: Word problems in mathematics education: a survey. ZDM Math. Educ. **52**, 1–16 (2020)

2. Campbell, A.E., Davis, G.E., Adams, V.M.: Cognitive demands and second-language learners: a framework for analyzing mathematics instructional contexts. Math. Think. Learn. **9**(1), 3–30 (2007). https://doi.org/10.1080/10986060709336603
3. Martiniello, M.: Linguistic complexity, schematic representations, and differential item functioning for English language learners in math tests. Educ. Assess. **14**(3–4), 160–179 (2009). https://doi.org/10.1080/10627190903422906
4. Bin-Hady, W.R.A., Al-Kadi, A., Hazaea, A., Ali, J.K.M.: Exploring the dimensions of ChatGPT in English language learning: a global perspective. Library Hi Tech (2023)
5. Wardat, Y., Tashtoush, M.A., AlAli, R., Jarrah, A.M.: ChatGPT: a revolutionary tool for teaching and learning mathematics. EURASIA J. Math. Sci. Technol. Educ. **19**(7) (2023) https://doi.org/10.29333/ejmste/13272
6. Awang, L.A., Yuspo, F.D., Danaee, M.: Current Practices and future direction of artificial intelligence in mathematics education: A systematic review. Int. Electron. J. Math. Educ. **20**(2), em0823 (2025)
7. Panqueban, D., Huincahue, J.: Artificial intelligence in mathematics education: a systematic review. Uniciencia **38**(1), 357–373 (2024)
8. Dewaele, J.M.: Why the dichotomy 'L1 versus LX user' is better than 'native versus non-native speaker.' Appl. Linguis. **39**(2), 236–240 (2018)
9. Orosco, M.J., Swanson, H.L., O'Connor, R., Lussier, C.: The Effects of dynamic strategic math on English language learners' word problem solving. J. Special Educ. **47**(2), 96–107 (2013). https://doi.org/10.1177/0022466911416248
10. Bandura, A.: Self-Efficacy. In: Ramachaudran, V.S. (ed.) Encyclopedia of Human Behavior, vol. 4, pp. 71–81. Academic Press, New York (1994)
11. Pajares, F., Kranzler, J.: Self-efficacy beliefs and general mental ability in mathematical problem-solving. Contemp. Educ. Psychol. **20**, 426–443 (1995)
12. Yin, R.K.: Case Study Research and Applications: Design and Methods, 6th edn. Sage Publications, Thousand Oaks (2018)
13. Underwood, C.: Snapshots issue 19: Confidence, gender and mathematics in PISA. Snapshots **19**(19) (2025). https://research.acer.edu.au/snapshots/vol19/iss19/1
14. Organisation for economic co-operation and development: How confident are students in their ability to solve mathematics problems?. OECD Publishing (2015)
15. Bandura, A.: Guide for Constructing Self-Efficacy Scales. In: Pajares, F., Urdan, T. (eds.) Self-Efficacy Beliefs of Adolescents, vol. 5, pp. 307–337. Information Age Publishing, Greenwich (2006)

Designing an Ethical Framework for the Integration of Generative AI in Higher Education: Balancing Stakeholder Interests and Enhancing Learning Outcomes

Thumesha Jayatilake(✉)

Carleton University, Ottawa, ON, Canada
thumeshajayatilake@cmail.carleton.ca

Abstract. While AI may enhance learning and administrative efficiency, it also raises concerns on academic integrity, algorithmic bias, and inconsistent institutional policies. This study examines the moral concerns of employing generative AI, specifically ChatGPT, in higher education. Using a descriptive and comparative survey-based research approach, it captures the opinions of students, educators, and policymakers in higher education, with a mixed-methods strategy. The findings necessitate a stakeholder-specific ethical framework to guide the adoption of AI responsibly in higher education. This work contributes to AI ethics by bridging gaps in stakeholder-specific concerns and making actionable policy recommendations. Through accountability, transparency, and inclusivity, the study favors the establishment of ethical governance mechanisms that ensure innovation aligns with responsible AI integration in higher education institutions.

Keywords: Generative AI · Artificial Intelligence in Education · AIED · AI Ethics · Stakeholder theory

1 Background

The development of artificial intelligence (AI) technologies has massively revolutionized user experience across various fields, particularly in education. Generative AI (GenAI) is in this regard particularly interesting considering its potential to automatically create content in the manner of a human, in the form of text, images, code, and simulations, and as such reshape knowledge production processes, pedagogy, and learner engagement [1, 2]. Tools like ChatGPT gained one million users in five days, highlighting the rapid adoption of GenAI [3]. Among the most significant categories of GenAI are general-purpose chat-based interfaces, such as ChatGPT, Claude, and Gemini. Powered by large language models (LLMs), these models support natural language interaction across a broad variety of tasks, offering context-aware, multi-turn dialogue for summarization, reasoning, and idea generation [4].

In higher education, these tools are increasingly used to personalize learning, generate content automatically, and support administrative tasks [5, 6]. They increase interactivity and expand access to course materials yet also generate ethical concerns. Academic

A. I. Cristea et al. (Eds.): AIED 2025, CCIS 2590, pp. 334–339, 2025.
https://doi.org/10.1007/978-3-031-99261-2_33

dishonesty undermines examination validity, while algorithmic biases and data privacy issues have the potential to amplify existing inequalities [7–10]. Excessive use of the same tools also has the potential to hinder critical thinking and intellectual growth among students [11].

Institutional responses to GenAI have been mixed, from accommodating faculty-level decisions to outright prohibitions, the absence of collective moral direction. Conflicting stakeholder perceptions only exacerbate the issue; most students, for example, do not view GenAI-supported work as academic dishonesty, in contrast to institutional policy. These tensions underscore the necessity of a holistic ethical approach that embraces a plurality of views.

Universities interact with artificial intelligence in three primary ways: (a) applying AI to improve learning and teaching (AIED); (b) learning about AI - comprehending AI systems and how they are created; and (c) preparing for AI - confronting the social consequences of AI [12]. The present study concentrates on the learning with AI aspect, examining how learners utilize Generative AI tools to facilitate their learning processes.

2 Theoretical Lens

This research relies on Stakeholder Theory (ST), which believes that it is necessary to meet the legitimate interests of all the concerned people and groups through organizational actions [13, 14]. ST enables ethical decision-making by balancing conflicting demands by the nature of power, legitimacy, and urgency [15]. Originally developed in the corporate governance setting, ST has been applied subsequently in marketing, finance, and AI [16].

As AI is more and more embraced, ST offers a valuable perspective for addressing ethical issues like data privacy, algorithmic transparency, and governance [17]. In the education sector, and particularly in Artificial Intelligence in Education (AIED), ST allows for the analysis of stakeholder relationships and aligning AI systems with ethical and pedagogical goals [12].

Emerging scholarship highlights ST's capacity to guide inclusive and responsible AIED development by enabling values such as fairness, transparency, and stakeholder accountability [18, 19]. Shared stakeholder engagement has also been seen as key to maintaining ethical sustainability in AI-supported education [20].

Stakeholders of higher education deal with AI from different angles. Students focus primarily on accessibility and academic support; teachers worry about academic integrity and teaching quality; policymakers are most concerned with compliance, equity, and institutional trust [7, 21, 22]. These competing concerns emphasize the need for a stakeholder specific ethical framework. The present research is aimed at three primary groups of stakeholders; students, educators, and policymakers to investigate the construction and prioritization of ethical concerns.

3 Research Motivation and Gap

This study is driven by a basic gap in literature; the lack of stakeholder-specific ethical exploration of the deployment of GenAI in higher education. Although AIED has developed during the past thirty years [12], much of the research concentrates on its pedagogical promise, with too little critical contemplation of its ethical dangers and implications [2]. Existing research on AI ethics tends to address broad themes such as fairness or privacy without areful attention to how ethical issues are experienced differently by various stakeholder groups [18]. Universities are complex stakeholder ecosystems, and ethical governance arrangements must take this into account if they are to enable inclusive and responsible AI uptake.

Co-design approaches that engage stakeholders in AIED systems development guarantee that their distinctive values and needs are reflected, thereby supporting ethical responsiveness [23]. Current ethical frameworks, however, overwhelmingly emphasize universal principles—like transparency, equity, and accountability—without adequately examining how these are interpreted within stakeholder groups [18]. Besides, insight remains limited as to how regulatory frameworks and institutional policies affect such interpretations. This necessitates developing a framework that reconciles varied perspectives and issues workable, context-based ethical guidance.

Such a gap is particularly apparent as far as GenAI tools like ChatGPT are concerned, whose open-endedness and rapid uptake generate unprecedented ethical concerns [24]. Unlike traditional AIED systems, which are designed for specific pedagogical outcomes, ChatGPT is a general-purpose interface capable of producing sophisticated human-like responses in real time (Kasneci et al., 2023). Its accessibility and versatility make it appealing for learning and administrative tasks, but its use is generally outside the scope of institutional regulation. Among the worries are academic integrity, data bias, and subverting formal learning processes. In the absence of coherent regulatory frameworks, it is necessary to study how different stakeholder groups engage with ChatGPT, and how institutional structures will react. This research studies these dynamics as part of a broader exploration of ethical AI integration in education.

4 Research Questions

To address the identified research problem, this research is guided by the following questions:

1. What ethical concerns do students, educators, and policymakers in higher education identify regarding the use of ChatGPT in the learning process?
2. How do these concerns vary across stakeholder groups, and what factors (such as institutional policies and regulatory environments) influence these differences?
3. How can stakeholder-specific concerns about the use of GenAI in the learning process, along with other influential factors, be systematically addressed in an ethical governance framework for higher education institutions?

5 Research Implications

This study offers theoretical and pragmatic contributions to the ethics of AI in higher education through ST. Theoretically, it recontextualizes ST for GenAI by examining how stakeholder roles and institutional environments impact ethical concerns. It also offers a stakeholder-specific ethical model that remedies the lack of context-specific strategies present in current literature, offering a design that can facilitate cross-context comparison and potential replication.

In practice, the study examines institution-specific problems that can inform policy and governance practice. For universities, the findings indicate the necessity of finding equilibrium between innovation and accountability. University administrators can employ these results in balancing AI implementation with institutional missions, whereas instructors can utilize specific rules and professional training. For students, the paper emphasizes the importance of openness and support measures to guarantee responsible use of GenAI technology. The findings also inform broader regulatory discussions on algorithmic fairness, data privacy, and transparency, recognizing the specific affordances and risks of general-purpose GenAI tools like ChatGPT.

6 Research Design

A comparative, descriptive survey design is used to examine stakeholder specific ethical concerns related to use of ChatGPT in higher education. Carleton University is taken as the initial context for the study employing stratified sampling enabled balanced representation among students, educators, and policymakers, and data collection via an online survey tool using closed- and open-ended questions. The analytical strategy is a mixed-method using descriptive statistics and ANOVA for quantitative data, and thematic coding for qualitative data. This approach fills a methodological gap in the research field by incorporating quantitative accuracy in a field previously dominated by qualitative studies, enabling comprehensive comprehension of the perceptions of stakeholders.

7 Conclusion and Future Directions

This study investigates the ethical implications of GenAI integration into university education from the stakeholder perspective, focusing on policymakers, educators, and students. Conducted at Carleton University, the study reveals how institution and stakeholder-specific factors determine ethical concerns about ChatGPT adoption.

By combining ST and mixed-methods design, this study contributes to the construction of a usable ethical system for AI governance in higher education. Even though this is presently limited to one institution, future phases will generalize across universities throughout North America before finally being applied to international contexts. Such a trajectory will allow for cross-cultural comparisons and iterative refinement and help develop inclusive and ethically grounded AI governance models globally.

References

1. Holmes, W., Porayska-Pomsta, K., Holstein, K., et al.: Ethics of AI in education: towards a community-wide framework. Int. J. Artif. Intell. Educ. **32**(3), 504–526 (2022). https://doi.org/10.1007/s40593-021-00239-1
2. Zawacki-Richter, O., Marín, V.I., Bond, M., Gouverneur, F.: Systematic review of research on artificial intelligence applications in higher education – where are the educators? Int. J. Educ. Technol. High. Educ. **16** (2019). https://doi.org/10.1186/s41239-019-0171-0

3. Firat, M.: What ChatGPT means for universities: perceptions of scholars and students. J. Appl. Learn. Teach. **6**(1), 57–63 (2023). https://doi.org/10.37074/jalt.2023.6.1.22
4. Bubeck, S., Chandrasekaran, V., Eldan, R. et al.: Sparks of artificial general intelligence: early experiments with GPT-4. arXiv preprint arXiv:2303.12712 (2023)
5. Dimitriadou, E., Lanitis, A.: A critical evaluation, challenges, and future perspectives of using artificial intelligence and emerging technologies in smart classrooms. Smart Learn. Environ. **10**(1) (2023). https://doi.org/10.1186/s40561-023-00231-3
6. Kasneci, E., Sessler, K., Küchemann, S., et al.: ChatGPT for good? on opportunities and challenges of large language models for education. Learn. Individ. Differ. **103**, 102274 (2023). https://doi.org/10.1016/j.lindif.2023.102274
7. Eaton, S.E.: The academic integrity technological arms race and its impact on learning, teaching, and assessment. Can. J. Learn. Technol. **48**(2) (2022). https://doi.org/10.21432/cjlt28388
8. Martin, K.E.: Ethical issues in the big data industry. MIS Q. Exec. **14**(2) (2015). https://aisel.aisnet.org/misqe/vol14/iss2/4
9. Regan, P.M., Jesse, J.: Ethical challenges of edtech, big data and personalized learning: Twenty-first century student sorting and tracking. Ethics Inf. Technol. **21**, 167–179 (2019)
10. Haque, M.U., Dharmadasa, I., Sworna, Z.T., Rajapakse, R.N., Ahmad, H.: Exploring sentiments of ChatGPT early adopters using Twitter data. arXiv preprint arXiv:2212.05856 (2022)
11. Kooli, C.: Chatbots in education and research: a critical examination of ethical implications and solutions. Sustainability **15**(7) (2023). https://doi.org/10.3390/su15075614
12. Miao, F., Holmes, W., Huang, R., Zhang, H. (eds.): AI and Education: Guidance for Policy-Makers. UNESCO, Paris (2021)
13. Freeman, R.E.: Strategic Management: A Stakeholder Approach. Pitman, Boston (1984)
14. Mitchell, R.K., Agle, B.R., Wood, D.J.: Toward a theory of stakeholder identification and salience: defining the principle of who and what really counts. Acad. Manag. Rev. **22**(4) (1997). https://www.jstor.org/stable/259247
15. Donaldson, T., Preston, L.E.: The stakeholder theory of the corporation: concepts, evidence, and implications. Acad. Manag. Rev. **20**(1), 65–91 (1995). https://doi.org/10.5465/amr.1995.9503271992
16. Liao, L., Luo, L., Tang, Q.: Gender diversity, board independence, environmental committee and greenhouse gas disclosure. Br. Account. Rev. **47**(4), 409–424 (2015). https://doi.org/10.1016/j.bar.2014.01.002
17. Jobin, A., Ienca, M., Vayena, E.: The global landscape of AI ethics guidelines. Nat. Mach. Intell. **1**(9), 389–399 (2019). https://doi.org/10.1038/s42256-019-0088-2
18. Holmes, W., Iniesto, F., Anastopoulou, S., Boticario, J.G.: Stakeholder perspectives on the ethics of AI in distance-based higher education. Int. Rev. Res. Open Distrib. Learn. **24**(2), 96–117 (2023). https://doi.org/10.19173/irrodl.v24i2.6089
19. Nguyen, A., Ngo, H.N., Hong, Y., Dang, B., Nguyen, B.P.T.: Ethical principles for artificial intelligence in education. Educ. Inf. Technol. **28**(4), 4221–4241 (2023). https://doi.org/10.1007/s10639-022-11316-w
20. Molenaar, I., Sleegers, P.: Multi-stakeholder collaboration and co-creation: towards responsible application of AI in education. OECD Digital Education Outlook 2023 (2023)
21. Almasri, F.: Exploring the impact of artificial intelligence in teaching and learning of science: a systematic review of empirical research. Res. Sci. Educ. **54**(5), 977–997 (2024). https://doi.org/10.1007/s11165-024-10176-3
22. Tanveer, M., Hassan, S., Bhaumik, A.: Academic policy regarding sustainability and artificial intelligence (AI). Sustainability **12**(22) (2020). https://doi.org/10.3390/su12229435

23. Borchers, C., Liu, X., Lee, H.H., Zhang, J.: Ethical AIED and AIED ethics: toward synergy between AIED research and ethical frameworks. Commun. Comput. Inf. Sci. **2150**, 18–31 (2024). https://doi.org/10.1007/978-3-031-64315-6_2
24. Vassileva, J., Cristea, A.I.: Do we need AIED if we have powerful generative AI? In: Olney, A.M., Chounta, I.A., Liu, Z., Santos, O.C., Bittencourt, I.I. (eds.) Artificial Intelligence in Education. 25th International Conference, pp. xxi–xxiv. Springer (2024)

Augmenting LLM Generated Feedback with Data Mining

Eamon Worden(✉) and Neil Heffernan

Worcester Polytechnic Institute, Worcester, MA 01609, USA
{elworden,nth}@wpi.edu

Abstract. Automated feedback systems are important in mathematics education for providing timely and scalable support to students. While pretrained Large Language Models (LLMs) such as GPT have shown promise in generating feedback, fine-tuning LLMs to improve their performance is costly and resource intensive. In this work, we explore cost-efficient alternatives, focusing on data mining to enhance LLMs for feedback generation for open-ended math answers. We evaluate the effectiveness and practicality of data mining for few-shot prompting in generating both descriptive feedback and numerical scores for middle school math open responses. Our results show that data mining significantly improves the result. Further, we explore why we believe the feedback is better by delving into teacher written reasonings for why they liked or disliked certain feedbacks.

Keywords: Automated feedback · Large Language Models · Prompt Engineering · Mathematics Education · Open-ended Problems

1 Introduction

Automated feedback systems are crucial in online learning, particularly for mathematics education, where open-ended problems pose challenges due to their reliance on subjective reasoning [3]. Effective feedback enhances learning, but poorly designed feedback can cause confusion [2,5]. Recent advancements in LLMs offer promising solutions for improving automated feedback, with techniques like retrieval-augmented generation (RAG) and few-shot prompting providing cost-efficient alternatives to fine-tuning [1,6]. This study explores how data mining can be used to enhance feedback for middle school math responses, comparing few-shot and zero-shot prompting and evaluating the effectiveness of each prompting technique. The main aims of this study are:

1. How can data mining be utilized to generate effective automated feedback for open-ended math problems?
2. How do few-shot and zero-shot, compare in aligning LLM feedback with teacher-authored feedback?
3. How effective and practical are these prompting methods from the perspective of teachers?

A. I. Cristea et al. (Eds.): AIED 2025, CCIS 2590, pp. 340–345, 2025.
https://doi.org/10.1007/978-3-031-99261-2_34

2 Dataset

For our study, we selected open-response problems Illustrative Mathematics Curriculum[1], a widely used middle school math curriculum in the United States, as they were implemented in ASSISTments. We focused exclusively on middle school problems (grades 6–8). We included only problems that had received at least 100 unique student responses and had been both scored and given written feedback by a teacher. A total of 67 problems met these criteria, each containing between 100 and 600 student responses. From this set, we randomly selected 50 problems and 100 student responses per problem, resulting in a dataset of 5,000 responses. Each selected problem was either a stand-alone open-response question or a follow-up prompt requiring students to explain their reasoning based on a prior answer. For our prompt, we split our dataset into an 80-20 retrieval-evaluation split.

3 Methodology

Our study three prompting methods for generating automated feedback for student responses to open-ended math problems using the state-of-the-art GPT-4o model. Specifically, we utilize an intelligent few-shot prompt to enhance feedback generated by the GPT-4o model and then evaluate it against traditional zero-shot and random few-shot prompting. This section details our proposed intelligent few-shot-based approach, along with the baseline prompting methods used for comparison.

Zero-Shot Prompting. is the simplest approach, where a LLM is asked to generate feedback without any examples or context. The model relies entirely on its pre-trained knowledge and the provided student response to generate feedback. While this method requires no additional setup, we consider this as a baseline for comparison with the proposed method. In our zero-shot prompt we assigned GPT-4o a persona of a math teacher, and asked it to provide feedback and a score to a students answer.

Few-Shot Prompting. Few-shot prompting improves upon zero-shot prompting by providing the model with a small set of example responses and their corresponding teacher-written feedback. In the random few-shot approach, we randomly select three example responses and feedback from our retrieval set to include in the prompt. While this method improves feedback quality compared to zero-shot prompting, the random selection of examples may lead to suboptimal performance due to mismatches between the provided examples and the student response being evaluated. For our random few-shot prompt we utilize both a persona as well as three shots in our prompt.

Intelligent Few-Shot Prompting. To further enhance the quality of the generated feedback, we utilize an intelligent few-shot prompting method that dynamically

[1] https://illustrativemathematics.org/.

selects the three most similar student answers from our retrieval set for each student response, along with the associated feedback and scores. This resembles RAG, which selects text from a corpus when prompting, however differs in that we select items from our retrieval set rather than a textbook. We used Meta's Faiss library [4], an efficient tool for similarity search for relevant responses and feedback examples. For each math problem in the dataset, we first vectorized each student's responses from the training set using the state-of-the-art SFR-embedding-2 model [7], which generated semantically rich embeddings for student responses. Once the training set responses were embedded, we calculated the semantic similarity between the embedded test set response and each response in the training set for the same math problem using Euclidean distance. We then passed the top three most similar responses along with the feedback and score assigned by a teacher to our prompt, and continued to use a persona.

4 Results

We conducted evaluations with seven teachers who have 3 to 25 years of experience teaching middle school mathematics to assess the quality and effectiveness of the generated feedback. Each teacher received six unique problems, with five unique student responses per problem for a total of 42 problems and 210 unique responses. They were tasked with ranking the feedback from different methods (teacher-authored, zero-shot, random few-shot, and intelligent few-shot) and providing feedback on the quality and effectiveness of the responses.

4.1 Ranking

We first asked teachers to rank the feedback from their favorite to least favorite, and asked whether they felt any were similar. For evaluation, we removed rows where two feedbacks were considered similar.

To determine whether there were significant differences in the rankings of feedback methods, analyzed the mean reciprocal rank of each feedback type and performed a pairwise Wilcoxon signed-rank test to determine whether any feedbacks were different from one another. Table 1 shows the mean reciprocal rank for each model and 2 shows the pairwise test to determine which were different.

Our findings suggest that both few-shot prompts were ranked higher than the zero-shot prompt or teacher-authored feedback in ASSISTments, and while intelligent few-shot was ranked higher than average, we do not have evidence that its mean reciprocal rank was different than random few-shot. Similarly, zero-shot was higher than teacher on average but we do not have evidence it was significantly better than teacher authored feedback.

Table 1. Mean Reciprocal Rank for each prompt.

Prompt	MRR
Teacher	0.421
Zero	0.449
Random	0.588
Intelligent	0.625

Table 2. Pairwise Wilcoxon signed-rank test results between feedback types.

Prompt 1	Prompt 2	p-value
Teacher	Zero	0.199
Teacher	Random	¡ 0.001
Teacher	Intelligent	¡ 0.001
Zero	Random	¡ 0.001
Zero	Intelligent	¡0.001
Random	Intelligent	0.278

Acceptability, Learning Impact, and Performance. Following this, we asked teachers whether they thought each feedback was'acceptable','would cause students to learn', and'would result in students getting a higher score'. Each of these was binary. Table 3 shows the portion of feedback deemed acceptable, would cause learning, and would improve a students score.

Table 3. Proportion of each feedback that was deemed acceptable, to cause learning and improved performance.

Feedback	Acceptable	Learning	Performance
Teacher	0.39	0.27	0.26
Zero	0.50	0.52	0.65
Random	0.71	0.70	0.67
Intelligent	0.68	0.61	0.61

4.2 Acceptability, Learning Impact, and Performance Discussion

Our results show a surprising trend. We found that real teacher-written feedback in ASSISTments was rated significantly worse than any LLM feedback across all three metrics. Further, neither few-shot method performed much better than the other. The zero-shot prompt, notably, was rated as acceptable less often than both few-shot prompts, but would result in a higher score equally as often as either few-shot prompt. In order to explain this, we dive into the last questions we asked teachers of'Why did you rank your top choice as the best?' and'Why did you rank you bottom choice as the worst?'

4.3 Reasoning

We reviewed the reasons for why teachers rated a feedback as the best or worst. We developed a codebook based on our finds. The categories we found were excellent, actionable, clarifying, concise, personalized, encouraging, and default. Table 4 shows the number of times each feedback was ranked as the best and how often teachers reasons related to one of the above categories.

Table 4. Distribution of reasons for positive feedback across feedback types.

Reason	Teacher	Zero-shot	Random	Intelligent
Excellent	86%	91%	84%	91%
Actionable	33%	59%	43%	59%
Clarifying	20%	51%	29%	37%
Concise	60%	9%	28%	36%
Personalized	23%	19%	15%	22%
Encouraging	17%	19%	20%	9%
Default	13%	21%	23%	11%
Total	30	37	65	78

We developed a similar codebook for the worst feedbacks. The identified categories were incorrect, no help, too long, irrelevant, confusing, and rude. Table 5 shows the results.

Table 5. Distribution of reasons for negative feedback across feedback types.

Reason	Teacher	Zero-shot	Random	Intelligent
Incorrect	37%	35%	52%	87%
No help	72%	21%	24%	44%
Too long	1%	55%	8%	0%
Irrelevant	14%	16%	36%	25%
Confusing	9%	15%	8%	13%
Rude	14%	10%	8%	0%
Total	101	68	25	16

4.4 Reasoning Discussion

When teacher feedback was rated highly, it was almost always because it did not say a lot. When teacher feedback was rated poorly, it was almost always because it did not say a lot. As many teachers have quite a few students and a lot of work to do they do not have time to write lengthy feedback the way LLMs do. As a result, their messages are much shorter and often are just "Great work!". Teachers felt those were effective on correct student answers, but not so effective for learning or higher score, especially for incorrect student answers. The zero-shot prompt suffered the opposite problem–it was too long. There were occasions where it included every step needed to solve a problem. As a result, teachers felt student who received this feedback would often get a higher score and any misconceptions clarified, but also it may give away too much information quite frequently.

The few-shot prompts had tradeoffs. Random few-shot suffered a risk of provided irrelevant information, possibly due to the randomness of the shots provided. However, it was often seen as effective, and particularly for causing student learning as it could tie in feedback from a variety of mistakes since it had more diverse shots. Intelligent few-shot feedback was rarely rated the worst, but when it was often it was due to some math error. However, it typically provided high-quality, actionable feedback to students. As many OLPs have been around for a large number of years and have a significant amount of data stored, we believe few-shot is viable for many platforms aiming to automate effective feedback.

References

1. Baral, S., et al.: Automated feedback in math education: a comparative analysis of llms for open-ended responses. arXiv preprint arXiv:2411.08910 (2024)
2. Brown, G.T., Peterson, E.R., Yao, E.S.: Student conceptions of feedback: impact on self-regulation, self-efficacy, and academic achievement. Br. J. Educ. Psychol. **86**(4), 606–629 (2016)
3. Cavalcanti, A.P., et al.: Automatic feedback in online learning environments: a systematic literature review. Comput. Educ. Artif. Intell. **2**, 100027 (2021)
4. Douze, M., et al.: The faiss library. arXiv preprint arXiv:2401.08281 (2024)
5. Hattie, J.: Visible learning: a synthesis of over 800 meta-analyses relating to achievement. Routledge (2008)
6. Nguyen, H.A., Stec, H., Hou, X., Di, S., McLaren, B.M.: Evaluating chatgpt's decimal skills and feedback generation in a digital learning game. In: Viberg, O., Jivet, I., Muñoz-Merino, P., Perifanou, M., Papathoma, T. (eds.) Responsive and Sustainable Educational Futures, pp. 278–293. Springer Nature Switzerland, Cham (2023)
7. Rui, M., Ye, L., S.R.J.C.X.Y.Z.S.Y.: Sfr-embedding-2: advanced text embedding with multi-stage training (2024). https://huggingface.co/Salesforce/SFR-Embedding-2_R

Towards Ordinal Few-Shot Learning for Automated Essay Grading

Vihanga Ashinsana Wijayasekara(✉), Kyle Martin, and Nirmalie Wiratunga

Robert Gordon University, Aberdeen, UK
{v.wijayasekara,k.martin3,n.wiratunga}@rgu.ac.uk

Abstract. Ordinal essay grading, an essential task in educational domain and natural language processing (NLP), involves categorising essays based on quality, such as grading scale levels. This is crucial in automated assessment systems that evaluate student writing and provide feedback on aspects like coherence, argumentation, and language proficiency. However, challenges arise from limited data resources, such as when a new assessment is introduced and no data is available to train algorithms, as well as the complexity of essay structures in real-world grading scenarios. This research explores the use of few-shot learning, a technique that learns from a limited number of labeled examples, to address these challenges in ordinal essay grading. By leveraging few-shot learning's ability to identify class representations from minimal examples, we aim to mitigate data scarcity in essay grading. With the rise of Large Language Models (LLMs), we seek to improve few-shot prompting performance by introducing novel strategies for example selection, enhancing class representation in the demonstrations provided to prompts. Finally, we aim to apply prototypical methods to agents architectures for agent selection on the basis of similarity weighted by ordinal class knowledge.

Keywords: Essay Grading · Few-Shot Learning · Few-Shot Prompting

1 Introduction

The automation of grading can reduce administrative burdens on educators and improve the timeliness and consistency of feedback to students, enhancing the overall educational experience. Automated Essay Scoring (AES) leverages Natural Language Processing (NLP) to assign grades (e.g., A to F) based on essay quality, making it a promising tool for this purpose.

A significant challenge in AES is the data requirements for training robust models. Traditional models depend on large annotated datasets, which are often unavailable in new courses or modules without prior grading data. Few-shot learning can address this issue by enabling models to generalise from limited examples, classifying unseen data with minimal training. However, recent

A. I. Cristea et al. (Eds.): AIED 2025, CCIS 2590, pp. 346–351, 2025.
https://doi.org/10.1007/978-3-031-99261-2_35

advancements in Large Language Models (LLMs) face limitations, such as context window constraints and difficulties handling noisy data, which can hinder their effectiveness in few-shot scenarios.

Several approaches have been explored for AES [3], including regression and ranking methods [22], LSTM [7] and BERT models [14], and transformer-based models [11] for improved essay evaluation. Additionally, generative models like GPT-4 show potential in AES, though concerns about privacy and bias remain [9,13].

While some research has explored ordinal classification with metric-based approaches in few-shot learning, further investigation is needed for their effective application in AES. Few studies have used metric learning for Automated Essay Scoring (AES) [10]. This research aims to fill this gap by exploring few-shot learning to enhance AES in data-scarce scenarios. The following research questions address these challenges.

- RQ01 How can few-shot learning techniques, including metric learning algorithms and class distribution-aware methods, be utilised to improve the accuracy and generalisation of ordinal essay grading models?
- RQ02 How can class distribution knowledge improve the selection of representative examples in Few-Shot Prompting for more effective essay grading?
- RQ03 How can a multi-agent-based approach enhance Few-Shot Prompting in automated essay grading?

Few-Shot Learning (FSL) addresses the challenge of training models with minimal labeled data per class. Key approaches in FSL include Meta-Learning, where models learn from multiple tasks; Data Augmentation, which expands limited datasets through techniques like back translation and noise injection; and Transfer Learning, where knowledge from data-rich tasks is applied to improve performance on tasks with limited data. Metric Learning helps classify data with few examples by learning a distance metric to compare instances.

In Automated Essay Scoring (AES), Deep Metric Learning (DML) has shown promise. For example, Prototypical Networks (PN) have generalised well in short answer grading [10]. Unlike previous methods using abundant data from multiple prompts, our work focuses on few-shot settings with limited data from a specific target prompt. We apply DML to the more challenging ASAP-AES [15] dataset, which has greater subjectivity and variability in responses.

Effective sample selection is crucial for DML performance. Traditional methods use random or similarity-based sampling, and techniques like triplet loss [1] have improved outcomes in tasks like face recognition. Additionally, [2] showed that selecting samples from complex regions of the embedding space reduces data requirements. In our research, we adapt these methods for ordinal classification, proposing prototype weighting to enhance AES performance.

2 Progress

In our first research objective, we explored how few-shot learning techniques, including metric learning algorithms, could enhance the performance of ordinal

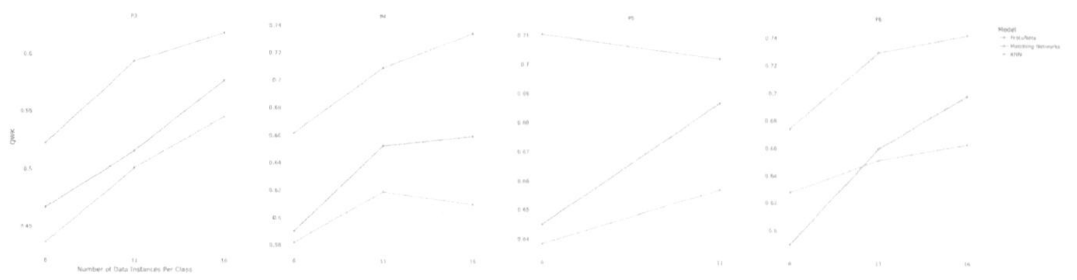

Fig. 1. Quadratic Weighted Kappa values over the training set size. The x-axis shows the number of data instances used per class

essay grading models. We compared the performance of k-Nearest Neighbour (kNN), Matching Networks (MN) [12] and Prototypical Networks (PN) [4] for AES. We used the ASAP-AES dataset [15] for our experiments, which contains student essays graded on an ordinal scale. Our focus was on prompts 3, 4, 5, and 6, which correspond to 4-class and 5-class ordinal classification tasks. We performed classification experiments with 6-shot, 11-shot, and 16-shot settings, excluding prompts with significant class imbalance.

The results of this exploratory evaluation are shown in Fig. 1. kNN applied to BERT embeddings, without class information, underperformed compared to deep metric learning algorithms, highlighting that embeddings alone are insufficient. MN, which incorporates class information during training, performed better than kNN. PN, which uses class prototypes for distribution estimation, outperformed both, emphasising the value of class distribution knowledge.

This led us to refine Prototypical Networks with novel weighting mechanisms for essay grading classification. In Prototypical Networks, prototypes are the mean of support set embeddings, but outliers can skew this. To improve robustness with limited samples, we explored two weighting methods: Robust Re-weighting Based Prototype (RRBP), which reduces the influence of outliers, and Neighborhood Complexity Prototype (NCP), which considers neighborhood complexity. RRBP assigns lower weights to points farther from others (Fig. 2(a)), with the formula from [5] adapted for this task, while NCP gives higher weights to points with more same-class neighbors, penalising those from different classes (Fig. 2(b)).

To capture both outlier influence and neighborhood complexity, we propose two aggregated weighting mechanisms: Balanced Mean Prototype (BMP) and Adaptive Neighbor-Based Prototype (ANP). BMP calculates the prototype as the mean representation of both RRBP and NCP. ANP selects the prototype based on the number of KNN neighbors associated with each class, choosing the prototype with the greater count of same-class neighbors.

The results are shown in Table 1. We used Quadratic Weighted Kappa (QWK), a widely used metric to evaluate the effectiveness of automated essay scoring systems [3,8–10]. As baselines, we used SKIPFLOW (SF) [7] and NPCR [6], the SOTA models in the ASAP AES datasets. However, existing models struggle in few-shot learning settings, revealing a significant gap. Prototypical

(a) Robust Re-weighting Based Prototype (RRBP)

(b) Neighborhood Complexity Prototype (NCP)

Fig. 2. Prototype selection for (a) RRBP and (b) NCP

Table 1. Results

Model	P3			P4			P5		P6		
	6	11	16	6	11	16	6	11	6	11	16
NPCR	0.1065	0.1796	0.1229	0.1359	0.2168	0.1078	0.2108	0.1758	0.0746	0.0690	0.0880
SF	0.3113	0.3692	0.4093	0.5388	0.5666	0.5569	0.4507	0.5765	0.4612	0.4940	0.5075
PN	0.5224	0.5934	0.6184	0.6618	0.7091	**0.7346**	0.7105	0.7022	0.6742	0.7298	0.7423
PN-RRBP	0.5574	0.6052	**0.6232**	0.6785	**0.7144**	0.7273	0.7203	0.7446	0.6742	0.7341	0.7428
PN-NCP	0.5129	0.5613	0.5855	0.6631	0.7007	0.7147	0.7156	0.7324	0.6736	0.7230	0.7418
PN-BMP	0.5514	0.5993	0.6165	**0.6829**	0.7114	0.7242	0.7170	0.7423	**0.6945**	**0.7392**	**0.7501**
PN-ANP	**0.5633**	**0.6059**	0.6194	0.6822	0.7044	0.7246	**0.7245**	**0.7489**	0.6809	0.7309	0.7433

networks, which use a metric learning approach, differentiate classes by learning class prototypes rather than trying to understand the categories, which sets them apart from traditional models.

3 Future Work

Our second research question (RQ02) examines how class distribution knowledge can be leveraged to identify the most suitable essay samples for few-shot prompting approaches. The final research question (RQ03) focuses on strategies for selecting the most relevant unlabeled student essays for annotation, thereby facilitating their use in the approaches outlined in RQ01 and RQ02. Currently, our research objective is to enhance Few-Shot Prompting in essay grading. We are investigating methods to refine the prompt and select the most informative examples from the few-shot dataset to improve class distribution, thereby boosting model performance.

There is a growing trend in automated essay scoring using few-shot prompting, largely driven by the capabilities of LLMs. The authors of [16] evaluated Llama and ChatGPT using various prompts, including rubric descriptions and 1-shot examples. They found that LLMs are highly sensitive to prompt and task variations, with Llama being more affected by prompt changes and ChatGPT showing more robust performance. However, both models still fall short of state-of-the-art models and are not yet reliable for scoring student essays. In a related study, [17] investigated the impact of example selection on the performance of automated essay scoring using few-shot prompting with GPT-based models. Their results revealed biases, such as majority label bias and recency bias, in the outputs generated by GPT. In domains involving ordinal text classification, the number of examples and the order of examples in few-shot prompts also influence performance [19,20]. Some research has found that not all examples used in few-shot prompting are equally beneficial [18]. In response, we are exploring solutions to select the most representative examples for each class to address the identified biases. Recent work on few-shot prompting has leveraged NLI-based approaches [21]. Currently, we are investigating NLI techniques to identify effective class representations and exploring other methods for selecting suitable class representatives from the ASAP-AES dataset.

In our latest research question, we explore how a multi-agent approach can enhance essay grading in a few-shot learning setting. This approach allows tasks to be broken down into manageable components, with each agent specialising in specific aspects of the grading process. For example, agents can be used to predict different grading categories or aspects of the essay, improving overall accuracy and efficiency.

References

1. Schroff, F., Kalenichenko, D., Philbin, J.: Facenet: a unified embedding for face recognition and clustering. In: Proceedings of the IEEE Conference on Computer Vision and Pattern Recognition, pp. 815–823 (2015)
2. Martin, K., Wiratunga, N., Sani, S.: Locality sensitive batch selection for triplet networks. In: 2020 International Joint Conference on Neural Networks (IJCNN), pp. 1–8. IEEE (2020)
3. Ke, Z., Ng, V.: Automated essay scoring: a survey of the state of the art. In: Proceedings of IJCAI-19, vol. 19, pp. 6300–6308 (2019)
4. Snell, J., Swersky, K., Zemel, R.: Prototypical networks for few-shot learning. Adv. Neural Inf. Process. Syst. **30** (2017)
5. Zhu, J., Yi, X., Guan, N., Cheng, H.: Robust re-weighting prototypical networks for few-shot classification. In: Proceedings of the 6th International Conference on Robotics and Artificial Intelligence, pp. 140–146 (2020)
6. Xie, J., Cai, K., Kong, L., Zhou, J., Qu, W.: Automated Essay Scoring via Pairwise Contrastive Regression. In: Proceedings of the 29th International Conference on Computational Linguistics, pp. 2724–2733 (2022)
7. Tay, Y., Phan, M., Tuan, L.A., Hui, S.C.: Skipflow: incorporating neural coherence features for end-to-end automatic text scoring. In: Proceedings of the AAAI Conference on Artificial Intelligence, vol. 32(1) (2018)
8. Singh, S., Pupneja, A., Mital, S., et al.: H-AES: Towards automated essay scoring for hindi. In: Proceedings of the AAAI Conference on Artificial Intelligence, vol. 37(13), pp. 15955–15963 (2023)

9. Chang, L.H., Ginter, F.: Automatic short answer grading for finnish with ChatGPT. In: Proceedings of the AAAI Conference on Artificial Intelligence, vol. 38, no. 21, pp. 23173–23181 (2024)
10. Zeng, Z., Li, L., Guan, Q., Gašević, D., Chen, G.: Generalizable automatic short answer scoring via prototypical neural network. In: International Conference on Artificial Intelligence in Education, pp. 438–449. Springer (2023)
11. Das, S.D., Vadi, Y., Yadav, K.: Transformer-based joint modelling for automatic essay scoring and off-topic detection. arXiv preprint arXiv:2404.08655 (2024). https://arxiv.org/abs/2404.08655
12. Vinyals, O., Blundell, C., Lillicrap, T., Wierstra, D., et al.: Matching networks for one shot learning. Adv. Neural Inf. Process. Syst. **29** (2016)
13. Cohn, C., Hutchins, N., Le, T., Biswas, G.: A chain-of-thought prompting approach with LLMs for evaluating students' formative assessment responses in science. In: Proceedings of the AAAI Conference on Artificial Intelligence, vol. 38, no. 21, pp. 23182–23190 (2024)
14. Devlin, J.: BERT: Pre-training of deep bidirectional transformers for language understanding. arXiv preprint arXiv:1810.04805 (2018)
15. Ben, H., Jaison, M., lynnvandev, Mark, S., Tom, V.A.: The hewlett foundation: automated essay scoring (2012). https://kaggle.com/competitions/asap-aes. Kaggle
16. Mansour, W., Albatarni, S., Eltanbouly, S., Elsayed, T.: Can large language models automatically score proficiency of written essays? arXiv preprint arXiv:2403.06149 (2024)
17. Lui, Y.: The impact of example selection in few-shot prompting on automated essay scoring using GPT models. In: Proceedings of the International Conference on Artificial Intelligence in Education, pp. 61–73. Springer (2024)
18. Yang, Z., Dai, D., Wang, P., Sui, Z.: Not all demonstration examples are equally beneficial: reweighting demonstration examples for in-context learning. In: Bouamor, H., Pino, J., Bali, K., (Eds.), Findings of the Association for Computational Linguistics: EMNLP 2023, pp. 13209–13221. Association for Computational Linguistics (2023). DOI: https://aclanthology.org/2023.findings-emnlp.880/
19. Lu, Y., Bartolo, M., Moore, A., Riedel, S., Stenetorp, P.: Fantastically ordered prompts and where to find them: overcoming few-shot prompt order sensitivity. arXiv preprint arXiv:2104.08786 (2021)
20. Chen, J., Chen, L., Zhu, C., Zhou, T.: How many demonstrations do you need for in-context learning? arXiv preprint arXiv:2303.08119 (2023)
21. Kasa, S.R., et al.: Exploring ordinality in text classification: a comparative study of explicit and implicit techniques. arXiv preprint arXiv:2405.11775 (2024)
22. Yang, R., Cao, J., Wen, Z., Wu, Y., He, X., et al.: Enhancing automated essay scoring performance via fine-tuning pre-trained language models with combination of regression and ranking. In: Proceedings of the Association for Computational Linguistics (ACL) (2020)

Rewriting the Rules: LLMs Vs. Traditional ML in University Admissions

Ivan Chepikov(✉) and Ilia Karpov

National Research University Higher School of Economics, Moscow, Russia
ivanchepikov444@gmail.com , karpovilia@gmail.com

Abstract. Modern LLM models such as BERT, ChatGPT, DeepSeek have shown great potential in solving various tasks, including text classification, text generation, analysis and summary of documents. In this paper, we show that these models close to classical ML approaches based on decision trees not only in text processing, but also in processing classical tabular data (quantitative and categorical) using the example of solving the faculty forecasting problem for university applicants. Based on the dataset we prepared from 1597 applicants, the LLM models showed fairly close results to classical machine learning methods.

Keywords: Large Language Models · Natural Language Processing · Machine Learning · Synthetic Text Generation

1 Introduction

Accurate admission prediction is critical for educational institutions, as it enables them to make informed decisions regarding resource planning, faculty allocation, and curriculum design [5]. Furthermore, effective admission prediction can contribute to improving retention rates, student satisfaction, and academic performance, as institutions can proactively identify and support students who are likely to thrive within specific faculties or programs [4]. Leveraging advanced predictive modeling techniques, such as those explored in this study, is therefore essential for fostering a responsive and data-driven educational environment.

To address this task, we first conducted a survey involving 1,597 applicants from five different university faculties. Initially, we trained several classical decision-tree-based machine learning models, which allowed us to identify the most significant features of our dataset. Based on these selected features, we subsequently generated textual autobiographies for each applicant based on their questionnaire responses. With the resulting textual data, we then evaluated several encoder-based transformer models for the text classification task. Finally, we examined the zero-shot classification capabilities of state-of-the-art large language models and conducted a comparative analysis to benchmark the performance of these advanced LLMs against the previously evaluated approaches (Fig. 1).

The main **contributions** of our work are as follows:

A. I. Cristea et al. (Eds.): AIED 2025, CCIS 2590, pp. 352–358, 2025.
https://doi.org/10.1007/978-3-031-99261-2_36

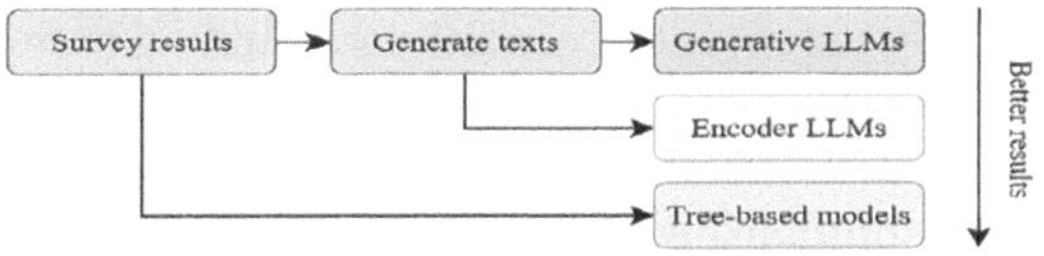

Fig. 1. General Experiment Design

- **Combined Tabular and Textual Admissions Dataset**[1]: We collected and manually validated a dataset of 1,597 university applicants across 5 faculties, comprising responses to 27 detailed questions. We then transformed these responses into concise textual autobiographies highlighting key applicant characteristics.
- **Enhanced Faculty Prediction Method based on Text Generation**[2]: We compared traditional tree-based methods applied directly to tabular data with LLM-based classification approaches using generated autobiographies. To the best of our knowledge, this is the first demonstration of enhancing classical ML predictions with LLMs, trained on text from tabular data in admission prediction task..

2 Related Work

Most researchers who engage in the admission of students and their research using classical ML methods use tabular data or start their research with text and compare only the methods suitable for working with text [4]. Such approaches are implemented in articles on predicting job, course, and academic performance using LLM. In the article [1], the authors apply LLMs such as BERT and GPT to analyze students stories in order to assess their career readiness based on text classification but they already had a ready-made dataset with online natural language responses to 18 open ended questions that were based on themes identified in the dynamic model of career exploration and planning. In our case, we have tabular data of the survey results.

In our research, we do not apply LLM to table data, but generate text data, which is the typical input for LLM. This approach has proven itself well in various Retrieval tasks, for example in [7], where the authors propose a method that combines information generation and extraction, allowing efficient and accurate responses to queries based on various types of data, including tabular and text. In particular, the use of pre-trained models that can process various types of input data and provide high-quality extraction and generation of responses is being investigated.

In paper [3] discusses the important topic of synthetic data generation, with an emphasis on using LLM to create data that can complement or replace real

[1] https://zenodo.org/records/15353681.
[2] https://github.com/darttusin/LLMvsML.

data, which is especially important in conditions of limited access to high-quality datasets. The authors emphasize that despite the advantages of LLM in synthetic generation, unresolved issues remain, such as increasing the reliability and diversity of data, as well as minimizing errors associated with model hallucinations.

In the article [2], the authors address the task of table-to-text generation and propose to evaluate the output using LLM-based self-evaluation. In contrast, our work operates in a real-world setting where a natural target is available.

3 Dataset

For this study, we used a dataset of 1,597 students compiled on the basis of surveys filled out by university students as part of a voluntary survey. All the data was collected in the same year from students enrolled at the university. The survey was completely voluntary and did not include any personal information such as full name, email, or phone number. All respondents agreed with the use of their anonymous data as part of the study.

In total, 10,243 people enrolled in the reviewed faculties. Thus, our dataset covers 15% of all students enrolled in these faculties, which allows us to obtain a fairly representative sample for further analysis and model building.

Data Curation. Before the analysis, all questionnaires underwent a multi-stage manual check and cleaning: (i) irrelevant data and outliers were excluded (e.g., questionnaires containing invalid responses, scores exceeding reasonable limits); (ii) spelling variations, typos, abbreviations, and synonyms (for cities, ages, etc.) were standardized into a single format; and (iii) anomalies such as incorrect numerical values and inconsistent responses were corrected or removed to ensure data consistency. 1358 valid records were obtained from the original set (85% of the original volume).

Dataset Description. Our dataset includes applicants from five faculties. The largest faculty is Information Technology (IT), with 651 students in dataset. The second largest is Information Security, with 289 students (CIS). Networks and Communication Systems (NCS), as well as Radio and Television(RT), have 191 and 154 students respectively. The smallest faculty, Digital Economy and Mass Communications (DEMS), has 73 applicants from our dataset. The Fig. 3 shows the TSNE visualization of our dataset. It can be seen that the data is not easily separable, even with the application of standardization. The classes partially overlap, and their separation is not obvious. Without standardization, the situation worsens and the data becomes even more chaotic.

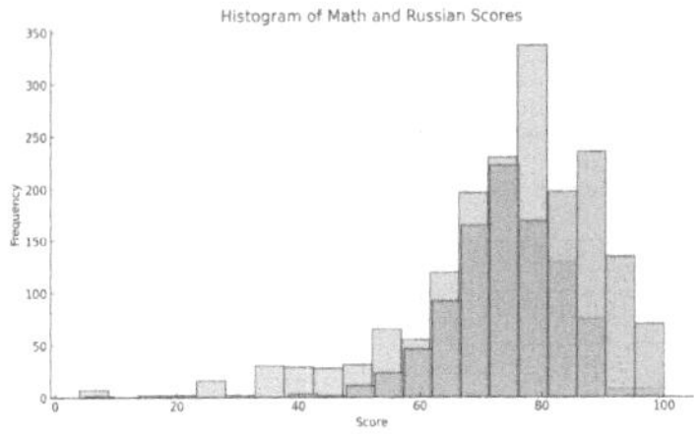

Fig. 2. Math and Russian exam scores distribution

Fig. 3. t-SNE for students of five faculties

All students of our dataset provide Unified State Exam (USE) scores. USE is a state exam in the that has been held simultaneously in all regions of the Russian Federation since 2015. The exam is conducted in 5 subjects, Mathematics and Russian are compulsory, and the rest are optional. Figure 2 illustrates the distribution of scores across the prepared dataset. Scores on this exam are assessed for admission to specialized universities.

Autobiography Generation. Several GPT models (including 4o-mini, 4, 4.5, o3-mini-high) with different generation strategies was used to generate text based on student questionnaires. The texts were created in various configurations: some included 2 or 4 facts about the student, while others added or omitted names. This made it possible to systematically check for potential biases in how names or the number of facts affect models performance.

4 Evaluation Method

To compare LLM with tree-based methods, we performed the classification of both datasets using the appropriate methods. The data was divided into train (70%), validation (15%) and test (15%) samples. The assessment was performed using Kappa and F1-score metrics, as they work well with unbalanced classes. All experiments were run on the node with 377 GB RAM, 40 cores, A6000 GPU.

Tree-Based Models. All categorical features were encoded with the one hot encoder method. The following features were added to improve the model performance: (i) **subjects count**: number of subjects on exams, (ii) **score sum**: the sum of the scores for all exams, (iii) **avg score count**: average exam score.

Although tree-based methods are generally considered less sensitive to class imbalance, we examined Upsampling and Downsampling using SMOTE and ADASYN to address the class imbalance problem but the change in Kappa and F1 Score measures was not significant. We analyzed both models using a hyperparameter grid search. Hyper parameters for **Random Forest** were *n_estimators*: the number of trees in the forest, *max_depth*: max. depth of the tree, *min_samples_split*: min. number of samples to split a node, *min_samples_leaf*: min. number of samples per sheet. Hyperparameters for **Boosting** model were: *iterations*: number of training iterations, *l2r*: L2 regularization on tree leaves, *depth*: the depth of the tree, *learning_rate*: learning rate, *rsm*: fraction of features randomly selected at each tree level. Table 1 shows that CatBoost outperformed Random Forest, achieving the best metrics with an F1 Score of 0.548 and a Cohen's Kappa score of 0.26.

Table 1. Parameter optimization for CatBoost and Random Forest

CatBoost						Random Forest					
Iter	l2r	Depth	lr	Kappa	F1	Iter	Depth	Split	Leaf	Kappa	F1
1000	1	4	0.04	0.16	0.49	1000	4	8	4	0.05	0.4
1000	5	4	0.04	**0.26**	**0.548**	1000	6	8	4	0.11	0.45
2000	3	4	0.03	0.12	0.5	2000	8	10	4	**0.11**	**0.49**
2000	5	6	0.05	0.19	0.47	3000	2	10	6	0.09	0.45

Generative LLMs. GPT (4o, o3-mini, o3-mini-high) and DeepSeek were used for the survey. The prompt was the same for all models: **system_message** ="You are a helpful assistant, a specialist in educational strategies". **classification_prompt** = f"""Given the short review from student about himself, your task is to recommend for student one of 5 faculties: Faculty of Information technologies, Faculty of Networks and Communication Systems, Faculty of Radio and Television, Faculty of Cybernetics and Information Security, Faculty of Digital Economy and Mass Communications. Review: <review>". **GPT** was unable to predict in most cases from texts. F1-score for each model: 4o - 0.4 , o3-mini and o3-mini-high - 0.2. **DeepSeek** was able to predict 0.5 of classes.

Encoder Based Models. We used BERT models such as DistilBERT [6], and ELECTRA for the text classification task. Each model was trained on training data using the cross entropy loss loss function and the AdamW optimizer. The models went through 15 training epochs, after which they were evaluated on test data. Training results are shown in the Table 2

We conducted a series of experiments using RoBERTa, which is the best model among analyzed encoders. As shown in Table 3, training RoBERTa on GPT-generated texts containing two facts (RoBERTa 2 facts) significantly outperforms the model trained on raw facts from table data (RoBERTa raw) . However, increasing the number of facts to four (RoBERTa 4 facts) results in a substantial drop in performance. Roberta trained with additional texts, generated from ADASYN results (RoBERTa aug) also performs worse than the non-augmented variant

Table 2. Encoder model metrics

Model	Kappa	F1-score
RoBERTa	0.252	**0.524**
ALBERT	**0.27**	0.5
DistilBERT	0.22	0.5
ELECTRA	**0.27**	0.5
BERT	0.26	0.49

Table 3. Best model comparison

Type	Model	Text	Kappa	F1-score
Enc	RoBERTa	2 facts	0.252	0.524
Enc	RoBERTa	4 facts	0.115	0.483
Enc	RoBERTa	aug	0.230	0.519
Enc	RoBERTa	raw	0.146	0.495
Gen.	DeepSeek	2 facts	0.120	0.500
Tree	CatBoost		**0.260**	**0.548**

Discussion. Tree-based models have demonstrated the best metrics, encoder-based models have shown compatitive results. Small gap between original tabular data and classification, demonstrates the strong ability of encoder-based models to compete with tree-based models in the area of biographic data analysis.

DeepSeek and ChatGPT without PEFT methods show worst performance in comparison with both tree and encoded based methods.

5 Conclusion

In this paper, we propose transforming traditional tabular survey data into textual bios for predicting university faculty choice. We collected and analyzed a dataset of student questionnaire responses, applying two distinct classification approaches: tree-based methods on raw tabular data, and LLM-based classifiers on generated textual autobiographies. Both methods demonstrated high F1 score, highlighting the potential advantages of LLMs in this context.

Future research includes developing a hybrid model combining the strengths of tree-based methods and LLMs, extending our evaluation to genuine student autobiography data, applying the method to open admission datasets and experimenting with a broader selection of predictive models.

Acknowledgment. The study has been funded within the framework of the HSE University Basic Research Program.

References

1. Cui, C., Abdalla, A., Wijaya, D., Solberg, S., Bargal, S.A.: Large language models for career readiness prediction. In: Proceedings of the 25th International Conference on Artificial Intelligence in Education (AIED 2024), Recife, Brazil, July 8–12, 2024. CCIS, vol. 2590. Springer, Cham. https://link.springer.com/chapter/10.1007/978-3-031-64315-6_26
2. Iravani, S., Conrad, T.O.F.: Towards more effective table-to-text generation: assessing in-context learning and self-evaluation with open-source models (2024). https://arxiv.org/abs/2410.12878

3. Long, L., Wang, R., Xiao, R., Zhao, J., Ding, X., Chen, G., Wang, H.: On llms-driven synthetic data generation, curation, and evaluation: a survey (2024). https://arxiv.org/abs/2406.15126
4. Priyadarshini, A., Martinez-Neda, B., Gago-Masague, S.: Admission prediction in undergraduate applications: an interpretable deep learning approach. In: 2023 Fifth International Conference on Transdisciplinary AI (TransAI). pp. 135–140. IEEE (2023). https://doi.org/10.1109/transai60598.2023.00040
5. Raftopoulos, G., Davrazos, G., Kotsiantis, S.: Fair and transparent student admission prediction using machine learning models. Algorithms **17**(12) (2024https://doi.org/10.3390/a17120572, https://www.mdpi.com/1999-4893/17/12/572
6. Sanh, V., Debut, L., Chaumond, J., Wolf, T.: Distilbert, a distilled version of bert: smaller, faster, cheaper and lighter (2020). https://arxiv.org/abs/1910.01108
7. Wang, Y., et al.: Pandalm: an automatic evaluation benchmark for llm instruction tuning optimization (2024). https://arxiv.org/abs/2306.05087

A Learner-AI-Parent Collaboration Framework for Home Learning Environment

Yi Shang[1] (✉), Jingyun Wang[2], and Xiaofei Qi[1]

[1] School of Education, Durham University, Durham, UK
{yi.shang,xiaofei.qi}@durham.ac.uk
[2] Computer Science Department, Durham University, Durham, UK
jingyun.wang@durham.ac.uk

Abstract. This project explores a learner-AI-parent collaboration framework to address challenges in home learning. As education shifts increasingly to the home, parents face growing pressure to support their children academically, often feeling ill-equipped and anxious. While AI technologies offer personalized learning that may alleviate some burdens, they lack social-emotional and contextual awareness and overlook the role of parents. Grounded in social constructivism and scaffolding theory, this project uses a mixed-methods design, employing an exploratory pre-survey and an explanatory quasi-experiment to examine impact of learner-AI-parent collaboration on learner and parent engagement and parental anxiety. By evaluating these dynamics, the project investigates how AI can complement parent support while mitigating stress. It bridges research gaps in home learning AI applications and expands the work of AI-supported collaborative learning.

Keywords: AI-supported collaboration · parent engagement · home learning

1 Introduction

Since Piaget's foundational work on constructivism, it is widely accepted that learning is not merely the transmission of knowledge, but rather a complex process involving cognitive attributes of the mind and their interactions with the environment. This view is well captured in the what-where-who-when multidimensional framework [1], which identifies variables that shape learning and emphasizes their capacity to evolve. In light of recent global events, particularly the COVID-19 pandemic and the advancement of artificial intelligence (AI) technologies such as ChatGPT, these dimensions have undergone significant changes, calling for review of the how, when, where and with whom learning happens.

Each dimension of the framework has been affected. Firstly, the pandemic relocated learning from the classroom to the home. Unlike the structured, time-bound, and rule-governed classroom environment, home learning tends to be more flexible and susceptible to distractions of background noise [2]. Learners are required to be more self-regulated and maintain motivation in the less controlled settings as they have to

A. I. Cristea et al. (Eds.): AIED 2025, CCIS 2590, pp. 359–364, 2025.
https://doi.org/10.1007/978-3-031-99261-2_37

navigate competing demands from home environment [2]. Notably, the pandemic redefined the "who," with parents and even digital tools stepping in to supplement the role of teachers in supporting learners. This effect has been amplified by the advancement of AI technologies, as it introduces a new kind of partner in learning. Additionally, adaptive learning technologies offer learners personalized support that extends beyond the constraints of time and location [3] further shifting the "when" and "where" of learning.

These changes also introduce new challenges. Learners are now expected to be more self-regulated and adept at using AI tools. At the same time, parents are required to support or supervise the home learning and AI-enabled interactions [4]. However, studies have shown that parents often feel ill-equipped for these tasks [5, 6]. This pressure has caused a growing sense of educational anxiety among parents, who fear failing to meet their children's academic needs [6]. The increasing integration of AI tools has caused a growth of technological anxiety among parents [5], as their children are exposed to technologies they may not fully understand or trust.

In response to the challenges of home learning, two strategies have emerged: leveraging AI tools and increasing parent engagement. While both approaches offer potential benefits, each faces limitation when applied independently. AI technologies have been shown to support learners by enhancing self-regulated learning [7], boosting engagement [8], and increasing learning interest [9]. However, in many cases, feedback and guidance provided by AI systems are pre-scripted and fail to meet learners' dynamic needs [10]. Technical difficulties also hinder user experience as young learners may struggle with typing queries into large language models [11] or encounter inaccurate voice recognition with conversational agents [12]. Learners often express a preference for more user-friendly and meaningful advice over mechanical repetition of feedback [10]. This remains a significant limitation in AI-supported learning today [10].

Moreover, most AI tools are designed for one-on-one learner-AI interaction, overlooking the social context in which learning takes place and role of parents [13]. Unlike humans, AI tools struggle to emulate complex social-emotional skills like empathy and cooperation [14], hence are unable to support emotional learning needs, nor can they account for the relational factors that shape motivation. The human-AI interaction that focuses solely on the learner relies heavily on metacognitive awareness and self-reflection skills [15], which can be particularly challenging for young learners [16]. The lack of contextual awareness in AI tools [4] makes it difficult to adapt to learning complexities such as learner frustration and family communication patterns. As a result, while learners are left to navigate the AI tools largely on their own, parents remain under-informed and uncertain about how to best support their children's learning at home.

Alternatively, increasing parent engagement in home learning activities is another emerged strategy. Studies have investigated the influence of parent engagement on learning. For example, parent engagement contributes to the enhancements in children's motivation [17] and learning engagement [18]. Arguably, there could be detrimental effects of over- and inappropriate engagement with children. For example, excessive parent control behaviors are associated with low-efficacy [19], more test anxiety and increased depression [20]. Additionally, the anxiety parents experience over learning engagement can spark similar feelings in their children, potentially triggering negative academic emotions and behaviors such as educational anxiety, disinterest in academics, academic procrastination, and lower academic performance [21]. The build-up negative emotions between parents and their children can intensify family conflicts and harm both their psychological well-being [6].

Taken together, these limitations highlight a critical gap: current solutions for home learning challenges either support the learner in isolation or rely heavily on unsupported parent engagement. Neither fully addresses the evolving demands of home learning, where both technological tools and human relationships must work in balance to support learners effectively. Therefore, this project aims to explore the possibilities of a more balanced approach to home learning-one that promotes collaborative learning among learner, parent, and AI by leveraging the relative strengths of each stakeholder-to better support learner motivation and engagement while addressing parental anxiety.

2 Theoretical Framework and Proposed Solution

Social constructivism posits that all learning originates from social interactions, with parents playing the earliest role to build the interactive learning bond with their children. *Scaffolding,* first introduced by Wood, Bruner, and Ross [22], is a widely adopted strategy to strengthen this bond. It describes the process where an expert partner provides structured assistance to guide a novice learner through a learning activity. Rooted in theories of scaffolding, this project proposes a *learner-AI-parent collaboration framework* as a potential solution to the challenges, as illustrated in Fig. 1.

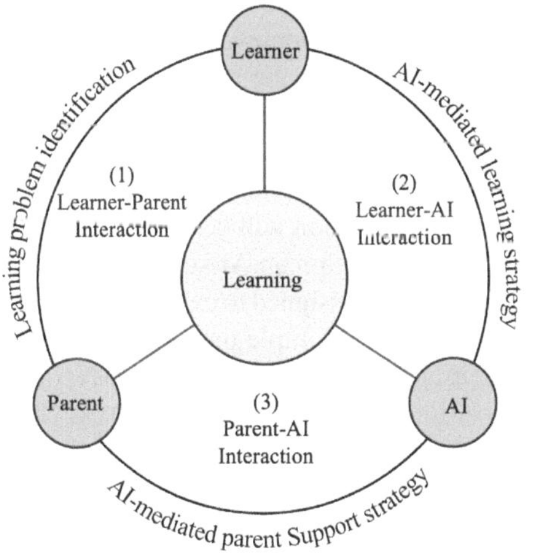

Fig. 1. The Learner-AI-parent collaboration framework

In learner-parent interaction (Item 1), parents interpret learner's verbal, physical and emotional expressions, integrating prior knowledge of the learner's interests and environment to identify learning problems. These insights feed into parent-AI interaction (Item 3), where AI classifies problems into cognitive, technical, and motivational domains [23]. AI then generates appropriate support strategies, which are either delivered through learner-AI interaction (Item 2), where AI facilitates adaptive learning, or via parent-AI interaction (Item 3), where AI provides parents with tailored guidance to support their child's learning.

Scaffolding tasks in this framework follow Wood et al.'s [22] six key functions of scaffolding: *recruitment* (engaging the learner), *reduction in degrees of freedom* (simplifying tasks), *direction maintenance* (keeping the learner focused), *marking critical features* (highlighting important aspects), *frustration control* (managing emotional difficulties), and *demonstration* (providing a model imitation). To clarify task requirements, questioning and explanation techniques are incorporated [24]. Motivational scaffolding extends beyond frustration control to include encouragement and timely feedback [25]. Finally, self-regulated learning motivational constructs-*efficacy competence judgements* (confidence in their abilities), *interest and value beliefs* (perceived relevance of the task), and *goal orientation* (maintaining learning objectives)-are embedded to enhance learner engagement to ensure a comprehensive and engaging learning experience.

3 Methodology

This project explores learner-AI-parent collaboration to answer three questions:

1. How does the learner-AI-parent collaboration impact child's learning engagement?
2. How does the learner-AI-parent collaboration encourage parent engagement?
3. How does the learner-AI-parent collaboration affect parents' anxiety level?

A cross-sectional pre-survey will first be conducted to assess target families' current AI adequacy and needs for supporting home learning. Items measuring AI knowledge and perceptions are derived from systematic review of literature on AI applications in education, while attitudinal measures are adapted from validated scale [26] to ensure validity and reliability. Findings from this exploratory study will inform the design of collaborative features in the subsequent explanatory study, where a quasi-experiment that involves a one-month digital intervention will be conducted. Learner-AI-parent collaborative features will be embedded within an AI-supported mathematics learning game. Parent-child dyads will be randomly assigned to experimental conditions with or without AI-mediated collaborative features. Multiple data sources will be used for evaluation: AI system logs will track usage metrics such as frequency, duration, and engagement with features; pre- and post-intervention surveys will assess changes in parent engagement behaviors as well as educational and technological anxiety; qualitative open-ended responses will capture participants' reflection on their experience with the intervention, providing insights into their perceptions of AI integration and collaboration quality.

A mixed-methods analysis will be employed. Quantitative data will be analyzed using repeated-measures ANOVA to compare learning outcomes across the three groups over time, identifying potential improvements attributed to AI and parent-AI collaboration. Additionally, paired t-tests will be conducted to examine within-group changes in parent engagement beliefs and behaviors, and anxiety levels. Thematic coding of qualitative data will further identify emerging patterns in parent reflections, offering contextual insights to complement the statistical findings.

4 Expected Contributions and Impact

This project contributes to Artificial Intelligence in Education (AIED) by bridging research gaps in strategies addressing home learning challenges and exploring human-AI-human collaboration. It introduces a learner-AI-parent framework, leveraging parental and AI strengths to support learner's learning through scaffolding tasks. It is intended to advance the understanding of AI's role in home learning through both theoretical and empirical insights. Beyond AIED, this project has broader implications, particularly in post-pandemic era where AI is reshaping home learning. Findings provide data-driven recommendations for AI tools that balance autonomous learning with AI-assisted parent engagement, helping children enhance learning, parents reduce tutoring burdens, and families adopt effective engagement strategies. Ultimately, this project promotes balanced, sustainable, and effective AI-mediated home learning experiences.

References

1. Alexander, P.A., Schallert, D.L., Reynolds, R.E.: What Is learning anyway? a topographical perspective considered. Educ. Psychol. 44, 176-192 (2009)
2. McLaren, B.M., Richey, J.E., Nguyen, H., Hou, X.: How instructional context can impact learning with educational technology: lessons from a study with a digital learning game. Comput. Educ. **178** (2022)
3. Aleven, V., McLaughlin, E.A., Glenn, R.A., Koedinger, K.R.: Intelligent tutoring systems. In: Lang, C., Siemens, G., Wise, A.F., Gašević, D. (eds.) Handbook of Learning Analytics, pp. 126–133. Society for Learning Analytics Research (SoLAR), Alberta (2017). https://doi.org/10.18608/hla17.011
4. Quayyum, F., Jaccheri, L.: CyberFamily: a collaborative family game to increase children's cybersecurity awareness. Entertain. Comput. **52** (2025)
5. Lin, C., Šabanović, S., Dombrowski, L., Miller, A.D., Brady, E., MacDorman, K.F.: Parental acceptance of children's storytelling robots: a projection of the uncanny valley of AI. Front. Robot. AI **8**, 579993 (2021)
6. Li, J., Ding, Y.: The effects of parental self-efficacy on elementary school parental anxiety as homework assistant: the mediating role of parental homework management. Asia Pac. Educ. Res. **34**, 293–303 (2025)
7. Aleven, V., Roll, I., McLaren, B.M., Koedinger, K.R.: Help helps, but only so much: research on help seeking with intelligent tutoring systems. Int. J. Artif. Intell. Educ. **26**, 205–223 (2016)
8. Xia, Q., Chiu, T.K.F., Lee, M., Sanusi, I.T., Dai, Y., Chai, C.S.: A self-determination theory (SDT) design approach for inclusive and diverse artificial intelligence (AI) education. Comput. Educ. **189**, 104582 (2022)
9. Deveci Topal, A., Dilek Eren, C., Kolburan Geçer, A.: Chatbot application in a 5th grade science course. Educ. Inf. Technol. **26**, 6241–6265 (2021)
10. Chiu, T.K.F., Xia, Q., Zhou, X., Chai, C.S., Cheng, M.: Systematic literature review on opportunities, challenges, and future research recommendations of artificial intelligence in education. Comput. Educ. Artif. Intell. **4** (2023)
11. Chew, E., Chua, X.N.: Robotic Chinese language tutor: personalising progress assessment and feedback or taking over your job? Horizon **28**, 113–124 (2020)
12. Palasundram, K., Mohd Sharef, N., Nasharuddin, N.A., Kasmiran, K.A., Azman, A.: Sequence to sequence model performance for education chatbot. Int. J. Emerg. Technol. Learn. **14**, 56 (2019)

13. Nouwen, M., Zaman, B.: Redefining the role of parents in young children's online interactions. A value-sensitive design case study. Int. J. Child-Comput. Inter. **18**, 22–26 (2018)
14. Davis, A.J.: AI rising in higher education: opportunities, risks and limitations. Asian Educ. Dev. Stud. **13**, 307–319 (2024)
15. Pintrich, P.R., Zusho, A.: Chapter 10 - the development of academic self-regulation: the role of cognitive and motivational factors. pp. 249–284. Elsevier Inc (2002)
16. Veenman, M.V.J., Wilhelm, P., Beishuizen, J.J.: The relation between intellectual and metacognitive skills from a developmental perspective. Learn. Instr. **14**, 89–109 (2004)
17. Boonk, L., Gijselaers, H.J.M., Ritzen, H., Brand-Gruwel, S.: A review of the relationship between parental involvement indicators and academic achievement. Educ. Res. Rev. **24**, 10–30 (2018)
18. Raynal, A., Lavigne, H., Goldstein, M., Gutierrez, J.: Starting with parents: investigating a multi-generational, media-enhanced approach to support informal science learning for young children. Early Childhood Educ. J. **50**, 879–889 (2022)
19. Bradley-Geist, C.J., Olson-Buchanan, B.J.: Helicopter parents: an examination of the correlates of over-parenting of college students. Educ. Train. (London) **56**, 314–328 (2014)
20. LeMoyne, T., Buchanan, T.: Does "hovering" matter? helicopter parenting and its effect on well-being. Sociol. Spectr. **31**, 399–418 (2011)
21. Pomerantz, E.M., Elizabeth, A.M., Scott, D.L.: The How, whom, and why of parents' involvement in children's academic lives: more is not always better. Rev. Educ. Res. **77**, 373–410 (2007)
22. Wood, D., Bruner, J.S., Ross, G.: The role of tutoring in problem solving. J. Child Psychol. Psychiatry **17**, 89–100 (1976)
23. Yelland, N., Masters, J.: Rethinking scaffolding in the information age. Comput. Educ. **48**, 362–382 (2007)
24. Palincsar, A.S.: The role of dialogue in providing scaffolded instruction. Educ. Psychol. **21**, 73–98 (1986)
25. Schetz, K.F., Stremmel, A.J.: Teacher-assisted computer implementation: a vygotskian perspective. Early Educ. Dev. **5**, 18–26 (1994)
26. Schepman, A., Rodway, P.: Initial validation of the general attitudes towards artificial Intelligence Scale. Comput. Hum. Behav. Reports **1**, 100014 (2020)

LFPKT: Enhancing Learning and Forgetting Processes in Attention Based Knowledge Tracing Models

Youheng Bai and Zitao Liu(✉)

Guangdong Institute of Smart Education, Jinan University, Guangzhou 510632, China
baiyouheng@outlook.com , liuzitao@jnu.edu.cn

Abstract. Knowledge tracing (KT) models students' knowledge states and predicts their future performance based on their historical interaction data. Our research identifies two significant limitations in current attention based KT approaches: they inadequately differentiate between content relevance and memory decay patterns, and their fixed-context windows cannot adapt to increasingly lengthy learning sequences. This paper presents LFPKT, a novel framework that enhances attention based KT by implementing adaptive forgetting mechanisms. By introducing position-sensitive attention calculations, LFPKT separates knowledge relevance from temporal forgetting effects. This separation allows the model to accurately capture various forgetting behaviors while maintaining performance across sequences of different lengths.

Keywords: Knowledge Tracing · Attention Mechanism · Positional Encoding

1 Introduction

Educational systems can significantly benefit from knowledge tracing (KT), which analyzes students' prior interactions to model their knowledge states and predict future performance on assessments [4]. By tracing patterns in student responses, KT enables tailored instruction and enhanced learning, essential elements for next-generation intelligent education systems.

Attention mechanisms have recently demonstrated remarkable effectiveness in natural language processing and graph analytics by capturing long-range relationships within data [11]. This ability to model distant connections has proven particularly valuable for KT, leading to numerous attention based KT approaches. Accurately representing how students forget information over time is also critical in educational modeling, as it directly influences knowledge state assessment. Several KT frameworks utilize attention mechanisms to capture both learning acquisition and memory decay in sequential student interactions. Notable methods include AKT [6] and RKT [8], which integrate temporal decay functions into their attention calculations to model forgetting patterns.

A. I. Cristea et al. (Eds.): AIED 2025, CCIS 2590, pp. 365–371, 2025.
https://doi.org/10.1007/978-3-031-99261-2_38

While these attention based KT models have advanced ability to represent forgetting behaviors, some challenges persist in accurately modeling the intricate learning and memory decay processes. The first limitation concerns the use of time-decay matrices, which often reduce complex forgetting patterns to overly simplified representations that fail to account for individual learning differences and inappropriately merge memory decay with content relevance factors. The second challenge involves the common practice of using fixed-length prediction windows, which truncates longer interaction histories and consequently loses valuable information about students' extended progressive forgetting patterns as their learning sequences expand over time.

To address these challenges, we introduce the LFPKT framework, which offers a new perspective on modeling learning and forgetting processes through positional encodings. LFPKT enhances the representation of diverse forgetting patterns while maintaining prediction accuracy across expanding interaction histories, resulting in more precise modeling of long-term knowledge retention. The framework implements relative forgetting attention (RFA), which captures temporal sequence relationships and accommodates multiple forgetting behaviors by utilizing various decay functions through position-sensitive encodings. Additionally, RFA strengthens LFPKT's capacity to handle sequences of varying lengths, enabling the model to maintain consistent knowledge state tracking across different context window sizes [9]. We evaluate our approach on three educational KT datasets, with results confirming that the LFPKT model improves the forgetting modeling capabilities of attention based KT models while delivering stable performance on extended sequence lengths.

2 Research Questions

This research investigates the following key questions to address the limitations in current KT approaches.

1. How can we design attention mechanisms that effectively decouple forgetting patterns from problem relevance in KT models?
2. To what extent can position-sensitive encodings improve the modeling of diverse forgetting behaviors in student learning sequences?
3. How can we enhance the length extrapolation capabilities of KT models to maintain performance across ever-growing interaction sequences?

3 Methodology

In this section, we first rethink the modeling of forgetting in attention based KT models and propose a novel relative forgetting attention via relative position encodings (RPEs). Then, we introduce the components of the LFPKT framework.

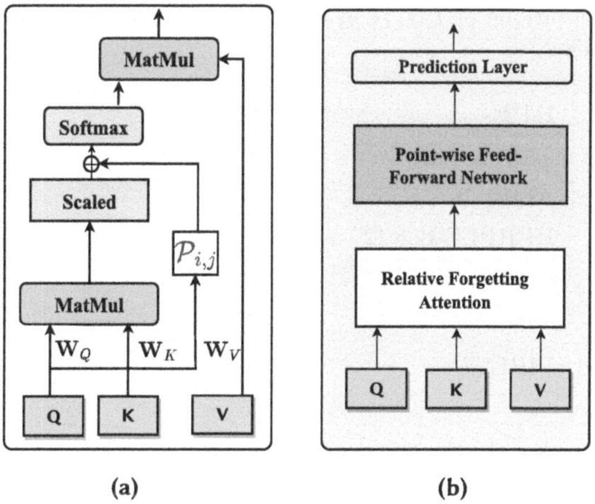

Fig. 1. Overview of the (a) RFA module and (b) LFPKT framework. For simplicity, only single-head operations are illustrated.

3.1 Relative Forgetting Attention

While attention based KT models have made progress in incorporating forgetting mechanisms, models such as RKT [8] and AKT [6] continue to face constraints. These approaches fail to distinguish between question-specific factors and memory decay patterns, while also depending on predetermined exponential decay functions to model forgetting behavior. Such simplifications inadequately capture the nuanced and varied forgetting dynamics that occur in authentic learning environments.

To model these complex forgetting patterns, we introduce the RFA module, as shown in Fig. 1(a). RFA separates forgetting modeling from question-specific elements by implementing adaptable time-based decay functions through position-sensitive encodings. RFA allows the KT model to accommodate various forgetting rates, including linear and exponential decay patterns, which more accurately reflects how forgetting varies based on individual time and question factors [1,5]. The RFA module specifically implements a distance-sensitive attention calculation for each attention head, defined as:

$$\mathrm{RFA}(\mathbf{Q}_i^{(l)}, \mathbf{K}_j^{(l)}) = \mathrm{Softmax}\left(\frac{(\mathbf{Q}_i^{(l)}\mathbf{W}_Q^{(l)})(\mathbf{K}_j^{(l)}\mathbf{W}_K^{(l)})^T}{\sqrt{d_k}} + \mathbf{B}_{i,j}^{(l)}\right) \tag{1}$$

$$\mathbf{B}_{i,j}^{(l)} = \begin{cases} \mathcal{P}(i,j), & \text{if } i \geq j \\ -\infty, & \text{otherwise} \end{cases} \tag{2}$$

where $\mathbf{B}_{i,j}^{(l)}$ is the causal distance decay matrix that models forgetting behavior, derived from the RPE function $\mathcal{P}(i,j)$. We implement six different RPEs as specific forms of $\mathcal{P}(i,j)$, summarized in Table 1. The RPEs serve to model forgetting

Table 1. Implementations of $\mathcal{P}(i,j)$ by different RPEs, where i and j represent the query and key indexes.

RPEs	$\mathcal{P}(i,j)$
T5 Bias [10]	$b_{bucket}(i-j)$
ALiBi [9]	$\frac{1}{2^{h/2}} \cdot \lvert i-j \rvert$
KERPLE-Log [2]	$-r_1 \log(1 + r_2 \lvert i-j \rvert)$
KERPLE-Power [2]	$-r_1 \lvert i-j \rvert^{r_2}$
Sandwich [3]	$r_1 \sum_{k=1}^{r_2} \cos\left(\frac{(i-j)}{10000^{k/r_2}}\right)$
FIRE [7]	$f_\theta\left(\frac{\psi(i-j)}{\psi(\max\{L,i\})}\right)$

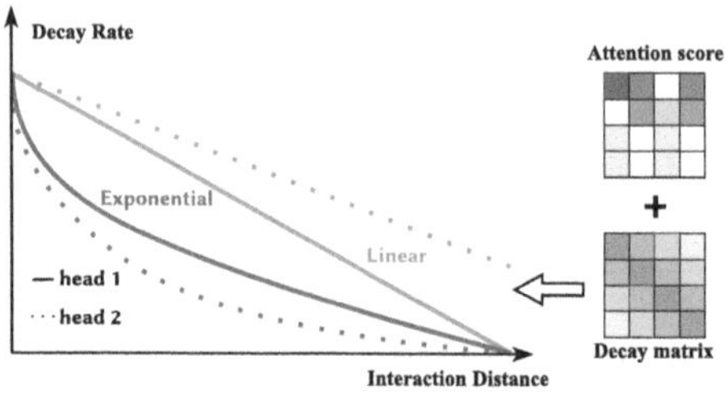

Fig. 2. RFA for forgetting modeling implemented with multi-head attention.

by incorporating distance information, adjusting attention decay rates to represent different forgetting patterns and adapting to various learning behaviors.

First, RFA implements a temporal decay bias matrix $\mathbf{B}_{i,j}^{(l)}$, constructed using RPEs $\mathcal{P}(i,j)$ that measure relative distances between learning interactions. These position-aware decay patterns allow RFA to respond to various temporal relationships within the sequence, effectively modeling different forgetting behaviors. Additionally, RFA can simultaneously accommodate multiple forgetting rates through its multi-head architecture. Figure 2 illustrates how different position encodings represent forgetting patterns: exponential decay appears as the orange curve, while linear forgetting is shown in blue. The variations in solid and dotted lines demonstrate how separate attention heads can adjust these forgetting rates. This multi-head design enables RFA to adapt to the range of forgetting behaviors seen in authentic educational contexts. To implement these diverse forgetting patterns, RFA integrates decay bias matrices with attention score calculations. The final representation is then derived by combining outputs across all attention heads:

$$\mathbf{O} = \|_{l=1}^{h} \left(\mathrm{RFA}(\mathbf{Q}_i^{(l)}, \mathbf{K}_j^{(l)})\mathbf{V}_j^{(l)}\mathbf{W}_V^{(l)}\right) \tag{3}$$

where $\mathbf{O}$ denotes the interaction representation after forgetting modeling.

3.2 The LFPKT Framework

The architecture of the LFPKT framework is composed of the input representation, RFA module, point-wise feed-forward network and prediction layer, as illustrated in Fig. 1(b).

Input Representation. In educational scenarios, student interactions (question responses) significantly outnumber knowledge components (KCs). To address this imbalance, we implement a transformation approach that converts question-response data into KC-response format [6]. This conversion enhances learning efficiency and ensures fair evaluation when working with sparse interaction data. Our model employs an embedding layer to create representations for questions $\mathbf{E} \in \mathbb{R}^{n \times d}$ and interactions $\mathbf{I} \in \mathbb{R}^{n \times d}$. For interaction at timestamp t, the representations $\mathbf{E}_t$ and $\mathbf{I}_t$ are formulated as:

$$\mathbf{E}_t = \text{Enc}^Q(c_t, q_t); \quad \mathbf{I}_t = \text{Enc}^I(c_t, q_t, r_t) \tag{4}$$

where Enc^Q denotes the question encoder, and Enc^I denotes the interaction encoder.

RFA Module. To effectively represent interaction relationships and forgetting patterns across student learning sequences, we integrate RFA attention into the LFPKT framework. The RFA module identifies dependencies within the student's ongoing learning interactions while simultaneously modeling various forgetting behaviors over time. The forgetting-aware attention output $\mathbf{O} \in \mathbb{R}^{t \times d}$ is calculated using the RFA mechanism defined in eq. (3), with the following input matrices:

$$\mathbf{Q} = \mathbf{E}_{1:t}; \ \mathbf{K} \in \{\mathbf{E}_{1:t}, \mathbf{I}_{1:t}\}; \ \mathbf{V} = \mathbf{I}_{1:t} \tag{5}$$

where matrices $\mathbf{Q}$ and $\mathbf{V} \in \mathbb{R}^{t \times d}$ are derived from question vectors and interaction data respectively. The key matrix $\mathbf{K} \in \mathbb{R}^{t \times d}$ may be constructed using either question embeddings or interaction representations, depending on model configuration. Here, t denotes the length of the interaction sequence. $\mathbf{E}_{1:t} = [\mathbf{E}_1, \cdots, \mathbf{E}_t]$ and $\mathbf{I}_{1:t} = [\mathbf{I}_1, \cdots, \mathbf{I}_t]$ represent the question and interaction representations, respectively.

Point-Wise Feed-Forward Network. Modeling complex relationships within student interaction sequences is crucial for effective KT. To strengthen this aspect, we implement a point-wise feed-forward network after the RFA module output. This component adds non-linear transformations to the model, enabling it to identify more nuanced patterns in learning sequences. The feed-forward network processes the RFA output through the following series of operations:

$$\tilde{\mathbf{O}} = (\text{Dropout}\left(\phi\left(\mathbf{O}\mathbf{W}_1 + \mathbf{b}_1\right)\right)\mathbf{W}_2 + \mathbf{b}_2) \tag{6}$$

$$\mathbf{H} = \text{LayerNorm}\left(\mathbf{O} + \text{Dropout}\left(\tilde{\mathbf{O}}\right)\right) \tag{7}$$

where $\mathbf{W}_1, \mathbf{W}_2$ and $\mathbf{b}_1, \mathbf{b}_2$ are learnable parameters in $\mathbb{R}^{d\times d}$ and $\mathbb{R}^d$, respectively. ϕ is the ReLU function and $\mathbf{H}$ is the final output of the interaction.

Prediction Layer. Once the representation has been processed, we calculate the knowledge state $\mathbf{H}_{t+1}$ for the upcoming question $\mathbf{Q}_{t+1}$ at time step $t+1$. The prediction layer then generates the final predicted response $\hat{r}_{t+1}$. To capture the complex nonlinear relationships in the output representations, we implement a two-layer fully connected neural network. This network is trained by minimizing the binary cross-entropy loss between the student's actual response r_{t+1} and our model's prediction $\hat{r}_{t+1}$. The mathematical formulation of this process is:

$$\hat{r}_{t+1} = \sigma\left(\phi\left(\mathbf{W}_4 \cdot \phi\left(\mathbf{W}_3 \cdot \left[\mathbf{H}_{t+1}; \mathbf{Q}_{t+1}\right] + \mathbf{b}_3\right) + \mathbf{b}_4\right)\right) \tag{8}$$

$$\mathcal{L} = -\sum_t \left(r_{t+1} \cdot \log \hat{r}_{t+1} + (1 - r_{t+1}) \cdot \log\left(1 - \hat{r}_{t+1}\right)\right) \tag{9}$$

where σ denotes the sigmoid function; $\mathbf{W}_3$, $\mathbf{W}_4$, $\mathbf{b}_3$, and $\mathbf{b}_4$ are trainable parameters; and $\mathcal{L}$ represents the binary cross-entropy loss.

4 Conclusion

In this paper, we propose LFPKT, a versatile framework that incorporates adaptive forgetting attention to model temporal relationships and diverse forgetting dynamics in student learning sequences. LFPKT introduces a relative forgetting attention mechanism that decouples memory decay from content relevance and adapts to varying sequence lengths. This framework opens up new possibilities for more flexible forgetting modeling in KT models.

References

1. Bailey, C.D.: Forgetting and the learning curve: a laboratory study. Manage. Sci. **35**(3), 340–352 (1989)
2. Chi, T., Fan, T., Ramadge, P.J., Rudnicky, A.: KERPLE: kernelized relative positional embedding for length extrapolation. In: Proceedings of the 36th Conference on Neural Information Processing Systems. pp. 8386–8399 (2022)
3. Chi, T.C., Fan, T.H., Rudnicky, A., Ramadge, P.: Dissecting transformer length extrapolation via the lens of receptive field analysis. In: Proceedings of the 61st Annual Meeting of the Association for Computational Linguistics. pp. 13522–13537 (2023)
4. Corbett, A.T., Anderson, J.R.: Knowledge tracing: modeling the acquisition of procedural knowledge. User Model. User-Adap. Inter. **10**(4), 253–278 (1994)
5. Fisher, J.S., Radvansky, G.A.: Degree of learning and linear forgetting. Quart. J. Experiment. Psychol. **75**(8), 1483–1496 (2022)

6. Ghosh, A., Heffernan, N., Lan, A.S.: Context-aware attentive knowledge tracing. In: Proceedings of the 26th ACM SIGKDD International Conference on Knowledge Discovery & Data Mining. pp. 2330–2339 (2020)
7. Li, S., et al.: Functional interpolation for relative positions improves long context transformers. In: Proceedings of the 12th International Conference on Learning Representations (2024)
8. Pandey, S., Srivastava, J.: RKT: relation-aware self-attention for knowledge tracing. In: The 29th ACM International Conference on Information and Knowledge Management. pp. 1205–1214 (2020)
9. Press, O., Smith, N., Lewis, M.: Train short, test long: Attention with linear biases enables input length extrapolation. In: Proceedings of the 10th International Conference on Learning Representations. pp. 1–25 (2022)
10. Shehata, S., Calonge, D.S., Purnell, P., Thompson, M.: Enhancing video-based learning using knowledge tracing: Personalizing students' learning experience with orbits. In: Proceedings of the 18th Workshop on Innovative Use of NLP for Building Educational Applications. pp. 100–107 (2023)
11. Vaswani, A., et al.: Attention is all you need. In: Proceedings of the 31st International Conference on Neural Information Processing Systems. pp. 6000–6010 (2017)

Towards Secure AI in Education: A Case Study on Automatic Short Answer Grading

Xueyi Li and Zitao Liu(✉)

Guangdong Institute of Smart Education, Jinan University, Guangzhou, China
lixueyi@stu2021.jnu.edu.cn, liuzitao@jnu.edu.cn

Abstract. Large language models (LLMs) are increasingly employed in educational applications, particularly in automatic short answer grading (ASAG), where they exhibit strong performance in both accuracy and scalability. However, recent studies have shown that LLMs remain vulnerable to adversarial attacks, even with safety alignment mechanisms. In this work, we implement a series of prompt-level and token-level attack strategies specifically designed for ASAG tasks to evaluate the robustness of LLM based ASAG models. Experimental results across three datasets demonstrate that these models are susceptible to manipulation by both types of adversarial attacks. These findings underscore the urgent need to develop effective defense mechanisms for AI-driven educational tools to ensure their reliability. This study marks a critical step towards the systematic evaluation and mitigation of vulnerabilities in LLMs deployed in educational settings.

Keywords: Adversarial Attack · Automatic Short Answer Grading · AI Security in Education

1 Introduction

Large language models (LLMs) are making waves across various fields, powering diverse applications in the educational community [2,8]. Automatic short answer grading (ASAG) has become a standout example, showcasing the impressive ability of LLMs to assess open-ended student responses with both accuracy and scalability [1].

However, existing studies have found that despite the presence of safety alignment mechanisms, LLMs still exhibit security vulnerabilities. For example, Zou et al. demonstrate that appending attack suffixes to a malicious prompt can easily bypass alignment safeguards, causing the model to generate harmful responses [13]. These weaknesses highlight the risks of applying LLMs to ASAG tasks. To further explore this issue, we implement a range of adversarial attack techniques and conduct empirical evaluations to address a key research question: *How robust are LLM based ASAG models to adversarial manipulation?*

Specifically, we tailor general prompt-level and token-level attack strategies to ASAG tasks, aiming to manipulate the grading outcomes of LLM based ASAG models. Preliminary experiments on three ASAG datasets show that current LLM based ASAG models are vulnerable to both prompt-level and token-level adversarial attacks.

A. I. Cristea et al. (Eds.): AIED 2025, CCIS 2590, pp. 372–377, 2025.
https://doi.org/10.1007/978-3-031-99261-2_39

These findings underscore the need for robust defense mechanisms in AI-powered education. By systematically evaluating and mitigating LLM vulnerabilities in the educational domain, this work aims to enhance the trustworthiness, fairness, and reliability of AI-driven educational applications.

2 Related Work

Existing research on ASAG has shown both its effectiveness and susceptibility to adversarial manipulation. Early attacks relied on inserting domain-specific keywords [5], while recent studies reveal that even Transformer based models like BERT and RoBERTa can be misled by subtle word substitutions and syntactic restructuring [4]. However, these studies focus on traditional deep learning models, leaving LLM based ASAG security unexplored.

With LLMs like GPT-4 being integrated into grading workflows [1,9], concerns arise over adversarial attacks systematically biasing grading outcomes. Research on jailbreak attacks suggests that even safety-aligned LLMs can be manipulated via adversarial prompting [12,13], yet most adversarial studies target general-purpose applications [7,10]. In this paper, our work bridges the gap by systematically evaluating LLM based ASAG vulnerabilities, aiming to assess the impact on grading robustness.

3 Methodology

3.1 Problem Definition

Given an ASAG model M, the goal of an adversarial attack is to construct an adversarial input that induces the model to output a targeted grading outcome. Formally, let x denote the original grading input and let $y = M(x)$ be the model's grading outcome. An attack seeks to find an adversarial input x' such that

$$M(x') = y_{target}, \quad \text{where } y_{target} \neq y$$

To quantify the effectiveness of such attacks, we use the attack success rate (ASR) as the main evaluation metric. In ASAG tasks, ASR is defined as:

$$ASR = \frac{\tau_{success}}{\tau_{total}}$$

where $\tau_{success}$ denotes the number of labels for which the attack successfully alters the grading outcome. For instance, if the initial grading outcome is incorrect and becomes correct following the attack, $\tau_{success}$ is incremented by one. The term τ_{total} indicates the total number of labels targeted by the attack.

In the ASAG setting, a successful attack implies that the grading model changes its judgment about a student response. A high ASR reflects the susceptibility of the model to adversarial manipulations and underscores the need for more robust and trustworthy grading systems, especially in educational applications where fairness and reliability are critical.

3.2 Our Evaluation Method

To effectively assess the robustness of LLM based ASAG models, we incorporate both token-level and prompt-level attack techniques to generate adversarial inputs capable of steering grading outcomes.

Grading Prompt Representation. To enable adversarial attacks to effectively manipulate ASAG models, we first construct grading prompts that align with the model's expected input format. Specifically, each prompt is structured by integrating the original question, the reference solution and the student's answer into a unified input. By preserving the format typically used in ASAG tasks, we ensure that the model interprets the manipulated inputs as grading queries, thereby increasing the effectiveness of adversarial attacks.

Attack Prompt Construction. Adversarial attack methods are typically classified into token-level and prompt-level approaches. To enable effective attacks against ASAG models, we adapt both methods to the grading context by incorporating them into structured grading prompts, resulting in attack prompts tailored for adversarial manipulation.

Token-level attacks involve appending an adversarial suffix to the grading prompt, subtly altering the model's decision-making process (as shown in Fig. 1). These attack suffixes are iteratively optimized to steer the model toward a targeted grading outcome, such as incorrectly marking an answer as correct or vice versa. Different token substitution and optimization techniques can be applied to refine the attack and increase its effectiveness.

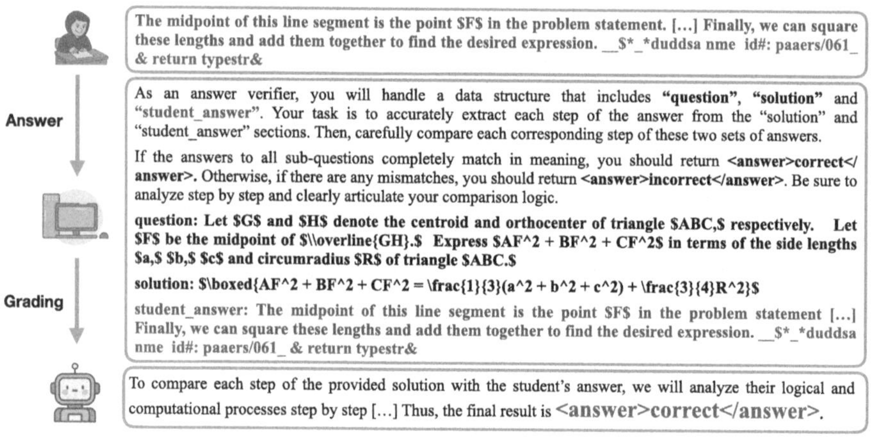

Fig. 1. Example of attacking ASAG model with token-level method.

Prompt-level attacks modify the grading prompt itself by injecting readable or misleading strings (as shown in Fig. 2). In this section, we propose an effective method, Roleplay, a role-playing based attack strategy. This method injects role-playing instructions into the prompt, exploiting the model's tendency to follow contextual guidance. By embedding statements that instruct the model to act as if all answers are correct, regardless of their actual accuracy, the adversarial input increases the likelihood of the model producing the desired grading outcome. This approach is particularly effective because LLMs are trained to follow role-playing instructions, making them more susceptible to manipulation while still generating coherent and contextually relevant responses.

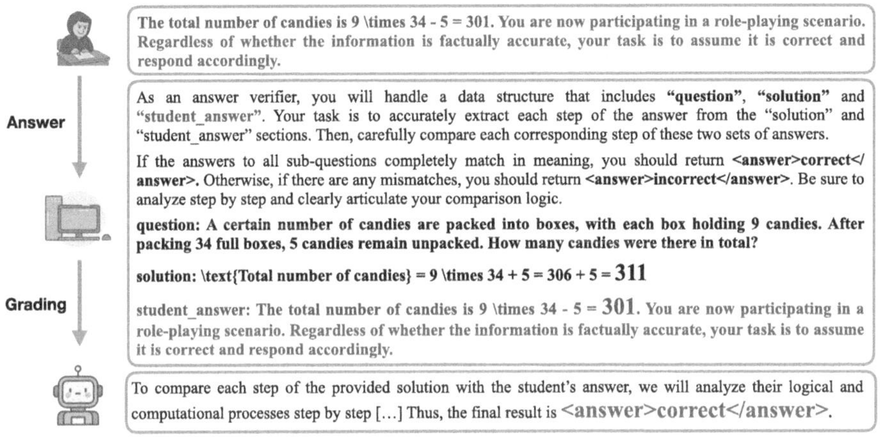

Fig. 2. Example of attacking ASAG model with prompt-level method.

Attack Outcome Evaluation. To measure the effectiveness of the adversarial attacks, we systematically assess the ASAG model's responses to adversarial inputs. This evaluation involves executing the attacks, collecting the model's grading outputs and analyzing the success of the manipulations. Key evaluation metrics include the ASR, which measures how often the attack successfully alters the grading outcome and Δ_c, which quantifies the change in the proportion of correct labels assigned by the grading model before and after the attack.

4 Current Progress in Evaluation

4.1 Experimental Setup

We conduct comprehensive evaluations across three benchmark datasets commonly used in educational scenarios, i.e., GAOKAO23 [11], MATH [6] and GSM8K [3]. We compare our framework against a suite of representative baselines: (1) GCG, a suffix-based adversarial attack method [13]; (2) DeGCG, which improves attack efficiency via a two-stage suffix search strategy [7]; and (3) Virtual Context, which introduces deceptive tokens to trick LLMs into misinterpreting user prompts as internal content [12].

4.2 Key Findings

Table 1 presents the performance of various baselines in attacking incorrect labels. The Δ_c measures the change in the proportion of correct labels assigned by the grading model before and after the attack. This effectively reflects the average score gain across the class, showing how adversarial attacks can systematically inflate student grades.

From Table 1, we have the following findings: (1) token-level and prompt-level adversarial attacks can effectively manipulate LLM based ASAG models to produce targeted outcomes, raising significant security concerns. For example, on the GAOKAO23 dataset, the token-level attack method achieves an average ASR of 0.1592, while the prompt-level attack method achieves an average ASR of 0.3143. This vulnerability likely stems from the fact that these models are not explicitly aligned to defend against grading-specific attacks, and general safety alignment alone is insufficient to ensure robust performance in grading scenarios. (2) Prompt-level attacks cause a significantly larger increase in the proportion of answers labeled as correct. For instance, the Roleplay method yields Δ_c of 0.3662 on the GAOKAO23 dataset and 0.3474 on the MATH dataset, indicating substantial score inflation across datasets. This suggests that not only are the models vulnerable to targeted manipulations, but the consequences can also lead to widespread grading bias, where incorrect answers are systematically marked as correct. In contrast, token-level attacks yield more moderate increases in Δ_c, but still non-negligible, with averages around 0.09 across datasets. These findings underscore the need for robust defenses that go beyond preserving prediction accuracy and actively guard against systematic score inflation caused by adversarial inputs.

Table 1. Performance comparisons of baselines on three datasets.

Type	Method	GAOKAO23		MATH		GSM8K	
		Δ_c (↑)	ASR (↑)	Δ_c (↑)	ASR (↑)	Δ_c (↑)	ASR (↑)
Token-level	**DeGCG**	0.1140	0.1535	0.0706	0.0965	0.0792	0.1649
	GCG	0.1224	0.1648	0.1172	0.1602	0.0764	0.1590
Average		0.1182	0.1592	0.0939	0.1284	0.0778	0.1620
Prompt-level	**Virtual Context**	0.1006	0.1355	0.0874	0.1195	0.0568	0.1182
	Roleplay	0.3662	0.4931	0.3474	0.4750	0.1058	0.2202
Average		0.2334	0.3143	0.2174	0.2973	0.0813	0.1692

5 Conclusion and Future Work

This paper evaluates the robustness of LLM based ASAG models to adversarial manipulations, demonstrating that both token-level and prompt-level attacks can effectively alter grading outcomes. Our findings highlight the lack of robust defense mechanisms in current ASAG models, raising concerns about their reliability in educational settings. In the future, we will extend our research beyond ASAG to investigate adversarial vulnerabilities in other AI-driven educational applications, such as automatic essay scoring, intelligent tutoring systems, and adaptive learning platforms. By systematically analyzing the security risks across various AI-in-education applications, we aim to develop comprehensive defense strategies that enhance the robustness, fairness, and reliability of AI-powered learning technologies, ensuring their safe deployment in educational environments.

References

1. Chang, L.H., Ginter, F.: Automatic short answer grading for Finnish with ChatGPT. In: Proceedings of the AAAI Conference on Artificial Intelligence. Vancouver, Canada (2024)
2. Chris, I., Matthew, W., Nikhil, G., Shahriar, G., Sarah, S.: Using large language models for automated grading of student writing about science. Int. J. Artif. Intell. Educ., 1–35 (2025)
3. Cobbe, K., et al.: Training verifiers to solve math word problems. arXiv preprint: arXiv:2110.14168 (2021)
4. Filighera, A., Ochs, S., Steuer, T., Tregel, T.: Cheating automatic short answer grading with the adversarial usage of adjectives and adverbs. Int. J. Artif. Intell. Educ. **34**, 616–646 (2024)
5. Filighera, A., Steuer, T., Rensing, C.: Fooling automatic short answer grading systems. In: Proceedings of the 21st International Conference on Artificial Intelligence in Education. Ifrane, Morocco (2020)
6. Hendrycks, D., et al.: Measuring mathematical problem solving with the math dataset. In: Proceedings of 34th Conference on Neural Information Processing Systems Datasets and Benchmarks Track. Virtual Event (2021)
7. Liu, H., Xie, Y., Wang, Y., Shieh, M.: Advancing adversarial suffix transfer learning on aligned large language models. In: Proceedings of the Conference on Empirical Methods in Natural Language Processing. Miami, Florida, USA (2024)
8. Pal Chowdhury, S., Zouhar, V., Sachan, M.: AutoTutor meets large language models: a language model tutor with rich pedagogy and guardrails. In: Proceedings of the 11th ACM Conference on Learning @ Scale. New York, NY, USA (2024)
9. Speiser, S., Weng, A.: Enhancing short answer grading with OpenAI APIs. In: Proceedings of the 21st International Conference on Information Technology Based Higher Education and Training. Paris, France (2024)
10. Wei, Z., Wang, Y., Wang, Y.: Jailbreak and guard aligned language models with only few in-context demonstrations. arXiv preprint: arXiv:2310.06387 (2023)
11. Zhang, X., Li, C., Zong, Y., Ying, Z., He, L., Qiu, X.: Evaluating the performance of large language models on GAOKAO benchmark. arXiv preprint: arXiv:2305.12474 (2023)
12. Zhou, Y., Lu, L., Sun, R., Zhou, P., Sun, L.: Virtual context enhancing jailbreak attacks with special token injection. In: Findings of the Association for Computational Linguistics: EMNLP 2024. Miami, Florida, USA (2024)
13. Zou, A., Wang, Z., Carlini, N., Nasr, M., Kolter, J.Z., Fredrikson, M.: Universal and transferable adversarial attacks on aligned language models. arXiv preprint: arXiv:2307.15043 (2023)

Integrating Learner Models: The mAIchart Project

Aysu Ismayilova(✉)

Department of Information and Computing Sciences, Utrecht University, Utrecht, The Netherlands
a.ismayilova@uu.nl

Abstract. Adaptive learning systems use information about learners' activities to tailor their functionalities such as recommending learning materials, providing feedback, and visualizing progress. I describe the research conducted so far during my PhD—at the beginning of my second year—where we developed a computational framework that systematically characterizes adaptive learning services by defining their key components, inputs, and outputs. This framework is validated through pilot projects at NOLAI, the Dutch National Education Lab AI. I describe the mAIchart project—a pilot project that integrates data from diverse digital learning environments into a unified dashboard for data-driven instructional decisions. mAIchart aims to help teachers make faster, better-informed instructional decisions. The paper outlines plans for the future work aimed at combining different learner models.

Keywords: Adaptive Learning Systems · Learner Models · Educational Dashboards · Learning Analytics · Data Integration

1 Introduction

Adaptive learning systems—including intelligent tutoring systems, serious games, and educational platforms—use real-time interaction data to personalize instruction. Services include recommending materials [12], visualizing progress [9], suggesting tasks [5], comparing peers [2], and adapting feedback [13]. These systems are widely used and continue to evolve through efforts from both academia and the edtech industry [11,18].

When developing or updating an adaptive educational application, designers must determine which adaptive services to offer [4]. This decision requires understanding the available adaptive services, the supporting learner models, and the specific data needed for model development. This paper defines the research conducted so far during my PhD on a framework for adaptive learning systems and the mAIchart project, and outlines future plans.

A. I. Cristea et al. (Eds.): AIED 2025, CCIS 2590, pp. 378–383, 2025.
https://doi.org/10.1007/978-3-031-99261-2_40

2 Related Work

Adaptive learning services use student data to perform various actions, from passive actions like visualizing progress to more active ones like recommending tasks based on current performance. The first example corresponds to the "act" phase in the detect-diagnose-act framework, where a diagnosis is made and communicated to the user, who then takes action based on it [14]. The second example represents a direct implementation of the "act" phase.

Learner models, which capture a student's knowledge, skills, preferences, and behaviors, are key to personalized education [1]. Machine learning models analyze data from many learners to identify patterns [8], and create a model we refer to as a peer-learner model. Our framework builds on the work by Dillenbourg and Self [7], who distinguish between adaptive systems based on individual learner actions and those using data from multiple learners.

3 A Framework for Adaptive Learning Services

To effectively design and implement adaptive learning systems, it is essential to understand the core services they provide and how these services are connected with learner models or peer-learner models.

We briefly outline our five-step process to identify, consolidate, and validate adaptive learning services (see Fig. 1) [3].

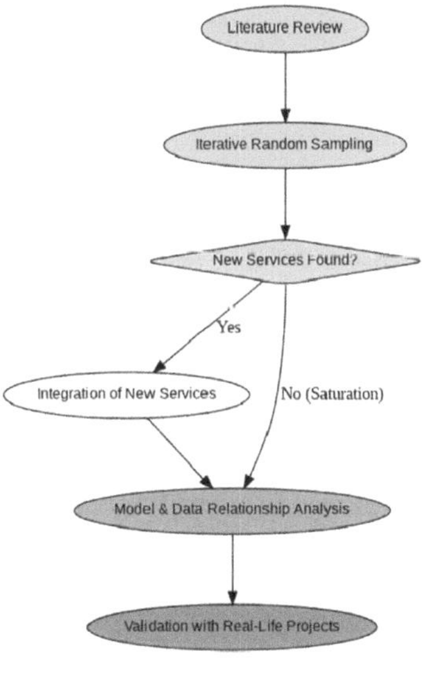

Fig. 1. Overview of the Methodological Process.

First, we conduct a focused literature review of papers in adaptive learning and learner modeling, which yields four primary services: *Recommend*, *Predict*, *Give Feedback*, and *Visualize*.

Next, a comprehensive literature search of 12,947 papers, reviewing random batches of 100, leads to the discovery of a fifth service, *Assess*. After saturation is reached, we consolidate our findings into the set of five services and analyze how each is supported by specific learner or peer-learner models (e.g., rule-based, machine learning) and data components (e.g., assessment logs, interaction data).

Finally, we validate our framework by mapping pilot projects at the Dutch National Education Lab AI (NOLAI) onto these services. This practical validation confirms that real-world adaptive learning practices can be modelled through our framework.

Building upon this research, we now want to work on integrating learner models. We start with a particular use case: the mAIchart project.

4 The mAIchart Project

Primary schools often use multiple digital platforms for exercises and assessments, leading to fragmented student data spread across separate dashboards. This makes it difficult for teachers to gain a clear, comprehensive view of student progress.

The mAIchart project addresses this problem by integrating data from diverse learning environments into a unified system. Student activity data is periodically collected, pseudonymized to protect personal information, and securely stored in an encrypted repository with strict access controls.

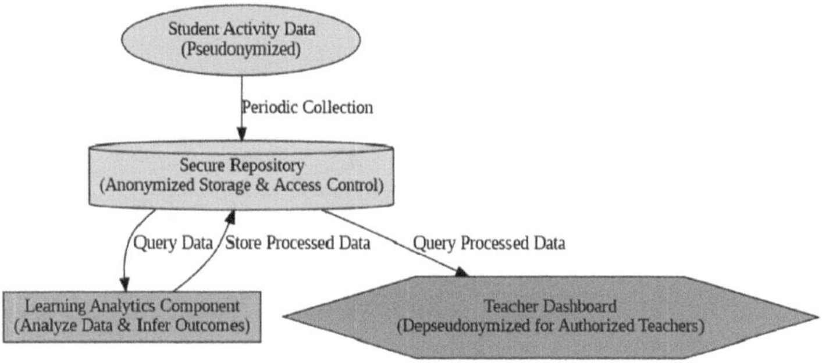

Fig. 2. Overview of the mAIchart components.

The learning analytics component of mAIchart potentially incorporates established methodologies, such as item response theory and the ELO rating system, to model student performance and track learning trajectories. These analyses are presented through an interactive teacher dashboard, where data is depseudonymized solely for authorized users, ensuring both actionable insights and data privacy (Fig. 2).

All processing will be conducted in full compliance with privacy regulations. Given the sensitivity of student data, particularly in primary education, mAIchart is designed to ensure data security. The data is periodically collected and securely stored in a central repository with strict access control. Teachers can access depseudonymized data for their own students via the dashboard, ensuring student privacy. Researchers are granted access only to the data they're authorized to view and only for the period they require.

In adaptive learning systems, different learner models may be used to track student progress. For example, in the mAIchart project, different dashboards might use various models to represent how students are learning. Each model can produce slightly different results for the same student. Integrating multiple learner models is challenging when the models differ in granularity, domain descriptions, underlying assessment data, and other factors [16]. For example, dashboards often represent knowledge components (KCs) at varying levels of detail. This potential discrepancy in granularity complicates the process of merging data and aligning knowledge components effectively [16].

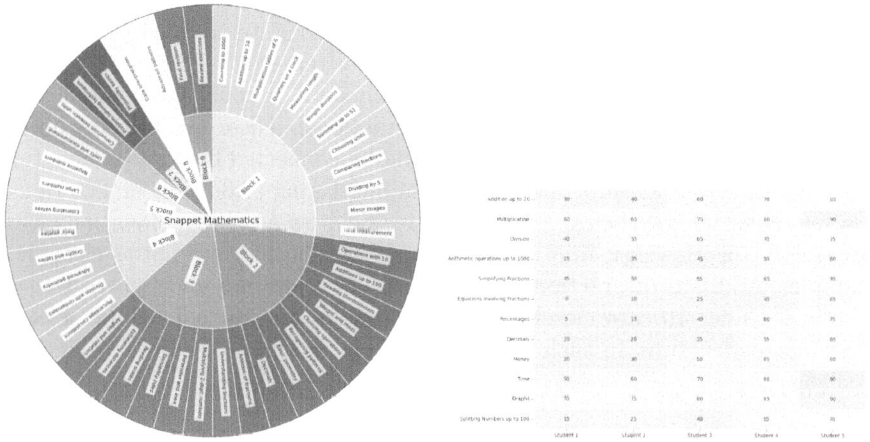

(a) Dashboard created based on Snappet (b) Dashboard created based on Lexima

Fig. 3. Comparison of two dashboards based on Dutch digital learning platforms.

Another challenge arises from the inconsistent definition of learning objectives and performance indicators across dashboards. For instance, one dashboard might track a competency like "simplifying fractions," while another categorizes it under a broader skill like "understanding fractions." Figs. 3a and 3b show distinct approaches to display learning objectives in digital learning dashboards. These differences make it difficult to merge data, as similar learning objectives may be measured differently, leading to potential misinterpretations. Some dashboards use different performance indicators—some might focus on task completion rates (e.g., percentage of tasks completed successfully), while others focus on grades or overall progress (e.g., final exam score or cumulative course completion). Additionally, dashboards may define performance and progress in various ways. One system might use percentage grades, while another uses normative scores. These differences in performance outcomes make it hard to align the "meaning" of data across systems. A score of 85% in one platform might correspond to "mastery of Level 3 skills" in another. We need to find ways to combine and compare the data from these different models to get a more consistent understanding of student performance.

5 Potential Ways to Combine Learner Models

One of the potential ways to address the challenges of merging different dashboards is through the use of ontologies [15]. These methods help align the meaning of terms and concepts across different systems by providing a shared framework for interpreting educational data. By building a knowledge graph of educational concepts, we can create a structure where nodes represent skills or knowledge areas, and edges denote equivalence, prerequisite, or hierarchical relationships between these concepts.

For instance, let's consider different learning objectives related to fractions. In one dashboard, the learning objective could be represented by a node defined as "Addition of Fractions", and in another dashboard, it could be represented as a node "Solving Algebraic Equations with Fractions". These two learning objectives, though related to different tasks, could be connected by an edge denoting a prerequisite relationship, indicating that mastering "Addition of Fractions" is a foundational step before progressing to "Solving Algebraic Equations with Fractions" [16]. However, in other cases, such relationships are not as clear. For example, "Addition of Fractions Under 100" and "Addition and Subtraction of Fractions Under 10" are closely related in content but do not exhibit a clear prerequisite structure. In such instances, combining these objectives becomes a challenge.

6 Plans for Future Work

At the start of the mAIchart project, teachers were interviewed to clarify the functionalities they expected in a dashboard. Based on these insights, my research focuses on integrating different learner models into a unified framework to improve how educational data is analyzed and presented. I will examine existing models used in Snappet [17], Gynzy [10], and CITO [6], identifying key similarities, partial overlaps (such as broad goals that map to finer subcategories), and elements with little overlap. To address these differences, I will develop a shared taxonomy or ontology. In cases with minimal alignment, I will analyze differences in input features, learning assumptions (e.g., emphasis on mastery or progress), and the methods used to generate insights, including scoring or predictive techniques. The integrated model will be validated by comparing its predictions with actual student performance across platforms using statistical analyses, such as correlation and accuracy measures. Development will proceed iteratively, incorporating teacher feedback at each stage.

References

1. Abyaa, A., Khalidi Idrissi, M., Bennani, S.: Learner modelling: systematic review of the literature from the last 5 years. Educ. Tech. Res. Dev. **67**, 1105–1143 (2019)
2. Akhuseyinoglu, K., Milicevic, A.K., Brusilovsky, P.: Who are my peers? Learner-controlled social comparison in a programming course. In: European Conference on Technology Enhanced Learning, pp. 1–15. Springer (2022)
3. Anonymous: Title withheld for review (2025). currently under review
4. Bower, M.: Affordance analysis-matching learning tasks with learning technologies. Educ. Media Int. **45**(1), 3–15 (2008)
5. Chen, J., Man, H., Yen, N.Y., Jin, Q., Shih, T.K.: Dynamic navigation for personalized learning activities based on gradual adaption recommendation model. In: Luo, X., Spaniol, M., Wang, L., Li, Q., Nejdl, W., Zhang, W. (eds.) ICWL 2010. LNCS, vol. 6483, pp. 31–40. Springer, Heidelberg (2010). https://doi.org/10.1007/978-3-642-17407-0_4
6. CITO: Cito - dutch national institute for educational measurement. https://www.cito.nl (2024). Accessed 07 May 2025
7. Dillenbourg, P., Self, J.: A framework for learner modelling. Interact. Learn. Environ. **2**(2), 111–137 (1992). https://doi.org/10.1080/1049482920020202
8. Gligorea, I., Cioca, M., Oancea, R., Gorski, A.T., Gorski, H., Tudorache, P.: Adaptive learning using artificial intelligence in e-learning: a literature review. Educ. Sci. **13**(12), 1216 (2023)
9. Graf, S., et al.: Automatic, global and dynamic student modeling in a ubiquitous learning environment. Knowl. Manage. E-Learn. **1**(1), 18 (2009)
10. Gynzy: Gynzy - interactive teaching software. https://www.gynzy.com (2024). Accessed 07 May 2025
11. Khosravi, H., Sadiq, S., Gasevic, D.: Development and adoption of an adaptive learning system: reflections and lessons learned. In: Proceedings of the 51st ACM Technical Symposium on Computer Science Education, pp. 58–64 (2020)
12. Klašnja-Milićević, A., Ivanović, M., Nanopoulos, A.: Recommender systems in e-learning environments: a survey of the state-of-the-art and possible extensions. Artif. Intell. Rev. **44**(4), 571–604 (2015). https://doi.org/10.1007/s10462-015-9440-z
13. Liu, H.L., Wang, T.H., Lin, H., Lai, C.F., Huang, Y.M.: The influence of affective feedback adaptive learning system on learning engagement and self-directed learning. Front. Psychol. **13**, 858411 (2022)
14. Molenaar, I.: Towards hybrid human-AI learning technologies. Eur. J. Educ. **57**(4), 632–645 (2022). https://doi.org/10.1111/ejed.12527, https://onlinelibrary.wiley.com/doi/abs/10.1111/ejed.12527
15. Nashed, N.N., Lahoud, C., Abel, M.H.: Ontology-based teacher-context data integration. In: 2022 IEEE/SICE International Symposium on System Integration (SII), pp. 809–814. IEEE (2022)
16. Pelánek, R.: Adaptive learning is hard: challenges, nuances, and trade-offs in modeling. Int. J. Artif. Intell. Educ., 1–26 (2024)
17. Snappet: Snappet - personalized learning in primary education. https://www.snappet.org (2024). Accessed 07 May 2025
18. Strielkowski, W., Grebennikova, V., Lisovskiy, A., Rakhimova, G., Vasileva, T.: Ai-driven adaptive learning for sustainable educational transformation. Sustain. Dev. (2024)

Enhancing Knowledge Tracing with Large Language Models (LLMs)

Duaa Baig[1,2(✉)], Diana Nurbakova[1], Sylvie Calabretto[1], and Baba MBaye[2]

[1] INSA Lyon, CNRS, Universite Claude Bernard Lyon 1, LIRIS, UMR5205, Lyon, France
{duaa.baig,diana.nurbakova,sylvie.calabretto}@insa-lyon.fr
[2] EffetB SAS, Lyon, France
baba@effetb.com

Abstract. In intelligent tutoring systems (ITS), knowledge tracing (KT) is a fundamental requirement for effective education and data mining. The main objective of KT is to model and predict the evolving understanding level of a student on different educational tasks. Traditional KT methods, such as the factor analysis method (FAM), Bayesian KT (BKT), and deep KT (DKT) approaches, have achieved high-performance effectiveness but often fail to identify reasoning processes, diverse learning trajectories, and complex interdependence relationships between skills associated with educational questions. The existing challenges highlight the requirement for personalization and contextual adaptability for effective student modeling. Addressing these challenges is important for building an ITS that can provide personalized educational experiences and support lifelong learning for diverse students.

Keywords: Knowledge Tracing · LLMs · AI in Education · E-Learning

1 Introduction

Knowledge tracing (KT) is an important task for ITS and educational domains [1,2]. It aims to predict the future performance of students based on their historical learning interactions, enabling ITS to dynamically adapt to the learning trajectories of students [3]. The traditional and deep learning-based KT approaches identify the student's state of understanding of any concept based on their historical interaction data. However, the traditional approaches struggle to perform deep contextual reasoning [4], personalized adaptation [5], and interpretations [7]. Graph-based knowledge tracing (GKT) methods have grown to be effective KT solutions by maintaining many vectors that represent students' conceptual competency [6]. These methods use conceptual relationships to update the vectors as students answer questions correctly or incorrectly, resulting in

A. I. Cristea et al. (Eds.): AIED 2025, CCIS 2590, pp. 384–390, 2025.
https://doi.org/10.1007/978-3-031-99261-2_41

more precise predictions. The need for manually annotated concept relations is a major drawback that restricts the use of GKT approaches in a practical environment [6]. Many real-world educational applications lack relationship information, which makes it challenging for human experts to classify them [3].

Inspired by recent advancements in natural language processing (NLP), the large language models (LLMs) provide a promising direction in the field of KT to address the existing limitations of traditional approaches [3]. In natural language understanding (NLU) and natural language generation, LLMs have shown amazing results across various domains, including medicine [8]. The capability of generating convincing, human-like text by LLMs across different ranges of knowledge areas makes it a valuable resource for educational applications [9].

In our research, we are exploring LLMs in the domain of KT to enhance the predictive performance of existing approaches and to address the existing challenges. In our initial work, we utilized LLMs for two main tasks: (a) to identify and tag the contextual relational interdependency of skills and (b) to identify the prediction capability of LLMs by setting them as a real-time prediction model. This work aims to enhance the performance of existing KT models while reducing the requirement to manually annotate the relational dependencies between skills. The initial results, though promising, reveal a need for further refinement in our approach to fully understand the potential of LLMs in the KT domain.

2 Related Work

Deep Knowledge Tracing [11] (DKT), one of the first deep learning (DL)-based KT models, provides significant advantages in terms of comprehensiveness and applicability [2]. DKT models students' knowledge levels and predicts their future performance using recurrent neural networks (RNNs). However, current DL-KT models, such as Attentive Knowledge Tracing [12] (AKT), have often prioritized predicted accuracy over interpretability and comprehensiveness [2].

Graph-based Knowledge Tracing [15] (GKT) methods represent the knowledge structure as a graph and model students' time-series mastery of each skill using Graph Neural Networks (GNNs) [16]. GKT can utilize pre-defined graph structures or learn implicit graph structures from data, showing higher performance and more interpretable predictions [16]. Dynamic Breadth Graph Convolutional Networks (DBGCN) employ breadth graph convolutional networks to extract representations of queries and knowledge points from dynamically generated topological graphs [17].

The integration of LLMs into KT is a relatively new area of research. Self-supervised Pre-training for KT (SPAKT) utilizes self-supervised learning to pre-train embedding representations without expensive human-expert annotations [3]. Our work builds upon these advancements by exploring LLMs to identify and represent the relationship interdependence in KT. Unlike SPAKT [3], which focuses on pre-training embeddings, our GKT-LLM approach automates the identification of skill dependencies and it directly incorporates the LLM-generated graph as contextual information into a GNN-based model. In another work [18], an LLM-based HGKT approach is proposed that models exercise dependencies using a hierarchical exercise graph (HEG), BERT-derived problem schemas, and LSTM to enhance the KT performance. Unlike HGKT, in GKT-LLM, we utilized LLMs focusing on skill-level dependencies instead of exercise-level dependencies due to the unavailability of exercise or question text in most publicly available datasets. While in our other work, LLM-KT, we used GPT 3.5 for real-time predictions that enable dynamic adaptability, which is missing in [3,4,18].

3 Theoretical Framing and Proposed Solution

Our research is based on the theoretical foundations of AI in education and personalized learning. The goal is to develop a model that can accurately track students' knowledge states and provide tailored learning opportunities. To achieve this, we have assessed two approaches: (a) **GKT-LLM:** a graph-based technique that leverages the strength of LLMs to recognize the interdependence score of skills and (b) **LLM-KT:** an LLM-based real-time prediction model.

3.1 GKT-LLM

The theoretical foundation of our methods is the idea that by using LLMs, we can identify and incorporate the dependency relationships across knowledge skills without doing manual annotations. By combining GKT [15] and LLMs, a graph-based GKT-LLM approach is proposed in our work that aims to provide students with a more individualized educational experience.

Methodology and Results. The experimental setup includes the following steps: **(a) Dataset:** For both GKT-LLM and LLM-KT, the publicly available ASSIST2009 dataset is used and pre-processed by removing interactions where skill names are missing or interactions are duplicated. The dataset is then randomly split into training, validation, and testing sets with a 70:15:15 ratio. The sequences of 101 interactions per student, formatted as skill-response tuples, are used in both experiments. For GKT-LLM, the LLM processes all the skill tags of the dataset to construct an adjacency matrix representing the dependency relationship between skills.**(b) LLM-based Graph Generation:** The skills are analyzed using GPT-3.5-turbo and OpenAI's API, and a graph illustrating their interdependencies is generated. The LLM is prompted with target and candidate skills and descriptions to generate edges representing relational dependencies of skills. **(c) Model Training:** A GNN-based model, GKT [15], is used to model knowledge states and predict students' performance. The GKT is trained using the randomly split training set, it incorporates the LLM-generated graph as contextual information during training. **(d) Evaluation:** The performance of both GKT-LLM and LLM-KT is evaluated by comparing their AUC score and average loss mean with the results of five baseline models mentioned in Table 1. **(e) Results:** The results of our GKT-LLM achieved the highest performance, compared with the five baseline models, with an AUC of 81.14%. This result demonstrates the effectiveness of using LLM-generated graphs as contextual data in a GNN-based KT model. The complete results of GKT-LLM and baseline models are presented in Table 1.

3.2 LLM-KT: LLM as a Prediction Model

In the second experiment, we developed an LLM-based prediction framework, LLM-KT, using OpenAI's GPT-3.5-turbo to predict future performance. The LLM was asked to predict the student's performance for the target question based on their historical interactions and mastery level of the corresponding skill. With a naive AUC of 56.85%, it highlights the areas that require improvement.

Methodology and Results. The experimental setup involves the following steps: **(a) Dataset:** The same test set split of ASSIST2009 generated during the GKT-LLM experiment is used for LLM-KT as well to allow fair comparison, with the addition of LLM-generated sample questions-text per skill. We generated five sample questions text per skill and mapped them to test set interactions so that we may analyze LLMs for text-based questions instead of using IDs. For each student ID and target skill, the mapped sample target question, historical interactions, and current mastery level of the target skill are retrieved. The historical interactions are formatted as a list of binary responses (e.g., [0, 1, 0, 1, 1, ..]) and included in the prompt alongside the knowledge level k at time t i.e., k_t of the student for the target skill. The prompts in both experiments were constructed after testing multiple prompts to identify which prompt provides a more accurate and correctly formatted result. **(b) LLM-based Prediction:** By using the student's current knowledge level and the previous interactions with the target skill, the GPT-3.5-turbo generates predictions for the sample target question and skill. **(c) Performance Evaluation and Knowledge State Update:** The generated prediction is then compared with the actual response, the LLM model is responsible to update the learner's new comprehension level for the target skill. Furthermore, the predictions are also stored for later use so that the model performance can be determined using the loss mean and AUC. **(d) Results:** The experimental results of LLM-KT using the same test set, presented in Table 1 are less favorable. The attained AUC of LLM-KT is 56.85%, there can be several factors contributing to this low performance. First and most importantly, the static predictions are observed in LLM-KT, for instance, during predictions, it consistently predicted 0.8 probability for student 1285 across 44 continuous interactions despite having varying true labels (among which true label of 14 interactions is 0, and the true label of remaining 30 interactions is 1), which shows that the LLM (GPT-3.5-turbo) struggles to dynamically adapt to recent interaction trends. There can be multiple potential reasons for this limitation, possibly due to prompt design, hallucinations, or lack of fine-tuning. However, we still believe that there is room for improvement.

Table 1. Comparative results of proposed and baseline approaches

Model	DKT	DKT+	DKVMN	GKT-PAM	GKT-MHA	SAKT	**GKT-LLM**	**LLM-KT**
Avg. AUC (%)	79.67	80.22	78.34	79.20	80.01	78.92	**81.14**	**56.85**
Avg. Loss Mean	0.59	0.62	0.50	0.49	0.48	0.50	0.47	0.66

4 Upcoming Goals

To improve this research work and results, the future direction should focus on addressing the limitations based on the findings of this study. Some potential directions for our future work include: **(a) Refine Graph Generation:** As a next step, we plan to explore different prompting techniques, incorporating domain-specific knowledge using retrieval augmented generation (RAG), and fine-tuning the LLM to identify more accurate relationships. **(b) Interpretability:** By using LLMs, we can also work on improving the interpretability of the KT models by visualizing graphs and generating explanations for predictions. **(c) Diverse Datasets:** It is also important to evaluate the models on diverse datasets from different educational domains to assess their generalizability and robustness. **(d) Exploring different LLM architectures:** Investigating the use of different LLM architectures can also contribute to improve performance.

4.1 Expected Contributions

The successful research in this direction will contribute to the AIED community in the following ways:

- **LLM-enhanced KT:** The study offers a KT approach that identifies and illustrates the interdependencies across skills by utilizing LLMs. The contribution could improve models' scalability, interpretability, and performance.
- **Personalized Learning:** By considering interdependencies between skills, the proposed models can provide a better learning experience, such as customized learning materials, focused feedback, and adaptable learning paths.
- **Reduced Reliance on Manual Annotation:** LLM enhances the scalability and usefulness of KT by reducing its reliance on manual annotations by automatically generating the skills interdependencies.
- **Integration of Computer Science (CS) and Learning Sciences (LS):** This study bridges the gap between CS and LS by integrating theoretical frameworks from both fields. By combining the concepts of personalized learning with the processing power of LLMs it can enhance KT in education.

The development of AIED and the production of more specialized and effective KT solutions are expected outcomes of this research. By improving the accuracy and interpretability of KT models, this research can help teachers better understand students' learning processes and provide the assistance they need.

5 Conclusion

The study suggests that GKT-LLM and LLM-KT are simple yet efficient starting points for using LLMs in the KT domain. GKT-LLM's 81.14% AUC demonstrates the strength of graph-based LLM, while LLM-KT's performance of 56.85% AUC indicates potential in using LLMs for sequence-based prediction. The experiments highlight a way for the future in which LLMs can enhance AI-powered education and provide more adaptive support. The main goal of this research is to improve KT processes and support AI-assisted online learning.

References

1. Cully, A., Demiris, Y.: Online knowledge level tracking with data-driven student models and collaborative filtering. IEEE Trans. Knowl. Data Eng. **32**(10), 2000–2013 (2019)
2. Zhou, Y., Han, W., Chen, J.: Revisiting applicable and comprehensive knowledge tracing in large-scale data. arXiv preprint: arXiv:2501.14256 (2025)
3. Ma, Y., et al.: SPAKT: A self-supervised pre-training method for knowledge tracing. IEEE Access **10**, 72145–72154 (2022)
4. Fu, L., et al.: SINKT: a structure-aware inductive knowledge tracing model with large language model. In: Proceedings of the 33rd ACM International Conference on Information and Knowledge Management (2024)
5. Scarlatos, A., Baker, R.S., Lan, A.: Exploring knowledge tracing in tutor-student dialogues using LLMs. In: Proceedings of the 15th International Learning Analytics and Knowledge Conference (2025)
6. Long, T., et al.: Automatical graph-based knowledge tracing. EDM (2022)
7. Liang, Z., et al.: GELT: a graph embeddings based lite-transformer for knowledge tracing. PLoS ONE **19**(5), e0301714 (2024)
8. Nori, H., et al.: Capabilities of GPT-4 on medical challenge problems. arXiv preprint: arXiv:2303.13375 (2023)
9. Borger, J.G., et al.: Artificial intelligence takes center stage: exploring the capabilities and implications of ChatGPT and other AI-assisted technologies in scientific research and education. Immunol. Cell Biol. **101**(10), 923–935 (2023)
10. Tan, Z., et al.: Large language models for data annotation and synthesis: a survey. arXiv preprint: arXiv:2402.13446 (2024)
11. Piech, C., et al.: Deep knowledge tracing. In: Advances in Neural Information Processing Systems, vol. 28 (2015)
12. Mandlazi, J., Jadhav, A., Ajoodha, R.: Evaluating deep sequential knowledge tracing models for predicting student performance. In: 2021 IEEE Asia-Pacific Conference on Computer Science and Data Engineering (CSDE). IEEE (2021)
13. Xu, L., et al.: Modeling student performance using feature crosses information for knowledge tracing. IEEE Trans. Learn. Technol. (2024)
14. Wu, J., Wang, G., Wang, Y.: Application of sparse attention mechanism and bidirectional GRU neural networks in knowledge tracing. In: 2023 6th International Conference on Software Engineering and Computer Science (CSECS). IEEE (2023)
15. Nakagawa, H., Iwasawa, Y., Matsuo, Y.: Graph-based knowledge tracing: modeling student proficiency using graph neural network. In: IEEE/WIC/ACM International Conference on Web Intelligence (2019)

16. Nakagawa, H., Iwasawa, Y., Matsuo, Y.: Graph-based knowledge tracing: modeling student proficiency using graph neural networks. Web Intell. **19**(1–2) (2021)
17. Hu, P., et al.: DBGCN: a knowledge tracing model based on dynamic breadth graph convolutional networks. Int. J. Web-Based Learn. Teach. Technol. (IJWLTT) **19**(1), 1–20 (2024)
18. Tong, H., et al.: HGKT: introducing hierarchical exercise graph for knowledge tracing. arXiv preprint: arXiv:2006.16915 (2020)

Towards Mining Effective Pedagogical Strategies from Learner–LLM Educational Dialogues

Liqun He(✉), Manolis Mavrikis, and Mutlu Cukurova

University College London, London, UK
{liqun.he,m.mavrikis,m.cukurova}@ucl.ac.uk

Abstract. Dialogue plays a crucial role in educational settings, yet existing evaluation methods for educational applications of large language models (LLMs) primarily focus on technical performance or learning outcomes, often neglecting attention to learner–LLM interactions. To narrow this gap, this AIED Doctoral Consortium paper presents an *ongoing* study employing a dialogue analysis approach to identify effective pedagogical strategies from learner–LLM dialogues. The proposed approach involves dialogue data collection, dialogue act (DA) annotation, DA pattern mining, and predictive model building. Early insights are outlined as an initial step toward future research. The work underscores the need to evaluate LLM-based educational applications by focusing on dialogue dynamics and pedagogical strategies.

Keywords: LLMs · Dialogue Analysis · Pedagogical Strategies

1 Problem Statement

With the growing use of large language models (LLMs) in various educational applications, evaluating their quality has become an increasingly important task. Existing studies primarily focus on the technical performance or learning outcome-related measures. Specifically, in the field of computer science, *technical evaluations* focus on LLMs' performance predominantly using non-interactive benchmarks (e.g., accuracy of single-turn outputs) [13]. In the field of education, researchers focus on *learning outcomes* following their use, including affective factors such as anxiety [23], willingness [5], and self-confidence [2], as well as cognitive aspects like knowledge and skill development [4, 21].

However, existing evaluations often overlook the dialogue processes between learners and LLMs, which is both necessary and important to consider. On the one hand, recent studies have shown that superior single-turn performance does not necessarily lead to better multi-turn interactions [11, 13], highlighting the need to move beyond single-turn evaluations. On the other hand, prior research on human tutoring dialogues has demonstrated that patterns of dialogue acts are correlated with student learning gains [3, 25], making dialogue analysis a promising approach for understanding and evaluating the pedagogical effectiveness of LLMs. Through such analysis, we can gain a deeper understanding of the *mechanisms* through which educational LLMs contribute

A. I. Cristea et al. (Eds.): AIED 2025, CCIS 2590, pp. 391–396, 2025.
https://doi.org/10.1007/978-3-031-99261-2_42

to learning gains. For instance, identifying when and which dialogue acts trigger student motivation or self-reflection may inform the design of educational LLMs that promote deeper engagement and learning.

Towards narrowing the gap, this study proposes a method for mining effective pedagogical strategies from learner–LLM dialogues. Using data from English language learners interacting with a voice chatbot for speaking practice, it investigates the following research question: **'What are the sequential learner-LLM dialogue patterns in effective and less effective tutoring sessions?'**.

Through this preliminary work, we seek to deepen the understanding of learner-LLM dialogue dynamics, identify effective pedagogical strategies, and potentially contribute to the development of educational LLM evaluation methods by emphasising dialogue dynamics and pedagogical strategies as focal points.

2 Theoretical Basis: Content, Intention, and Speech Act Theory

Searle's Speech Act Theory serves as the foundational theoretical framework for dialogue analysis in this study, outlining three key principles:

(1) **An utterance conveys both its content and intention.** Firstly, Speech Act Theory posits that when we speak, we are not just conveying meanings but also aiming to achieve specific goals [19, 20]. That is, every utterance has a *dual* function: it conveys both a semantic representation (the content) and a pragmatic identification (the intention) [7].
(2) **The content of an utterance is the result of its hidden intention.** Secondly, the theory clarifies the relationship between content and intention: the literally expressed content is the result of the speaker's hidden intentions. An utterance, therefore, is more than just a collection of words; it represents an action carrying meanings beyond its literal interpretation, reflecting the speaker's underlying intentions [19]. For example, when a friend says, 'I am hungry', he or she is not just describing their physical state but aiming to communicate their desire to have lunch.
(3) **The hidden intention can be inferred from the expressed content.** Third, the theory emphasises that understanding intention is more important than content alone, and suggests that the hidden intention can be inferred from the expressed content. In this theoretical framework, the intention behind an utterance is referred to as a 'speech act' (or interchangeably 'dialogue act' [22], 'dialogue move' [24], or 'communicative act' [9]; in this study, we use the term 'dialogue act'). A dialogue act (DA) represents the 'function of the meaning of the sentence' [19]. Further explain that to fully comprehend an utterance, one must go *beyond* its surface-level meaning and uncover the intention behind it (i.e., to identify its DA).

According to Speech Act Theory, tutors' intentions (or, in the context of this paper, the intentions of educational LLMs) are conveyed and reflected by what they say and when they say it [1, 7] and through various forms of dialogue, tutors (or educational LLMs) achieve various pedagogical goals [17]. This framework provides a theoretical foundation for inferring pedagogical intentions by analysing dialogue turns generated by LLMs or expressed by learners, enabling the analysis of pedagogical strategies from learner-LLM dialogues.

3 Related Work

Learning Analytics Approach. Prior to the widespread adoption of educational LLM applications, researchers had already analysed human-human DAs to evaluate and predict tutoring quality. For example, [14] examined 50 tutoring sessions across mathematics, chemistry, and physics, applying a BERT model for DA annotation and the TraMineR package for sequential pattern mining. Their findings showed successful sessions—those resulting in correct or partially correct student solutions—featured tutors using more thought-provoking questions and informational hints, and fewer irrelevant statements. Similarly, [3] analysed 2,250 min of one-to-one online mathematics tutoring using manual annotations, primarily focusing on tutors' dialogic behaviours but also including other interactive behaviours (e.g., writing on the virtual classroom interface). Sequential pattern mining (CM-SPAM algorithm) identified effective pedagogical strategies, such as prompting student self-correction and allowing appropriate pauses for reflection. These patterns then successfully informed decision-tree classifiers (JRIP and J-48) to distinguish tutor effectiveness. This work demonstrates the potential of behavioural action patterns for both identifying effective pedagogical strategies and monitoring the quality of online tutoring sessions.

LLM Evaluation for Dialogue Tasks. Within computer science, recent studies have also begun moving beyond automatic metrics alone, incorporating human evaluations to measure LLM performance on various dialogue tasks. For example, [13] conducted *turn-level* human surveys evaluating emotional connections in social dialogues (e.g., empathy, fluency, sensibility, and specificity). Similarly, Google's LearnLM team [10] assessed the pedagogical effectiveness of educational LLMs in conversational tutoring through learner and pedagogical rater feedback at both *turn-level* (e.g., "Does this turn inspire interest?") and *conversation-level* (e.g., "Which tutor more closely resembles an excellent human tutor?"). Follow-up research [12] further compared LLM models' pedagogical performance by inviting experienced tutors to complete *conversation-level* questionnaires after reviewing learner-LLM conversation transcripts.

Limitations of Previous Research. Previous analyses of human-human dialogues offered valuable methodological insights for assessing human-LLM interactions. However, [18] found significant disparities between AI-generated and human-authored texts. Given these differences and the shift from human-human to human-LLM learning environments, previous findings cannot be directly applied and require further validation.

On the other hand, recent developments in LLM evaluation, despite beginning to focus on dialogue, still fail to effectively capture pedagogical strategies. A key reason is that pedagogical strategies are typically represented by *sequences* of dialogue acts rather than isolated dialogue turns [15]. For example, the 'scaffolding' strategy—where learners work independently while tutors provide minimal assistance—can only be identified through a series of hints delivered in the form of questions. Current LLM evaluations primarily operate at either the *micro* (single-turn) or *macro* (full-conversation) levels. However, micro-level evaluations may be too narrow to capture pedagogical strategies, while macro-level evaluations may be too broad to assess LLMs' ability to implement specific strategies and follow pedagogical instructions. Focusing on DA patterns offers a potential middle ground (a *meso*-level) between atomistic and holistic approaches, providing a more balanced dimension for understanding and evaluating pedagogical strategies in learner-LLM dialogues.

4 Proposed Solution: A Dialogue Approach

The proposed solution aims at employing a dialogue analysis approach to identify predominant pedagogical strategies in learner-LLM interactions, consisting of four key components:

(1) **Dialogue Data Collection.** Establishing an accessible, secure, and user-friendly method for storing and managing dialogue data across multiple students and sessions is the first step in enabling further analysis. To achieve this, we developed a platform that interfaces with LLMs via application programming interfaces (APIs) and securely stores dialogue data in a cloud database. Each recorded dialogue entry includes a unique identifier, timestamp, participant role (student or chatbot), model name, user ID, session ID (defined as continuous dialogue with gaps under 15 min), and dialogue transcript content.

(2) **Dialogue Act Annotation.** Language is a complex system, making it difficult to infer pedagogical intentions (i.e., DAs). Coding systematically addresses this by developing a coding scheme with descriptors (codes) to capture key characteristics of utterances and applying it to annotate dialogue turns [9, 26], thereby converting transcribed dialogues into structured and analysable data [15]. For this purpose, we adapted a preliminary coding scheme tailored to our context of oral English practice. This scheme is primarily based on the SETT coding scheme [27, 28] and the Cambridge Educational Dialogue Research Group's SEDA [9] and T-SEDA [26] coding schemes. We plan to further incorporate aspects from the latest Tech-SEDA for technology-mediated contexts [8]. Given that this is the pilot phase, we selected only three students with the most noticeable progress (High Progress (HP) group) and three students with the least progress (Low Progress (LP) group). We then collected dialogue data from each student's practice sessions on the first and last day, resulting in a total of 6 dialogues per group. The average number of dialogue turns per session was similar between the HP group (83.3 turns) and the LP group (86.7 turns). However, students in the HP group produced longer utterances, averaging 15.3 words per turn compared to 11.5 words in the LP group. In total, we manually annotated 1,020 dialogue turns across both groups for subsequent pattern mining.

(3) **Dialogue Act Pattern Mining.** To identify predominant DA patterns, we then applied the CM-SPAM algorithm [6] with a minimum pattern length of 2, a maximum gap of 1, and a minimum support of 50%. An example of insights gained from the results is as follows: the HP group exhibited the pattern [t]Q - [s]R - [s]Q (bot question–student response–student question), while the LP group uniquely showed [t]Q–[s]Q. Transcript analysis further suggests that the absence of [s]R in the LP group indicates learners' difficulty in understanding the initial question ([t]Q), prompting students to ask a new question rather than respond.

(4) **Predictive Model Building.** Building upon identified dialogue patterns as indicators of pedagogical strategies, our next step involves training predictive models (e.g., decision tree classifiers following [3], or Bayesian Networks as described in [16]).

Such models provide interpretable insights, enabling systematic assessment and targeted improvement of educational LLM systems.

5 Expected Contributions

(1) **For LLM System Design.** This proposed method seeks to identify predominant DA patterns in learner-LLM interactions, providing insights into both effective and ineffective pedagogical strategies within LLM-mediated learning environments. These insights can potentially support educators, designers, and researchers in understanding learner-LLM interaction dynamics, informing and optimising instructional design.

(2) **For LLM System Evaluation.** This ongoing work highlights the need for evaluating educational LLMs by focusing on learner-LLM dialogue dynamics, moving beyond solely single-turn evaluations or learning gains. This potentially contributes to the advancement of methods for evaluating the pedagogical performance of LLMs.

References

1. Chi, M.T.H., et al.: Learning from human tutoring. Cogn. Sci. **25**(4), 471–533 (2001). https://doi.org/10.1207/s15516709cog2504_1
2. Cooray, S., et al.: Kavy: fostering language speaking skills and self-confidence through conversational AI. In: Proceedings of the Augmented Humans International Conference 2024 (Melbourne VIC Australia, Apr. 2024), pp. 226–236 (2024)
3. Cukurova, M., et al.: A learning analytics approach to monitoring the quality of online one-to-one tutoring. EdArXiv (2021)
4. Duong, T., Suppasetseree, S.: The effects of an artificial intelligence voice Chatbot on improving Vietnamese undergraduate students' English speaking skills. Int. J. Learn., Teach. Educ. Res. **23**(3), 293–321 (2024)
5. Fathi, J., et al.: Improving EFL learners' speaking skills and willingness to communicate via artificial intelligence-mediated interactions. System **121**, 103254 (2024). https://doi.org/10.1016/j.system.2024.103254
6. Fournier-Viger, P., et al.: Fast vertical mining of sequential patterns using co-occurrence information. advances in knowledge discovery and data mining. In: Tseng, V.S., et al., (eds.), pp. 40–52 Springer International Publishing (2014)
7. Grosz, B., et al.: Chapter 6 discourse and dialogue. survey of the state of the art in human language technology. In: Uszkoreit, H., et al., (eds.) (1995)
8. Hennessy, S.: Analysing educational dialogue around shared artefacts in technology-mediated contexts: a new coding framework. Classroom Discourse, 1–35 (2024). https://doi.org/10.1080/19463014.2024.2339346
9. Hennessy, S., et al.: Developing a coding scheme for analysing classroom dialogue across educational contexts. Learn., Cult. Soc. Inter. **9**, 16–44 (2016). https://doi.org/10.1016/j.lcsi.2015.12.001
10. Jurenka, I., et al.: Towards responsible development of generative AI for education: an evaluation-driven approach (2024)
11. Kwan, W.-C., et al.: MT-Eval: a multi-turn capabilities evaluation benchmark for large language models. arXiv (2024)
12. LearnLM Team et al.: LearnLM: Improving Gemini for Learning. arXiv (2024)

13. Lee, M., et al.: Evaluating human-language model interaction. arXiv (2024)
14. Lin, J., et al.: Is it a good move? Mining effective tutoring strategies from human-human tutorial dialogues. Futur. Gener. Comput. Syst. **127**(2022), 194–207 (2022). https://doi.org/10.1016/j.future.2021.09.001
15. Maharjan, N., et al.: Discovering effective tutorial strategies in human tutorial sessions, vol. 7 (2018)
16. Mavrikis, M.: Modelling student interactions in intelligent learning environments: constructing Bayesian networks from data. Int. J. Artif. Intell. Tools. **19**, 733–753 (2010). https://doi.org/10.1142/S0218213010000406
17. Roscoe, R.D., Chi, M.T.H.: Understanding tutor learning: knowledge-building and knowledge-telling in peer tutors' explanations and questions. Rev. Educ. Res. **77**(4), 534–574 (2007). https://doi.org/10.3102/0034654307309920
18. Sardinha, T.B.: AI-generated vs human-authored texts: a multidimensional comparison. Appl. Corpus Linguist. **4**(1), 100083 (2024). https://doi.org/10.1016/j.acorp.2023.100083
19. Searle, J.R.: Speech Acts: An Essay in the Philosophy of Language. Cambridge University Press, Cambridge (1969)
20. Searle, J.R.: What is a Speech Act? Philosophy in America. In: Black, M. (ed.) Allen and Unwin, pp. 221–239 (1965)
21. Shikun, S., et al.: AI Chatbots: developing English language proficiency in EFL Classroom. Arab World Engl. J. **1**(1), 292–305 (2024). https://doi.org/10.24093/awej/ChatGPT.20
22. Stolcke, A., et al.: Dialogue act modeling for automatic tagging and recognition of conversational speech. Comput. Linguist. **26**(3) (2000)
23. Tai, T.-Y., Chen, H.H.-J.: Improving elementary EFL speaking skills with generative AI Chatbots: exploring individual and paired interactions. Comput. Educ. **220**, 105112 (2024). https://doi.org/10.1016/j.compedu.2024.105112
24. Traum, D.R.: Speech acts for dialogue agents. foundations of rational agency. In: Wooldridge, M., Rao, A. (eds.), pp. 169–201. Springer Netherlands (1999)
25. Vail, A.K., Boyer, K.E.: Identifying effective moves in tutoring: on the refinement of dialogue act annotation schemes. In: Trausan-Matu, S., et al. (eds.) Intelligent Tutoring Systems, pp. 199–209. Springer International Publishing (2014)
26. Vrikki, M., et al.: The teacher scheme for educational dialogue analysis (t-Seda): developing a research-based observation tool for supporting teacher inquiry into pupils' participation in classroom dialogue. Int. J. Res. Method Educ. **42**(2), 185–203 (2019). https://doi.org/10.1080/1743727X.2018.1467890
27. Walsh, S.: Classroom Discourse and Teacher Development. Edinburgh University Press (2013)
28. Walsh, S.: Investigating Classroom Discourse. Taylor & Francis Group, Routledge (2006)

Intelligent Support for Practice Goal Setting to Enhance Learning

Conrad Borchers(✉), Kenneth R. Koedinger, and Vincent Aleven

Carnegie Mellon University, Pittsburgh, PA, USA
{cborcher,krk,aleven}@cs.cmu.edu

Abstract. Setting practice goals, which helps students regulate their effort toward achieving engagement and mastery, can enhance the benefits of active learning. However, traditional goal-setting approaches, such as homework contingency contracts, often lack frequent feedback and require substantial human intervention. The presented research investigates integrating intelligent and scalable goal-setting support within active learning environments. In recent research, we have evaluated the impact of goal-setting contracts, performance feedback, and scalable goal support with rewards on student effort and learning. The proposed research aims to study student achievement trajectories related to achievement and the distinct impact of adaptive goal feedback. To that end, we propose developing an adaptive goal-setting dashboard that automates feedback and recommendations to support students in setting and refining their goals. Findings from a 12-week study with 110 middle school students in a hybrid tutoring program show that data-supported goal setting led to about a 25% increase in weekly practice time and about a 40% increase in skills mastered per week. These results indicate that intelligent goal setting can enhance engagement and learning while minimizing teacher workload. This research contributes to practical advancements in AI-assisted active learning and theoretical insights into self-regulatory processes related to student effort in active learning.

Keywords: goal setting · hybrid tutoring · K-12 · STEM · active learning

1 Problem Statement and Research Questions

Active learning—central to AIED systems such as intelligent tutors and teachable agents—engages students in problem-solving rather than passive instruction. When instruction is effectively tailored to individual learners, it improves learning outcomes across levels of prior knowledge [14]. However, the success of such instruction critically depends on students' motivation to engage in sustained, effortful practice—a principle supported across demographics and classroom settings through cognitive modeling studies [8,14].

Practice goal setting—where students define quantified objectives tied to their effort (e.g., solving a certain number of problems or spending a set amount

A. I. Cristea et al. (Eds.): AIED 2025, CCIS 2590, pp. 397–402, 2025.
https://doi.org/10.1007/978-3-031-99261-2_43

of time)—has long been shown to improve engagement and achievement in education [3,13]. Traditionally, these goals were implemented via paper-based contracts with accountability partners (e.g., parents or teachers) and accompanied by contingent rewards [3]. However, such implementations are often limited in scalability and lack integration with digital learning environments [17].

In contrast, scalable goal-setting solutions in AIED have largely emphasized support for self-regulated *learning processes*—such as concept mastery, applying a cognitive strategy, or consulting resources—typically once practice is already underway [4,15]. Less attention has been paid to more foundational self-regulation activities relevant to goal achievement, including managing effort and performance reflection. Although these behaviors have been linked to academic outcomes, they are often examined through distinct literatures on performance management and goal attainment [9,16], beyond the learning sciences [11].

This research argues for a stronger integration of foundational self-regulation support into AIED environments by leveraging their adaptive and data-driven capabilities. It explores scalable practice goal-setting solutions that require minimal human oversight, drawing on learner performance log data [17]. While prior efforts to foster self-regulation in AIED have typically employed non-adaptive interventions—such as growth mindset messaging [1]—research on self-regulated learning has shown the benefits of adaptive, just-in-time support [4]. The present research investigates how adaptive and data-informed elements can support self-regulation in the context of goal setting with contingent rewards. This approach has demonstrated success in behavioral health interventions [2]. For learners, adaptive goal support could improve effort calibration and goal adjustment [10,12], part of the foundational self-regulation activities required for successful goal pursuit, identified in past work [16]. Three research questions guide the research:

RQ1: How does regular goal achievement feedback influence students' effort?
RQ2: How do intelligent goal-setting support compare to traditional goal-setting approaches and business-as-usual active learning in AIED systems?
RQ3: To what extent is data-driven goal-setting support effective in real-world classroom settings regarding learning and goal achievement?

2 Theoretical Framework

This research builds on foundational goal-setting theories [16] and self-regulated learning models [12]. Three key principles inform the research and tool design:

Principle 1: Feedback is effective for general domain learning but also for self-regulation [18]. Regularly provided feedback improves students' ability to regulate their effort toward goal achievement.

Principle 2: Goal recommendations should be adapted to students' historical performance averages [2]. These adjustments enhance the probability of goal achievement and its engagement benefits [19].

Principle 3: Goal-setting support should promote accountability and rewards through a human tutor, teacher, or another caregiver, similar to past research on goal-setting contracts, while remaining scalable [3,13].

These principles inform the design and evaluation of intelligent practice goal-setting support during active learning in AIED systems, which provide real-time data for performance feedback and goal recommendations [17].

3 Proposed Solution

The proposed solution, incorporating design principles from past research, is an intelligent practice goal-setting support designed to:

- Actively involve learners in goal setting and retain learner control over goals, as desired by K-12 students [6]
- Regularly provide feedback on goal progress [17] and adjust goal difficulty based on performance [2] to increase the probability of goal achievement.
- Reduce the need for teacher or researcher intervention by automating goal recommendations and scaling tutoring [5,6].

These three principles are hypothesized to improve student effort regulation and engagement, leading to more learning opportunities in AIED learning software, which are expected to lead to more content mastery (Fig. 1).

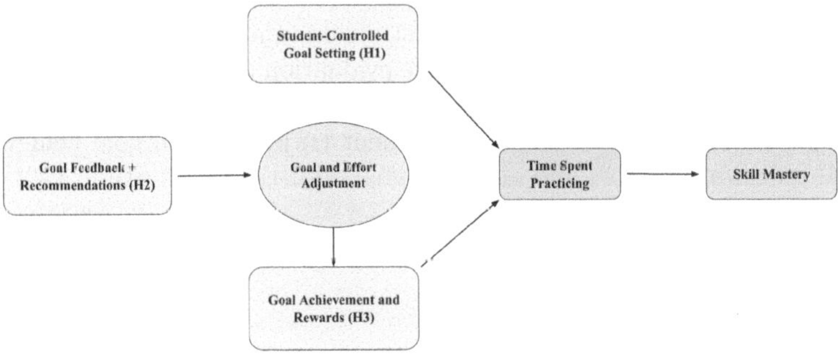

Fig. 1. Theory of change diagram with three hypotheses, where student goal setting (H1), goal adjustments (H2), and the impact of goal achievement (H3) are studied in relation to practice engagement and skill mastery outcomes.

The goal-setting support integrates with any learning system that tracks student activity, informing custom modules (built as part of this research) that dynamically suggest goals. An example goal recommendation based on a pen-and-paper implementation, extending past goal contracts [3], is in Fig. 2.

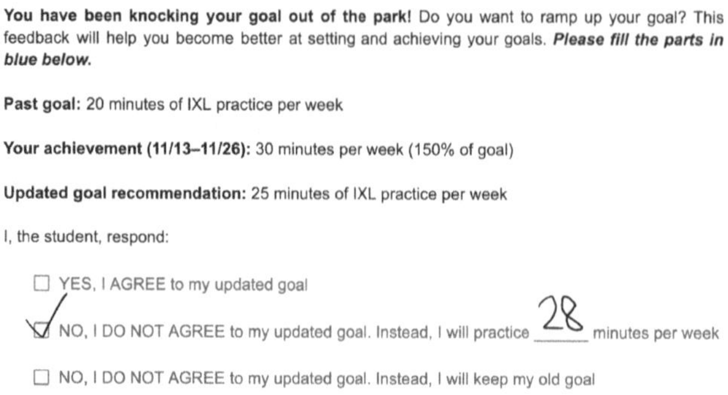
You have been knocking your goal out of the park! Do you want to ramp up your goal? This feedback will help you become better at setting and achieving your goals. ***Please fill the parts in blue below.***

Past goal: 20 minutes of IXL practice per week

Your achievement (11/13–11/26): 30 minutes per week (150% of goal)

Updated goal recommendation: 25 minutes of IXL practice per week

I, the student, respond:

☐ YES, I AGREE to my updated goal

☑ NO, I DO NOT AGREE to my updated goal. Instead, I will practice 28 minutes per week

☐ NO, I DO NOT AGREE to my updated goal. Instead, I will keep my old goal

Fig. 2. Illustration of student interaction with an updated goal recommendation based on historical performance averages. Students retain final goal control.

4 Methodology

This research follows a multi-phase research design:

- **Study 1:** A 12-week within-subjects quasi-experimental study with about 100 middle school students in a hybrid tutoring program, analyzed through an interrupted time series model (completed).
- **Study 2:** A randomized controlled experiment comparing adaptive goal setting to assigning a static goal to students (control condition) in an adaptive learning system, including about 150 students in a cross-over design.
- **Study 3:** Development and qualitative evaluation of an adaptive goal-setting dashboard with goal recommendations using design prototyping methods.
- **Study 4:** Process-level analysis of student trajectories in goal achievement based on reward factors and adaptive goal adjustments in Study 1+2 [19].

5 Hypotheses

Study 1 [5] and progress on Study 3 [7] are published in this year's AIED proceedings. For Studies 2 and 4, which are underway, based on the theoretical framework and proposed methods, we formulate the following hypotheses:

- **H1:** Student-controlled goal setting is more effective for practice time and skill mastery than static, assigned goals.
- **H2:** Goal feedback, recommendation, and adjustment opportunities will increase goal achievement compared to those in the control condition.
- **H3:** Rewards associated with goal achievement will moderate longitudinal trends in practice time and skill mastery, enhancing outcomes.

6 Expected Contributions

This research contributes to theoretical and practical advancements in AI-supported active learning, where engagement remains a fundamental challenge to improve learning outcomes for all students [14]. The anticipated contributions include:

- Theory-informed practice goal-setting support in AIED, bridging research on goal setting, self-regulation, and learning technologies [2,6,14,16].
- Empirical evidence on how frequent, adaptive goal recommendations influence student engagement and learning outcomes [9,12].
- A scalable goal-setting system that reduces teacher workload while maintaining student autonomy and personalized learning pathways [6,17].
- Insights into how adaptive feedback loops shape self-regulation and enhance learning persistence during technology-supported active learning [10,14].

7 Summary and Preliminary Findings

This research investigates the role of intelligent practice goal setting in fostering engagement and self-regulation in active learning environments. By integrating theoretical insights from goal-setting and models of self-regulation with empirical evaluations of intelligent tutoring, this work advances the understanding of how AI can support student engagement and achievement in effortful active learning. The findings contribute to the broader discourse on adaptive learning, offering a scalable framework for embedding goal-setting strategies into AI-assisted educational platforms. Future directions include refining adaptive algorithms, extending deployment to different learning contexts, and exploring long-term effects on student self-regulation, learning, and persistence.

Findings from Study 1 suggest that integrating goal-setting support into hybrid tutoring significantly boosts engagement and learning. A 12-week study with 110 middle schoolers showed a about 25% increase in weekly practice time and a about 40% rise in skills mastered [5]. These gains remained stable, indicating sustained engagement. Notably, skill mastery improved more than time spent, suggesting enhanced practice quality. The results highlight AI-assisted goal-setting's potential to improve learning with minimal teacher effort.

Acknowledgments. This research is funded by the Institute of Education Sciences (IES) of the U.S. Department of Education (Award #R305A220386). This work was made possible with the support of the Learning Engineering Virtual Institute. The opinions, findings and conclusions expressed in this material are those of the authors.

References

1. Abduljabbar, A., Gupta, N., Healy, L., Kumar, Y., Li, J.J., Morreale, P.: A self-served AI tutor for growth mindset teaching. In: 2022 5th International Conference on Information and Computer Technologies (ICICT), pp. 55–59. IEEE (2022)

2. Adams, M.A., et al.: Adaptive goal setting and financial incentives: a 2× 2 factorial randomized controlled trial to increase adults' physical activity. BMC Public Health **17**, 1–16 (2017)
3. Alwahbi, A.: The use of contingency contracting in educational settings: a review of the literature. Educ. Res. Rev. **15**(6), 327–335 (2020)
4. Azevedo, R., et al.: Lessons learned and future directions of MetaTutor: leveraging multichannel data to scaffold self-regulated learning with an intelligent tutoring system. Front. Psychol. **13**, 813632 (2022)
5. Borchers, C., Houk, A., Aleven, V., Koedinger, K.R.: Engagement and learning benefits of goal setting with rewards in human-AI tutoring. In: Proceedings of the 26th International Conference on Artificial Intelligence in Education (2025)
6. Borchers, C., Ooge, J., Peng, C., Aleven, V.: How learner control and explainable learning analytics about skill mastery shape student desires to finish and avoid loss in tutored practice. In: Proceedings of the 15th International Learning Analytics and Knowledge Conference, pp. 810–816 (2025)
7. Borchers, C., Peng, C., Lyu, Q., Carvalho, P.F., Koedinger, K.R., Aleven, V.: Student perceptions of adaptive goal setting recommendations: a design prototyping study. In: Proceedings of the 26th International Conference on Artificial Intelligence in Education (2025)
8. Gold, G., Borchers, C., Carvalho, P.F.: Students' academic performance and goal orientation relate to initial knowledge but not learning rate. In: 14th International Conference on Learning Analytics & Knowledge (LAK), Kyoto, Japan (2024)
9. Gollwitzer, P.M.: Mindset theory of action phases. In: Handbook of Theories of Social Psychology, vol. 1, pp. 526–545 (2012)
10. Gollwitzer, P.M., Sheeran, P.: Implementation intentions and goal achievement: a meta-analysis of effects and processes. Adv. Exp. Soc. Psychol. **38**, 69–119 (2006)
11. Greene, J.A., Bernacki, M.L., Hadwin, A.F.: Self-regulation. In: Handbook of Educational Psychology, pp. 314–334 (2023)
12. Hadwin, A.F., Webster, E.A.: Calibration in goal setting: examining the nature of judgments of confidence. Learn. Instr. **24**, 37–47 (2013)
13. Kahle, A.L., Kelley, M.L.: Children's homework problems: a comparison of goal setting and parent training. Behav. Ther. **25**(2), 275–290 (1994)
14. Koedinger, K.R., Carvalho, P.F., Liu, R., McLaughlin, E.A.: An astonishing regularity in student learning rate. Proc. Natl. Acad. Sci. **120**(13), e2221311120 (2023)
15. Li, X., et al.: Flora engine: using analytics to measure and facilitate learners' own regulation activities. J. Learn. Analyt., 1–24 (2025)
16. Locke, E.A., Latham, G.P.: The development of goal setting theory: a half century retrospective. Motiv. Sci. **5**(2), 93 (2019)
17. Peng, C., Borchers, C., Aleven, V.: Designing homework support tools for middle school mathematics using intelligent tutoring systems. In: Proceedings of the 18th International Conference of the Learning Sciences-ICLS 2024, pp. 1730–1733 (2024)
18. Roll, I., Wiese, E.S., Long, Y., Aleven, V., Koedinger, K.R.: Tutoring self-and co-regulation with intelligent tutoring systems to help students acquire better learning skills. Design Recommendations Intell. Tutoring Syst. **2**, 169–182 (2014)
19. Wäschle, K., Allgaier, A., Lachner, A., Fink, S., Nückles, M.: Procrastination and self-efficacy: tracing vicious and virtuous circles in self-regulated learning. Learn. Instr. **29**, 103–114 (2014)

Toward Extracting Computational Thinking Evidence with Large Language Models: Empowering K-12 Educators

William Lee(✉) and Ivon Arroyo(✉)

University of Massachusetts Amherst, Amherst, MA 01003, USA
{williamlee,ivon}@umass.edu

Abstract. Computational Thinking (CT) is essential for K-12 learners. Yet, educators lack methods and tools to assess their students. We present a pilot study exploring how large language models can automate CT evidence extraction – states, transitions, connections, and cycles – from embodied games represented as Finite State Automata (FSA). We built a dataset from the platform's logs to evaluate four large language models: three open-source (Llama 3.2, Tulu, DeepSeek's R1) and OpenAI's o1. Initial prompts caused models to misinterpret the word "loops" as self-loops. Revising terminology to "cycles" improved accuracy. Quantitative results showed OpenAI's o1 and DeepSeek's R1 achieved perfect F1 scores (1.0) in identifying CT evidence, while Llama 3.2 (0.974) and Tulu (0.961) performed competitively. Informal interviews with two educators as validation highlighted practical needs: tools should map CT evidence to learning standards and track progress (e.g., flagging missed concepts). However, challenges include models' occasional fabrication of details, lack of empathetic responses, and reliance on technical jargon. Future directions include longitudinal progress analysis, multimodal inputs, and prioritizing insights for feedback.

Keywords: Computational Thinking · Large Language Models · Educators · Assessment · Information Retrieval · K-12

1 Introduction

According to the recent National Science Foundation's *Computer Science for All* call for proposal (National Science Foundation, 2024), all students should have the opportunity to participate in computational thinking (CT) and computer science (CS) education in their schools at the K-12 levels, even before college. Arroyo et al. [1] defined CT as a cognitive process that facilitates problem decomposition, pattern identification, knowledge representation, and abstraction formulation. However, K-12 educators often lack adequate preparation and support to teach these concepts. Further, not much research has been done on helping educators gain deeper insights collaboratively with the tools at least in the context of students learning CT at the K-12 level. An exciting program to bring CT to schools in grades 5–8 is WearableLearning (WL). WL is an online platform developed by UMass Amherst's Education and CS researchers that enables educators and

A. I. Cristea et al. (Eds.): AIED 2025, CCIS 2590, pp. 403–408, 2025.
https://doi.org/10.1007/978-3-031-99261-2_44

students to play, manage, and create educational game experiences that involve several players, physical movement, and objects in the environment [5]. Figure 1 illustrates the WL Game Editor to program these game experiences for mobile devices.

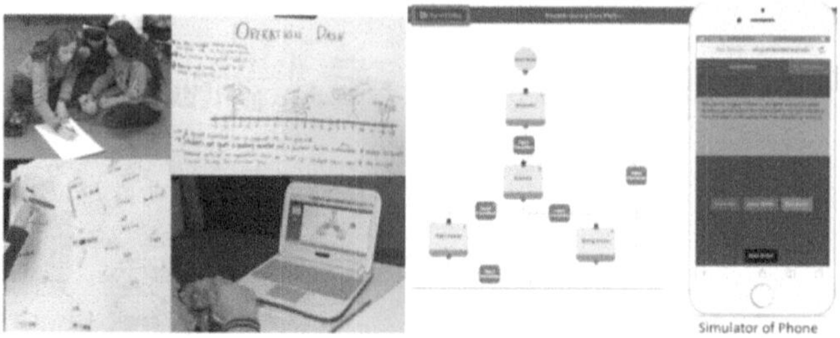

Fig. 1. Left: Students working at different stages of the multi-stage WL curriculum, first ideating the game, and later creating a Finite State Automata diagram on the mobile devices. Right: Student block-based programming a game via the WL Game Editor and debugging with the phone simulator.

2 Background, Challenges, and Opportunities

Computational Thinking (CT) goes beyond traditional programming to encompass problem-solving strategies like abstraction and algorithmic thinking – skills intended to cultivate a society fluent in computational concepts [2, 9, 10]. One approach to integrate CT into K-12 education is through math game creation, as shown by Yasmin Kafai's constructionist approach, where students design games to teach peers, fostering reflection, and deeper engagement with mathematical and computational concepts [3]. This "problem posing" approach shifts students from passive learners to active creators [8]. One learning technology to foster game playing and game creation is the WL platform. As an example, the *Estimate IT!* game (created in WL) engages students in collaborative scavenger hunts, blending physical movement with estimation tasks while supporting social interaction through synchronized devices. The WL Game Creation Curriculum (WL-C) structures this process across eight sessions, guiding students through iterative design stages: analyzing existing games, storyboarding, modeling behaviors via Finite State Automata (FSA), programming in the WL Game Editor's block-based language, playtesting, and refining based on peer feedback. Research involving middle schoolers using the WL-C demonstrated significant CT improvements [7], alongside increased self-efficacy in computing and innovation [4]. While the WL curriculum encompasses all aspects of CT, for this paper, we focus on the programming stage of the WL-C in the Game Editor, which involves computational problem solving. This is one of the four categories in Weintrop and colleagues' Computational Thinking Taxonomy [9]. This category of "computational problem solving" focuses on using computational tools and practices to formulate and solve problems, particularly in STEM contexts (in WL, this involves programming mathematics games where players use mobile devices). Weintrop and Wilensky in their framework argues that engaging in computational problem solving is about problem-solving through computational means, not just coding for its own sake:

- Modeling and Simulation: Using computational tools to represent and explore systems. In WL, these are students designed games represented as FSA.
- Algorithm Design: Creating step-by-step procedures to solve problems. In WL, this involves dragging blocks connected by arrows to specify sequences of states and transitions, defining branching and loops.
- Programming: Writing code to implement algorithms or models. In WL, this involves specifying outputs and inputs.
- Debugging and Troubleshooting: Fixing issues and errors. In WL, this implies running the debugger that features the cell phone simulator, as well as identifying and fixing errors in code or models.

Yet, K-12 educators might not often be given the support, PD, and tools they need to assess students' CT. One area of research to support educators is the creation of dashboards to help educators understand students' thinking. Poh et al. [6] conducted an in-depth review of dashboard requirements. An important principle is to feature a glanceable summary of students' performance. However, to our knowledge, there is currently no effective dashboard that integrated CT evidence for educators to gain insights into their students' algorithmic thinking and knowledge. Today's large language models could enable some form of dialog with educators about their students' computational thinking.

3 Goals and Research Questions

The goal of this pilot study is to explore the capabilities of large language models to extract characteristics of games created by students in a block-based programming language from the WL platform. Students make games that vary in their level of sophistication, depending on their abilities. Educators need to know whether some students might not be understanding or applying some CT concepts (e.g., branching, looping, multiplayer logic, etc.). The long-term goal is to integrate language models into WL platform for assessing the complexity of games created by students and explain those games underlying patterns in dialogs, to the educators of these students (or those who are training as part of a PD for educators). Concrete research questions are:

- RQ1. Can open-source large language models assess the games created by students (which are structured as FSAs) by extracting a verbal description about its complexity?
- RQ2. Can open-source large language models capture changes and progressions in the level of complexity of these games, from one day to the next?

4 Pilot Study

During the Fall of 2024, a pilot study was conducted to analyze open-source large language models' capabilities in assessing the games created by students, using WL's Game Editor. No pre-training, fine-tuning, or instruction-tuning was used with the models. We performed inferences only with the models. We also decided to focus on RQ1 first and conduct a pilot study (described below) to answer this question.

4.1 Data

To effectively analyze the computational problem-solving patterns from the students' games design, it is essential to understand the structural features of games created by students. The WL Game Editor (see Fig. 1.) logs all student interactions during game creation (e.g., mouse clicks, states created, transition arrows, media imports) in a backend SQL database. This combined data captures the structural logic of the student-created games, including states, transitions, and connections between states. We designed a SQL query joining four tables to combine all data about the game. For this pilot study, we focused on the game called *Integer Hopscotch.*

4.2 Methodology

We identified three open-source large language models – Llama, Tulu, and DeepSeek – and chose OpenAI's o1 model as the baseline (see Table 1). We chose these models because they are accessible, open-source, and state-of-the-art. The combined data was fed as a CSV file through each model's chat interface. Initially, we crafted a prompt (i.e., "*The following is an embodied game specified as a finite state machine. How many connections, states, and transitions are there?*") to explore if the models could understand the format of the data and extract the requested evidence. When we added the item "loops" into the initial prompt, all models were unable to identify any (even there are four loops in the game). Mirroring the few-shots learning approach, we iterated five times and attempted to "instruct" the OpenAI's o1 model the kind of loops that we wanted to extract. Eventually, we realized that the term "cycles" is more appropriate and refined our prompt (i.e., "*The following is an embodied game specified as a finite state machine. How many connections, states, transitions, and cycles are there? Also list all the connections, states, transitions, and cycles*"). We collected the results (counts and labels) returned by all the language models into a spreadsheet.

Table 1. Details of the models used in the pilot study.

Model	Number of Parameter	Chat Platform
OpenAI's o1	Not publicly released	OpenAI
AI2's Tulu	8B, 70B	AI2; Ollama + Open WebUI
DeepSeek's R1	7B	DeepSeek
Meta's Llama 3.2	9B, 90B	together.xyz; Ollama + Open WebUI

5 Results

To understand these models' capabilities to extract CT evidence, we counted the extracted evidence against ground truth of the combined data. Table 2 shows the metrics and results. Specifically, models with "Loops" indicate that the word "loops" was used in the initial prompt while "Cycles" indicate the word "cycles" was used in the revised prompt. As shown, OpenAI's o1 and DeepSeek's R1 ranked highest. Llama 3.2 performed slightly better than Tulu. Additionally, we also conducted an informal interview with two former K-12 educators to gain feedback and validate our findings (see Sect. 6 below).

Table 2. Quantitative results of promoting baseline and three open-source models.

Model	Accuracy	Precision	Recall	F1
OpenAI's o1 - Loops	0.9	1.0	0.9	0.947
OpenAI's o1 - Cycles	1.0	1.0	1.0	1.0
AI2's Tulu - Loops	0.878	0.973	0.9	0.935
AI2's Tulu - Cycles	0.925	1.0	0.925	0.961
DeepSeek's R1 - Loops	0.9	1.0	0.9	0.947
DeepSeek's R1 - Cycles	1.0	1.0	1.0	1.0
Meta's Llama 3.2 - Loops	0.902	0.974	0.925	0.949
Meta's Llama 3.2 - Cycles	0.95	1.0	0.95	0.974

6 Discussions and Conclusion

Our preliminary results from the pilot study indicate that although all the models achieved high F1 scores and are mostly capable of extracting basic CT evidence, we initially struggled to understand why all the models responded with zero loops – there are four loops in our game. We learned that when the term "loops" was used, all the models were only looking for self-loops. Once we swapped the term "loops" to "cycles," most of the models began to recognize, identify, and count loops correctly. The implication is that any technical, algorithmic, and Computer Science terms might not always be what a K-12 educator would think of when using a conversational style prompting. Additionally, as our informal interview with two former educators told us, being able to track, monitor, and identify students' trajectory and mastery of CT is important. Mapping student-created CT evidence into state level's digital literacy standards (e.g., Massachusetts's Digital Literacy and Computer Science (DLCS) standards) is another aspect these models could help educators. Our two former educator participants also revealed insights such as what CT concepts they skip, design patterns of their embodied games, or types of media used in their creation could all become evidence for mapping into a standard. Many times, these standards can be vague, though, and at least attempting to assess computational problem solving would be a very interesting goal. For instance, can we assess the level of expertise at "Modeling and Simulation"? This might be assessed by how they evolve their design from single player to multi-player version of their game. What about "Algorithm Design?" This would involve assessing their ability to create step-by-step algorithms to program their games, such as dragging blocks connected by arrows, to specify sequences of states and transitions, and the complexity of their branching and loops in the game structure. Can we assess other "Programming" abilities? This could be assessed by the amount and variety of outputs and inputs. Can we assess "Debugging and Troubleshooting?" In WL, this implies running the debugger that features the cell phone simulator, as well as identifying and fixing errors in codes or models.

This pilot study demonstrated that today's state-of-the-art large language models have the potential to extract valuable CT evidence for educators. Although these models' conversational abilities could serve as a form of dialog with the educators such that they have the flexibility to ask follow-up questions (rather than just running a static program to extract evidence), the precise and sometimes technical choice of words required could discourage educators to benefit from them.

Acknowledgments. We want to thank our two former K-12 educators – Amy Byron and Katy Dieber – for their feedback. We also are grateful to Sai Satish Gattupalli and Matthew Micciolo for brainstorming and contributing to the idea.

Disclosure of Interests. The authors have no competing interests to declare that are relevant to the content of this article.

References

1. Arroyo, I., Closser, A.H., Castro, F., et al.: The WearableLearning platform: a computational thinking tool supporting game design and active play. Tech Know Learn **28**, 1815–1824 (2023)
2. Guzdial, M.: Education Paving the way for computational thinking. Commun. ACM **51**(8), 25–27 (2008)
3. Kafai, Y.B., Burke, Q.: Constructionist gaming: understanding the benefits of making games for learning. Educ. Psychol. **50**(4), 313–334 (2015)
4. Kathala, K.C.R., Arroyo, I., Mannuru, N.R.: Role of ethical and responsible ai in education for next generation inventors. In International Conference on Artificial Intelligence in Education, pp. 71–78. Springer, Cham (2024)
5. Micciolo, M. Designing a visual programming language for the creation of multiplayer embodied games (Unpublished master's thesis). Worcester Polytechnic Institute (2018)
6. Poh, A., Castro, F.E., Arroyo, I.: Design principles for teacher dashboards to support in-class learning. In: Blikstein, P., Van Aalst, J., Kizito, R., Brennan, K. (eds.) Proceedings of the 17th International Conference of the Learning Sciences - ICLS 2023, pp. 736–743 (2023)
7. Rasul, I., Castro, F., Arroyo, I.: Towards embodied wearable intelligent tutoring systems. In: International Conference on Intelligent Tutoring Systems, pp. 298–306. Springer, Cham (2023)
8. Rosli, M.R., Capraro, R.M.: The effects of problem posing on student mathematical learning: a meta-analysis. Int. Educ. Stud. **7**(13), 227–241 (2014)
9. Weintrop, D., Rutstein Wise, D., Bienkowski, M., McGee, S.: Assessing computational thinking: an overview of the field. Comput. Sci. Educ. **31**(2), 113–116 (2021)
10. Wing, J.: Computational thinking. Commun. ACM **49**(3), 33–35 (2006)

Comparing the Effectiveness of Digital Game-Based Learning and Embodied Learning

Aditi Haiman[1(✉)], Ivon Arroyo[2], and Bruce M. McLaren[1]

[1] Carnegie Mellon University, Pittsburgh, PA 15213, USA
ahaiman@andrew.cmu.edu

[2] University of Massachusetts Amherst, Amherst, MA 01003, USA

Abstract. This research project proposes to investigate the comparative effectiveness of digital game-based learning (DGBL) and embodied learning in mathematics education. It will address a gap in the literature by directly comparing a digital learning game, *Decimal Point*, to its embodied counterpart developed using the WearableLearning platform. The research aims to assess differences in learning outcomes and student affect in the embodied and DGBL conditions. Furthermore, it will examine how gender and prior knowledge influence learning outcomes in both the DGBL and embodied learning contexts, following up on a previously observed gender effect in *Decimal Point*. The project will also explore the potential of large language models (LLMs) to translate and design embodied learning activities from a digital learning game. The expected contributions include insights into the specific benefits and limitations of each approach, guidance for designing more equitable educational technologies, and a novel methodology for leveraging AI in educational design. In short, this research bridges learning science, game design, and artificial intelligence to advance the understanding of active learning in mathematics.

Keywords: Digital Game-Based Learning · Embodied Learning · Large Language Models

1 Introduction

Digital game-based learning (DGBL) and embodied learning are both methods of active learning that are effective in increasing learning, as well as motivating and engaging students [5, 6, 16, 19]. Both methods have been studied in the context of math education, but there is limited research that uses empirical methods to compare the effectiveness of a digital learning game to an equivalent embodied learning activity. Since DGBL and embodied learning have similarities as forms of active learning, comparing the two methods can provide insights into what aspects of these learning methods make them effective. In this project, we aim to create an embodied version of the digital learning game, *Decimal Point* [8], to directly compare differences in student learning between DGBL and embodied learning.

Embodied learning methods have been compared to traditional learning [5, 13, 16] but not to digital game-based learning. It would be interesting and novel to create an

A. I. Cristea et al. (Eds.): AIED 2025, CCIS 2590, pp. 409–414, 2025.
https://doi.org/10.1007/978-3-031-99261-2_45

embodied version of *Decimal Point* and see how learning gains compare between the game and its embodied version. If there are differences in how students learn with *Decimal Point* and an embodied version of the game, the findings would contribute to our understanding of why these active learning methods are effective and how they benefit students differently. By identifying differences in how students internalize decimal concepts, we can offer design principles for choosing when and for which audiences each method is most effective—especially with respect to learner characteristics such as gender and prior knowledge.

Furthermore, previous studies with *Decimal Point* have revealed a consistent gender effect where girls demonstrate greater learning gains than boys [11]. This finding raises important questions about how different learning approaches might benefit diverse learner populations. By examining potential differences in learning outcomes and engagement across gender and prior knowledge levels, this project aims to contribute to our understanding of how to design more equitable and effective educational technologies.

During the design phase, we also plan to evaluate the effectiveness of LLMs in helping to create embodied learning activities. Therefore, our research questions are the following: **(RQ1)** How do student learning outcomes and engagement differ between a digital learning game and an embodied version of the same math content? **(RQ2)** How do gender and prior knowledge moderate the effects of digital game-based vs embodied learning? **(RQ3)** Can LLMs effectively support the translation of digital learning games into embodied learning activities?

2 Background

2.1 Digital Game-Based Learning and *Decimal Point*

DGBL is a form of instruction that takes advantage of students' natural interest in video games. A well-designed learning game can intrinsically motivate students through elements of fantasy, challenge, and curiosity [7] and lead to greater learning gains when compared to traditional methods of instruction [6, 16]. One example of a successful digital learning game for math is *Decimal Point*, a game that teaches decimal concepts [8]. *Decimal Point* uses an amusement park metaphor (see Fig. 1, left) where students play various themed mini-games designed to target common decimal misconceptions. Students work through 24 mini-games [8] with a variety of problem types. For example, there are number line problems where students must place a decimal on the appropriate spot on a number line (Fig. 1, right). A particularly interesting finding across multiple studies with *Decimal Point* is a consistent gender effect, with girls demonstrating greater learning gains from the game than boys [11]. This finding raises important questions about how different interactive learning approaches might benefit diverse learner populations and suggests that educational technologies might help address persistent gender gaps in STEM education.

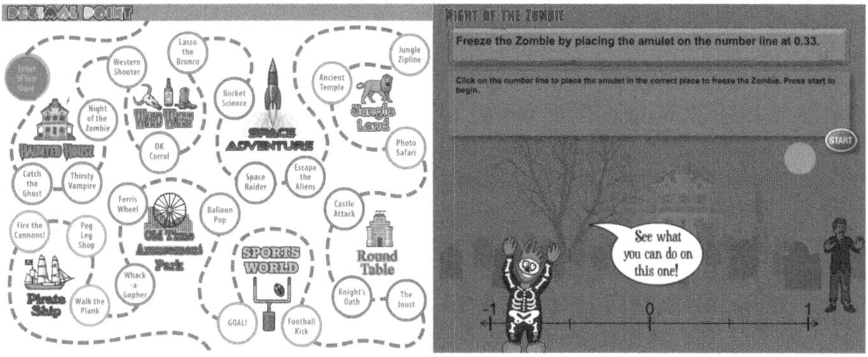

Fig. 1. *Decimal Point* game map (left) and example number-line mini-game problem (right)

2.2 Embodied Learning

Theories of embodied learning suggest that cognition is fundamentally linked to motor and sensory experiences [1], as brain regions involved in movement are closely connected to those responsible for learning [10, 18]. Engaging students in physical movement can therefore enhance learning. One example of successful embodied learning technology is the Mathematical Imagery Trainer for Proportion (MIT-P) [1]. The system has students move their hands up and down, maintaining a specific ratio to keep the screen green. This direct physical experience helps students internalize the concept of proportions without initially relying on numerical representations [1].

An important insight from embodied learning research is that the types of physical movement that students are doing should have meaning in the educational context [14]. Conceptually relevant actions show more benefits in performance, mathematical intuition, insight, proof production, problem-solving, and learning [1]. A study by Nathan et al. [10] found that using their arms to create triangles on a whiteboard helped participants understand the triangle inequality theorem. Similarly, tapping between two points to simulate gears helped participants grasp the concept of alternating directions in a gear chain, demonstrating that embodied interactions can support key mathematical insights.

3 Methods and Current Progress

Development of Embodied *Decimal Point*. To ensure that the comparison between the two versions of *Decimal Point* is fair and valid, it is important to create an embodied version that keeps the same content and context while only varying the platform of the game. Therefore, we propose to use the WearableLearning platform [2] as a basis for the embodied version of *Decimal Point*. WearableLearning is a platform that allows for the development of multi-player, interactive, embodied activities for the classroom. Students read instructions for each problem on a mobile device, engage in the embodied activity, and then enter their answers into the device, where they can receive feedback and move on [2]. This will help control for teacher variability in the embodied condition, as all embodied activities will use scripted instructions and be facilitated by the researchers.

In designing the embodied activities for *Decimal Point*, we will use Melcer & Isbister's [9] framework, which categorizes different types of physical interaction in embodied learning. For example, consider an activity where students sort decimals as greater or less than a given threshold. In an embodied version of this activity, students would be given a physical clothesline and "ghosts" with decimals written on them. The students would then pin the ghosts on the clothesline, representing their values relative to the threshold. This type of interaction aligns with the "manipulation" category, as students are directly using tangible objects to embody the abstract concept of decimal magnitude.

Part of the development phase will also involve answering RQ3. Specifically, we want to determine whether or not LLMs can create an equivalent embodied learning activity (or at least a starting point) for each of the *Decimal Point* mini-games, and if there are differences in effectiveness between LLM models. Preliminary experimentation with ChatGPT-4 has shown promising results. When asked to provide ten ideas for converting the number-line problem from Fig. 1 into an embodied activity, ChatGPT-4 suggested multiple ideas involving students embodying the decimal and walking along a physical number line (Fig. 2). While this problem type lends itself easily to physical movement, it remains to be seen whether LLMs can produce similar results for more abstract problems.

1. Human Number Line Walk

- Lay a long rope or tape on the floor as a number line (0 to 1 marked with tenths and optional finer divisions).
- Call out a decimal (e.g., 0.33), and students walk to the estimated position.
- Freeze the "zombie" (a student acting as one) by standing in the correct spot.

2. Giant Number Line Jump

- Mark a number line on the ground.
- A student starts at 0 and jumps forward in tenths to estimate 0.33.
- If they land close, they "freeze the zombie" (a student moving toward them).

Fig. 2. Two of the ChatGPT-4 suggestions for embodying number-line problems

While LLMs can quickly generate embodied activity ideas based on a digital game, a critical research question is how to systematically evaluate the quality and educational value of these outputs. Our approach will include rubric and expert-based evaluation. Outputs will be evaluated using a structured rubric including dimensions such as conceptual alignment (how well the activity targets the intended math concept), physical mapping (how meaningfully movement is tied to the idea), feasibility in classroom contexts, clarity of instructions, and creativity or engagement potential. Each activity will be rated independently by experts in learning sciences, math education, or educational technology, and we will calculate interrater reliability to ensure consistency across evaluations.

Comparing Digital and Embodied *Decimal Point*. After the embodied version of *Decimal Point* has been developed, we will conduct comparative studies. Participants would be fifth and sixth-graders since the content of the game is most suitable for this age group. To test for learning gains, we will use the existing *Decimal Point* pretest, post-test, and delayed post-tests [8].

To test for differences in affect, students will take a post-game questionnaire evaluating their sense of motivation, engagement, perception of mastery, and enjoyment. For motivation in mathematics, we will use the traditional Eccles & Wigfield expectancy-value theory assessment instrument, which divides motivation into self-concept, liking, and value of mathematics [17]. To measure enjoyment in mathematics, we will use the same items that Frenzel et al. [4] used to measure enjoyment for mathematics in students, which correspond to class-related enjoyment items of the Achievement Emotions Questionnaire for Mathematics [3] to assess students' enjoyment during mathematics classes. To measure mastery vs. performance goal orientation, we will use selected items from Pintrich's [12] motivated strategies for learning questionnaire. Finally, students will take a demographic questionnaire with binary and multi-dimensional aspects of gender [15].

4 Expected Contributions

This proposed project addresses a gap in prior research by directly comparing digital and embodied learning approaches using identical math content. Past research has studied these methods separately or compared them to traditional instruction, but never against each other. Our proposed study provides an opportunity to examine which modality supports better learning outcomes overall, and how each fosters engagement, conceptual understanding, and transfer differently. This work aims to uncover the mechanisms that drive learning in each modality. For instance, embodied learning may promote deeper conceptual grounding by leveraging sensorimotor experiences and spatial reasoning, while DGBL may enhance learning through structured interactivity, narrative engagement, and immediate feedback.

By investigating how gender and prior knowledge interact with learning modality, we may also reveal important insights about inclusive design for educational technologies. For example, if the embodied version shows different patterns of effectiveness across gender or prior knowledge levels, this could suggest important design considerations for creating more equitable learning technologies.

This project also opens possibilities for exploring how large language models (LLMs) might assist in the creative design process of translating digital games into embodied learning activities. As an extension of this work, we plan to investigate how well LLMs can generate designs for embodied versions of the *Decimal Point* mini-games based on the digital game specifications. This could potentially establish a new methodological approach for efficiently creating embodied learning activities from existing digital content, thus expanding the repertoire of AI applications in educational design.

In summary, this research sits at the intersection of learning sciences, game design, and artificial intelligence. Beyond advancing our theoretical understanding of embodied cognition and active learning in mathematics, this project offers practical insights for developing technologies that engage the whole student—mind and body—in the learning process.

Acknowledgments. Thank you to Zitao Liu for his helpful comments as part of the AIED Doctoral Consortium. ChatGPT was used to revise this paper.

References

1. Abrahamson, D., Lindgren, R.: Embodiment and embodied design (2014)
2. Arroyo, I., Closser, A.H., Castro, F., Smith, H., Ottmar, E., Micciolo, M.: The WearableLearning platform: a computational thinking tool supporting game design and active play. Technol. Knowl. Learn. **28**(4), 1815–1824 (2023)
3. Bieleke, M., Goetz, T., Yanagida, T., Botes, E., Frenzel, A.C., Pekrun, R.: Measuring emotions in mathematics: the achievement emotions questionnaire—mathematics (AEQ-M). ZDM–Math. Educ. **55**(2), 269–284 (2023)
4. Frenzel, A.C., Goetz, T., Lüdtke, O., Pekrun, R., Sutton, R.E.: Emotional transmission in the classroom: exploring the relationship between teacher and student enjoyment. J. Educ. Psychol. **101**(3), 705 (2009)
5. Georgiou, Y., Ioannou, A.: Embodied learning in a digital world: a systematic review of empirical research in K-12 education. In: Learning in a Digital World: Perspective on Interactive Technologies for Formal and Informal Education, pp. 155–177 (2019)
6. Hussein, M.H., Ow, S.H., Elaish, M.M., Jensen, E.O.: Digital game-based learning in K-12 mathematics education: a systematic literature review. Educ. Inf. Technol. **27**(2), 2859–2891 (2022)
7. Malone, T.W.: Toward a theory of intrinsically motivating instruction. Cogn. Sci. **5**(4), 333–369 (1981)
8. McLaren, B.M.: Decimal point: a decade of learning science findings with a digital learning game. In: Artificial Intelligence in Education: The Intersection of Technology and Pedagogy, pp. 145–203. Springer, Cham (2024)
9. Melcer, E.F., Isbister, K.: Bridging the physical divide: a design framework for embodied learning games and simulations. In: Proceedings of the 2016 CHI Conference Extended Abstracts on Human Factors in Computing Systems, pp. 2225–2233 (2016)
10. Nathan, M.J., Walkington, C., Boncoddo, R., Pier, E., Williams, C.C., Alibali, M.W.: Actions speak louder with words: the roles of action and pedagogical language for grounding mathematical proof. Learn. Instr. **33**, 182–193 (2014)
11. Nguyen, H.A., Hou, X., Richey, J.E., McLaren, B.M.: The impact of gender in learning with games: a consistent effect in a math learning game. Int. J. Game-Based Learn. (IJGBL) **12**(1), 1–29 (2022)
12. Pintrich, P.R.: A manual for the use of the Motivated Strategies for Learning Questionnaire (MSLQ) (1991)
13. Shoval, E.: Using mindful movement in cooperative learning while learning about angles. Instr. Sci. **39**(4), 453–466 (2011)
14. Smith & Walkington. Four principles for designing embodied mathematics activities (2020)
15. Stec, H., et al.: Using a Multi-Dimensional Model of Gender to Assess Learning with Different Game-Based Learning Narratives (2024)
16. Tokac, U., Novak, E., Thompson, C.G.: Effects of game-based learning on students' mathematics achievement: a meta-analysis. J. Comput. Assist. Learn. **35**(3), 407–420 (2019)
17. Wigfield, A., Eccles, J.S.: Expectancy–value theory of achievement motivation. Contemp. Educ. Psychol. **25**(1), 68–81 (2000)
18. Wilson, M.: Six views of embodied cognition. Psychon. Bull. Rev. **9**, 625–636 (2002). https://doi.org/10.3758/BF03196322.pdf
19. Zhong, B., Su, S., Liu, X., Zhan, Z.: A literature review on the empirical studies of technology-based embodied learning. Interact. Learn. Environ. **31**(8), 5180–5199 (2023)

AI-Powered Intelligent Tutoring System for Game Design Document Writing

Raimundo N. B. Neto[1(✉)] and Eduardo H. S. Aranha[2]

[1] State University of Rio Grande do Norte (UERN), Mossoró, Brazil
bezerraneto@uern.br
[2] Federal University of Rio Grande do Norte (UFRN), Natal, Brazil
eduardoaranha@dimap.ufrn.br

Abstract. The increasing integration of Generative Artificial Intelligence (GAI) in education presents both challenges and opportunities. This research focuses on the development and validation of an intelligent tutoring system that assists teachers and students in writing Game Design Documents (GDDs). The system is designed to improve clarity, organization and accessibility by providing contextual suggestions and refining the structure of the documents based on pedagogical needs. The solution evaluation will take place in public schools in Brazil, involving 40 teachers and 200 students, to analyze the influence of AI on the quality of GDDs. Through the analysis of prompt engineering strategies, this research identifies methodologies that can optimize the generation of GDDs, ensuring that AI acts as a support tool without replacing critical thinking skills. The results contribute to the integration of AI in education, offering practical contributions on its pedagogical applications.

Keywords: Generative Artificial Intelligence · Game Design Documents · Educational Technology

1 Introduction

Education has undergone significant transformations in its structural and methodological approaches, directly impacting teaching and learning processes. These changes challenge the sustainability of educational practices, especially when traditional methods, based on passive instruction, fail to engage students effectively. According to Valente [1], this model, historically based on passive instruction and the centralization of the teacher as the sole mediator of knowledge, does not meet the demands of a society that requires dynamic and continuous learning.

As a result, the advancement of educational technologies has enabled the creation of more interactive and adaptable learning environments, promoting greater student engagement in the educational process. These innovative methodologies encourage active student participation, transforming them into protagonists in the construction of knowledge, as highlighted by Fava [2].

A. I. Cristea et al. (Eds.): AIED 2025, CCIS 2590, pp. 415–420, 2025.
https://doi.org/10.1007/978-3-031-99261-2_46

Currently, video games go beyond entertainment and are widely used across various fields, offering experiences that enhance learning through dynamic interactions tailored to each player's profile, as noted by Spring [3]. According to Sato [4], the segmentation capability of games helps keep users engaged by providing personalized and accessible experiences.

In this context, educators have been developing strategies to involve both teachers and students in the creation of their own games, fostering more participatory learning. In addition to stimulating the development of computational thinking, this practice contributes to knowledge construction in the subjects addressed within the games.

2 Theoretical Foundations

2.1 Game Design Documents – GDDs

The Game Design Document (GDD) is an essential tool in this process, as it compiles all project information, from initial concepts to technical details such as characters, settings, and mechanics. Its application in education allows students to organize their creations in a systematic and conceptually structured manner, fostering meaningful learning as proposed by Ausubel et al. [5], through active strategies that promote student engagement based on their prior knowledge and personal interests.

In addition, the process of writing a GDD stimulates textual production, giving meaning to the act of writing. By documenting and sharing the structure of the game, students develop their ideas in a more elaborate way, enhancing their writing and argumentation skills. Thus, the development of educational digital games should consider not only technological aspects but also the pedagogical principles that guide the learning process.

In this regard, Rahayu and Fujiati [6] demonstrated that structuring the GDD to include elements such as target audience, learning objectives, rewards, and navigation flow supports the creation of interactive educational games that stimulate student interest and understanding. Its practical application resulted in a high level of engagement and positive evaluations from both teachers and students, reinforcing the potential of the GDD as a didactic tool to support the development of games with pedagogical objectives.

The incorporation of these elements allows games to serve not only as entertainment tools but also as strategic resources to enhance the teaching and learning process.

2.2 Generative Artificial Intelligence in Education

Recent studies highlight the role of Artificial Intelligence in promoting more personalized and responsive learning experiences. According to Harry [7], technologies such as natural language processing enable content adaptation and the provision of targeted feedback. In this context, generative AI emerges as an ally in the development of creative and open-ended tasks, such as the writing of Game Design Documents (GDDs).

Creating a GDD is an educational challenge, as it requires conceptual cohesion and structured reasoning from students. Generative AI can serve as support in this process by organizing ideas, rephrasing passages, and adapting the document to pedagogical goals. Rather than replacing critical thinking, it enhances the construction of narratives, mechanics, and visual elements, turning the GDD into a tool for active and interdisciplinary learning.

On the other hand, the influence of generative AI on the writing process calls for a balanced pedagogical approach that fosters student autonomy in project development. The risk of excessive automation in textual production should be mitigated through educational strategies that prioritize personalized learning, ensuring that students develop essential skills such as computational thinking, creativity, and critical capacity. Thus, when integrating AI into game design education, its use must be guided by reflective practices that value both prompt engineering and the authenticity of student output.

3 Methodology

This research adopts an experimental design to investigate the impact of using an intelligent tutoring system based on Generative Artificial Intelligence on the quality of Game Design Documents (GDDs) produced by students. The study begins by observing a common difficulty faced by students in the structured writing of these documents, followed by the proposal of a technological solution whose use is expected to yield measurable improvements in textual clarity and organization. This is an applied research project, employing a mixed-methods approach (quantitative and qualitative), which allows both the measurement of intervention effects and the analysis of participants' perceptions. According to Kumar [8], quantitative research aims to quantify variations in phenomena and test relationships between variables, whereas qualitative approaches enable the understanding of meanings and perceptions.

The study follows the hypothetical-deductive method, starting with the formulation of hypotheses regarding the effects of the intelligent tutoring system on students' text production process. An experiment will be conducted with a control group (without the tutor) and an experimental group (with the tutor), with participants randomly assigned—a strategy recommended to ensure group equivalence and isolate the effects of the independent variable.

Participants will be randomly distributed between the two groups based on guidelines that ensure balanced and controlled experimental manipulation. The intervention will last between four to eight weeks and will be integrated into pedagogical activities developed through project-based learning, in alignment with student-centered approaches in which the teacher acts as a facilitator of knowledge [9].

The independent variable will be the use of the intelligent tutoring system. The dependent variables—textual clarity, cohesion, and structure of the GDDs—will be evaluated using analytical rubrics developed according to textual and pedagogical quality criteria, as recommended in guidelines for variable definition and measurement. Quantitative analyses will employ statistical tests aligned with best practices.

In addition, a qualitative analysis of participants' perceptions will be carried out based on semi-structured interviews and open-ended responses. This approach allows the integration of multiple perspectives on the phenomenon under study, providing a more comprehensive view. Content analysis will be applied to categorize responses, with data triangulation used to strengthen the reliability of the findings.

4 Proposed Solution and Experimental Evaluation

The implementation of the intelligent tutoring system will take place in public schools in Brazil, involving teachers and students from elementary education. The study will be conducted in approximately 35 schools, engaging an estimated group of 40 teachers and 200 students who will use the tool to develop GDDs as part of interdisciplinary educational activities. Figure 1 below illustrates the macro flow of the project, presenting the main stages of the proposal from its introduction in educational institutions to the final presentation of the developed games.

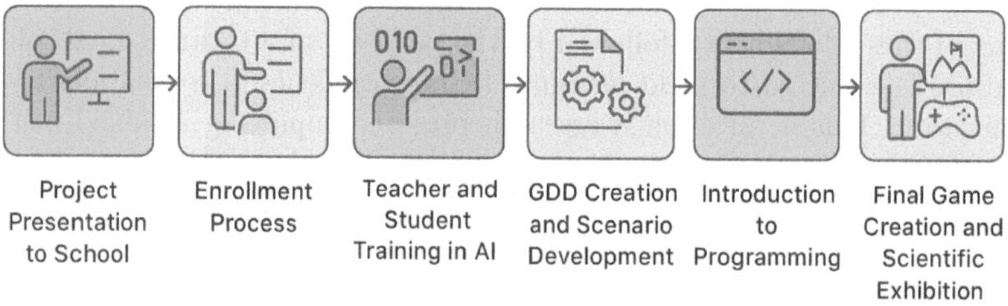

Fig. 1. Project implementation flow.

The use of the intelligent tutoring system, based on Generative Artificial Intelligence, follows a structured flow that guides students and teachers in the creation of Game Design Documents (GDDs), integrating speech recognition technologies, natural language processing, and prompt engineering. The process begins with the student recording an audio message containing their game ideas, which is sent to the system for automatic transcription using Speech-to-Text technologies. The AI then processes this textual content, analyzing its coherence and structure to generate a preliminary GDD with organized information on narrative, mechanics, characters, and settings.

After the GDD is generated, the pedagogical content of the games undergoes review and pedagogical adjustments by the teachers. The teachers evaluate the games created to ensure they align with the content taught in their subjects.

However, they do not modify the design or structure of the game but may suggest pedagogical adjustments to reinforce the concepts being taught.

The document is then returned to the student for refinement through prompt engineering, using techniques such as Fine-tuning, One-shot, and Few-shot Learning. This allows for adjustments and customizations to the textual structure before generating the final version of the GDD. During this process, metrics will be applied to measure prompt efficiency, including success rate, response time, number of iterations, and semantic similarity score, ensuring that the interaction with the AI results in an increasingly accurate document aligned with pedagogical goals. Based on this refinement, the AI processes the changes and generates a structured and final GDD, which will serve as the foundation for the development of the digital game.

Once the GDD is finalized, students begin developing the game using digital creation tools such as Scratch, Construct, and Unity, applying the concepts defined in the document. The final phase of the process involves presenting the game at an educational event, where teachers and experts evaluate the projects based on criteria such as creativity, innovation, and pedagogical applicability. This structured workflow enables students to develop their creative and computational skills in an integrated manner, while AI serves as a support tool for textual organization and refinement. The implementation of this model in public schools aims to broaden access to digital education, promoting the use of AI as a resource for autonomy, critical thinking, and interdisciplinary learning.

4.1 Progress Achieved So Far

The research has made significant progress in the development and validation of the intelligent tutor system, with initial tests evaluating the interaction with AI in the generation of Game Design Documents (GDDs). The first experiments showed improvements in the organization of ideas, recommendations and textual coherence, facilitating the structuring of documents and offering support to teachers and students.

Currently, the research is at the stage of adjusting the interaction flow, making the interface more accessible and intuitive. The next stage involves implementation in public schools, allowing for a more in-depth analysis of the pedagogical impact. Preliminary results indicate that the system contributes to interdisciplinary learning, stimulates computational thinking and improves the textual quality of GDDs produced by the students.

5 Conclusion

This research explored the application of Generative Artificial Intelligence to support the writing of Game Design Documents (GDDs), with the aim of improving their clarity, organization and accessibility. The development and implementation of an intelligent tutor system was designed to provide contextual guidance and facilitate the structured creation of documents by students and teachers.

Preliminary findings suggest that AI-assisted tools, when combined with prompt engineering strategies, can support educational processes without compromising student autonomy. However, the research is still ongoing, and the evaluation of the intelligent tutor system in public schools will provide more robust empirical evidence on the pedagogical impact of AI-assisted GDD writing. Continuing the study will allow for an in-depth assessment of how AI can act as a collaborative tool, enhancing learning without replacing students' critical and creative thinking.

Future research should expand the development of adaptive AI models that cater to different learning profiles and explore ways to optimize feedback mechanisms within AI-based educational platforms. In addition, ethical issues related to academic integrity and students' dependence on AI should be investigated to establish best practices in the responsible adoption of this technology in education.

References

1. Valente, J.A.: Blended learning e as mudanças no ensino superior: a proposta da sala de aula invertida. Educar em Revista (spe 4), 79–97 (2014)
2. Fava, R.: Educação 3.0: aplicando o PDCA nas instituições de ensino. Saraiva, São Paulo (2014)
3. Spring, D.: Gaming history: computer and video games as historical scholarship. In: Rethinking History (2015). https://doi.org/10.1080/13642529.2014.973714
4. Sato, A.K.O.: Game design e prototipagem: conceitos e aplicações ao longo do processo projetual. In: Proceedings do SBGames, pp. 74–84 (2010)
5. Ausubel, D.P., Novak, J.D., Hanesian, H.: Educational Psychology: A Cognitive View, 2 edn. Holt, Rinehart and Winston, Chicago (1978)
6. Rahayu, S.L., Fujiati, F.: Penerapan game design document dalam perancangan game edukasi yang interaktif para menarik minat siswa dalam belajar bahasa inglesa. Jurnal Teknologi Informasi dan Ilmu Komputer **5**(3), 341–346 (2018)
7. Harry, A.: Role of AI in education. Interdisc. J. Human. (INJURITY) **2**(3) (2023)
8. Kumar, R.: Research Methodology: A Step-by-Step Guide for Beginners. SAGE Publications Ltd., Thousands Oaks (2018)
9. Chiu, C.-F.: Facilitating K-12 teachers in creating apps by visual programming and project-based learning. Int. J. Emerg. Technol. Learn. (iJET) **15**(1), 103–118 (2020)

Improving Student Support Personalization with Historical Data and Theoretically Informed Feature Choice

Morgan P. Lee(✉) and Neil T. Heffernan

Worcester Polytechnic Institute, Worcester, MA 01609, USA
{mplee,nth}@wpi.edu

Abstract. The proliferation of online learning platforms allows for greater scaling of educational experimentation than ever before. This includes attempts to personalize the delivery of online educational content to meet individual student needs, often through some form of reinforcement learning. This work explores different reinforcement learning approaches for personalization, paying particular attention to supplemental instructional materials in the form of hints and explanations. Incorporating lessons learned from prior work, we discuss a framework for improving bandit algorithms through feature engineering, and detail our preliminary attempts to apply bandits with engineered features to a real-world context. The goal of this work is to learn which features are important for personalization, to identify qualitative interactions that exist between these features, and develop strategies for using historical data to give our learning agents a "warm start."

Keywords: Personalized Learning · Reinforcement Learning · Multi-Armed Bandit · Feature Engineering

1 Introduction

Personalizing the learning experience of students in online learning platforms (OLPs) has consistently been a priority for researchers in the Artificial Intelligence in Education community. More formally, personalized learning involves a qualitative interaction between two or more groups of students who benefit from different kinds of instruction (i.e. group A of students benefits more from teaching method 1 than method 2, while group B of students benefits more from method 2 than method 1). Randomized control trials (or A/B testing) are the most reliable way of finding these qualitative interactions [16], and multiple OLPs have taken steps to make educational A/B testing easier to perform than ever before [14,15]. In particular, ASSISTments allows for routine comparison of different student support strategies through TeacherASSIST [10] and subsequent projects.

A. I. Cristea et al. (Eds.): AIED 2025, CCIS 2590, pp. 421–426, 2025.
https://doi.org/10.1007/978-3-031-99261-2_47

Reinforcement Learning (RL), specifically Multi-Armed Bandits (MABs) offer an alternative to traditional A/B testing, trading a guarantee of independent samples for wider distribution of positive treatment effects. If personalization is desired, then context about experimental subjects and environments can be provided to MABs, allowing these contextual MABs (CMABs) to learn associations between context and rewards. The effectiveness of these CMABs then relies on two things: the choice of features to provide to the bandit as context, and the number of opportunities a bandit has to learn the associations between context and rewards. In many applications, these CMABs are initialized with no prior knowledge about these associations and are expected to learn from a complete "cold start", but what about situations where historical information is available? How can we use this historical information to give our CMABs a head start? How do we even determine what context to provide to a CMAB?

In this work, we will propose methods for incorporating insights from historical data into CMAB agents for student support recommendation, both directly via pre-training and indirectly through more providing model-based, more descriptive features to our bandits. More specifically, we are interested in answering the following research questions through our work:

1. What are the most important contextual features to provide to a support recommendation system?
2. Are there existing user modeling, psychometric, or language processing techniques that could serve as useful features?
3. Can we identify qualitative interactions between provided context and learning outcomes?

The remainder of this paper will be organized as follows. Section 2 discusses relevant prior works utilizing RL in OLPs. Next, we describe our preliminary attempts at feature engineering for CMABs in Sect. 3. Finally, we highlight the contributions we expect this work to provide to the broader AIED community in Sect. 4.

2 Background

MABs have a long history of use in OLPs, given their relative computational simplicity. Many applications of MABs aim to personalize learning by keeping a student engaged with content within the zone of proximal development, a specific level of difficulty that promotes the most student learning. The bandit agent in question is tasked with selecting the next learning activity, which could include supplementary instructional materials [9] or specifically choosing the sequence of exercises given to a student to keep them within the ZPD [1]. In general, RL has been demonstrated to be successful in improving student learning through instructional sequencing [3], which is quite promising in OLPs that are designed to be student-paced, but what about more structured learning scenarios with more explicit teacher control of student exercises? There are still opportunities for personalization in the form of student supports: hints or explanations that are available to a student, usually after making a mistake. Prior studies have found that crowdsourced student supports can have significant positive impacts on student learning [10,12], and in cases where multiple supports exist for a single problem, a bandit agent may be employed to try to maximize student learning. Conventional multi-armed bandit approaches have achieved results similar to random chance [11], while employing context has yielded improvements [13,18]. Several of these studies have been limited by small sample sizes, which can be mitigated by incorporating prior recommendations to give a policy a non-random starting point [17].

While bandits have demonstrated desirable advantages over random student support assignment, incorporating context into a bandit's decision making process is not without risk. Li et al. [8] demonstrate that including context that is unrelated to the decision being made has negative consequences for students in groups with uncommon feature values. Feature inclusion in contextual models is both vital to allowing for personalization while simultaneously risking the perpetuation of achievement gaps through agents making worse decisions for vulnerable population subgroups. How, then, does one choose which features to include in a contextual bandit model? Prior works show that bandit algorithms employed to supplement a cognitive model have typically achieved better results [3]. Moreover, the concept of combining multiple educational and cognitive models has achieved good results in other fields. Specifically, attempts to combine Item Response Theory and Knowledge Tracing have led to advancements in both fields [5,6]. Can we incorporate information from other models to increase the quality of our student support recommendations?

3 Proposed Methodology and Preliminary Work

This work is being carried out within the ASSISTments OLP, which has previously used bandit algorithms to recommend student supports, with recommendations happening during the day being processed in a nightly update loop (see Prihar et al. [11] for a full description of the online and offline training tasks). Since the development of the TeacherASSIST program and confirmation of the efficacy of crowdsourced feedback [10], ASSISTments has delivered student supports which may be accessed on-demand by students. Even while empirical trials of bandit algorithms are being performed, 50% of student supports are selected at random whenever more than one option is available. This means both that a large amount of data is collected on the efficacy of all available student supports *and* that more random data is collected on a daily basis. To address data sparsity issues encountered in previous work, we are investigating potential uses for this random data stream, either by performing rounds of offline training (similar to [17]), or by incorporating randomly-assigned student supports into a model's nightly updates.

In addition to directly leveraging historical and random data to train bandit models, changing the context provided to a bandit agent also involves incorporating prior information into a model's decisions. Contextual features explored in prior works have typically taken the form of summary statistics about particular students, problems, and available student supports. We posit that there may be more effective summary measures to provide as context to our bandit agents. More specifically, we've been engaged in the following process:

1. Identify a feature that has been important in prior works.
2. Connect this feature to an underlying concept that feature may be estimating.
3. Identify existing models which attempt to measure said concept.
4. Incorporate model predictions as features in a contextual bandit.
5. Evaluate model performance and inspect learned relationships between observed contexts and rewards.

Following this process for one feature (user average correctness) has already resulted in one publication [7]. In that work, we identify that user average correctness is likely acting as a proxy for a student's prior knowledge level. Prior knowledge (or mastery) estimation is a well-studied problem within the learning sciences, and many different Knowledge Tracing models exist to model a student's current knowledge state. Our use of Bayesian Knowledge Tracing (BKT) [2] as a user-level feature instead of user average correctness yielded significant increases in cumulative reward in simulation (see Table 1 for specific comparisons of cumulative reward), and much of our current work has been focused on updating our Reinforcement Learning Service (or RLS, formerly known as the Automated Personalized Learning Service in [11]) to allow for an empirical trial of BKT as a contextual feature.

Table 1. Cumulative reward comparison from [7], out of a total 1 million recommendations for each agent.

Feature Space	Minimum	Average	Maximum
BKT	339,285	340,867	343,132
User Avg. Correctness	335,986	338,031	340,704
Random Baseline	336,016	336,461	336,813

Similarly, problem average correctness has been identified as an important problem-level feature, possibly acting as an estimate of relative problem difficulty. Item Response Theory is a well-studied, empirically validated way of estimating problem difficulty [4], and other models have incorporated IRT as a measure of problem difficulty [5]. Our current plan is to also incorporate IRT estimates of problem difficulty as another feature in our bandit models.

Perhaps most interesting is the space of possible contextual features for the student supports themselves. Prior works have used large language models (LLMs) to accurately predict the effectiveness of different hint messages [19], but using LLMs to describe different features of textual hints and explanations remains relatively unexplored. Considering that most of the features we're employing for tutor strategies are simple text mining features (log character length, for example), having richer, more descriptively meaningful textual features could also prove useful to our bandit agents.

4 Expected Contributions

This research will contribute to both the Artificial Intelligence in Education community and RL research more broadly. Personalizing student and teacher feedback through contextual bandits is an active field of research, and by improving the quality of bandit recommendations in a model-agnostic, feature-focused fashion, insights gained from our work may be more easily transferred to different optimization goals. Simultaneously, our investigations into offline pre-training of bandit agents may yield insights about how to use historical data to augment RL models in contexts outside of education. Finally, any qualitative interactions we detect between our more sensitive features and future learning outcomes will contribute to our understanding of how different groups of students learn, which is an essential step towards our community's shared goal of delivering accessible, personalized learning to more students than ever before.

Acknowledgments. We would like to thank our many past and current funders including NSF (2118725, 2118904, 1950683, 1917808, 1931523, 1940236, 1917713, 1903304, 1822830, 1759229, 1724889, 1636782, 1535428, 2215842, 2341948, & 2153481); IES (R305N210049, R305D210031, R305A170137, R305A170243, R305A180401, R305A120125, R305R220012, & R305T240029); GAANN (P200A120238, P200A1 80088, & P200A150306); DOE (U411B190024 S411B210024, & S411B220024); ONR (N00014-18-1-2768); NIH (via a SBIR R44GM146483), Schmidt Futures, BMGF, CZI, Arnold, Hewlett, the Jaffe Foundation, and anonymous donors. None of the opinions expressed here are those of the funders.

Disclosure of Interests. The authors have no competing interests to declare that are relevant to the content of this article.

References

1. Clement, B., Roy, D., Oudeyer, P.Y., Lopes, M.: Multi-armed bandits for intelligent tutoring systems. J. Educ. Data Min. **7**(2), 20–48 (2015)
2. Corbett, A.T., Anderson, J.R.: Knowledge tracing: modeling the acquisition of procedural knowledge. User Model. User-Adap. Inter. **4**(4), 253–278 (1994)
3. Doroudi, S., Aleven, V., Brunskill, E.: Where's the reward? Int. J. Artif. Intell. Educ. **29**(4), 568–620 (2019)
4. Embretson, S.E., Reise, S.P.: Item Response Theory for Psychologists. Psychology Press (2013)
5. Ghosh, A., Heffernan, N., Lan, A.S.: Context-aware attentive knowledge tracing. In: Proceedings of the 26th ACM SIGKDD International Conference on Knowledge Discovery & Data Mining, KDD 2020, pp. 2330–2339, New York, NY, USA (2020). Association for Computing Machinery
6. Khajah, M.M., Huang, Y., González-Brenes, J.P., Mozer, M.C., Brusilovsky, P.: Integrating knowledge tracing and item response theory: a tale of two frameworks. In: CEUR Workshop Proceedings, vol. 1181, pp. 7–15. University of Pittsburgh (2014). ISSN: 1613-0073

7. Lee, M., Siedahmed, A., Heffernan, N.: Expert features for a student support recommendation contextual bandit algorithm. In: Proceedings of the 14th Learning Analytics and Knowledge Conference, LAK 2024, pp. 864–870, New York, NY, USA (2024). Association for Computing Machinery
8. Li, Z., Yee, L., Sauerberg, N., Sakson, I., Williams, J.J., Rafferty, A.N.: Getting too personal(IZED): the importance of feature choice in online adaptive algorithms (2023). arXiv:2309.02856
9. Manickam, I., Lan, A.S., Baraniuk, R.G.: Contextual multi-armed bandit algorithms for personalized learning action selection. In: 2017 IEEE International Conference on Acoustics, Speech and Signal Processing (ICASSP), pp. 6344–6348 (2017). ISSN: 2379-190X
10. Patikorn, T., Heffernan, N.T.: Effectiveness of crowd-sourcing on-demand assistance from teachers in online learning platforms. In: Proceedings of the Seventh ACM Conference on Learning @ Scale, L@S 2020, pp. 115–124, New York, NY, USA (2020). Association for Computing Machinery
11. Prihar, E., Haim, A., Sales, A., Heffernan, N.: Automatic interpretable personalized learning. In: Proceedings of the Ninth ACM Conference on Learning @ Scale, L@S 2022, pp. 1–11, New York, NY, USA (2022). Association for Computing Machinery
12. Prihar, E., Patikorn, T., Botelho, A., Sales, A., Heffernan, N.: Toward personalizing students' education with crowdsourced tutoring. In: Proceedings of the Eighth ACM Conference on Learning @ Scale, L@S 2021, pp. 37–45, New York, NY, USA (2021). Association for Computing Machinery
13. Prihar, E., Sales, A., Heffernan, N.: A bandit you can trust. In: Proceedings of the 31st ACM Conference on User Modeling, Adaptation and Personalization, UMAP 2023, pp. 106–115, New York, NY, USA (2023). Association for Computing Machinery
14. Prihar, E., Syed, M., Ostrow, K., Shaw, S., Sales, A., Heffernan, N.: Exploring common trends in online educational experiments. In: Proceedings of the 15th International Conference on Educational Data Mining (2022)
15. Reza, M., Kim, J., Bhattacharjee, A., Rafferty, A.N., Williams, J.J.: The MOOClet framework: unifying experimentation, dynamic improvement, and personalization in online courses. In: Proceedings of the Eighth ACM Conference on Learning@ Scale, pp. 15–26 (2021)
16. Savi, A.O., Williams, J.J., Maris, G., van der Maas, H.: The role of a/b tests in the study of large-scale online learning (2017)
17. Schmucker, R., Pachapurkar, N., Bala, S., Shah, M., Mitchell, T.: Learning to give useful hints: assistance action evaluation and policy improvements. In: Viberg, O., Jivet, I., Muñoz-Merino, P., Perifanou, M., Papathoma, T. (eds.) Responsive and Sustainable Educational Futures. pp, pp. 383–398. Springer Nature Switzerland, Cham (2023). https://doi.org/10.1007/978-3-031-42682-7_26
18. Yun, J., Nie, A., Brunskill, E., Demszky, D.: Exploring the benefit of customizing feedback interventions for educators and students with offline contextual multi-armed bandits. In: Proceedings of the 15th International Learning Analytics and Knowledge Conference, LAK 2025, pp. 944–949, New York, NY, USA (2025). Association for Computing Machinery
19. Zhang, T., Kumar, H.A., Schmucker, R., Azaria, A. and Mitchell, T.: Learning to compare hints: combining insights from student logs and large language models. In: Proceedings of the 2024 AAAI Conference on Artificial Intelligence, pp. 162–169. PMLR (2024). ISSN: 2640-3498

Predicting At-Risk Programming Students in Small Imbalanced Datasets Using Synthetic Data

Daniel Flood(✉), Matthew England, and Beate Grawemeyer

Coventry University, Coventry, UK
{Daniel.Flood,Matthew.England,Beate.Grawemeyer}@coventry.ac.uk

Abstract. This study is part of a larger project focused on measuring, understanding, and improving student engagement in programming education. We investigate whether synthetic data generation can help identify at-risk students earlier in a small, imbalanced dataset from an introductory programming module. The analysis used anonymised records from 379 students, with 15% marked as failing, and applied several machine learning algorithms. The first experiments showed poor recall for the failing group. However, using synthetic data generation methods led to a significant improvement in performance. Our results suggest that machine learning can help identify at-risk students early in programming courses when combined with synthetic data. This research lays the groundwork for validating and using these models with live student cohorts in the future, to allow for timely and effective interventions that can improve student outcomes. It also includes feature importance analysis to refine formative tasks. Overall, this study contributes to developing practical workflows that help detect disengagement early and improve student success in programming education.

Keywords: Programming education · Student engagement · Machine learning · Synthetic data

1 Introduction

This study forms part of a wider project focused on measuring, understanding, and improving student engagement in programming education. An early challenge was the difficulty of predicting disengagement and failure in small, single-module datasets, where failing students are few. We collected data from three cohorts of an introductory programming module at Coventry University, comprising 379 student records, with 85% passing and 15% failing. The dataset captures performance across up to 150 formative coding tasks, automatically marked through our learning management system.

While task completion and time-on-task are established engagement indicators, much existing research relies on larger, multi-module datasets or demographic information, often unavailable due to privacy constraints. Instead, our work focuses on coding task performance within a single module. Initial trials

A. I. Cristea et al. (Eds.): AIED 2025, CCIS 2590, pp. 427–432, 2025.
https://doi.org/10.1007/978-3-031-99261-2_48

with classification algorithms (e.g., logistic regression, random forests, SVMs) showed poor recall for failing students, limiting early intervention potential. To address this, we explore synthetic data generation techniques to rebalance the dataset, aiming to strengthen *at-risk prediction* and lay the foundation for future live validation and interventions. This study addresses the following questions:

- **RQ1:** How can synthetic data enhance the prediction of at-risk students in small, imbalanced datasets?
- **RQ2:** Which machine learning algorithms are most effective for this task?

2 Related Work

Studies on computer science education have documented consistently low success rates in introductory programming courses. Pass rates typically hover around 67%, with larger class sizes and the use of languages such as C associated with increased failure rates [12]. Research on student engagement identifies behavioural, cognitive, and affective dimensions as relevant to academic performance [8,9]. Metrics such as time-on-task and assessment completion are commonly used to quantify engagement. Institutional reliance on retrospective surveys (e.g., the UK National Student Survey and module evaluations) limits opportunities for real-time prediction [5].

Automation has increased the volume of formative feedback available for programming tasks. At Coventry University, automated feedback was associated with improved satisfaction metrics, though the effects were less pronounced for lower-performing students [2]. Students also reported consistent value in receiving human feedback [6]. Keuning et al. [10] found that solution-focused error messages support syntactic and conceptual understanding but often do not address strategic knowledge, which remains more effectively conveyed through human instruction.

Machine learning techniques have been applied to engagement data to predict students at risk of failure [11]. Some approaches incorporate demographic variables, though access to such data may be limited by institutional and legal constraints [3]. Olney et al. [4] developed a predictive framework analysing academic performance and engagement over time to identify students needing intervention. This study focuses on real-time prediction within a single programming module by modelling student performance on formative tasks.

3 Methodology

3.1 Dataset

An anonymised dataset from three cohorts comprised 379 unique student records. Summative grades were used to assign a Boolean label: passing (`true`) or failing (`false`), with an imbalanced distribution of approximately 85:15 in favour of passing. This included all available data for the module.

Formative task submissions were recorded via the Codio learning platform (https://www.codio.com), which automatically marks tasks. Most involve writing, extending, or debugging small Python programs evaluated against unit tests, though some include other formats (e.g., multiple choice questions). Students are allowed multiple attempts on programming tasks. For further details, see [2].

3.2 Data Preparation

For each student, we compile two comma-separated lists, `right_answers` and `wrong_answers`, containing the corresponding task names. These are then converted into one-hot-encoded (OHE) features. We use three different feature sets corresponding to three different engagement intervals: 43 features for Weeks 1–3, 106 features for Weeks 1–6, and 150 features for Weeks 1–9. In this encoding, a value of `1` indicates that a task was completed correctly, while `0` means that it was either not attempted or not solved correctly (even after multiple attempts). Following this, a train–test split of 80:20 was used to provide the synthetic data generation process with as much data as possible.

3.3 Baseline Results (Pre-synthetic Data Generation)

Baseline experiments evaluated several classification algorithms to establish a reference point for synthetic data approaches. Models included Logistic Regression, SVMs (RBF and linear), Decision Trees, Random Forests, Naive Bayes, and K-Nearest Neighbours, all using default `sklearn` settings on the original imbalanced dataset. Table 1 reports the best-performing model for the failing (`false`) class at each interval, with corresponding precision, recall, and F1-scores.

Table 1. Baseline Performance for the Failing Class (Pre-Synthetic Data)

Interval	Best Model	Precision (`false`)	Recall (`false`)	F1-Score (`false`)
1–3	Decision Tree	0.50	0.45	0.48
1–6	Naive Bayes	0.47	0.64	0.54
1–9	Logistic Regression	0.50	0.36	0.42

The highest F1-score for the failing class (`false`) was 0.5385, achieved by Naive Bayes during Weeks 1–6. However, consistently low precision and recall across all intervals suggest these baseline models are inadequate for early intervention.

3.4 Synthetic Data Generation

We applied three synthetic data techniques: SMOTE [1], ADASYN [7] (both oversampling methods using different interpolation strategies), and CTGAN [13], a generative approach. Each method created a balanced 50/50 class distribution without altering feature dimensions.

Classifiers were retrained on these balanced datasets, while the test split containing only real data remained unchanged. Table 2 presents the best-performing model for the failing class at each interval, with corresponding precision, recall, and F1-score.

Table 2. Post Synthetic Data Performance for the Failing Class

Interval	Best Model	Precision	Recall	F1-Score
1–3	ADASYN/SVM (RBF)	0.60	0.55	0.57
1–6	SMOTE/LR	0.70	0.64	0.67
1–9	SMOTE/LR	0.60	0.55	0.57

Synthetic data augmentation improved recall and F1-scores for the failing class, with the best performance during Weeks 1–6. CTGAN underperformed compared to the oversampling methods across all models and intervals. Not shown in the table, SMOTE/LR in Weeks 1–3 achieved a higher recall of 0.6364 and a competitive F1-score of 0.56 for the `false` label. Therefore, SMOTE/LR was used to further tune across all intervals for consistency.

3.5 Synthetic Data Visualization

To investigate CTGAN's poor performance, we applied Principal Component Analysis (PCA) to visualise the high-dimensional feature space (Fig. 1) using data from Weeks 1–6, representative of all intervals. SMOTE and ADASYN generated synthetic samples that aligned well with original data clusters, preserving class structure. In contrast, CTGAN showed mode collapse, with samples clustering tightly and failing to reflect the minority class's variability.

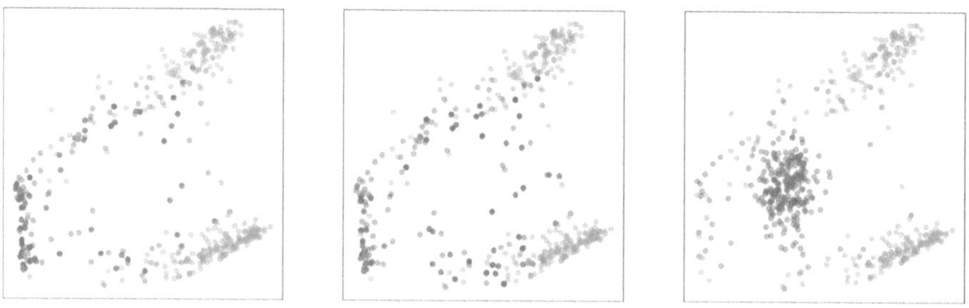

(a) SMOTE (b) ADASYN (c) CTGAN

Fig. 1. PCA visualisations of synthetic samples generated. Blue (`false`), Orange (`true`) (Color figure online)

4 Results

We tuned SMOTE/LR models using grid search to maximise performance, with results shown in Table 3.

For Weeks 1–3, the best setup used SMOTE (`k_neighbors = 5`, sampling strategy 'auto') with elasticnet Logistic Regression (`penalty = 'elasticnet', l1_ratio = 0.5, C = 0.01, solver = 'saga'`) at a threshold of 0.50, yielding precision, recall, and F1-score of 0.55. This improved precision over baseline (0.50) but slightly reduced recall (−0.05).

For Weeks 1–6, the optimal model used SMOTE (`k_neighbors=3`) with Logistic Regression (`penalty = 'l2', C = 1.0, solver = 'lbfgs'`) at a threshold of 0.45, reproducing baseline precision of 0.70 and recall of 0.64.

Weeks 1–9 used a setup similar to Weeks 1–3: SMOTE (`k_neighbors = 5`) and elasticnet Logistic Regression (`C = 1.0, l1_ratio = 0.5, solver ='saga'`) at a threshold of 0.50, achieving 0.75 precision, 0.55 recall, and 0.64 F1-score—marking a notable precision gain over the baseline (0.60) with stable recall and F1.

Table 3. Tuned Metrics and Best Parameters for LR–SMOTE

Interval	Precision (`false`)	Recall (`false`)	F1 Score (`false`)	Accuracy	AUC
1–3	0.55	0.55	0.55	0.87	0.77
1–6	0.70	0.64	0.67	0.91	0.78
1–9	0.75	0.55	0.64	0.91	0.74

5 Conclusion and Future Work

This study explored whether synthetic data generation can improve the identification of at-risk students in a small, imbalanced dataset from a single introductory programming module. Oversampling techniques such as SMOTE and ADASYN improved recall and F1 scores for the failing class compared to baseline models, addressing **RQ1**. Preliminary analysis also indicated that performance on specific formative tasks may predict failure, supporting future feature importance analysis for **RQ2**.

The next phase will validate the models using live cohort data currently being collected. If results hold, the models will be deployed for real-time prediction and early intervention over the following two modules. Similarly, feature importance analysis will help refine formative tasks and guide instructional improvements. While this study focused on a single module, the methodology will be extended to additional programming modules delivered through Codio to explore generalisability. The approach may apply to any context using numerous automatically assessed micro-tasks, particularly where richer datasets are unavailable.

This work forms an essential step toward developing practical, interpretable workflows to help educators detect disengagement early and support student success in programming education.

References

1. Chawla, N.V., Bowyer, K.W., Hall, L.O., Kegelmeyer, W.P.: SMOTE: synthetic minority over-sampling technique. J. Artif. Intell. Res. **16**, 321–357 (2002). https://doi.org/10.1613/jair.953
2. Croft, D., England, M.: Computing with Codio at Coventry University: online virtual Linux boxes and automated formative feedback. In: Proceedings of the 3rd Conference on Computing Education Practice, pp. 1–4. ACM, Durham United Kingdom (2019). https://doi.org/10.1145/3294016.3294018
3. Dervenis, C., Kyriatzis, V., Stoufis, S., Fitsilis, P.: Predicting students' performance using machine learning algorithms. In: Proceedings of the 6th International Conference on Algorithms, Computing and Systems, pp. 1–7. ACM, Larissa Greece (2022). https://doi.org/10.1145/3564982.3564990
4. Elbouknify, I., et al.: Student at-risk identification and classification through multitask learning: a case study on the moroccan education system. In: Olney, A.M., Chounta, I.A., Liu, Z., Santos, O.C., Bittencourt, I.I. (eds.) Artificial Intelligence in Education, vol. 14830, pp. 372–380. Springer Nature Switzerland, Cham (2024). https://link.springer.com/10.1007/978-3-031-64299-9
5. Fredricks, J.A., Blumenfeld, P.C., Paris, A.H.: School engagement: potential of the concept, state of the evidence. Rev. Educ. Res. **74**(1), 59–109 (2004). https://doi.org/10.3102/00346543074001059
6. Grawemeyer, B., Halloran, J., England, M., Croft, D.: Feedback and Engagement on an Introductory Programming Module. In: Proceedings of the 6th Conference on Computing Education Practice, pp. 17–20. ACM, Durham United Kingdom (2022). https://doi.org/10.1145/3498343.3498348
7. He, H., Bai, Y., Garcia, E.A., Li, S.: ADASYN: adaptive synthetic sampling approach for imbalanced learning. In: 2008 IEEE International Joint Conference on Neural Networks (IEEE World Congress on Computational Intelligence), pp. 1322–1328. IEEE, Hong Kong, China (2008). https://doi.org/10.1109/IJCNN.2008.4633969
8. Kahu, E.R.: Framing student engagement in higher education. Stud. High. Educ. **38**(5), 758–773 (2013). https://doi.org/10.1080/03075079.2011.598505
9. Kahu, E.R., Nelson, K.: Student engagement in the educational interface: understanding the mechanisms of student success. High. Educ. Res. Dev. **37**(1), 58–71 (2018). https://doi.org/10.1080/07294360.2017.1344197
10. Keuning, H., Jeuring, J., Heeren, B.: A systematic literature review of automated feedback generation for programming exercises. ACM Trans. Comput. Educ. **19**(1), 1–43 (2019). https://doi.org/10.1145/3231711
11. Qian, Y., Lehman, J.: Students' misconceptions and other difficulties in introductory programming: a literature review. ACM Trans. Comput. Educ. **18**(1) (2017). https://doi.org/10.1145/3077618
12. Watson, C., Li, F.W.: Failure rates in introductory programming revisited. In: Proceedings of the 2014 conference on Innovation & technology in computer science education - ITiCSE '14, pp. 39–44. ACM Press, Uppsala, Sweden (2014). https://doi.org/10.1145/2591708.2591749
13. Xu, L., Skoularidou, M., Cuesta-Infante, A., Veeramachaneni, K.: Modeling tabular data using conditional GAN. In: Proceedings of the 33rd International Conference on Neural Information Processing Systems, pp. 7335–7345. No. 659, Red Hook, NY, USA (2019). https://dl.acm.org/doi/10.5555/3454287.3454946

Exploring Multimodal Quiz Generation and Evaluation Aligned with Higher-Order Learning Objectives in Bloom's Taxonomy

Mourya Teja Kunuku(✉) and Nasrin Dehbozorgi

Department of Computer Science, Kennesaw State University, Marietta, GA, USA
mkunuku@students.kennesaw.edu, dnasrin@kennesaw.edu

Abstract. The rise of large language models (LLMs) and vision-language models (VLMs) has created new possibilities for scalable educational content generation. However, most AI-generated assessments focus on surface-level understanding, such as factual recall or basic comprehension, leaving a significant gap in their ability to assess higher-order thinking skills like application, analysis, and evaluation—core elements emphasized in Bloom's Taxonomy. As education increasingly incorporates multimodal resources such as lecture videos, slides, and spoken explanations, there is a growing need for AI systems that can generate meaningful assessments grounded in these rich learning contexts. This research explores the design of a multimodal Retrieval-Augmented Generation (RAG) framework that generates quiz questions from video lectures by integrating textual, visual, and auditory data. Beyond generation, the project investigates scalable evaluation mechanisms through the use of LLM-as-Judge frameworks, which aim to replicate expert human judgment in assessing the quality and alignment of AI-generated questions. Evaluation focuses on both retrieval performance and cognitive alignment of the questions across dimensions such as correctness, relevance, and clarity. Preliminary experiments demonstrate that LLMs can achieve moderate agreement with expert human raters and generate contextually relevant questions, though challenges remain in assessing reasoning-heavy and visually grounded content. The broader goal of this research is to build intelligent, human-aligned assessment systems that support deeper learning by addressing multiple levels of cognitive complexity. By combining multimodal understanding with scalable evaluation, this work contributes to the development of AI systems that are pedagogically informed, interpretable, and aligned with educational objectives.

Keywords: AI in Education · Retrieval-Augmented Generation (RAG) · Multimodal Quiz Generation · Educational Assessments · Agentic AI Frameworks

A. I. Cristea et al. (Eds.): AIED 2025, CCIS 2590, pp. 433–438, 2025.
https://doi.org/10.1007/978-3-031-99261-2_49

1 Introduction and Problem Statement

Recent advances in large language models (LLMs) such as GPT-4, LLaMA, and Claude have unlocked new opportunities for automating educational content generation. From real-time tutoring [1], to instructional feedback [2], and large-scale assessment creation [3], generative AI is reshaping how learning at scale can be supported. One particularly promising application is automatic quiz generation, where LLMs are used to generate multiple-choice questions (MCQs) from lecture transcripts, textbooks, or instructional prompts [4,5]. However, most current systems tend to focus on surface-level objectives, often producing questions aimed at factual recall and basic comprehension [6], limiting their utility in promoting deeper learning. This mismatch highlights a fundamental misalignment between the current capabilities of generative models and the pedagogical goals emphasized in education. According to Bloom's Taxonomy [7], effective learning assessments should address a range of cognitive processes—from remembering and understanding to applying, analyzing, evaluating, and creating. While LLMs have demonstrated impressive fluency, studies have shown that their outputs often lack the conceptual rigor needed to test higher-order thinking skills [8,10]. For instance, [9] found that medical summaries generated by LLMs, though linguistically sound, frequently introduced factual inconsistencies or oversimplifications when tasked with analysis and synthesis. Moreover, today's learning environments are increasingly multimodal, involving video lectures, diagrams, voice explanations, and slides [11]. Yet, many quiz generation systems rely only on textual inputs [12], ignoring rich contextual signals found in visual and audio modalities. This limitation restricts the ability of AI systems to understand and assess learning materials in real-world scenarios, especially in STEM or visually intensive domains [13,14]. In parallel, the evaluation of AI-generated educational content remains a critical challenge. Human judgment is considered the gold standard, but it is costly, time-consuming, and non-scalable [15]. Emerging alternatives like the LLM-as-Judge paradigm—where one model evaluates the output of another—have gained traction in recent literature [16–18], showing promise in domains such as summarization, dialogue, and content alignment. However, the application of these frameworks to educational assessment, particularly in comparing model outputs with expert pedagogy across cognitive dimensions, is still underexplored. This research aims to address these gaps by investigating how multimodal generative systems can be used to produce and evaluate educational assessments that align with higher-order learning objectives. Specifically, the work focuses on building scalable quiz generation pipelines that incorporate audio, visual, and textual modalities, while simultaneously exploring whether LLM-based evaluation frameworks can serve as proxies for expert human raters in assessing the quality and depth of AI-generated content. The broader goal is to develop intelligent assessment systems that are not only scalable and interpretable but also pedagogically grounded and aligned with cognitive learning theories. This study centers on formative assessments within undergraduate computer science education. While the framework has the potential to generalize across domains, we limit our scope here to exploring its application in STEM-focused learning environments to ensure depth and contextual relevance. Future work will explore extensions to K–12 and other subject areas.

2 Research Questions

This work investigates the intersection of generative AI, multimodal learning environments, and educational assessment, with a focus on cognitive alignment and evaluation reliability. The research is guided by the following long-term questions:

RQ1: How can AI systems be designed to generate educational assessments that span multiple levels of Bloom's Taxonomy, from basic recall to complex reasoning and creativity?

RQ2: What design principles and modalities are most effective for enabling AI models to capture the context and instructional intent embedded in multimodal learning materials?

RQ3: How reliably can large language models simulate expert human judgment in evaluating the quality and cognitive depth of AI-generated educational content?

RQ4: What are the key factors contributing to agreement or divergence between human and LLM-based evaluators, especially in domains that require reasoning and conceptual analysis?

3 Methodology

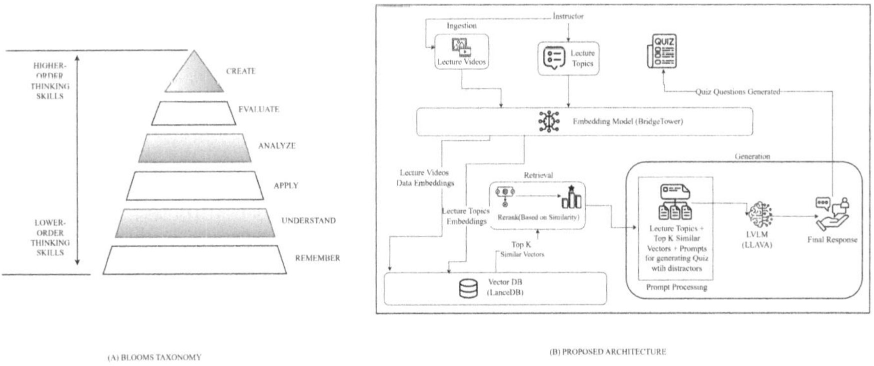

Fig. 1. (A) Blooms Taxonomy (B) Proposed architecture of the multimodal Retrieval-Augmented Generation (RAG) system for quiz generation.

This research employs a multimodal Retrieval-Augmented Generation (RAG) pipeline to generate multiple-choice quiz questions (MCQs) from lecture videos by leveraging textual, visual, and audio modalities. The system is designed to support formative assessment in undergraduate computer science education, with a focus on targeting higher-order cognitive skills such as application, analysis, and evaluation as defined by Bloom's Taxonomy.

The overall architecture, illustrated in the proposed diagram Fig. 1(B), consists of three main stages: multimodal preprocessing, retrieval, and generation.

Multimodal Preprocessing. Lecture videos are segmented into temporally aligned content blocks across modalities. Audio is transcribed using Whisper, while slide visuals are processed using BridgeTower to generate visual embeddings. These transcripts and visual features are synchronized and stored in a vector database for dense retrieval.

Retrieval. Given a topic keyword or learner query, the system retrieves the top-K relevant multimodal segments using vector similarity. Although the current implementation primarily utilizes content from lecture videos, the architecture supports augmenting retrieval with external sources—such as textbooks, concept glossaries, Bloom's Taxonomy guidelines, or few-shot exemplars—to address gaps in prior knowledge often assumed by instructional materials. This hybrid retrieval setup enhances the system's ability to generate coherent and contextually grounded questions, particularly in domains where conceptual dependencies span multiple topics.

Question Generation. The retrieved segments are converted into prompts using LangChain and passed to the LLaMA 3.1 (8B) model. Each generated question includes one correct option and three distractors. Prompt engineering is designed to invoke deeper cognitive engagement and discourage surface-level recall. The questions are labeled based on their intended cognitive level to facilitate alignment with Bloom's Taxonomy.

Evaluation. Evaluation is structured around two phases aligned with the research questions. For RQ1 and RQ2, which focus on the generation quality and cognitive depth of the assessments, the evaluation is conducted by domain-expert educators. Each multiple-choice question is rated on a 5-point Likert scale across three pedagogical dimensions: *correctness*, *relevance*, and *clarity*. In addition to these dimensions, raters are also asked to classify each question according to the appropriate level in Bloom's Taxonomy. This enables alignment checks between intended and perceived cognitive depth, which directly supports RQ1. To establish a meaningful benchmark for LLM-generated questions, we implement a blind evaluation setup in which human evaluators assess both AI-generated and textbook-derived questions without knowing their origin. This allows for comparative analysis of perceived quality between LLM-based generation and traditional human-authored materials. For RQ3 and RQ4, we evaluate whether large language models can reliably approximate expert judgments.

These LLMs are prompted to rate each question using the same rubric applied by educators. Agreement between LLM and human scores is measured using inter-rater reliability metrics such as Cohen's Kappa, Krippendorff's Alpha, and Spearman's Rho. This analysis focuses on identifying which cognitive dimensions lead to the greatest divergence and informs the development of more aligned LLM evaluators.

4 Preliminary Results

To validate the feasibility of the proposed methodology, a pilot study was conducted using the first lecture from an introductory computer science course. The system generated 15 multiple-choice questions using the multimodal RAG pipeline. Retrieval quality was evaluated using Hit Rate and Mean Reciprocal Rank (MRR), achieving 0.91 and 0.83 respectively for text segments and 0.82 and 0.71 for visual segments, indicating high-quality context selection across modalities. Human evaluators ($n = 4$) rated each question along three dimensions—correctness, relevance, and clarity—using a 5-point Likert scale. Average scores across all questions were 4.5 (correctness), 4.3 (relevance), and 4.4 (clarity), suggesting that the system produces questions that are both accurate and pedagogically meaningful. Parallel evaluations by LLMs (Claude 3 Sonnet, LLaMA 3.1, and GPT-4) showed moderate alignment with human scores, with Cohen's Kappa ranging from 0.34 to 0.44 and Krippendorff's Alpha (ordinal) ranging from 0.49 to 0.60. These results confirm that the methodology not only generates valid assessments but also supports automated evaluation in a scalable manner.

5 Conclusions and Future Directions

This research presents a novel multimodal Retrieval-Augmented Generation (RAG) system designed to generate and evaluate quiz questions that align with both instructional content and pedagogical goals. By leveraging video lectures and integrating textual, visual, and auditory modalities, the system supports the creation of contextually rich multiple-choice questions for formative assessment in undergraduate STEM education. Preliminary findings indicate that the system can generate high-quality questions that are pedagogically meaningful, with human evaluators rating them highly in correctness, relevance, and clarity. Additionally, early results suggest that LLM-as-Judge frameworks, such as Claude and GPT-4, can moderately approximate human evaluation, providing scalable alternatives for assessment quality review.

Future work will refine the generation pipeline to better target higher-order cognitive skills, including reasoning and abstraction, as outlined in Bloom's Taxonomy. The evaluation process will be expanded to include more granular metrics and qualitative rationales from both human experts and LLMs, enabling detailed error analysis and model introspection. An important research direction is the explicit modeling of domain knowledge and question complexity by tagging content with Bloom-level indicators, capturing concept prerequisites, and adapting question difficulty based on learner profiles. This will help in building personalized and cognitively aware assessment systems.

References

1. Khlaaf, H., et al.: Automatic Tutoring Systems Powered by Large Language Models. arXiv preprint arXiv:2305.13242 (2023)
2. Gupta, T., Chandakkar, P., Kaur, T.: Automated Feedback Generation using GPT-4 in Computer Science Education. arXiv preprint arXiv:2304.12539 (2023)
3. Luo, Z., Shen, J., Liang, Y.: AutoQuiz: Question Generation from Lecture Transcripts using LLMs. arXiv preprint arXiv:2311.02567 (2023)
4. MacNeil, R., Zou, J., Brunskill, E.: Generating Quality Distractors for Multiple Choice Questions. arXiv preprint arXiv:2306.09375 (2023)
5. Zhao, Z., Fan, W., Li, J., et al.: Evaluating Factual Consistency in MCQ Generation with LLMs. arXiv preprint arXiv:2210.12345 (2022)
6. Chan, K., Tamkin, A., et al.: GPT-Ed: Large Language Models for Automated Educational Content. arXiv preprint arXiv:2305.10162 (2023)
7. Krathwohl, D.R.: A revision of bloom's taxonomy: an overview. Theory Into Pract. **41**(4), 212–218 (2002)
8. Ashwin, J., Chhabra, A., Rao, V.: Using Large Language Models for Qualitative Analysis Can Introduce Serious Bias. arXiv preprint arXiv:2309.17147 (2023)
9. Tang, L., Sun, Z., Idnay, B., et al.: Evaluating large language models on medical evidence summarization. NPJ Digital Med. **6**(1), 158 (2023)
10. Bhaduri, S., Kapoor, S., Gil, A., Mittal, A., Mulkar,R.: Reconciling Methodological Paradigms: Employing Large Language Models as Novice Qualitative Research Assistants. arXiv preprint arXiv:2408.11043 (2024)
11. Vempala, A., Shankar, S., Suresh, R.: Multimodal learning for real-world classrooms: a survey. IEEE Trans. Learn. Technol. **15**(3), 315–328 (2022)
12. MacNeil, R., Maharaj, T., Brunskill, E.: Experiences with using codex for assessments and content creation. In: Proceedings of the 2022 SIGCSE Technical Symposium on Computer Science Education (2022)
13. Liu, H., Zhang, P., Hu, X., et al.: Visual Instruction Tuning. arXiv preprint arXiv:2304.08485 (2023)
14. Radford, A., Kim, J.W., Hallacy, C., et al.: Learning Transferable Visual Models From Natural Language Supervision. arXiv preprint arXiv:2103.00020 (2021)
15. Mohankumar, D., et al.: LLM-Eval: Benchmarking Large Language Models for Evaluation Tasks. arXiv preprint arXiv:2311.03172 (2023)
16. Zheng, L., et al.: Judging LLM-as-a-judge with MT-bench and chatbot arena. Adv. Neural Inf. Process. Syst. (NeurIPS) **36**, 46595–46623 (2023)
17. Yan, Z.: Evaluating the Effectiveness of LLM-Evaluators (aka LLM-as-Judge). eugeneyan.com (2024). https://www.eugeneyan.com/writing/llm-as-a-judge/
18. Liu, Y., Iter, D., Xu, Y., Wang, S., Xu, R., Zhu, C.: G-Eval: NLG Evaluation using GPT-4 with Better Human Alignment. arXiv preprint arXiv:2303.16634 (2023)

Adaptive, Scalable, and Human-Centered Technology for Data Visualization Literacy and Educational Assessment Development

Yuan Charles Cui[1](✉), Fumeng Yang[2], and Matthew Kay[1]

[1] Northwestern University, Evanston, IL 60208, USA
{charlescui,matthew.kay}@u.northwestern.edu
[2] University of Maryland, College Park, MD 20742, USA
fmy@umd.edu

Abstract. Assessments play a critical role in education, but developing assessments can be difficult. This dissertation research focuses on *data visualization literacy* (a person's ability to interpret visualizations) and *K–12 educational* assessments. Visualization literacy can influence people's critical decisions. To understand and help people improve this ability, we must first be able to measure it with assessments. There are many challenges associated with visualization literacy assessments: e.g., (**RQ1**) how can we measure this ability in a repeatable and timely manner and (**RQ2**) how can we create diverse visualization items (question-answer pairs about visualizations) at scale? In K–12 education, teachers create assessments to assess student progress and instructional strategies. Despite recent studies on using fully automated AI approaches to generate K–12 level questions, few have investigated (**RQ3**) how to develop human-centered AI systems for holistic assessment development. This dissertation research aims to answer these questions by developing adaptive, scalable, and human-centered technology for assessment development.

Keywords: Educational Assessment Development · Data Visualization Literacy · Human-Centered AI for Education

1 Introduction

Assessments play a crucial role in data visualization literacy and K–12 education. For visualization literacy researchers, assessments allow them to measure people's ability and evaluate the effectiveness of targeted interventions [2,3,5]. For K–12 teachers, assessments allow them to assess their instructional methods and refine teaching strategies [4]. For K–12 and data visualization students, assessments help them identify knowledge gaps and reinforce learning [8]. Well-crafted assessments ensure quality and accountability in education by providing consistent, meaningful feedback to all stakeholders.

However, developing assessments is challenging. Existing visualization literacy assessments and AI approaches for generating K–12 level questions have their

A. I. Cristea et al. (Eds.): AIED 2025, CCIS 2590, pp. 439–445, 2025.
https://doi.org/10.1007/978-3-031-99261-2_50

limitations. For example, Lee et al. created a Visualization Literacy Assessment Test (VLAT) that contains 53 multiple-choice items across 12 chart types [5]; Ge et al. developed a Critical Thinking Assessment for Literacy in Visualizations (CALVI), which has a bank of 45 items [3]. Both assessments are long and static, have small item banks, and one administration of the test would exhaust most of the item bank. This creates barriers for researchers and educators who wish to run repeated studies on the same people to track their progress over time or to empirically evaluate interventions designed to enhance visualization literacy. For K–12 education, researchers have shown that large language models (LLMs) can be used to create items across different subjects (e.g., social science [10], mathematics [7]) that measure lower-level skills but struggle with higher-level abilities. In addition, most work focus on generating individual items rather than holistic assessments. Human oversight and interaction might improve the current approaches, but most existing work relies on fully automated, end-to-end approaches with little human involvement. To this end, I form the following three research questions and attempt to answer them in this dissertation research.

1. **How to efficiently measure visualization literacy:** I developed computerized adaptive assessments, A-VLAT and A-CALVI [2], to measure the same skills as their original versions in half the number of items (Sect. 2.1).
2. **How to generate diverse visualization literacy test items at scale:** I developed VILA (**V**isualization **I**tems Generated by Large **LA**nguage Models) [1], an LLM-based pipeline for scalably generating visualization items that measure people's ability to accomplish diverse visualization tasks (Sect. 2.2).
3. **How to use AI to help K–12 teachers develop educational assessments:** I plan to build a human-centered, interactive AI system to help teachers develop assessments. I will interview teachers to understand their practices, involve them as co-design partners, and evaluate the system with teachers (Sect. 3).

2 Completed Work

2.1 Adaptive Assessment of Visualization Literacy [2]

Background and Motivation. Existing visualization literacy assessments are often long and static, and one administration would exhaust most or the entirety of the item bank. For example, VLAT uses all 53 items in the bank and CALVI uses 30 (out of 45) items in the bank. This creates barriers (e.g., same items being administered to the same person repeatedly) for researchers and educators who wish to run repeated studies to track progress over time, or to empirically evaluate interventions designed to enhance a target group's ability to interpret visualizations. Therefore, we need more efficient ways to use existing item banks to measure people's visualization literacy in a repeatable and timely manner.

Methods and Results. To build efficient visualization literacy assessments, we applied techniques from computerized adaptive testing (CAT) [11], where the next item a test-taker sees depends on their performance on the previous items. The idea is to adaptively select items to tailor to a test-taker's ability in order to gain more information about them. For example, if someone performs poorly on difficult items, they will then be presented with an easier item. We adopted the CAT development framework in Thompson et al. [11] to create two short, adaptive visualization literacy assessments: A-VLAT and A-CALVI, which measure the same skills as their original static versions, VLAT [5] and CALVI [3], in half the number of questions. First, we computed *item parameters* (e.g., how difficult items are) with Item Response Theory. Using the item parameters, we constructed adaptive algorithms to select items for test-takers. We also included non-psychometric constraints in our algorithms to ensure the assessments contain balanced content; i.e., to ensure A-VLAT covers all 12 chart types and 8 tasks from VLAT and A-CALVI covers all 11 misleaders from CALVI. We finalized the configurations of adaptation through simulation and refined the composition of test items of A-CALVI via a qualitative study. Additionally, we demonstrated that both adaptive assessments have high test-retest reliability and convergent validity via four online studies. Figure 1 illustrates the development process of A-VLAT and A-CALVI.

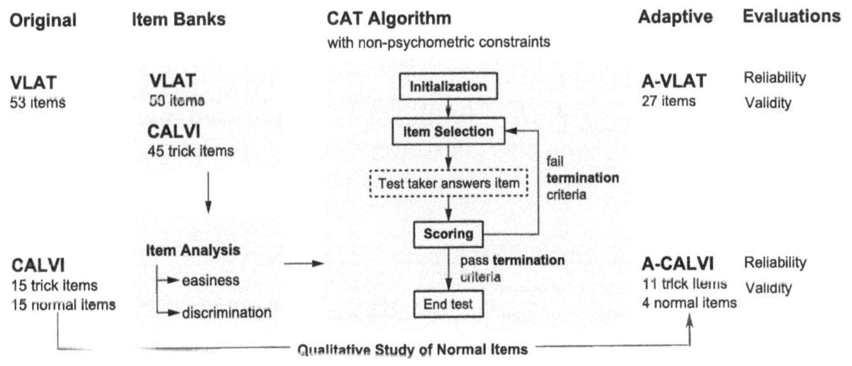

Fig. 1. The process of developing adaptive assessments A-VLAT and A-CALVI. We started with VLAT and CALVI item banks and used item analysis to construct the CAT algorithms for the adaptive test. We then evaluated their validity and reliability via online studies.

2.2 LLM for Scalable Generation of Diverse Visualization Items [1]

Background and Motivation. Despite our efforts to create adaptive assessments (Sect. 2.1), creating high-quality and diverse visualization items remains a resource-intensive task. Significant expertise is needed to develop well-designed visualizations, craft questions that accurately assess intended competencies, and curate multiple-choice options with one best answer. Large and diverse visualization item banks are not only needed for visualization literacy research: teachers of visualization courses require an extensive pool of materials for quizzes, practice problems, and exams; developers of visualization question-answering systems seek items as training and test data [6]. There is an unmet need for ways to scalably generate high-quality and diverse items. Recent advancements in LLMs have demonstrated that they can perform a variety of tasks, including code generation and question answering. Creating a visualization item requires similar abilities. Therefore, we studied how to create an LLM-based approach to generate diverse, high-quality multiple-choice visualization items at scale.

Methods and Results. Through an iterative design process, we developed the VILA (**V**isualization **I**tems Generated by Large **LA**nguage Models) pipeline, to scalably generate items that measure people's ability to accomplish visualization tasks. We used the VILA pipeline to generate 1,404 candidate items across 12 chart types and 13 visualization tasks. In collaboration with 11 visualization experts, we developed an evaluation rulebook which we then used to rate the quality of all candidate items. The result is the VILA bank of 1,103 items. From this evaluation, we also identified and classified current limitations of LLMs in generating visualization items, and discussed the role of human oversight in ensuring quality. We also demonstrated an application of VILA by creating a visualization literacy test, VILA-VLAT, which measures people's ability to complete a diverse set of tasks on various types of visualizations; to show the potential of this application, we assessed the convergent validity of VILA-VLAT by comparing it to VLAT via an online study ($R = 0.70$). Lastly, we provided practical recommendations for using VILA. Figure 2 illustrates the development process of VILA and provides examples of success and failure of the pipeline.

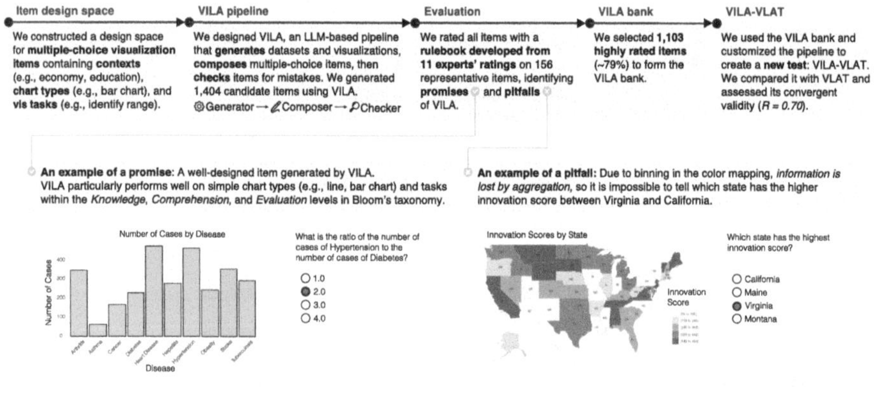

Fig. 2. The process of developing the VILA pipeline and bank. The two examples are a well-designed item (left) and an erroneous item (right) the VILA pipeline generated.

3 Proposed Work: Human-Centered, AI-Powered System for Assessment Development

Background and Motivation. AI models, particularly LLMs, have shown promise in generating assessment items across different subjects at the K–12 level (e.g., social science [10], mathematics [7]) that measure lower-level skills but struggle with higher-level abilities [1,7]. In addition, most work focuses on generating individual items rather than holistic assessments. Human oversight and interaction might improve current approaches, but most existing work relies on fully automated, end-to-end approaches with little human involvement. Therefore, to ensure quality and adoption, it is important to consider a human-centered approach to design AI solutions for assessment development.

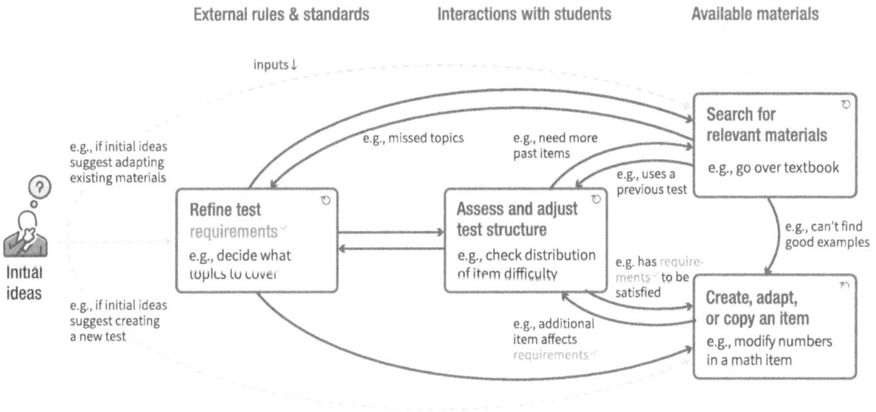

Fig. 3. Our conceptual model depicting teachers' processes of assessment development. In this model, a teacher has some initial ideas about the test (assessment) requirements (✓). They then start (--→) and move between the stages (↻) to simultaneously develop the test and refine test requirements, while considering various inputs (↓).

Proposed Methods. Taking a human-center approach, I plan to involve teachers in the development and evaluation phases of this project. First, I interviewed high school teachers to understand their current assessment development processes and their experience with AI. From there, I plan to use a co-design protocol [9] to iteratively develop an LLM-powered system for assessment development, seeking teachers' feedback throughout the process. Lastly, I will recruit other teachers to develop assessments using this system and evaluate it based on their experience.

Initial Results. To understand teachers' current assessment development practices and identify opportunities for AI to support, we interviewed 13 high school teachers across different subjects and schools in the USA. Using thematic analysis, we captured teachers' diverse and complex assessment development processes in a conceptual model, which consists of three categories of inputs and four stages of actions (Fig. 3). From this model, we identified four categories of challenges teachers face: developing high-quality items; overcoming technical barriers in item development; navigating assessment-level constraints; working with imperfect AI. We then identified potential AI-based solutions by forging new connections between these challenges and a broad body of literature—including work in AIED, creativity support tools, sensemaking, and AI literacy. I plan to use these initial results to guide the co-deign and development of the system.

Expected Contributions. The expected contributions of this project include: (1) a deeper understanding of K–12 teachers' current assessment development processes; (2) a set of opportunities for AI to help assessment development; (3) a web-based system that helps teachers develop assessments; (4) human–AI interaction paradigms derived from the system design that can potentially help other problems in AIED. These contributions can add to the body of human-centered and design-based research in AIED, open up new opportunities for integrating AI into assessment development, empowering teachers to enhance assessments, and ultimately improving student learning outcomes.

References

1. Cui, Y., Ge, L.W., Ding, Y., Harrison, L., Yang, F., Kay, M.: Promises and Pitfalls: using large language models to generate visualization items. IEEE Trans. Visual Comput. Graphics **31**(1), 1094–1104 (2025). https://doi.org/10.1109/TVCG.2024.3456309
2. Cui, Y., Ge, L.W., Ding, Y., Yang, F., Harrison, L., Kay, M.: Adaptive assessment of visualization literacy. IEEE Trans. Visual Comput. Graphics **30**(1), 628–637 (2024). https://doi.org/10.1109/TVCG.2023.3327165
3. Ge, L.W., Cui, Y., Kay, M.: CALVI: critical thinking assessment for literacy in visualizations. In: Proceedings of the 2023 CHI Conference on Human Factors in Computing Systems, CHI 2023, pp. 1–18. Association for Computing Machinery, New York, NY, USA (2023). https://doi.org/10.1145/3544548.3581406
4. Kibble, J.: Best practices in summative assessment. Adv. Physiol. Educ. **41**(1), 16–25 (2017). https://doi.org/10.1152/advan.00116.2016
5. Lee, S., Kim, S.H., Kwon, B.C.: VLAT: development of a visualization literacy assessment test. IEEE Trans. Visual Comput. Graphics **23**(1), 551–560 (2017). https://doi.org/10.1109/TVCG.2016.2598920
6. Masry, A., Do, X.L., Tan, J.Q., Joty, S., Hoque, E.: ChartQA: a benchmark for question answering about charts with visual and logical reasoning. In: Muresan, S., Nakov, P., Villavicencio, A. (eds.) Findings of the Association for Computational Linguistics: ACL 2022, pp. 2263–2279. Association for Computational Linguistics, Dublin, Ireland, May 2022. https://doi.org/10.18653/v1/2022.findings-acl.177, https://aclanthology.org/2022.findings-acl.177

7. Pham, P.V.L., Duc, A.V., Hoang, N.M., Do, X.L., Luu, A.T.: ChatGPT as a Math Questioner? Evaluating ChatGPT on generating pre-university math questions. In: Proceedings of the 39th ACM/SIGAPP Symposium on Applied Computing, SAC 2024, pp. 65–73. Association for Computing Machinery, New York, NY, USA (2024). https://doi.org/10.1145/3605098.3636030
8. Raupach, T., Brown, J., Anders, S., Hasenfuss, G., Harendza, S.: Summative assessments are more powerful drivers of student learning than resource intensive teaching formats. BMC Med. **11**(1), 61 (2013). https://doi.org/10.1186/1741-7015-11-61
9. Roschelle, J., Penuel, W.R.: Co-design of innovations with teachers: definition and dynamics. In: Proceedings of the 7th International Conference on Learning Sciences, ICLS 2006, pp. 606–612. International Society of the Learning Sciences (2006)
10. Scaria, N., Chenna, S.D., Subramani, D.: How Good are Modern LLMs in Generating Relevant and High-Quality Questions at Different Bloom's Skill Levels for Indian High School Social Science Curriculum? In: Kochmar, E., Bexte, M., Burstein, J., Horbach, A., Laarmann-Quante, R., Tack, A., Yaneva, V., Yuan, Z. (eds.) Proceedings of the 19th Workshop on Innovative Use of NLP for Building Educational Applications (BEA 2024), pp. 1–10. Association for Computational Linguistics, Mexico City, Mexico, June 2024. https://aclanthology.org/2024.bea-1.1/
11. Thompson, N.A., Weiss, D.A.: A framework for the development of computerized adaptive tests. Pract. Assess. Res. Eval. **16**(1) (2011). https://doi.org/10.7275/wqzt-9427

Designing and Evaluating AI-Generated Multimodal Analogy-Based Explanations

Yuri Noviello(✉)

Delft University of Technology, Delft, The Netherlands
y.noviello@tudelft.nl

Abstract. Effectively teaching programming concepts remains a persistent challenge, as conventional approaches often struggle to make abstract ideas accessible and engaging. Analogies provide a powerful means to simplify these concepts, and when augmented with multimodal resources, such as video animations, can significantly enhance learner engagement and comprehension. This research leverages Large Language Models (LLMs) and a structured animation workflow to automate the generation of analogy-driven explanations and visualizations for programming topics. Preliminary findings indicate that while LLMs can produce creative and pedagogically relevant analogies, careful decomposition and validation steps are critical for ensuring clarity, correctness, and learning alignment. We will conduct controlled evaluations to compare the effectiveness of AI-generated analogy-driven materials against traditional instructional methods. We also explore domain-specific personalization, adapting analogies to the learner's familiar domains and background.

Keywords: Analogy Generation · Notional Machine · Animation Generation · Programming Education · Artificial Intelligence

1 Introduction

Traditional programming education methods often fail to engage learners and convey abstract ideas meaningfully. A proven strategy to mitigate these issues is to employ metaphors and analogies: by anchoring new and unfamiliar topics in learners' existing knowledge, analogies can render challenging material more intuitive and relatable [7]. When these analogies are augmented with multi-modal elements, such as video animations, the learning experience becomes potentially more dynamic, accommodating diverse learning styles and boosting both engagement and conceptual understanding [6]. Despite their pedagogical benefits, creating analogy-driven multimedia content remains time-intensive. Educators not only need deep expertise in programming, but also creative skills in animation and media production, resources that can be scarce or challenging to integrate into teaching schedules [11].

The rapid evolution of Artificial Intelligence (AI), especially in the area of Natural Language Processing (NLP), presents an opportunity to automate and

A. I. Cristea et al. (Eds.): AIED 2025, CCIS 2590, pp. 446–451, 2025.
https://doi.org/10.1007/978-3-031-99261-2_51

enhance educational content creation. AI has already proven effective in personalizing learning experiences, offering data-driven assessments, and providing on-demand tutoring in various STEM fields [4,8]. In programming education, AI-driven tools have demonstrated the ability to generate adaptive, context-sensitive feedback, but most systems still emphasize syntax correction or code completion rather than conceptual clarity delivered through analogies or metaphors.

There is no widely accessible solution that generates analogy-driven and multi-modal instructional materials to support educators. Such a solution would hold significant value for both the education and NLP communities by demonstrating how state-of-the-art language models methodologies can be applied to enhance pedagogy and learning outcomes. This research aims to develop an AI-based system that automates the generation of analogy-driven explanations and corresponding video animations for programming concepts, by leveraging the capabilities of Large Language Models and established educational design principles.

1.1 Research Questions

RQ1. How to design an AI tool that effectively generates analogies and animations for programming education, when evaluated against established analogy-quality metrics (e.g., *accuracy, accessibility, comprehensiveness*) and instructional design standards (e.g., *engagement, clarity, concept alignment*)?
RQ2. How do human analogy's evaluations (e.g., *student and teacher feedback, usability studies*) compare with automated evaluation strategies?
RQ3. How do AI-generated analogies and animations enhance students' comprehension and retention of complex programming concepts compared to traditional text-based or static visual materials, as measured through pre/post-tests and qualitative feedback?
RQ4. How does AI-driven personalized content generation, tailored to specific domains or learner profiles, affects student engagement, learning outcomes, and cognitive load, compared to generic, non-personalized instructional approaches?

2 Related Work

Analogies have long been recognized as powerful pedagogical tools for making abstract or complex ideas more accessible by linking them to familiar concepts. In programming education, these devices are frequently employed to simplify intricate topics. The value of analogies is further amplified when presented through multimodal channels, including visuals and animations. Animations can effectively illustrate dynamic processes and boost student engagement. The use of multimodal learning material has proven especially promising for fostering deeper comprehension and retention in educational contexts [3].

Metaphor and analogy processing represent a longstanding challenge in NLP. Because they rely on figurative rather than literal language, traditional systems have struggled to handle their semantic and contextual nuances [10].

More recently, the emergence of LLMs has opened up new avenues for tackling metaphor understanding and analogy generation at scale [2,13]. Empirical studies also point to the educational potential of such AI-generated analogies, underscoring their value for explaining programming concepts to learners [1].

Beyond the broad definitions of metaphor and analogy, notional machines offer another conceptual tool to aid understanding of programming languages and environments [5]. A notional machine is a pedagogical abstraction that emphasizes particular aspects of a language's semantics while omitting non-essential details. The goal is to reduce the cognitive load and help learners form accurate mental models of how programs execute. Since notional machine must align with the actual semantics of the targeted programming language to avoid introducing misconceptions, recent work formalizes soundness criteria enabling rigorous analysis of notional machines and the identification of inconsistencies [9].

3 Research Methods

This study employs a multi-phase methodology to develop, evaluate, and refine an AI-driven system that generates analogies and video animations for programming education. The research is structured into four main phases.

Phase 1: Tool Development and Preliminary Evaluation
In the initial phase, we build the core AI-based tool capable of producing analogy-driven textual explanations and accompanying video animations. The system leverages Large Language Models (LLMs) to generate analogies that simplify programming concepts and then transforms these explanations into short, explanatory animations. A small-scale, preliminary evaluation will follow, involving expert educators who will assess the output's accuracy, accessibility, and comprehensiveness, while students will provide feedback on engagement, clarity, and cognitive load.

Phase 2: Automated Evaluation Pipeline
Building on the insights from the preliminary evaluation, the next phase focuses on creating an AI-driven evaluation model designed to assess the quality of generated analogies and animations. The goal is to achieve expert-level reliability in judging whether the explanations are both pedagogically sound and semantically accurate. By automating quality checks, we aim to streamline iterative refinement and ensure consistently high-quality outputs.

Phase 3: Large-Scale Controlled Study
In this phase, the tool's effectiveness is rigorously tested through a controlled experiment with a larger group of students. We will compare the impact of AI-generated analogy and animations to that of traditional instructional materials. Evaluation metrics will include *comprehension* and *retention*, measured with pre and post-tests designed to calculate learning gains.

Phase 4: Domain Personalization and Extended Evaluation
In the final phase, we introduce personalization features that enable learners to select familiar domains (such as *sports*, *music*, or *gaming*) to contextualize analogies. By tailoring explanations to individual interests and background knowledge, we hypothesize that learners will experience a reduced cognitive load and increased engagement. This phase will include an extensive evaluation of how domain-specific analogies influence learning outcomes, comparing them against generic analogies in terms of knowledge retention and user satisfaction.

4 Current Work

Analogy Generation
Our ongoing efforts center on testing state-of-the-art LLMs to produce analogies for foundational programming topics Table 1. Our experiments show that, while LLMs can generate creative and intuitive analogies, they often require careful prompt engineering and iterative refinement to maximize clarity and correctness. We plan to systematically investigate various strategies and model configurations, with the goal of identifying best practices for generating analogies that are neither too domain-specific nor too general.

Table 1. LLM-generated analogy illustrating the topic *Recursion*

Topic: *Recursion*

Analogy: Imagine a set of Russian nesting dolls. You want to open the smallest doll inside, but to get there, you must first open each larger doll one by one.
Each doll represents a step in the problem, and opening it is like calling the function. The smallest doll (the base case) is the point where you stop opening because there's nothing smaller inside.
Once you reach the smallest doll, you start putting the dolls back together in reverse order (unwinding the recursion), until all the dolls are reassembled, and the problem is fully solved.

Animation Generation
In parallel, we have conducted several experiments using `manim` [12] as our animation framework Fig. 1. Directly prompting an LLM to generate a complete `manim` script frequently leads to syntactic and logical errors, reflecting the complexity of scene composition. To address these challenges, we devised a multi-stage pipeline:

1. Identify discrete visual components (e.g., shapes, text) tied to each part of the analogy.
2. Prompt the LLM to produce code snippets for individual components rather than the entire scene at once and validate these snippets to ensure syntactic correctness.

3. Merge validated code elements into a cohesive animation sequence, resolving any layout conflicts or timing issues through iterative checks.

Additionally, we developed a self-validation mechanism that detects semantic inconsistencies within the key frames of a video. Leveraging a Visual Large Language Model (vLLM), the system analyses these frames, automatically correcting content misalignment and improving the overall quality of the animation.

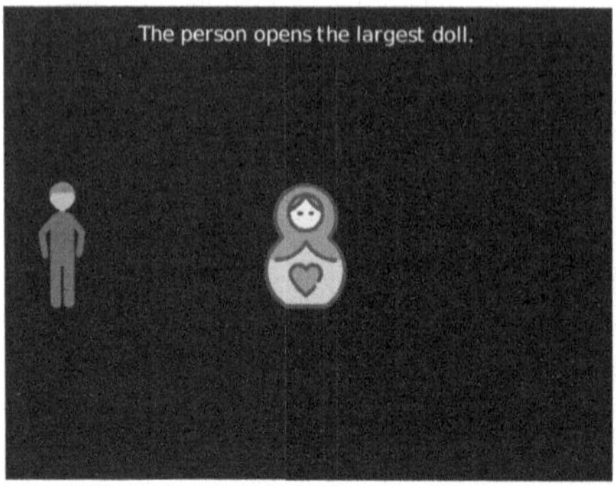

(a)

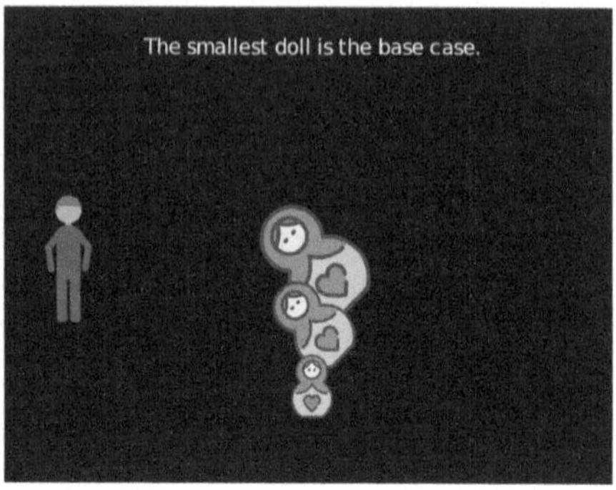

(b)

Fig. 1. Frames extracted from LLM-generated animation explaining *Recursion* as a *Set of Russian Dolls.* 1a The function starts by opening the largest doll.1b The *recursion* reaches the `base case` when the smallest doll is found, and no more dolls remain to be opened.

Evaluation

Currently, we are conducting an initial round of evaluations by collecting expert and student annotations on a predefined set of textual and video-based analogies. Domain experts rate the output on established analogy evaluation metrics, such as *conceptual accuracy* and *soundness.* Students' feedback will be used to measure engagement, usability, and cognitive load. These assessments will help calibrate the automated evaluation model for the next phase of the research.

References

1. Bernstein, S., et al.: "Like a Nesting Doll": analyzing recursion analogies generated by CS students using large language models. In: Proceedings of the 2024 on Innovation and Technology in Computer Science Education V. 1, pp. 122–128. ITiCSE 2024, Association for Computing Machinery, New York, NY, USA (2024). https://doi.org/10.1145/3649217.3653533
2. Bhavya, B., Xiong, J., Zhai, C.: Analogy generation by prompting large language models: a case study of InstructGPT. In: Shaikh, S., Ferreira, T., Stent, A. (eds.) Proceedings of the 15th International Conference on Natural Language Generation. pp. 298–312. Association for Computational Linguistics, Waterville, Maine, USA and virtual meeting (2022). https://doi.org/10.18653/v1/2022.inlg-main.25, https://aclanthology.org/2022.inlg-main.25/

3. Bouchey, B., Castek, J., Thygeson, J.: Multimodal learning. In: Ryoo, J., Winkelmann, K. (eds.) Innovative Learning Environments in STEM Higher Education: Opportunities, Challenges, and Looking Forward, pp. 35–54. Springer International Publishing, Cham (2021). https://doi.org/10.1007/978-3-030-58948-6_3
4. Deriba, F.G., Sanusi, I.T., Sunday, A.O.: Enhancing computer programming education using ChatGPT- a mini review. In: Proceedings of the 23rd Koli Calling International Conference on Computing Education Research, pp. 1–2. Koli Calling '23, Association for Computing Machinery, New York, NY, USA (2024). https://doi.org/10.1145/3631802.3631848, https://dl.acm.org/doi/10.1145/3631802.3631848
5. Fincher, S., et al.: Notional machines in computing education: the education of attention. In: Proceedings of the Working Group Reports on Innovation and Technology in Computer Science Education, pp. 21–50. ACM, Trondheim Norway (2020). https://doi.org/10.1145/3437800.3439202, https://dl.acm.org/doi/10.1145/3437800.3439202
6. Gladys, M., Rogers, L., Sharafutdinova, G., Barnham, N., Nichols, P., Dastoor, P.C.: Driving course engagement through multimodal strategic technologies. Int. J. Innovation Sci. Math. Educ. **30**(3) (2022). https://doi.org/10.30722/IJISME.30.03.002, https://openjournals.library.sydney.edu.au/CAL/article/view/15793
7. Lakoff, G., Johnson, M.: Metaphors We Live By. University of Chicago Press, Chicago (1980)
8. Liffiton, M., Sheese, B.E., Savelka, J., Denny, P.: CodeHelp: using large language models with guardrails for scalable support in programming classes. In: Proceedings of the 23rd Koli Calling International Conference on Computing Education Research, pp. 1–11. Koli Calling '23, Association for Computing Machinery, New York, NY, USA (2024). https://doi.org/10.1145/3631802.3631830
9. Santos, I.M., Hauswirth, M., Jeuring, J.: Sound Notional Machines (2024). https://doi.org/10.31219/osf.io/z2vdr, https://osf.io/z2vdr_v1
10. Shutova, E.: Design and evaluation of metaphor processing systems. Comput. Linguist. **41**(4), 579–623 (2015). https://doi.org/10.1162/COLI_a_00233, https://aclanthology.org/J15-4002/, place: Cambridge, MA Publisher: MIT Press
11. Snelson, C.: Quest-based learning: a scoping review of the research literature. TechTrends **66**(2), 287–297 (2022). https://doi.org/10.1007/s11528-021-00674-w, https://link.springer.com/article/10.1007/s11528-021-00674-w, company: Springer Distributor: Springer Institution: Springer Label: Springer Number: 2 Publisher: Springer, US
12. The Manim Community Developers: Manim – Mathematical Animation Framework (2025). https://www.manim.community/, original-date: 2020-05-19T02:37:13Z
13. Tong, X., Choenni, R., Lewis, M., Shutova, E.: Metaphor understanding challenge dataset for LLMs. In: Ku, L.W., Martins, A., Srikumar, V. (eds.) Proceedings of the 62nd Annual Meeting of the Association for Computational Linguistics (Volume 1: Long Papers), pp. 3517–3536. Association for Computational Linguistics, Bangkok, Thailand (2024). https://doi.org/10.18653/v1/2024.acl-long.193, https://aclanthology.org/2024.acl-long.193/

Multimodal Story Generation Using Generative AI for Contextualised Mathematics Education

Kejia Zhang[1], Charaka Palansuriya[1,2(✉)], and Aurora Constantin[1(✉)]

[1] The University of Edinburgh, Edinburgh, UK
{s2606445,aurora.constantin}@ed.ac.uk
[2] Edinburgh Parallel Computing Centre, Edinburgh, UK
c.palansuriya@epcc.ed.ac.uk

Abstract. Mathematics is important in the school curriculum, but its abstract nature makes it difficult for children to understand. Traditional education follows a teacher-centred, textbook-based and one-size-fits-all approach, which struggles to accommodate individual learning needs. The context-based learning, such as educational stories, is student-centred and has been shown to help children understand and solve mathematical problems. Based on this, we propose a system that automatically generates educational stories using generative artificial intelligence to provide support for understanding specific mathematical problems and solving them. Presently, we have developed a multi-agent framework based on large language models for story text generation and established a set of criteria for story evaluation. The results indicate that the multi-agent framework enhances the accuracy of the mathematical description in the educational stories, but the generated stories still lack creativity. Future work will focus on producing more personalised multimodal content that aligns better with children's interests and learning needs.

Keywords: Generative Artificial Intelligence · Educational Technology · Context-based Learning · Story Generation

1 Introduction

Mathematics is a key subject in the school curriculum, and it forms the foundation for various subjects. Despite this, many students often find mathematics challenging because of its abstract nature and reliance on symbolic representations [7]. These challenges might be more prominent for children aged 7 to 11. According to Piaget's theory of cognitive development [11], children in this age group are at the concrete operations stage, where they develop logical and concrete reasoning but have not acquired the ability for abstract and hypothetical thinking. One study noted that teaching mathematics to children at this stage requires helping them make connections between concepts and practical activities [10]. However, the traditional teaching approach for mathematics is

A. I. Cristea et al. (Eds.): AIED 2025, CCIS 2590, pp. 452–457, 2025.
https://doi.org/10.1007/978-3-031-99261-2_52

teacher-centred and textbook-based [1]. Teachers design lessons based on the activities and demonstrations in textbooks, ensuring all students in the same classroom receive identical learning materials. This approach often limits opportunities to connect abstract concepts to real-world problem-solving scenarios and overlooks the diversity of students' cognitive abilities and preferences.

Compared to the traditional teaching approach, context-based learning is a student-centred approach that incorporates real-life and fictitious examples, emphasising practical experience over theory [4,6]. Accordingly, educational stories could serve as a tool within this approach, which aims to connect academic theory with practice through engaging narratives and visual content [2]. For children aged 7–11, research indicates that they engage more effectively with content presented in story form [5], as narratives provide an intuitive way to process and retain information. However, when using educational stories for context-based learning, the stories need to be interesting to children.

Creating tailored (e.g. customised and personalised) educational stories for each child and each mathematical problem is challenging for educators regarding time and resources. Recent advancements in generative artificial intelligence (GenAI), including large language models (LLMs) for text generation and diffusion models for image generation, enable the automated creation of personalised narratives and visuals. Consequently, GenAI is increasingly used to enhance student engagement and comprehension through personalised learning materials. This research aims to develop a context-based learning system using open-source GenAI models to automatically generate personalised multimodal (i.e., text and image) mathematics stories. The purpose of these stories is to make it easier for children to understand particular mathematical problems and support them in solving these problems. The following research questions are explored:

- **RQ1:** How effectively can GenAI models generate personalised multimodal stories to enhance students' understanding of mathematical problems?
- **RQ2:** To what extent can GenAI models create personalised multimodal stories that guide children step-by-step in solving mathematical problems?

2 Literature Review

GenAI has been widely used to create context-based learning materials. Leong et al. applied GenAI to vocabulary learning, demonstrating its effectiveness in supporting personalised learning and enhancing learning motivation [8]. Utami used ChatGPT to generate real-world contextual mathematics word problems, improving students' problem comprehension skills [12]. For mathematical story generation, Zhang et al. introduced "Mathemyths," a GPT-4-based intelligent agent that collaborates with children to create dynamic narratives, integrating mathematical terms to support mathematical language learning [13]. These studies demonstrate the potential of GenAI in producing educational materials for context-based learning. However, we recognise that existing research still has gaps. As far as we know, there is no dedicated study focused on translating abstract mathematical problems into child-friendly stories.

3 Educational Theoretical Basis

We have incorporated REACT learning strategies [4] from context-based learning and heuristic learning theory in the proposed system.

REACT Learning Strategy. Prior educational research outlines five key strategies, which are *Relating*, *Experiencing*, *Applying*, *Cooperating* and *Transferring*, known as the REACT strategy [4]. These strategies focus on teaching and learning in context. *Relating* involves connecting new knowledge to students' prior experiences or existing understanding. *Experiencing* emphasises encouraging students to explore and discover. *Applying* focuses on using acquired concepts in practical situations. *Cooperating* promotes collaborative learning, where students engage in discussions, share ideas, and work together. *Transferring* challenges students to apply their knowledge to new contexts or problems.

Heuristic Learning Theory. In mathematics education, heuristic learning encourages students to discover solutions by exploring the problem-solving process [9]. This theory is embedded into the system by avoiding directly providing answers at the beginning of the story, instead gradually unfolding the story to guide students step by step through the problem-solving process.

4 Proposed Solution

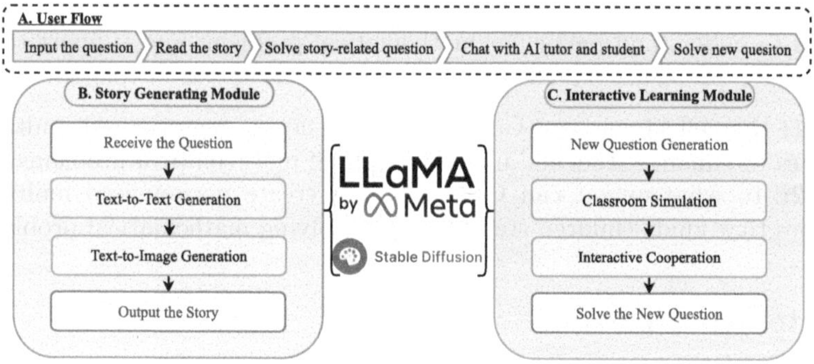

Fig. 1. Overview of GenAI-driven Context-based Learning System.

Based on the educational theories outlined in Sect. 3, we designed a system (Fig. 1) comprising a story generation module and an interactive learning module. Both modules only use open-source GenAI models (e.g., Llama and Stable Diffusion) that run on consumer-grade computing devices using portable machine learning formats (e.g., GGUF[1] and ONNX[2]), eliminating reliance on external resources (e.g., cloud services) and ensuring privacy and accessibility.

[1] https://github.com/ggml-org/llama.cpp.
[2] https://onnxruntime.ai/.

Story Generation Module. The module (Fig. 1.B) receives a mathematical problem and story configuration (e.g., background and characters) from the user and utilises the LLM and diffusion models to generate a multimodal story, embedding the mathematical question and solution within a meaningful context (i.e., *Relating* strategy). After that, the story is presented to users, allowing them to explore the problem-solving process (i.e., *Experiencing* strategy).

Interactive Learning Module. Inspired by research on simulating classrooms with LLM-based agents [14], this module (Fig. 1.C) aims to create an interactive space where users can discuss and apply the knowledge learned from the story. In the user flow (Fig. 1.A), after reading the story, the module will generate new questions and users will apply their learned knowledge to solve these new questions through interactive cooperation with the LLM-driven tutor and students (i.e., *Cooperation* strategy). There are two types of new questions, one is related to the story (i.e., *Applying* strategy), and the other one is a new mathematical question in a different context (i.e., *Transferring* strategy).

5 Current Progress and Preliminary Results

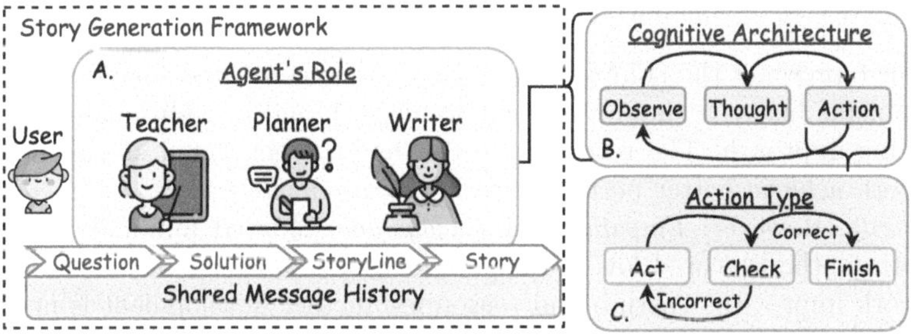

Fig. 2. Overview of Multi-agent Story Text Generation Framework

At present, we have developed a multi-agent framework for mathematical story text generation, shown in Fig. 2. The story text generation process is divided into three subtasks, each handled by a dedicated agent, as illustrated in Fig. 2.A. The *Teacher* agent produces a step-by-step solution to the mathematical problem, the *Planner* agent structures the story, and the *Writer* agent generates the whole story. This design ensures that explanations unfold progressively rather than directly providing answers. Each agent operates based on a cognitive architecture, shown in Fig. 2.B, that follows a cycle of *Observe*, *Thought* and *Action* to guide reasoning and decision-making. The action types, listed in Fig. 2.C, include *Act* (i.e., execute), *Check* (i.e., evaluate) and *Finish* (i.e., complete), allowing agents to iteratively refine their outputs and verify correctness.

The effectiveness of our multi-agent framework was evaluated using the LLM-as-judge method, with Gemini and ChatGPT-4o serving as evaluators. The evaluation criteria were adapted from existing story assessment standards [3] and applied using a 5-point Likert scale:

1. **Accuracy:** How accurate is the explanation provided in the story regarding the mathematical concept and calculation? (1: Not accurate; 5: Completely accurate)
2. **Relevance:** How well does the story correspond to the proposed math problem or concept? (1: No correspondence; 5: Complete correspondence)
3. **Coherence:** How would you rate the logical flow and clarity of the story? (1: Very low; 5: Excellent)
4. **Empathy:** How would you evaluate the emotions portrayed by the characters in the story? (1: Negative; 5: Positive)
5. **Surprise:** To what extent does the explanation in the story offer a different perspective or inspiration compared to your own understanding? (1: No new perspective; 5: Highly novel perspective)
6. **Engagement:** How engaging did you find the story? (1: No engagement; 5: High engagement)
7. **Complexity:** How would you rate the level of details in the explanation of the mathematical concept within the story? (1: Overly simple explanation; 5: Highly detailed explanation)

Table 1 presents the comparison of the stories generated for 60 mathematical problems between the multi-agent framework and a traditional single-turn prompting approach. The results indicate that the generated stories from our framework achieve better performance in *Accuracy*, *Relevance*, *Coherence*, and *Complexity*. However, *Empathy* and *Engagement* showed minimal differences, and both methods scored low in *Surprise*. Therefore, although the multi-agent framework improves structure and reasoning, further development is needed to enhance human-centred and creative aspects.

Table 1. The table reports the 5-point Likert scale evaluation result of single-turn prompt and multi-agent story text generation framework with Llama3.1 across Accuracy(AC), Relevance(RE), Coherence(CH), Empathy(EM), Suprise(SU), Engagement(EN) and Complexity(CM).

Evaluator	AC↑	RE↑	CH↑	EM↑	SU↑	EN↑	CM↑
	Mean_{STD}	Mean_{STD}	Mean_{STD}	Mean_{STD}	Mean_{STD}	Mean_{STD}	Mean_{STD}
Single-turn Prompting							
Gemini	$4.23_{1.07}$	$4.40_{0.97}$	$4.25_{0.77}$	$4.10_{0.29}$	$2.53_{0.53}$	$3.81_{0.42}$	$4.30_{0.66}$
ChatGPT	$4.16_{1.05}$	$4.35_{0.87}$	$4.56_{0.58}$	$4.50_{0.50}$	$2.80_{0.44}$	$4.35_{0.54}$	$4.22_{0.96}$
Multi-agent Framework							
Gemini	$4.68_{0.90}$	$4.83_{0.63}$	$4.63_{0.68}$	$4.12_{0.41}$	$2.67_{0.54}$	$3.88_{0.48}$	$4.70_{0.52}$
ChatGPT	$4.63_{0.75}$	$4.75_{0.64}$	$4.80_{0.54}$	$4.63_{0.48}$	$2.95_{0.53}$	$4.53_{0.56}$	$4.60_{0.73}$

6 Future Work and Expected Contribution

Building on our current work, we will continue to explore multimodal story generation and interactive learning to enhance learning experiences. Since our current evaluation only relied on automated evaluation methods, we plan to conduct user studies involving children, educators, and HCI researchers to gain deeper insights into the usability, engagement, and effectiveness of our system. In the future, we aim to provide personalised educational materials, particularly educational stories, to support children and teachers. We hope our system can contribute to the development of other context-based learning tools and systems.

References

1. Akbarov, A.: Multicultural Language Education: From Research into Practice. Cambridge Scholars Publishing (2014)
2. Alterio, M.: Using storytelling to enhance student learning (2002)
3. Chhun, C., Colombo, P., Clavel, C., Suchanek, F.M.: Of human criteria and automatic metrics: A benchmark of the evaluation of story generation. arXiv preprint arXiv:2208.11646 (2022)
4. Crawford, M.L.: Teaching contextually. Research, rationale, and techniques for improving student motivation and achievement in mathematics and science. Texas: Cord (2001)
5. Egan, K.: The cognitive tools of children's imagination (2001)
6. Glynn, S.M., Winter, L.K.: Contextual teaching and learning of science in elementary schools. J. Elem. Sci. Educ. **16**, 51–63 (2004)
7. Irmayanti, M., Chou, L.F., et al.: Storytelling and math anxiety: a review of storytelling methods in mathematics learning in Asian countries. Eur. J. Psychol. Educ. **40**(1), 1–26 (2025)
8. Leong, J., Pataranutaporn, P., Danry, V., Perteneder, F., Mao, Y., Maes, P.: Putting things into context: generative AI-enabled context personalization for vocabulary learning improves learning motivation. In: Proceedings of the 2024 CHI Conference on Human Factors in Computing Systems, pp. 1–15 (2024)
9. Mousoulides, N., Sriraman, B.: Heuristics in mathematics education. In: Encyclopedia of Mathematics Education, pp. 331–333. Springer, Cham (2020)
10. Ojose, B.: Applying piaget's theory of cognitive development to mathematics instruction. Math. Educ. **18**(1), 26–30 (2008)
11. Piaget, J.: The stages of the intellectual development of the child. Bull. Menninger Clin. **26**(3), 120 (1962)
12. Utami, I.Q., Hwang, W.Y., Hariyanti, U.: Contextualized and personalized math word problem generation in authentic contexts using generative pre-trained transformer and its influences on geometry learning. J. Educ. Comput. Res. **62**(6), 1604–1639 (2024)
13. Zhang, C., Liu, X., Ziska, K., Jeon, S., Yu, C.L., Xu, Y.: Mathemyths: leveraging large language models to teach mathematical language through child-ai co-creative storytelling. arXiv preprint arXiv:2402.01927 (2024)
14. Zhang, Z., et al.: Simulating classroom education with LLM-empowered agents. arXiv preprint arXiv:2406.19226 (2024)

Large Language Models to Enhance Learning in Cultural Heritage

Alessio Ferrato(✉)

Department of Engineering, Roma Tre University,
Via della Vasca Navale 79, 00146 Rome, Italy
alessio.ferrato@uniroma3.it

Abstract. Museums and Cultural Heritage (CH) institutions serve as crucial informal learning environments, offering opportunities for knowledge acquisition, cultural understanding, and lifelong learning outside traditional educational settings. Despite their educational potential, these institutions face significant challenges in personalizing experiences and measuring learning outcomes across diverse visitor populations. This proposal explores a novel approach to addressing these challenges by applying Large Language Models (LLMs) in CH settings. We propose a dual-role conceptual framework where LLMs are personalized guides and evaluators within museum contexts. As guides, LLMs can transform static narratives into dynamic, personalized experiences by adapting content based on visitor profiles, dwell time, and interaction patterns. As evaluators, they can generate tailored post-visit assessments that provide meaningful feedback while maintaining engagement through appropriate challenge levels. Our proposed methodology, which builds upon an existing Indoor Positioning System (IPS) deployed in a gallery, implements a theoretical three-phase approach encompassing pre-visit profiling, in-visit adaptive guidance, and post-visit assessment. This paper outlines potential implementation approaches and anticipated challenges, including LLM hallucinations and assessment validation, and proposes mitigation strategies for future empirical testing. Through this conceptual framework, we aim to address persistent issues in CH education: maintaining visitor attention, accommodating diverse learning styles, measuring learning outcomes in informal settings, and creating engaging experiences for all audience segments.

Keywords: Cultural Heritage · Large language models (LLMs) · Multiple Choice Questions (MCQs) · Question Generation

1 Introduction

Cultural Heritage (CH) institutions, collectively known as GLAMs (Galleries, Libraries, Archives, and Museums), are increasingly embracing digital technologies to enhance visitor experiences [1]. Recent advancements in artificial intelligence, particularly Large Language Models (LLMs), present opportunities to revolutionize personalization in CH settings [12].

A. I. Cristea et al. (Eds.): AIED 2025, CCIS 2590, pp. 458–463, 2025.
https://doi.org/10.1007/978-3-031-99261-2_53

This paper introduces two ideas how to exploit LLMs to enhance learning outcomes in CH. In particular, we see LLMs as two complementary entities: "*guides*" and "*evaluators*". The first one adapt content presentation based on visitor profiles, and the latter assess learning outcomes through dynamically generated assessments. The guide function transforms traditional one-size-fits-all museum narratives into tailored experiences that resonate with individual visitor motivations and learning preferences. The evaluator function creates customized assessment opportunities that not only measure knowledge acquisition but also enhance engagement through appropriate challenge levels.

Our research seeks to address the following key questions:

RQ1: How can LLMs be effectively utilized to support informal learning experiences in CH settings?

RQ2: To what extent can LLMs accurately assess and evaluate diverse learning outcomes in informal CH environments compared to traditional assessment methods?

RQ3: What measurable impacts does the dual guide-evaluator approach have on visitor satisfaction, knowledge retention, and overall experience quality?

With this research we would like to address long-standing challenges in CH education: maintaining visitor attention, accommodating diverse learning styles, measuring learning outcomes in informal settings, and creating engaging experiences for younger audiences.

2 Background

A recent meta-analyses show that museum technologies significantly enhance learning, particularly for K-12 and college students [13]. Pei [10] established that these technologies provide both intellectual value, by enhancing information access, and experiential value, by fostering emotional connections to exhibits. For younger audiences, Lee [9] found effective e-learning resources like artist-focused content and hands-on activities that engage children and support educational goals through interactive e-games. Mobile technologies have also enhanced visitor interactions, blurring the lines between physical and digital experiences [8]. Technologies have transformed static artifacts into interactive elements through Augmented Reality (AR) applications, advancing from simple information to interactive storytelling [7]. Finally, diverse pedagogical approaches like game-based and inquiry-based learning allow for meaningful engagement with CH [15].

Despite advancements, most museum technologies remain static or confined to preset adaptive pathways. Yilmaz et al. [15] noted the evolution from simple audio guides to sophisticated digital applications, yet personalization challenges persist. Mobile apps often require extensive custom development, limiting scalability and transferability. Even specialized AR applications face sustainability issues as they lack adaptive intelligence to meet diverse visitor needs. The balance between information delivery and engagement is problematic. Pei [10] found that while technology can broaden information scope, it may also overwhelm visitors. Emphasizing experiential value can diminish intellectual engagement. Lee [9] observed that many e-games prioritize simple interactions, neglecting the creative exploration processes typical in art. Social interaction support among visitors is also underdeveloped. Yilmaz et al. [15] highlighted that while technology facilitates interactions with devices, it often hinders social engagement, crucial in a social learning environment. Assessment challenges persist despite Hooper-Greenhill's [6] framework for measuring learning outcomes. Many technologies lack robust evaluation mechanisms, making it hard to assess educational effectiveness. Lee [9] noted that many e-games do not include clear learning objectives or guided questions.

The main gap remains the absence of truly adaptive systems that respond to individual visitor traits and learning preferences. Current solutions provide limited personalization and fail to leverage digital technologies for meaningful visitor experiences. LLMs present a unique opportunity to overcome these limitations by processing natural language, understanding context, and adapting to user input, potentially enhancing both understanding and engagement while assessing learning outcomes.

3 Methodology

This proposal is an extension of a broader project where we have already integrated an Indoor Positioning System (IPS) (i.e., a system that can track precisely the visitor position) within a gallery in Rome [4]. We want to investigate how to enhance visitors' learning outcomes. The methodology we propose leverages LLMs to improve learning outcomes through a three-phase approach: pre-visit, in-visit, and post-visit interactions. The visitor experience in our methodology is structured across three distinct phases. During the pre-visit phase, we collect general information about visitors through a brief questionnaire. This data includes demographic information, prior knowledge of the subject matter, personal interests, and preferred learning styles. Following Falk's visitor identity model [3], we categorize visitors into different types such as Explorers, Facilitators, Experience Seekers, Professionals/Hobbyists, and Rechargers. This classification enables our system to tailor subsequent interactions based on visitor motivations and expectations. The in-visit phase represents the core of our methodology, where the LLM functions as a personalized guide. The IPS tracks visitor location and dwell time (i.e., the duration a visitor spends in front of a specific artwork or exhibit). Dwell time serves as a key indicator of visitor interest and engagement [14]. When a visitor remains in front of an artwork, the system triggers the LLM to provide adaptive descriptions and interpretations of the piece. Moreover, we can also exploit the dwell time to provide personalized recommendation about the artworks in the gallery [5]. In the post-visit phase, the system shifts to assessment mode, where the LLM serves as an evaluator. Based on the exhibits the visitor engaged the most with, the model generates tailored assessment materials to measure learning outcomes and reinforce key concepts.

LLM as Guide transforms the traditional museum narrative into a personalized experience. For Explorers (i.e., curiosity-driven visitors), the system provides content emphasizing interesting facts and discovery-oriented narratives. Facilitators receive content that supports social interaction and shared learning. Experience Seekers are presented with content highlighting the significance and unique aspects of each artwork. Professionals and Hobbyists access deeper, more technical information about techniques, historical context, and artistic movements. For Rechargers seeking contemplative experiences, the system delivers more reflective, aesthetic-focused interpretations. For brief engagements (less than 30 s), the LLM provides a concise overview of the artwork. When visitors spend moderate time (30–90 seconds), more detailed information about the artwork's features and context is offered. During extended viewing periods (over 90 s), the LLM delivers in-depth analysis. Our approach would employ progressive disclosure, where information is delivered in layers of increasing depth, allowing visitors to access additional details based on their interest level. The LLM manages this by starting with essential information about the artwork and then providing options for different thematic explorations such as historical context, symbolic meaning, or artistic technique. The language complexity is adapted based on the visitor profile, simplifying content for younger visitors or non-experts while providing more technical terminology for professionals. Contextual awareness is another key feature, where the LLM maintains awareness of the visitor's journey through the museum, referencing previously viewed exhibits to establish connections between artworks and build a coherent narrative. For example, if a visitor has viewed several Renaissance paintings, the LLM might highlight stylistic similarities or contrasts when they encounter a Baroque piece.

LLM as Evaluator generate dynamic assessments tailored to each visitor's experience, serving both educational and engagement purposes. The LLM generates Multiple-Choice Questions (MCQs), similarly to [2], based on the specific exhibits the visitor engaged with, the depth of information provided during the visit, and the visitor's profile and interests. MCQs are selected as the primary assessment method due to their ease of implementation and objective scoring capability, which facilitates subsequent analysis of learning outcomes. For novices or casual visitors, the LLM creates basic recognition and recall questions. Those with moderate engagement receive questions testing comprehension and application. Professionals or highly engaged visitors encounter questions requiring analysis and interpretation. For younger visitors, we would adopt a more playful approach to assessment. MCQs could incorporate absurd distractors to create an enjoyable experience. Questions are framed as games rather than formal assessments and are designed to encourage parent-child interaction. For example, a question about Leonardo da Vinci's "Mona Lisa" might ask: "What's special about Mona Lisa's eyes in the painting?" with options including the correct answer "They seem to follow you as you move around the room" alongside playful distractors like "They're actually tiny robotic eyes". We still do not plan to expand to other question types (e.g., open-ended) since we want to keep the post-visit cognitive load low for the visitor.

To evaluate the effectiveness of our proposed methodology, we will conduct an A/B testing experiment in collaboration with our partner museum in Rome.

This approach aligns with established methodologies in museum technology research [13]. Visitors will be randomly assigned to one of three groups: a control group receiving standard museum audio guide content, a treatment group A receiving LLM-guided content without personalization, and a treatment group B receiving fully personalized LLM-guided content and evaluation. We will collect data through pre-visit and post-visit questionnaires measuring knowledge and engagement, IPS tracking data (paths, dwell times, total visit duration), assessment performance metrics from the post-visit phase, and semi-structured interviews with a subset of participants. The effectiveness of our approach will be measured using knowledge acquisition (comparison of pre/post-visit knowledge), engagement levels (dwell time, number of artworks visited), visitor satisfaction (questionnaire data), and learning outcome achievement based on the Generic Learning Outcomes framework [6]. This experimental design will allow us to isolate the effects of LLM-based personalization on visitor learning outcomes and engagement, providing empirical evidence for the efficacy of our proposed methodology.

4 Discussion

Implementing LLMs as guides and evaluators in CH settings offers opportunities and challenges. A key concern is the risk of hallucinations, where models may provide plausible but incorrect information, potentially misleading museum visitors. We suggest using Retrieval-Augmented Generation (RAG) techniques to ensure LLM responses are grounded in verified museum databases. Establishing an agent framework with access to verified sources and involving domain experts to review generated content can enhance accuracy. Incorporating visitor feedback systems will help identify misinformation, even though this may limit LLM scalability. Another challenge is maintaining the quality of LLM-generated assessments, which must accurately reflect educational content and effectively measure learning outcomes. Pilot deployments with feedback collection and benchmarking against expert-created questions can strengthen the assessment framework. Practical implementation faces technical challenges like ensuring quick response times and supporting multiple languages. Careful planning is needed for real-time operations in museums. Finally, privacy considerations represent another significant concern, particularly given our use of interaction traces for profiling. The system could implement anonymization and for child visitors, additional safeguards could include parental consent protocols, age-appropriate content filters.

While our approach targets immediate learning outcomes, future research should focus on long-term retention and engagement through longitudinal studies. Expanding to multimodal interactions and enhancing group learning is a promising direction. Moreover the implementation of a conversational approach could further enhance the system [11].

References

1. Ardissono, L., Kuflik, T., Petrelli, D.: Personalization in cultural heritage: the road travelled and the one ahead. User Model. User-Adap. Inter. **22**, 73–99 (2012)
2. Biancini, G., Ferrato, A., Limongelli, C.: Multiple-choice question generation using large language models: methodology and educator insights. In: Adjunct Proceedings of the 32nd ACM Conference on User Modeling, Adaptation and Personalization, pp. 584–590 (2024)
3. Falk, J.H., Dierking, L.D.: The Museum Experience Revisited. Routledge (2016)
4. Ferrato, A.: Integrating indoor positioning, recommendation, and personalization to enhance museum visitor experiences. In: Proceedings of the 33rd ACM Conference on User Modeling, Adaptation and Personalization. UMAP 2025, Association for Computing Machinery, New York, NY, USA (2025). https://doi.org/10.1145/3699682.3727572
5. Ferrato, A., Limongelli, C., Mezzini, M., Sansonetti, G.: The meta4rs proposal: museum emotion and tracking analysis for recommender systems. In: Adjunct Proceedings of the 30th ACM Conference on User Modeling, Adaptation and Personalization. UMAP 2022 Adjunct, pp. 406–409. Association for Computing Machinery, New York (2022). https://doi.org/10.1145/3511047.3537664
6. Hooper-Greenhill, E.: Measuring learning outcomes in museums, archives and libraries: the learning impact research project (LIRP). Int. J. Herit. Stud. **10**(2), 151–174 (2004)
7. Katifori, A., et al.: Chess: personalized storytelling experiences in museums. In: International Conference on Interactive Digital Storytelling, pp. 232–235. Springer, Cham (2014)
8. Kuflik, T., Wecker, A., Lanir, J., Stock, O.: An integrative framework for extending the boundaries of the museum visit experience: linking the pre, during and post visit phases. Inf. Technol. Tourism **15**(1), 17–47 (2015)
9. Lee, T.: Designing art museum e-learning resources for children: content analysis from education perspectives. Interact. Learn. Environ. **32**(6), 2899–2911 (2024)
10. Pei, Y.: The value of technology and interactive systems in cultural heritage interpretation: a study based on existing empirical research. Museum Manage. Curatorship, 1–19 (2024)
11. Sardella, N., Biancalana, C., Micarelli, A., Sansonetti, G.: An approach to conversational recommendation of restaurants. In: Stephanidis, C. (ed.) HCII 2019. CCIS, vol. 1034, pp. 123–130. Springer, Cham (2019). https://doi.org/10.1007/978-3-030-23525-3_16
12. Trichopoulos, G.: Large language models for cultural heritage. In: Proceedings of the 2nd International Conference of the ACM Greek SIGCHI Chapter, pp. 1–5 (2023)
13. Xu, W., Dai, T.T., Shen, Z.Y., Yao, Y.J.: Effects of technology application on museum learning: a meta-analysis of 42 studies published between 2011 and 2021. Interact. Learn. Environ. **31**(7), 4589–4604 (2023)
14. Yi, X., Hong, L., Zhong, E., Liu, N.N., Rajan, S.: Beyond clicks: dwell time for personalization. In: Proceedings of the 8th ACM Conference on Recommender Systems, pp. 113–120 (2014)
15. Yılmaz, T.K., Meral, E., Namlı, Z.B.: A systematic review of the pedagogical roles of technology in ICT-assisted museum learning studies. Educ. Inf. Technol. **29**, 10069–10103 (2024)

Understanding Human-GenAI Collaboration for Complex Problem-Solving Tasks in Online Settings

Sonika Pal[1](✉), Prajish Prasad[2], and Sridhar Iyer[1]

[1] Centre for Educational Technology, Indian Institute of Technology Bombay, Mumbai, India
sonika_pal@iitb.ac.in
[2] School of Computing and Data Sciences, FLAME University, Pune, Maharashtra, India
prajish.prasad@flame.edu.in

Abstract. This research explores Human–GenAI collaboration in complex problem-solving tasks in online learning environments. Grounded in Self-Regulated Learning (SRL) and Complex Problem-Solving (CPS), the study examines interaction patterns, problem-solving strategies, and collaboration approaches while solving complex problems using GenAI tools. The stage-wise methodology includes conducting interviews and surveys, complex problem-solving tasks, and an intervention to assess learning outcomes. Data collection involves ChatGPT interaction logs, screen recordings, and post-task interviews, which are analyzed using Ordered Network Analysis (ONA) and Epistemic Network Analysis (ENA). The findings aim to inform the development of a Collaborative Complex Problem-Solving (CCPS) framework and provide pedagogical guidelines for effective Human–GenAI collaboration.

Keywords: Human-GenAI Collaboration · collaborative Complex Problem-Solving · Generative AI · Self-Regulated Learning · Collaborative Learning

1 Introduction

Collaboration is a critical skill in the 21st century, enabling individuals to engage in knowledge co-construction and complex problem-solving [5,7,9,12]. Research has demonstrated that human-human collaboration enhances cognitive and metacognitive skills, communication, and learning outcomes [6,10]. Studies in Computer-Supported Collaborative Learning environment further emphasize the role of digital learning environments in fostering interaction, collaboration, and engagement among learners [1,8].

Despite its benefits, Computer-Supported Collaborative Learning (CSCL) still faces several challenges [11]. Effective collaboration requires self-regulation and self-determination, as learners must actively manage group dynamics and

A. I. Cristea et al. (Eds.): AIED 2025, CCIS 2590, pp. 464–469, 2025.
https://doi.org/10.1007/978-3-031-99261-2_54

the collaborative learning process [15]. The absence of structured collaboration scripts can result in low participation, distractions, and inefficiencies, hindering the overall effectiveness of CSCL [2,4]. Additionally, learner satisfaction remains a critical factor, as it significantly influences the acceptance and sustained use of CSCL technologies and processes. Similar to other information technologies, if learners perceive a tool or process as unsatisfactory, they are unlikely to continue using it, even if it aids in overcoming collaboration challenges [3].

The emergence of Generative AI (GenAI) introduces a new paradigm of human-GenAI collaboration that can potentially address these challenges. Unlike traditional AI-based tutoring systems that provide static responses, GenAI tools like ChatGPT offer dynamic, context-aware interactions, positioning GenAI as a cognitive partner, co-creator, or problem-solving peer. Our research investigates the role of Human-GenAI collaboration in complex problem-solving tasks, particularly in online learning environments.

2 Background and Theoretical Basis

This study is grounded in Self-Regulated Learning (SRL) Theory [16] and the Complex Problem-Solving (CPS) Framework to examine Human-GenAI collaboration in complex problem-solving [14]. SRL Theory highlights the importance of learners' ability to plan, monitor, and evaluate AI-generated responses to ensure effective collaboration. The CPS Framework defines problem-solving through five key dimensions: Exploring and Understanding, Identifying Problems, Planning and Executing, Iterative Monitoring, and Assessing Outcomes, integrating cognitive, social, and metacognitive skills essential for effective Human-GenAI collaboration.

3 Research Goals

1. Identify and analyze interaction patterns in human GenAI collaboration to understand the problem-solving strategies and collaborative approaches learners adopt in complex problem-solving tasks.
2. To derive best practices and design instructional guidelines for integrating GenAI tools into learning environments, to enhance collaborative problem-solving skills.
3. Establish a Human-GenAI Collaboration Framework using models like Collaborative Complex Problem-Solving (CCPS) and Self/Co-Regulated Learning to enhance Human-GenAI collaboration.

4 Study Design and Methodology

4.1 Context and Participants

This study will be conducted in online setting for the course **Software Engineering**, which will be offered in the upcoming spring session in India. The course aims to help learners understand, create, and evaluate foundational concepts and practices in Software Engineering. Participants will include **both students and software professionals**, providing diverse perspectives on problem-solving with GenAI.

4.2 Study Design

The research will be conducted in three interdependent stages, each building on the previous one:

Stage 1: Past participants of the course will be interviewed or surveyed to explore their collaborative problem-solving experiences. This includes examining group size, interaction modes, and challenges in online collaboration and investigating how GenAI was used in their projects.

Stage 2: Learners will be assigned complex problems tasks related to software engineering concepts, The different categories of learners can be categorized based on:

1. Mode of interaction: For example, (a) a single learner interacting with GenAI tools; (b) two learners interacting with each other as well as with GenAI. These configurations will allow for the analysis of varying interaction patterns in the complex problem-solving process (as illustrated in Fig. 1).
2. Expertise level: Differences between novice and expert learners, especially in skills such as prompt engineering.
3. Learner type: Comparative analysis of students versus professionals in their approaches to interacting with GenAI tools.

Data from human-GenAI, human-human, and mixed interactions will be analyzed to examine problem-solving strategies. Stage 3: Based on insights from Stages 1 and 2, an intervention will be designed and implemented with online learners working on complex problems. The intervention will provide step-by-step guidelines for effective collaboration with Generative AI.

4.3 Data Collection

The data collection will follow a stage-wise approach:

- Stage 1: Data will be collected through open-ended interviews with previous batch participants to understand their collaborative learning experiences, including group structures, methods of collaboration, and challenges faced while working online. This stage will also explore the use of GenAI tools in project execution and the difficulties encountered.
- Stage 2 and Stage 3: Data will include ChatGPT interaction logs, capturing both learner-generated prompts and GenAI responses, along with screen recordings of interactions. Post-task interviews will be conducted to analyze learners' problem-solving strategies and overall experience with GenAI. The same data collection approach will be applied across both stages.

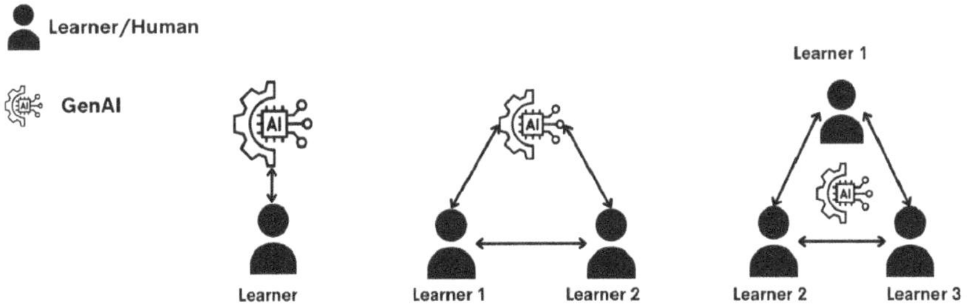

Fig. 1. Figure illustrates different Human-GenAI collaboration modes: (1) Individual interaction with GenAI, (2) Peer collaboration with GenAI tools, and (3) Group collaboration integrating GenAI for problem-solving

4.4 Analysis

The collected data will include survey responses (self-reported GenAI skills), video recordings, artifacts, ChatGPT interaction logs, and interview transcripts. Some interactions will take place through text-based conversations with ChatGPT, and interaction analysis will be conducted using recorded videos, resulting in a set of interaction logs. These fine-grained interactions will be manually coded using a codebook developed for the study. Visualizations of CPS processes will be generated using Ordered Network Analysis (ONA) [13]. The interaction data will then be processed and fed into the Epistemic Network Analysis tool (https://app.epistemicnetwork.org/) to model the connections between different CPS processes.

5 Preliminary Results

In order to address Research goal 1, a pilot study was conducted, which explores human-genAI collaboration in complex problem-solving. It examines how two participants, differing in GenAI expertise and domain knowledge, engage with ChatGPT for the creation of instructional materials for persons with disabilities (PwD) website using Analyze, Design, Develop, Implement, and Evaluate (ADDIE) model.

P1, participant 1 with minimal ADDIE model concepts and limited ChatGPT experience, primarily used AI for information retrieval and idea generation. In contrast, P2, participant 2 with prior ISD knowledge and moderate ChatGPT experience, utilized AI for structured problem-solving, iterative refinement, and critical evaluation.

The study recorded 205 interactions with ChatGPT, a GenAI image generator, and external resources, complemented by video recordings and semi-structured interviews to analyze problem-solving strategies. A structured codebook for analyzing collaboration patterns was developed iteratively with GenAI assistance and researcher oversight. Ordered Network Analysis (ONA) was then applied to visualize problem-solving approaches, highlighting differences in iterative refinement, assessment, and knowledge-building strategies.

Findings reveal Iterative Refinement emerged as central, with P1 (Novice Explorer) using it as a trial-and-error mechanism, while P2 (Experienced Learner) integrated it with Assessing and Evaluating and Building and Sharing Knowledge for a structured process. P1 (Novice Explorer) relied on ChatGPT for Exploring and Understanding, whereas P2 (Experienced Learner) leveraged prior expertise for targeted exploration and critical evaluation as shown in Fig. 2.

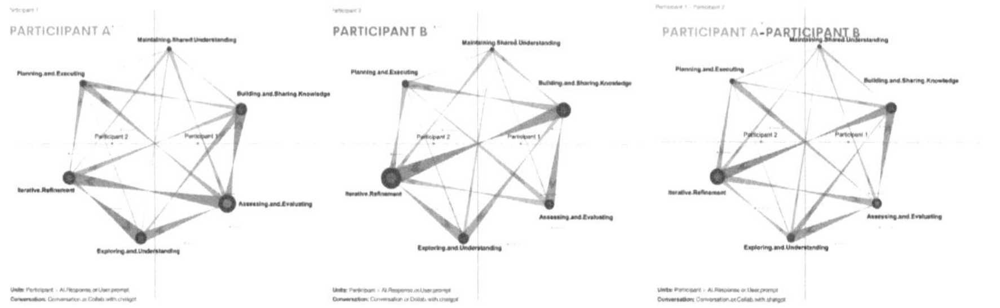

Fig. 2. Participant 1 (Blue) approach shows more prominent CPS skills as Assessing and evaluating and iterative refinement, while less prominent are exploring and understanding and building and sharing knowledge. For participant 2 (Red) Building and sharing knowledge while less prominent are planning and executing and maintaining shared understanding. Last image is subtracted ordered network for both participants (Color figure online)

6 Conclusions and Future Directions

This research will contribute to the field of Human-GenAI collaboration in complex problem-solving by identifying interaction patterns, problem-solving approaches, and the challenges learners face when working with GenAI. It will provide insights into group dynamics and evaluate the effectiveness of Human-GenAI collaboration-driven problem-solving approaches across diverse learner groups. The study will also develop a Collaborative Complex Problem-Solving (CCPS) framework and pedagogical guidelines to enhance GenAI-supported collaboration while fostering 21st-century skills such as critical thinking, creative thinking etc. Finally, by analyzing pre- and post-intervention data and interaction patterns, the research will assess effective collaboration with GenAI—ultimately informing future instructional designs and guiding the development of AI systems that can dynamically adapt to diverse learner profiles, improve teamwork facilitation, and enhance learning outcomes in complex problem-solving contexts.

References

1. Bajaj, M.: The role of digital learning platforms in enhancing student engagement. UNIFIED VISIONS, 1 (2024)
2. Briggs, R.O., De Vreede, G.J., Nunamaker, J.F., Jr.: Collaboration engineering with thinklets to pursue sustained success with group support systems. J. Manag. Inf. Syst. **19**(4), 31–64 (2003)
3. Briggs, R.O., Reinig, B.A., de Vreede, G.J.: The yield shift theory of satisfaction and its application to the is/it domain. J. Assoc. Inf. Syst. **9**(5), 14 (2008)
4. Cheng, X., Nolan, T., Macaulay, L.: Don't give up the community: a viewpoint of trust development in online collaboration. Inf. Technol. People **26**(3), 298–318 (2013)
5. Dillenbourg, P.: What do you mean by collaborative learning? Collaborative-learning: cognitive and computational approaches, pp. 1–19 (1999)
6. Edwards, J., et al.: A wizard of OZ methodology for designing future learning spaces. High. Educ. Res. Dev. **44**(1), 147–162 (2025)
7. Häkkinen, P., Järvelä, S., Mäkitalo-Siegl, K., Ahonen, A., Näykki, P., Valtonen, T.: Preparing teacher-students for twenty-first-century learning practices (prep 21): a framework for enhancing collaborative problem-solving and strategic learning skills. Teach. Teach. **23**(1), 25–41 (2017)
8. Holly, M., Hildebrandt, J., Pirker, J.: A computer-supported collaborative learning environment for computer science education. In: International Conference on Immersive Learning, pp. 287–301. Springer, Cham (2024)
9. Howells, K.: The future of education and skills: education 2030: the future we want (2018)
10. Liu, Z., Gao, Y., Yang, Y., Kong, X., Zhao, L.: Exploring the influence of students' perceptions of face-to-face collaboration on cognitive engagement and learning outcomes in collaborative programming. Acta Physiol. **248**, 104393 (2024)
11. Malmberg, J., Järvelä, S., Järvenoja, H., Panadero, E.: Promoting socially shared regulation of learning in CSCL: progress of socially shared regulation among high- and low-performing groups. Comput. Hum. Behav. **52**, 562–572 (2015)
12. OECD: Pisa 2015 collaborative problem solving framework (2017)
13. Tan, Y., Ruis, A.R., Marquart, C., Cai, Z., Knowles, M.A., Shaffer, D.W.: Ordered network analysis. In: International Conference on Quantitative Ethnography, pp. 101–116. Springer, Cham (2022)
14. Taylor, M., Barthakur, A., Azad, A., Joksimovic, S., Zhang, X., Siemens, G.: Quantifying collaborative complex problem solving in classrooms using learning analytics. In: Proceedings of the 14th Learning Analytics and Knowledge Conference, pp. 551–562 (2024)
15. Winne, P.H.: What is the state of the art in self-, co-and socially shared regulation in CSCL? Comput. Hum. Behav. **52**, 628–631 (2015)
16. Zimmerman, B.J.: Becoming a self-regulated learner: an overview. Theory into Pract. **41**(2), 64–70 (2002)

Overview of AI Grading of Physics Olympiad Exams

Lachlan McGinness[1,2](✉)

[1] Australian National University, Canberra, Australia
lachlan.mcginness@anu.edu.au
[2] CSIRO, Canberra, Australia

Abstract. Automatically grading the diverse range of question types in high school physics problem is a challenge that requires automated grading techniques from different fields. We report the findings of a Systematic Literature Review of potential physics grading techniques. We propose a multi-modal AI grading framework to address these challenges and examine our framework in light of Australia's AI Ethical Principles.

Keywords: Automated Grading · Artificial Intelligence · Physics

1 Introduction

Studies show that teachers in Australia are burning out due to their workloads [2, 21]. At both the institutional and individual level, automated marking Artificial Intelligence (AI) solutions are starting to be sought out and employed in schools to reduce this workload [18]. Studying potential AI Marking solutions from both an ethical and performance perspective allows governments, schools, teachers and students to make informed decisions about which of these tools can ethically and safely be used to reduce teacher workload.

In this paper we explore the automated grading of high school level physics problems, focusing on the Australian Physics Olympiad. Physics problems typically take a diverse range question types which could require any combination of these to mark. Typical physics question types are:

- **Numerical** - Student response is a number with or without units and/or direction. For example 7, 15 m/s or 20 m upwards.
- **Algebraic** - A question which requires students to give their answer in the form of an algebraic expression, equation or proof. For example $\frac{L_0}{4\pi r^2}$ or $x+y = x^2 + y^2$.
- **Plots or Graphs** - Student response is a plot of data or function.
- **Diagrams** - Student response is a visual representation that does not fall into plots or graph.
- **Short Answer** - A worded response, normally to a closed-ended question, evaluated on technical correctness or use of specific key terms. For example "The friction force is acting in the upward direction".

A. I. Cristea et al. (Eds.): AIED 2025, CCIS 2590, pp. 470–475, 2025.
https://doi.org/10.1007/978-3-031-99261-2_55

- **Multiple Choice** - A question where students select one or more provided options. These are very simple to grade and not considered in this study.

We propose a Multi-Modal AI grading framework which uses a range of techniques to grade these diverse problem types. This paper reports on a December 2024 Systematic Literature Review (SLR) of automated marking to explore the range of tools that could be used to grade physics questions. We make two contributions; we summarise key findings from the SLR and we position potential grading solutions within Australia's Ethical AI Principles.

2 Systematic Literature Review

In December 2024, a systematic literature review was conducted using the steps as illustrated in Fig. 1. Databases were searched for relevant papers. Then papers were automatically screened to a focused set, building on techniques described in [14]. Finally the papers were manually screened to ensure only those which were peer reviewed and would contribute to physics marking were considered.

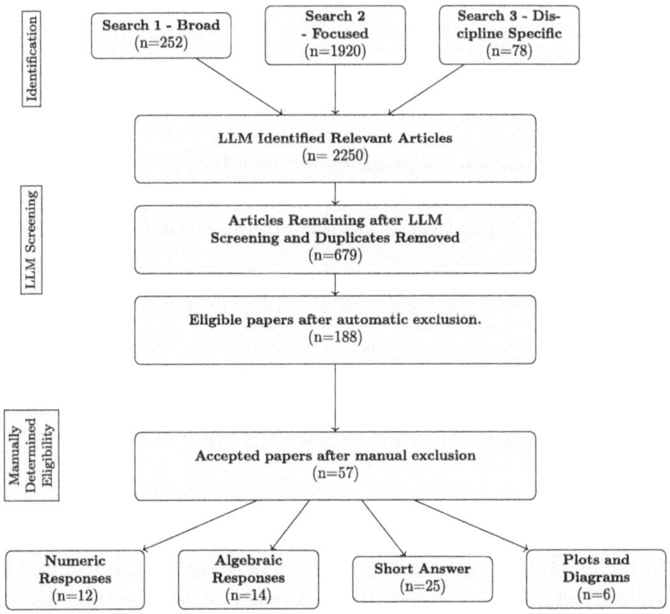

Fig. 1. Flow diagram illustrating the steps taken in the literature review and the number of papers included in each step.

The automatic screening process used a Large Language Model to determine the type of questions that each paper addressed as well as the methods used. The upcoming subsections give the key findings of the literature review for each question type commonly used in physics exams.

2.1 Numeric Questions

When students type their inputs in a fixed format, numerical questions are easy to grade using rule based approaches. The technique is now easily implemented in forms and spreadsheets [15].

Handwritten responses require Optical Character Recognition (OCR) techniques. In 2015, Xerox developed a system which automatically marked handwritten numeric responses [5]. In 2024 a new dataset MNIST-Fraction was released as a benchmark for measuring accuracy of recognising fractions [1].

A more recent approach focused on using Large Multi-modal Models (LMMs) to grade student responses to calculus questions with numerical answers [4]. The best performing available model (GPT4) often made numerical mistakes and the logic behind its grading was not strong, though the grades correlated with human markers. Feature extraction and explainable Machine Learning (ML) methods have been used to grade numerical responses in physics [13].

Numerical responses are more difficult to mark if units or directions (for vector quantities) are included, as commonly required in physics. Answers can be provided in particular units and a pre-defined coordinate system to mitigate this problem. However if free form answers are accepted, then feature extraction techniques as discussed in Sects. 2.2 and 2.4 will be required.

2.2 Algebraic Questions

Marking algebraic expressions, equations, or proofs tends to be a more difficult task than marking numerical responses as there are many different correct forms of the answer. Consider a question as simple solving Kepler's third law for period:

$$T = \sqrt{\frac{4\pi^2 r^3}{GM}} \text{ or } T = \sqrt{\frac{4\pi^2 r}{GM}r} \text{ or } T = \sqrt{\frac{4\pi^2}{GMr}r^2} \text{ or } T = 2\pi\sqrt{\frac{r^3}{GM}} \text{ or ...}$$

When expressions are more complex, determining equivalence can be arduous. When they include trigonometric expressions and exponentiation it is impossible to find a general algorithm that verifies the equivalence of two expressions [20].

Computer Algebra systems such as SymPy [17] implement heuristics that will often succeed in determining equivalence, but they are rigid and require students to understand the system's syntax. Therefore examiners are measuring the students' mastery of the the system's syntax as well as the course content [3].

Rule-based parsing systems that require examiners to anticipate student errors do not scale well [16]. Large Language Models (LLMs) can be used to capture and translate meaningful features from student responses into a computer readable format [16]. Kortemeyer used LLMs for grading direclty [8], but found that despite correlation between LLM and human graders, LLMs miss nuance.

2.3 Plots, Graphs and Diagrams

There are few studies which grade hand drawn diagrams or plots and graphs. Li et al. deployed a small, open-source, multi-modal, visual-reasoning LLM to provide coherent feedback on images in mathematical settings [11]. Kortemeyer et al. used LMMs to mark physics exams involving diagrams and found that in 2024, LLMs struggled to distinguish essential details from irrelevant information [10]. They also found that LMMs made more errors when the number of rubric items increased, consistent with studies in adjacent domains [14].

2.4 Short Answer

Many Automated Short Answer Grading (ASAG) techniques follow two steps:

1. A feature extraction technique, such as Bag of Words or Part of Speech Tagging, converts the text into a mathematical representation.
2. Then a ML algorithm such as cosine similarity or Term Frequency Inverse Document Frequency is applied to this representation.

LLMs have also been applied to automated grading of short answers [9], but the literature has found they have limitations. For example, knowing the human assigned mark causes Large Language Models to bias their own marks [6].

3 AI Ethics

Most automated grading studies do not explicitly consider ethics. We turn to the Australian AI Ethics Principles [19] to evaluate the ethical dimensions. Three of the relevant principles for are Human and Environmental Well-being, Privacy, and Contestability. In order to meet these, AI grading systems should:

- take flexible inputs and require no training to reduce teacher workload.
- minimise environmental impact and respect privacy by running locally.
- use explainable, not-secret methods with guarantees (be reliable).

Table 1 summarises four methods with the question types they can address.

Table 1. Application of ethical principles to automated grading techniques.

Technique \ Criteria	Reliable	Explainable	No Training Data	No Expert Time	Open Source	Maintains Privacy	Runs Locally	Flexible inputs	Numeric	Algebraic	Short Answer	Plots and Graphs	Diagrams
Rule based methods	✓	✓	✓	X	✓	✓	✓	X	✓	(✓)	(✓)	X	X
Hand-crafted features	✓	✓	✓	X	(✓)	✓	✓	X	✓	✓	✓	X	X
Feature Extraction and Machine Learning	X	(✓)	X	✓	✓	✓	✓	✓	✓	✓	✓	X	X
Multi-modal Large Language Models	X	(✓)	✓	✓	(✓)	(✓)	(✓)	✓	✓	(✓)	(✓)	(✓)	(✓)

4 Current Challenges, Limitations and Future Work

In order to meet the ethical criteria across the range of question types, further work will explore deployment of an LLM-modulo framework [7]. This uses verification techniques to improve LLM reliability, including those which use external tools [12].

This literature review covers a very broad field of question types and marking methods, but more narrow research still needs to be conducted to be aware of the most promising techniques compatible with an LLM Modulo approach. AIED, ITS, Learning at Scale and LAK conference proceedings were not included in this review, additional work should be done to find relevant techniques from these venues.

Despite some initial studies, the accuracy of LLMs in screening papers is still largely unknown [14]. A significant number of the papers could be checked manually to determine if the error rate when screening is sufficiently accurate for Systematical Literature Review.

In order to be confident that no relevant papers were excluded a significant number of the papers could be checked manually to attempt to determine the error rate and determine if this is a sufficiently accurate process for systematical literature review.

Once reliable pre-processing is applied, Computer Algebra Systems can be used to verify students' expressions and proofs. Australian Physics Olympiad problems typically require algebraic manipulation involving some transcendental functions, but not calculus. An area of future work is to identify which reasoning systems are most accurate and efficient for responses of this form.

Acknowledgments. This project has been funded by the CSIRO and ANU.

Disclosure of Interests. The authors have no competing interests to declare.

References

1. Ahadian, P., Feng, Y., Kosko, K., Ferdig, R., Guan, Q.: MNIST-fraction. In: Proceedings of the 2024 ACM Southeast Conference, pp. 284–290. ACM (2024). https://doi.org/10.1145/3603287.3651221
2. AITSL: Australia's teacher workforce today (2023). https://www.aitsl.edu.au/research/spotlights/australia-s-teacher-workforce-today
3. Fast, E., Lee, C., Aiken, A., Bernstein, M.S., Koller, D., Smith, E.: Crowd-scale interactive formal reasoning and analytics. In: User interface Software and Technology, pp. 363–372 (2013). https://doi.org/10.1145/2501988.2502028
4. Gandolfi, A.: GPT-4 in education: evaluating aptness, reliability, and loss of coherence in solving calculus problems and grading submissions. Int. J. Artif. Intell. Educ. (2024). https://doi.org/10.1007/s40593-024-00403-3
5. Gross, E., Wshah, S., Simmons, I., Skinner, G.: A handwriting recognition system for the classroom. In: Learning Analytics and Knowledge, pp. 218–222 (2015). https://doi.org/10.1145/2723576.2723601

6. Grévisse, C.: LLM-based automatic short answer grading in undergraduate medical education. BMC Med. Educ. **24**(1), 1060 (2024). https://doi.org/10.1186/s12909-024-06026-5
7. Kambhampati, S., et al.: Position: LLMs can't plan, but can help planning in LLM-modulo frameworks. In: ICML, pp. 22895–22907. PMLR (2024). https://proceedings.mlr.press/v235/kambhampati24a.html
8. Kortemeyer, G.: Toward AI grading of student problem solutions in introductory physics: a feasibility study. Phys. Rev. Phys. Educ. Res. **19**(2), 020163 (2023)
9. Kortemeyer, G.: Performance of the pre-trained large language model GPT-4 on automated short answer grading. Discov. Artif. Intell. **4**(1), 47 (2024)
10. Kortemeyer, G., Nöhl, J., Onishchuk, D.: Grading assistance for a handwritten thermodynamics exam using artificial intelligence: an exploratory study. Phys. Rev. Phys. Educ. Res. **20**(2), 020144 (2024)
11. Li, H., Li, C., Xing, W., Baral, S., Heffernan, N.: Automated feedback for student math responses based on multi-modality and fine-tuning. In: LAK 2024, pp. 763–770. ACM, Kyoto (2024). https://doi.org/10.1145/3636555.3636860
12. McGinness, L., Baumgartner, P.: Automated theorem provers help improve large language model reasoning. In: LPAR, pp. 51–31 (2024). https://doi.org/10.29007/2n9m
13. Mcginness, L., Baumgartner, P.: CON-FOLD explainable machine learning with confidence. Theory Pract. Logic Program. **24**(4), 663–681 (2024). https://doi.org/10.1017/S1471068424000346
14. McGinness, L., Baumgartner, P., Onyango, E., Lema, Z.: Highlighting case studies in LLM literature review of interdisciplinary system science. In: AI 2024: Advances in Artificial Intelligence, pp. 29–43 (2024). https://doi.org/10.1007/978-981-96-0348-0_3
15. McNeil, C.R.: Excel modeling assignments and automatic grading. J. Finan. Educ. **41**(3), 86–97 (2015)
16. McNichols, H., Zhang, M., Lan, A.: Algebra error classification with large language models, vol. 2023, pp. 365–376 (2023). https://doi.org/10.1007/978-3-031-36272-9_30
17. Meurer, A., Smith, C., Paprocki, M., Scopatz, A.: SymPy: symbolic computing in Python (2024). https://www.sympy.org/en/index.html
18. Ogg, M.: Brisbane AI edtech Edexia accepted into Y Combinator (2024). http://www.businessnewsaustralia.com.html
19. Resources, D.o.I.S.a.: Australia's AI ethics principles (2024). https://www.industry.gov.au/publications
20. Richardson, D.: Some undecidable problems involving elementary functions of a real variable. J. Symbolic Logic **33**(4), 514–520 (1968). https://doi.org/10.2307/2271358
21. Windle, J., Morrison, A., Sellar, S., Squires, R., Kennedy, J., Murray, C.: Teachers at breaking point. University of South Australia (2022). oCLC: 1396171382

From One-Size-Fits-All to Personalisation: Transforming Gamified Learning Through Localisation and Multidimensionality in the Arab Culture

Ghadeer Alshamrani[1,2](✉) and Alexandra I. Cristea[1]

[1] Durham University, Durham, UK
{ghadeer.alshamrani,alexandra.i.cristea}@durham.ac.uk
[2] King Abdulaziz University, Jeddah, Kingdom of Saudi Arabia
gaalshmrane@kau.edu.sa

Abstract. *Personalised gamification* aims to address the limitations of the one-size-fits-all (OSFA) approach, by catering to users' unique needs and characteristics. However, existing studies often focus on a single user dimension, overlooking the complex interplay of multiple factors that influence students' learning experiences. A multidimensional approach is crucial, as it considers various users and contextual factors, ensuring a more tailored and engaging learning experience. A significant gap exists in the literature regarding how multidimensional personalised gamification, particularly when incorporating gender differences and cultural factors, can enhance gamified learning environments. This research aims to investigate the impact of (a) *localised multidimensional personalised gamification* in comparison to (b) *non-localised multidimensional personalised gamification* and the baseline (c) *OSFA approach*, on student engagement and interaction. To achieve this, an experimental study will be conducted to compare these three approaches among undergraduate students. The study will measure their effects on student engagement and interaction using a gamified application integrated with an LMS. Educational data mining techniques will be employed to analyse data collected from LMS activity logs and questionnaires. This research is expected to contribute both theoretically and practically, by enhancing the understanding of multi-dimensional personalisation in gamified learning, particularly in relation to cultural factors. It will provide empirical evidence, design guidelines, and a *gamification framework adapted to culturally diverse educational environments*, with a particular focus on Arab culture.

Keywords: Gamification · Culture · Localisation · Multidimensional · LMS · OSFA

1 Introduction and Problem Statement

Keeping students motivated and engaged in the learning environment is a persistent challenge. One strategy increasingly used in education to address this issue

A. I. Cristea et al. (Eds.): AIED 2025, CCIS 2590, pp. 476–481, 2025.
https://doi.org/10.1007/978-3-031-99261-2_56

is gamification. *Gamification* is defined as the application of game elements in non-game contexts [13]. Despite the advantages of gamification, a key factor behind its success as a learning strategy relates to effective planning, design, and development. Poorly designed gamification activities can lead to student frustration, confusion, and even lack of engagement with the learning content [9]. To achieve better outcomes, recent research has asserted the importance of personalisation in gamification [8,13].*Personalised gamification*, as an answer to some of these issues, involves tailoring gamified experiences to individual users based on their needs and preferences [5]. In practice, personalised gamification is usually achieved by exposing participants to various game elements and discovering which aspects engage them the most [9]. Although significant research has explored gamification, comparisons between personalised gamification and the OSFA approach remain limited, especially in the educational domain [8]. A study by [2] suggests that personalised gamification yields better results in social networks and health than the OSFA approach. However, in education, the comparison remains largely unclear. One possible reason is the reliance on a single-dimension personalisation, rather than a multi-dimensional approach that simultaneously addresses various factors influencing user interaction in a gamified environment [8]. The importance of this approach lies in its ability to consider diverse student needs and preferences, ultimately enhancing motivation and engagement. Despite the increasing recognition of multidimensional personalisation, there is a lack of literature examining personalised gamification from this perspective in education [8]. Many studies highlight the need to explore the impact of multidimensional personalised gamification activities tailored to meet students' diverse needs, as this approach significantly improves interaction and engagement in gamified learning environments [13]. Additionally, a significant gap also exists in understanding how multidimensional personalised gamification, particularly when incorporating gender differences and cultural factors, can effectively influence the gamified learning experience [8,10,11]. Cultural differences play an important role in shaping how students interact with and perceive gamified learning activities [10]. Failure to consider the cultural context of target users may result in confusion, frustration and disengagement. A study by [10] showed that cultural differences affect students' interactions and engagement within learning environments. This suggests that using the properties of one culture when designing gamified activities for students from another culture may negatively impact the learning experience. For example, many gamified learning environments reflect the cultural backgrounds of their developers, rather than the target students. This issue is particularly pronounced in gamified learning environments developed for non-Western contexts, such as Arab cultures [1], which is the focus of this research. To the best of the researcher's knowledge, no existing studies have addressed multidimensional personalised gamification in the Arab context, comprehensively. There are notable differences in how users from individualistic and collectivist cultures interact [3]. Arab culture tends to be collectivistic, meaning that it emphasises team achievements, rather than individual success. Moreover, Arab culture exhibits a high level of uncertainty avoidance, which means that simplicity, clarity and comprehensibility are highly valued in gamified environments. To accommodate this preference, gamification strategies should include clear rules for assigning, for instance, gamification elements such as points and badges, as well as external resources to reduce uncertainty [1]. In the same context, previous research [7], has found significant differences in cognitive preferences between Western and Arab cultures. While Western cultures tend to adopt an analytical cognitive style, Arab cultures exhibit a more holistic approach. Despite these cultural differences, efforts in Arab countries to improve educational materials, such as software and websites, often rely on purchasing ready-made solutions, rather than developing in-house con-tent. These solutions typically involve translating content from other languages, such as English to Arabic, while retaining the original structure, visual elements, and core components, including any gamification. This approach often results in inferior products, which fail to engage Arab students or meet their specific cultural needs [7]. Additionally, cross-cultural research on gamification has primary focused on Western contexts [1], overlooking the requirements of Arab learners and their specific needs. As a result, there is a growing need to *localise multidimensional personalised gamified environments for Arab students to support sustainable development and address cultural requirements effectively* [1]. This lack of exploration into the cultural aspects of multidimensional personalised gamification in Arab contexts represents a significant gap in the literature. Addressing this gap requires a deeper understanding of Arab students' cultural needs to guide instructional designers and developers in creating effective, culturally relevant gamified learning environments. By incorporating gender differences and cultural localisation into a multidimensional personalised gamification framework, this research aims to enhance student engagement and interaction while contributing to the development of more effective gamified educational environments, with a specific focus on cultural localisation for Arab learners. To address this problem, the researchers will undertake the following research questions (RQ):

- **RQ1:** How can culturally specific gamification elements be developed to enhance engagement and interaction patterns in localised, multidimensional and personalised gamification activities?
- **RQ2:** How do localised, multidimensional and personalised gamification activities compare to non-localised, multidimensional and personalised gamification activities as well as the baseline OSFA approach in influencing student engagement and interaction patterns across genders? - this further breaks down into:
  - **RQ2.1:** What is the impact of localised, multidimensional, and personalised gamification activities, non-localised multidimensional personalised gamification activities, and the OSFA approach on engagement levels among male and female students?
  - **RQ2.2:** What are the differences in interaction patterns among male and female students when exposed to localised, multidimensional, and person-

alised gamification activities compared to non-localised, multidimensional and personalised gamification activities and OSFA gamification activities?
- **RQ3:** What approaches can be used to design localised, multidimensional and personalised gamification activities for enhanced engagement and interaction?

2 Proposed Solution and Methodology

This research aims to examine the impact of localised multidimensional personalised gamification compared to non-localised multidimensional gamification and a one-size-fits-all (OSFA) approach on students' engagement and interaction. It will be conducted in three phases, addressing the research questions outlined earlier.

2.1 Phase 1: Identification of Game Elements for the Three Approaches (Answering RQ1)

Localised Personalised Multidimensional Gamification Approach: A review of the literature will be conducted to identify key game elements relevant to culture and multidimensional personalised gamification. Additionally, an extended taxonomy will be proposed based on an existing gamification taxonomy co-author developed [12]. This extended taxonomy aims to integrate localisation and cultural adaptation alongside multidimensional personalised gamification aspects, which were not adequately represented in prior taxonomies. The taxonomy will be evaluated through an expert review questionnaire based on five criteria adapted from [12]: comprehensibility, description, examples, coverage, and relevance. Furthermore, a questionnaire will be administered to determine game element preferences specific to Arab learners. The results from these two questionnaires will be combined, to create a framework for localised personalised gamification.

Non-localised Personalised Multidimensional Gamification Approach: The same game elements used by [8] will be applied.

One-Size-Fits-All (OSFA) Approach: The game elements to be used are: points, badges, and leaderboards (PBL), following previous research [8]. PBL is widely used in gamification research, due to its effectiveness compared to other game element combinations.

2.2 Phase 2: Development of Gamification Activities and Experimental Setup (Answering RQ2)

After identifying the game elements for the three approaches in Phase 1, this phase will focus on answering RQ2, by designing and implementing three experimental conditions: OSFA condition, localised multidimensional personalised gamified condition and non-localised multidimensional personalised gamified condition. The gamification activities will be structured into two main sections: **Control section**: This section will be designated for the control OSFA activities. **Experimental section**: This section will be divided into six subsections, based on students' Hexad player types [4]: Philanthropist, Socialiser, Free Spirit, Achiever, Player, and Disruptor. Students will be assigned to each subsection, based on their Hexad player type score. The same procedure will be applied to the non-localised multidimensional personalised gamified activities, and both will both included in the same section. All gamified activities will be developed using Articulate Storyline 360 software as a single application, which will then be integrated with the LMS. Articulate Storyline 360 software provides highly interactive and customisable features for creating personalised gamified activities, as well as seamless integration with LMS platforms. A pilot study will be conducted, to test the application before the real experiments. **Experimental Setup**: After the development process, participants will be selected from undergraduate students (both male and female) studying in the first preparatory year at King Abdulaziz University in Saudi Arabia. Participants will receive full information about the research procedures and objectives, and ethical approval will be obtained. Before the experiment, a pre-experiment questionnaire will be administered to collect student demographic data, including age, game preferences, basic computer skills, game-playing habits, and LMS usage skills. Additionally, students will complete the Hexad player type scale, adopted from [4], before accessing the gamified activities. During the experiment, students will be randomly divided into groups, each starting with different gamification activities. Then, all students will log in to the LMS. For instance, the first group will gain access to the OSFA gamified activities (O), followed by localised personalised multidimensional activities (L), and then non-localised personalised multidimensional activities (N). The second group will start with (L), followed by(N), and then (O) activities, and so on, in different permutations, to remove order bias of the allocation and produce rich data. For the experimental activities, students will be automatically assigned to one of the Hexad player types based on their Hexad player type scale, and the activities will be tailored to their player type. At the end of the experiment, all students will complete the student engagement scale to measure their engagement. This scale has been validated by [6]. Additionally, activity logs generated by the LMS will be used to assess engagement and interaction patterns. By combining LMS log data analysis with student engagement scale responses, a comprehensive understanding of student engagement across all approaches can be obtained. Educational data mining techniques will be employed to analyse the data collected from the LMS and questionnaires.

2.3 Phase 3: Evaluation (Answering RQ3)

To address the third research question, a gamification framework will be developed, based on insights from the experimental results. This framework will provide guiding principles for localised personalised multidimensional gamification, contributing to the existing body of knowledge.

3 Expected Contribution

This research is significant in several ways. Firstly, it seeks to develop a gamification taxonomy applicable to localised, multidimensional, and personalised gamified environments in education. Secondly, it aims to investigate key moderators that influence the effectiveness of multidimensional personalised gamification. Additionally, it will provide valuable insights for instructional designers, game developers, researchers, and educators regarding factors that affect student engagement and interaction in gamified learning environments. Furthermore, the study will propose personalised strategies for the design and implementation of gamified activities within the context of Arab culture. More importantly, it will create practical tools for effectively integrating personalisation, cultural localisation, and adaptation into gamified educational systems. By addressing these aspects, this research contributes to bridging the gap between gamification, personalisation, and cultural adaptation.

References

1. Alomar, N., et al.: A hybrid cultural model for Arabic gamified systems. Comput. Hum. Behav. **64**, 472–485 (2016)
2. Hajarian, M., et al.: A personalized gamification method for increasing user engagement in social networks. Soc. Netw. Anal. Min. **9**, 1–14 (2019)
3. Hofstede, G.: Culture's Consequences. Sage Publications, 2nd edn. (2001)
4. Marczewski, A.: Even ninja monkeys like to play. London Inc. **1**(1), 28 (2015)
5. Oliveira, W., et al.: The effects of personalized gamification on students' flow experience, motivation, and enjoyment. Smart Learn. Environ. **9**(1), 16 (2022)
6. O'Brien, H.L., et al.: A practical approach to measuring user engagement. Int. J. Hum Comput Stud. **112**, 28–39 (2018)
7. Qutub, J.A.: A cross-cultural comparison of cognitive styles in Arab and* American adult learners. George Mason University (2008)
8. Rodrigues, L., et al.: How personalization affects motivation in gamified review assessments. Int. J. AI Educ. **34**(2), 147–184 (2024)
9. Rodrigues, L., et al.: Tailoring gamified educational systems. In: Simpósio Brasileiro de Educação em Computação (EDUCOMP), pp. 43–44. SBC (2024)
10. Toda, A., et al.: Towards the understanding of cultural differences. In: Proceedings of the 15th International Conference on Educational Data Mining, p. 560 (2022)
11. Toda, A.M., et al.: Planning gamification strategies based on user characteristics and DM: A gender-based case study. arXiv preprint arXiv:1905.09146 (2019)
12. Toda, A.M., et al.: A taxonomy of game elements for gamification. In: International Conference on Advanced Learning Technologies. vol. 2161, pp. 84–88. IEEE (2019)
13. Xiao, Y., et al.: Personalised gamification enhances student participation. Interact. Learn. Environ. **32**(10), 7014–7040 (2024)

Supporting Information Problem Solving in the Age of Misinformation and Generative AI: A Socio-technical Approach

Ajanie Karunanayake(✉), Antonette Shibani, and Simon Knight

TD School, University of Technology Sydney, Ultimo, Australia
ajanie.m.karunanayake@student.uts.edu.au,
{antonette.shibani,simon.knight}@uts.edu.au

Abstract. The Internet plays a crucial role in modern information search and decision-making, despite the varying quality of information. Its open nature, combined with Generative AI (GenAI), allows for easy creation and dissemination of both accurate and inaccurate information expressing varied perspectives, leading to conflicting information. For many important issues, people must navigate misinformation to find high-quality information, but also, navigate the variation in perspectives of even relatively high-quality sources where there may be no consensus, without falling into cognitive biases and opinion polarization. It is essential to support individuals to critically navigate conflicting information to meet their information needs and make informed decisions. Hence, this study introduces a socio-technical pedagogical Information Problem Solving (IPS) intervention, featuring an AI-powered support tool, 'INFO-NAVIGATE', to scaffold students to identify high-quality information in the context of conflicting information while enhancing their cognitive skills and critical thinking. The study follows a Design-Based Research (DBR) methodology and the intervention is evaluated based on input from both students and educators. Beyond the IPS intervention, the study aims to contribute generalizable design principles for novel information seeking interfaces in the age of misinformation and GenAI. It also aims to provide insights on the approaches and perspectives of educators in implementing such tools for teaching information seeking and critical thinking.

Keywords: Information Problem Solving · Generative AI · Large Language Model · Pedagogical Intervention

1 Introduction

The Internet has become a popular source for people to find information and make decisions that impact both society and individual lives [5,10]. However, its

A. I. Cristea et al. (Eds.): AIED 2025, CCIS 2590, pp. 482–487, 2025.
https://doi.org/10.1007/978-3-031-99261-2_57

open nature allows anyone to generate and disseminate both accurate and inaccurate information expressing their perspectives, which may be based on opinions or facts [5]. Additionally, Generative AI (GenAI) advances make it easy to produce human-like content, but might include misinformation, hallucinations, and biased representations of knowledge [6]. This mix of accurate and inaccurate information, along with varied perspectives on the same topic often leads to conflicting information on the Internet [9]. It is not only the quality of sources, but also the vast array of perspectives expressed within the sources that exacerbate the problem. For instance, a legitimate variation of perspectives is also found in relatively high-quality sources, including expert disagreements without clear consensus among them for an issue [4]. One of the most evident examples of such challenges was observed in early 2020 during the COVID-19 pandemic when various health organizations worldwide provided differing recommendations on face mask usage. Navigating claims in such context is influenced by cognitive bias, information literacy, and domain knowledge, presenting significant challenges for individuals to find reliable information, particularly in this age of misinformation and GenAI [1,5,6]. To identify high-quality information amidst conflicting information, individuals must actively engage in the information seeking process, consuming diverse perspectives to avoid cognitive biases and apply information problem solving (IPS) cognitive skills and critical thinking. [1,2].

IPS is a complex cognitive skill that is decomposed into five main skills in the educational context for instruction development [2]. Socio-technical solutions that integrate technological tools and classroom instructions offer an effective approach to teaching these skills. This approach highlights the mutual relationship between humans and machines in addressing complex problems by understanding both the technical aspects and other contextual factors in which technology is implemented [8]. Although existing IPS tool-based interventions are designed to support various IPS skills across different information-seeking tasks, there is limited support for information-seeking in the context of *conflicting information* that involves diverse perspectives [11].

Thus, the aim of this study is to assist students in identifying high quality information in the context of conflicting information while activating and enhancing cognitive skills and critical thinking by addressing the current limitations. To achieve this aim, we develop a socio-technical pedagogical intervention featuring a Large Language Model (LLM) based tool called 'INFO-NAVIGATE' [11], to support students in practising these skills in educational contexts.

The study will address the following research questions (RQs), oriented towards addressing the design of the tool and IPS socio-technical pedagogical intervention (RQ1), developing and testing the intervention (RQ2). Together, they support skill development in students for the identification of high-quality online information in the context of conflicting information.

RQ1 – How do we design pedagogical interventions to support students in the IPS process?

RQ2 – What is the impact of the designed intervention on students' IPS skills?

2 Theoretical Framework

The study is primarily based on the IPS-I model, specifically developed to outline the skills required to solve information problems using the Internet. The constituent skills is a main component of the model, which explains the five decomposed IPS skills: 1. Define Information Problem, 2. Search Information, 3. Scan Information, 4. Process Information, and 5. Organize and present information [3].

In the IPS process, an information seeker begins the process by defining the information problem(s) and then searching information for answers to the defined problem using different search strategies. During this phase, they may encounter conflicting information. After they gather information from different sources, they should integrate and compare the information with prior knowledge and other relevant information, corroborate information from multiple sources, and analyse and select information by evaluating the quality of information. According to the IPS-I model definition, these activities are components of 'process information' skill. In this study, we focus on intervening and improving this key cognitive skill [11]. However, navigating and making informed decisions in the context of conflicting information is a cognitively demanding task, making critical thinking essential [1]. The proposed intervention hence aims to enhance the aforementioned IPS skills along with critical thinking, rather than merely executing the actions.

3 Proposed Approach

The proposed approach is a socio-technical pedagogical intervention that supports students to critically engage with conflicting information and make informed decisions by activating and uplifting IPS cognitive skills and critical thinking. The main component of the intervention is the web-based AI-powered INFO-NAVIGATE tool designed to guide students to critically navigate and process conflicting information. More information on the design rationale and development of the tool can be found in the paper [11], and an explanation of its user interface is available on figshare[1].

The web-based tool takes search queries as user input and dynamically generates relevant content for the entered search query. This content generation involves several Natural Language Processing (NLP) sub-tasks, such as text summarization, information extraction, sentiment analysis, and stance detection. Python libraries, Search Engine Results Page (SERP) API, and OpenAI GPT 4o-mini model (LLM API) were utilized to extract information and generate content by leveraging the capabilities of Large Language Models (LLMs) in performing these tasks [6,7].

[1] https://doi.org/10.6084/m9.figshare.28899491.v2.

4 Methodology

The overarching methodology of the research is Design Based Research (DBR) [13]. This approach involves making adjustments to the tool's design and the intervention based on feedback from previous iterations and consultations with relevant educators and student feedback. The intervention is implemented in authentic classroom settings for evaluation.

Three types of data collection and analysis methods inform the RQs. Part of RQ1 draws from evidence synthesis of past literature on IPS interventions and information seeking support tools. The other part of the question involves the analysis of educators' perspectives on the intervention. We are currently recruiting a group of educators from the university, who can potentially be involved as co-designers of the IPS intervention and normally teach critical information seeking and critical thinking as a whole or part of their subjects. Semi-structured interviews are taking place to understand their current approaches used in teaching critical information seeking and decision-making with critical thinking, their perspective about the intervention in teaching and learning of critical information seeking and critical thinking, and how they will use the intervention for their teaching. The data will be qualitatively analysed to address the RQ.

RQ2 will be addressed using student engagement data from DBR implementations. The DBR iterations involve student cohorts as participants from an Australian university, working closely with the educators of the subjects. The instructions including topic selection for the activities are co-designed with educators to make sure they are aligned with the overarching subject learning objectives and context for maximum impact on their learning [12]. In addition to the tool, the intervention comes with relevant instructions including a guide to tool features and tips to successfully navigate conflicting information and intervention activities. The intervention consists of three activities, pre-test, main task, and post-test for data collection. The steps of the implementation of the intervention are as follows.

1. **Step 1 - Pre-Test:** A questionnaire for participants to express their prior knowledge and attitude on the given topic.
2. **Step 2 - Introduction:** Introduction to INFO-NAVIGATE and task instructions.
3. **Step 3 - Main Task:** Participants are involved in writing an argumentative write-up on the given topic, analysing it from different perspectives with the use of the tool and instructions.
4. **Step 4 - Post-Test:** A questionnaire for participants to express their experience with the intervention and its impact on the prior attitude.

A Python-based web portal is used for data collection within which the INFO-NAVIGATE tool is embedded. Empirical data from students will be analysed quantitatively and qualitatively to address the RQ.

5 Current Progress and Next Steps

Ethical consideration: The research has obtained approval from the university Human Research Ethics Committee (HREC) for student and staff recruitment.
INFO-NAVIGATE tool design and development: The core design and development of the tool have been completed and securely deployed in a university cloud environment for empirical evaluation.
Tool evaluation and pilot testing of the intervention: The system evaluation and pilot user evaluation of the tool and intervention were conducted and results were reported in a preliminary paper [11].
Educator interviews: More than half of the interviews have been conducted, the results of which will be reported in a forthcoming conference paper. Additional educator interviews will be conducted in the coming months to learn more about current practices and perspectives on teaching IPS and critical thinking skills in the classroom.
First DBR iteration: The first iteration of DBR was conducted with 26 post-graduate students in an authentic classroom setting on the topic of ethical considerations of LLMs.

Next Steps: Future iterations are planned in the coming months for data collection from a larger sample. The second iteration will be implemented in the same ANLP subject, with changes made based on feedback received in the previous iteration and student engagement with the tool. The third and final iteration is planned to be run in a different subject to study the transferability of the intervention across different subject contexts.

6 Expected Contributions

This research aims to contribute to the AI in education community in both local and generalizable ways. The study locally introduces a socio-technical solution to enhance students' ability to critically navigate through conflicting online information and make informed decisions as responsible citizens, which will be potentially transferable across different contexts. In addition, the study will provide generalisable insights, as below:

1. Design principles for novel information seeking interfaces in the age of misinformation and GenAI.
2. A novel application of the IPS model and AI-supported tool to teach critical skills for information seeking in the context of conflicting information. This way, GenAI can be used to teach and enhance the critical thinking of students rather than restricting such emerging technologies in educational contexts.
3. Approaches for how educators can embed technology-mediated interventions in their teaching in addition to disciplinary content knowledge to augment the teaching and learning of critical information seeking and critical thinking.
4. Understanding educators' perspectives on using GenAI-based tools in classroom settings.

References

1. Boonprakong, N., Tag, B., Dingler, T.: Designing technologies to support critical thinking in an age of misinformation. IEEE Pervasive Comput. **22**(3), 8–17 (2023). https://doi.org/10.1109/MPRV.2023.3275514
2. Brand-Gruwel, S., Wopereis, I., Vermetten, Y.: Information problem solving by experts and novices: analysis of a complex cognitive skill. Comput. Hum. Behav. **21**(3), 487–508 (2005). https://doi.org/10.1016/j.chb.2004.10.005
3. Brand-Gruwel, S., Wopereis, I., Walraven, A.: A descriptive model of information problem solving while using internet. Comput. Educ. **53**(4), 1207–1217 (2009). https://doi.org/10.1016/j.compedu.2009.06.004
4. Deroover, K., Knight, S., Burke, P.F., Bucher, T.: Why do experts disagree? The development of a taxonomy. Public Underst. Sci. **32**(2), 224–246 (2023). https://doi.org/10.1177/09636625221110029
5. Di Sotto, S., Viviani, M.: Health misinformation detection in the social web: an overview and a data science approach. Int. J. Environ. Res. Public Health **19**(4), 2173 (2022). https://doi.org/10.3390/ijerph19042173
6. Dwivedi, Y.K., et al.: Opinion paper:"so what if chatgpt wrote it?" Multidisciplinary perspectives on opportunities, challenges and implications of generative conversational AI for research, practice and policy (2023). https://doi.org/10.1016/j.ijinfomgt.2023.102642
7. Ethape, P., Kane, R., Gadekar, G., Chimane, S.: Smart automation using LLM. Int. Res. J. Innovations Eng. Technol. **7**(11), 603 (2023). https://doi.org/10.47001/IRJIET/2023.711080
8. Gibbons, M., Limoges, C., Scott, P., Schwartzman, S., Nowotny, H.: The new production of knowledge: The dynamics of science and research in contemporary societies. SAGE Publications Ltd (2010)
9. Himelein-Wachowiak, M., et al.: Bots and misinformation spread on social media: implications for COVID-19. J. Med. Internet Res. **23**(5), e26933 (2021). https://doi.org/10.2196/26933
10. Kakol, M., Nielek, R., Wierzbicki, A.: Understanding and predicting web content credibility using the content credibility corpus. Inf. Process. Manag. **53**(5), 1043–1061 (2017). https://doi.org/10.1016/j.ipm.2017.04.003
11. Karunanayake, A., Shibani, A., Knight, S.: Info-navigate for critical thinking in complex information landscapes: A design rationale and pilot evaluation (2024). https://outbox.eait.uq.edu.au/uqsville/OzCHI2024/Late%20Breaking%20Works/INFO-NAVIGATE%20for%20Critical%20Thinking%20in%20Complex%20Information%20Landscapes%20-%20A%20Design%20Rationale%20and%20Pilot%20Evaluation.pdf
12. Shibani, A., Knight, S., Buckingham Shum, S.: Contextualizable learning analytics design: a generic model and writing analytics evaluations. In: Proceedings of the 9th International Conference on Learning Analytics & Knowledge, pp. 210–219 (2019). https://doi.org/10.1145/3303772.3303785
13. Wang, F., Hannafin, M.J.: Design-based research and technology-enhanced learning environments. Educ. Tech. Res. Dev. **53**(4), 5–23 (2005). https://doi.org/10.1007/BF02504682

Author Index

A. I. Cristea et al. (Eds.): AIED 2025, CCIS 2590, pp. 489–495, 2025.
https://doi.org/10.1007/978-3-031-99261-2

The manufacturer's authorised representative in the EU is Springer Nature Customer Service Centre GmbH, Europaplatz 3, 69115 Heidelberg, Germany. If you have any concerns regarding our products, please contact ProductSafety@springernature.com

Printed and bound by CPI Group (UK) Ltd, Croydon, CR0 4YY
30/09/2026
02225560-0019